PHYSIOLOGICALLY BASED PHARMACOKINETIC MODELING

PHYSIOLOGICALLY BASED PHARMACOKINETIC MODELING

Science and Applications

EDITED BY

MICAELA B. REDDY, PH.D.
Colorado State University
Fort Collins, Colorado

RAYMOND S. H. YANG, PH.D.
Colorado State University
Fort Collins, Colorado

HARVEY J. CLEWELL III, M.A.
CIIT Centers for Health Research
Research Triangle Park, North Carolina

MELVIN E. ANDERSEN, PH.D.
CIIT Centers for Health Research
Research Triangle Park, North Carolina

WILEY-INTERSCIENCE

A JOHN WILEY & SONS, INC., PUBLICATION

Published by John Wiley & Sons, Inc., Hoboken, New Jersey.
Published simultaneously in Canada.

Library of Congress Cataloging-in-Publication Data:
Physiologically based pharmacokinetic modeling : science and application / edited by Micaela B. Reddy . . . [et al.].
p. cm.
Includes index.
ISBN 0-471-47814-8 (cloth)
1. Pharmacokinetics. 2. Xenobiotics. 3. Toxicology. I. Reddy, Micaela B.
RM301.5.P53 2005
615′.7—dc22

2005005177

Printed in the United States of America
10 9 8 7 6 5 4 3 2 1

CONTENTS

PREFACE

In recent years, there has been an enormous expansion of uses of physiologically based pharmacokinetic (PBPK) modeling in areas related to environmental chemicals and drugs. For individuals interested in PBPK modeling, it is relatively easy to locate and use the contributions of previous authors on a specific chemical of interest. However, it is more difficult to locate broader sets of contributions containing useful modeling techniques and applications. Our purpose was to provide a broad review of the PBPK modeling literature, before the size of the body of work grew large enough to make such an effort prohibitive, and to provide a resource to contain comprehensive coverage of the PBPK modeling literature from its beginnings in the mid-1900s through the first few years of the twenty-first century. This monograph is meant to be a useful reference and educational tool for those professionals and graduate students in toxicology, pharmacology, computational biology, and risk assessment interested in PBPK modeling as a tool for quantifying tissue doses and for describing the response of organisms to chemical exposures.

Our initial literature search in 2001 and updated in 2002, conducted using the Web of Science, Medline, and Toxline databases and incorporating keywords such as physiologically based pharmacokinetic/PBPK model, physiologically based toxicokinetic/PBTK model, and physiologically based pharmacodynamic/PBPD model, uncovered over 1000 references. As the term PBPK model did not become popular until the 1980s, for earlier contributions we relied on literature searches using the names of authors known by the editors to have made early contributions in the field, followed up by searches on other authors and articles cited in these articles. We chose to organize this diverse body of work based on classes of chemicals (e.g., volatile organics and environmental contaminants) and modeling purposes (e.g., perinatal transfer models and dermal absorption models). Our goal was to be fairly comprehensive, but to stress primary contributions in PBPK model development and in applications of these models to investigate factors that regulate chemical distribution within the body. We have also attempted to include articles that appeared over the past few years during completion of this volume. While we have made attempts to be inclusive in our coverage of the PBPK modeling literature, some important contributions may have been missed in our review process. We apologize to authors whose work may have been inadvertently overlooked in these various chapters and not captured by the editorial review.

This monograph describes the development of PBPK modeling for toxic compounds over the past eight decades and their current uses, providing background on the basics of PBPK modeling for understanding the physical chemical, and biological properties that determine absorption, distribution, metabolism, and elimination of xenobiotics. Early PBPK modeling applications with volatile anesthetics and chemotherapeutics paved the way for applying these techniques to a wide range of volatile compounds of occupational and environmental significance. The past 15 years have witnessed extensive application with many other classes of chemicals: metals, inorganic chemicals, pesticides, persistent organics pollutants, drugs, and the metabolites of these classes of chemicals. PBPK models have played important roles in unraveling dose–response behaviors based on estimates of tissue dose and have revolutionized low dose and interspecies extrapolations in risk assessment. Following an introductory chapter on PBPK modeling, a series of chapters reviews PBPK model results for various classes of compounds with coverage of historical development, modeling challenges specific to classes of chemicals, and current practices. Comments are also provided regarding the use of these PBPK models to support pharmacodynamic modeling for various toxic responses and future directions where modeling approaches will be helpful.

This monograph arose through efforts of graduate students, postdoctoral fellows, and professors at Colorado State University to review literature in specific areas and produce a series of chapters. These individuals worked in the Quantitative and Computational Toxicology Program at the Center for Environmental Toxicology and Technology in the Department of Environmental and Radiological Health Sciences. Many of these individuals have graduated from Colorado State and left for other positions. The editors wish to express their sincere appreciation for all the assistance provided by these individuals in developing this monograph. Each of these individuals is cited as the authors on the chapters where they contributed.

Micaela B. Reddy
Raymond S. H. Yang
Harvey J. Clewell III
Melvin E. Andersen

ACKNOWLEDGMENTS

The concepts and work discussed in this presentation were partially contributed by many colleagues associated or collaborating with the Quantitative and Computational Toxicology group at Colorado State University; we are grateful for their contributions and intellectual stimulation. Any advances in science require funding support from many agencies. We thank NIH (Superfund Basic Research Program Project P42 ES05949; research grants R01 ES09655; 3 RO1 ES-09655-01S1, and RO3 ES10116 ZES1; training grants T32 ES 07321, F32 ES 11425, and 1 F32 ES 05901; and Career Development Awards 1 K08 CA72396, 1 K01 CA75955-01A1, and 01 K25 ES11146), ATSDR (Cooperative Agreement U61/ATU881475), NIOSH/CDC (research grant 1 RO1 OH07556-01), US Air Force (research grants F33615-91-C-0538; F49620-94-1-0304), US Navy (research contract, F3360195MSA05), and USEPA (Cooperative Agreement CR 821922-01-0; Contract No. 3C-R102-NTEX). Without the generous support of these agencies, the development of some of the research described herein and the writing of this monograph could have never been possible.

Two of the authors, MEA and HJC, gratefully acknowledge support from the Long-range Research Initiative program of the American Chemistry Council (LRI-ACC) for their continuing support of development of PBPK tools to improve human health risk assessment. We also want to acknowledge our debt to colleagues from our Wright-Patterson AFB years, Drs. Michael MacNaughton, Michael Gargas, Jeff Fisher, and James McDougal, who were all critical parts of a team of scientists that initially developed the experimental methods and modeling tools to support PBPK modeling in toxicology and risk assessment.

Several individuals made important contributions in this project. We thank Dr. Ying Ou, who initiated the project. The authors also thank Mr. Roger Paxton, Miss Christina Barinque, Miss Lisa Casuto, and Mrs. Linda Monum for their valuable assistance with the monograph.

CONTRIBUTORS

Andersen, Melvin, E., Ph.D.
Division of Computational Biology
CIIT Centers for Health Research
Research Triangle Park, NC 27709-2137

Bae, Dong-Soon, Ph.D.
Laboratory of Cellular Carcinogenesis and Tumor Promotion
National Cancer Institute
National Institutes of Health
Bethesda, MD 20892-4255

Belfiore, Carol J., Ph.D.
Quantitative and Computational Toxicology Group
Center for Environmental Toxicology and Technology
Department of Environmental and Radiological Health Sciences
Colorado State University
Fort Collins, CO 80523-1690

Campain, Julie, Ph.D.
Center for Environmental Toxicology and Technology
Department of Environmental and Radiological Health Sciences
Colorado State University
Fort Collins, CO 80523-1690

Clewell, Harvey J., III, M.A.
Center for Human Health Assessment
CIIT Centers for Health Research
Research Triangle Park, NC 27709-2137

Davidson, Robert T., Ph.D.
Department of Nutrition, Dietetics, and Food Science
Brigham Young University
Provo, UT 84602

Dennison, James E., Ph.D.
Quantitative and Computational Toxicology Group
Center for Environmental Toxicology and Technology

Department of Environmental and Radiological Health Sciences
Colorado State University
Fort Collins, CO 80523-1690

Dobrev, Ivan D., Ph.D.
Quantitative and Computational Toxicology Group
Center for Environmental Toxicology and Technology
Department of Environmental and Radiological Health Sciences
Colorado State University
Fort Collins, CO 80523-1690

Lee, Sun K., Ph.D.
Quantitative and Computational Toxicology Group
Center for Environmental Toxicology and Technology
Department of Environmental and Radiological Health Sciences
Colorado State University
Fort Collins, CO 80523-1690
Present Address:
Hoffman–LaRoche
Nonclinical Drug Safety
Nutley, NJ 07110

Lu, Yasong, M.S.
Quantitative and Computational Toxicology Group
Center for Environmental Toxicology and Technology
Department of Environmental and Radiological Health Sciences
Colorado State University
Fort Collins, CO 80523-1690

McMullin, Tami S., B.S.
Department of Environmental and Radiological Health Sciences
Colorado State University
Fort Collins, CO 80523-1690

Pott O'Brien, Wendy, D.V.M., Ph.D.
United States Environmental Protection Agency, Region 8
Denver, CO 80202

Reddy, Micaela B., Ph.D.
Quantitative and Computational Toxicology Group
Center for Environmental Toxicology and Technology
Department of Environmental and Radiological Health Sciences
Colorado State University
Fort Collins, CO 80523-1690

Yang, Raymond S. H., Ph.D.
Quantitative and Computational Toxicology Group
Center for Environmental Toxicology and Technology
Department of Environmental and Radiological Health Sciences
Colorado State University
Fort Collins, CO 80523-1690

CHAPTER 1

INTRODUCTION: A HISTORICAL PERSPECTIVE OF THE DEVELOPMENT AND APPLICATIONS OF PBPK MODELS

Melvin E. Andersen, Raymond S. H. Yang, Harvey J. Clewell III, and Micaela B. Reddy

1.1 INTRODUCTION

Pharmacokinetics is the quantitative study of factors that control the time course for absorption, distribution, metabolism, and excretion of chemicals within the body. Pharmacokinetic (PK) models provide sets of equations that simulate the time courses of chemicals and their metabolites in various tissues throughout the body. The interest in PK modeling in toxicology and pharmacology arose from the need to relate internal concentrations of active compounds at their target sites with the doses of chemical given to an animal or human subject. The reason, of course, is the fundamental tenet in pharmacology or toxicology that both beneficial and adverse responses to compounds are related to the concentrations of active chemicals reaching target tissues rather than the amounts of chemical at the site of absorption.

Physiologically Based Pharmacokinetic Modeling, edited by Micaela B. Reddy, Raymond S. H. Yang, Harvey J. Clewell III, and Melvin E. Andersen.

The relationships between tissue dose and administered dose can be complex, especially in high-dose toxicity testing studies, with multiple, repeated daily dosing, or when metabolism or toxicity at routes of entry alter uptake processes for various routes of exposure. PK models of all kinds are primarily a tool to assess chemical dosimetry at target tissues for a wide range of exposure situations.

In physiologically based pharmacokinetic (PBPK) modeling, compartments correspond to discrete tissues or to groupings of tissues with appropriate volumes, blood flows, and pathways for metabolism of test chemicals (Bischoff and Brown 1966). These PBPK models include pertinent biochemical and physicochemical constants for metabolism and solubility in each compartment. Routes of dosing (routes of administration) are included in their proper relationship to the overall physiology. For instance, dermally absorbed compounds penetrate the skin, enter the mixed venous blood, and then travel through the heart and lungs to the arterial blood for distribution. Orally absorbed compounds move through intestinal tissues and portal blood to the liver before moving to the mixed venous blood for distribution to the remainder of the body. The equations that form the basis of the PBPK model also account for the time sequence of dose input into test subjects and permit input by multiple routes, if necessary, for specific exposure situations. Each compartment in the model is described with a mass-balance differential equation (MB-DE) whose terms mathematically represent biological processes. The set of equations is solved by numerical integration to simulate tissue time-course concentrations of chemicals and their metabolites.

Some PBPK models account for interactions of circulating compounds with specific receptors or the covalent interactions of chemicals with tissue constituents. Modeling these reversible and irreversible molecular interactions with cell constituents is the initial step in developing physiologically based pharmacodynamic (PBPD) models for effects of chemicals on biological processes. This monograph emphasizes progress in PBPK rather than PBPD modeling. A number of short reviews are available that focus on earlier stages of the development of PBPK modeling approaches (e.g., Himmelstein and Lutz 1979; Gerlowski and Jain 1983; Leung 1991), including a volume on PBPK modeling in chemical risk assessment (National Research Council 1987). More recent progress in PBPK modeling has not yet been thoroughly reviewed. Some of the aims of this volume are to provide an overview of the range of applications of PBPK modeling, the classes of compounds evaluated with these tools, and the insights derived from the application of PBPK models to the distribution of chemicals in intact animals. This first chapter traces the history and background of PBPK modeling over the last century, providing the background that places in perspective the rather astonishing expansion of PBPK modeling in toxicology and risk assessment over the past decade. The subsequent chapters focus either on specific chemical classes or on specific model applications.

1.2 A HISTORICAL PERSPECTIVE

The history of PBPK modeling for drugs and environmental compounds provides an interesting look at the interface between posing scientific questions and the technologies necessary to solve them. It also highlights the interdisciplinary contributions—from medicine, engineering, toxicology, pharmaceutics, and risk assessment—that made the present-day use of these PBPK modeling tools a reality. The scientific question posed almost from the earliest PK studies was, What physiological processes are important in creating and maintaining sufficient tissue concentrations of a test compound to ensure a biological effect? This question, in one form or another, motivates almost all PBPK modeling contributions whether in pharmacology or toxicology.

1.2.1 Responses to Inhaled Compounds

Inhalation anesthesiologists have maintained a long tradition of understanding the role of ventilation rates, blood flow rates, and tissue solubility on the uptake and distribution of volatile anesthetics to the central nervous system. In the 1920s, Haggard (1924a,b) quantitatively described the importance of physiological factors for the uptake of ethyl ether into the body during the first few breaths. Accomplishing this analysis required writing an equation for the relationship between inhaled ether and the concentration of ether in blood. Tools for solving this equation over time were not available, so the mathematical analysis was limited to the first few breaths when venous concentrations remained small. The American Chemical Society Monograph Series, Vol. 35 by Henderson and Haggard (1942) represents, to the authors of this chapter, the first detailed discussion of toxicology of inhaled compounds in the context of the principles that control exposure, absorption, and physiological actions. It is the first articulation of a PBPK modeling strategy in occupational and environmental toxicology.

More complete PBPK models for inhalation were provided by Kety (1951), Mapleson (1963), and Riggs (1963). In these models, body tissues were lumped together based on blood perfusion rates, giving sets of tissues referred to as richly perfused or poorly perfused. Mapleson (1963) solved the set of equations using an analog computer to give solutions to the complete time course within the various tissue groups. These analog computer PBPK models for inhaled gases and vapors were extended by Fiserova-Bergerova and colleagues (1975, 1979, 1980) to focus on compounds in the occupational environment and to describe metabolism of these compounds in liver. The extension to include metabolism was particularly important for subsequent work in toxicology because most compounds of interest in occupational toxicology are metabolized and metabolites are often involved in toxic responses.

1.2.2 Pharmaceutical Applications

In the 1930s, Teorell (1937a,b) provided a set of equations for uptake, distribution, and elimination of drugs from the body. These articles are rightly regarded as providing the first physiological model for drug distribution. However, computational methods were not available to solve the sets of equations at this time. Exact mathematical solutions for distribution of compounds in the body could only be obtained for simplified models in which the body was reduced to a small number of compartments that did not correspond directly with specific physiological compartments. Over the next 30 years, PK modeling focused on these simpler descriptions with exact solutions rather than on developing models more concordant with the structure and content of the biological system itself. These approaches are sometimes referred to as "data-based" compartmental modeling since the work generally took the form of a detailed collection of time-course blood/excreta concentrations at various doses (Fig. 1.1). Time-course curves were analyzed by assuming particular model structures and estimating a small number of model parameters by curve-fitting. In the earliest of these models, all processes for metabolism, distribution, and elimination were treated as first-order (i.e., they increased in direct proportion to the concentration of the chemical species). Two areas of concern that particularly affected data-based compartmental PK modeling arose in the 1960s and early 1970s: (1) the saturation of elimination pathways and (2) the possibility that blood flow rather than metabolic capacity of an organ might limit clearance. Saturation led to models that were not first-order, making it difficult to derive exact solutions to the sets of equations. Blood-flow-limited metabolism in an organ meant that the removal rate constant (k_{out}) from a central compartment (Fig. 1.1) could not increase indefinitely as the metabolic capacity increased.

1.2.3 Occupational and Environmental Applications

Data-based compartmental models were brought to toxicology and risk assessment in a series of innovative studies by the late Dr. Perry Gehring (1938–2003) and his colleagues at the Dow Chemical Company in Midland, MI, in order to examine PK behavior where specific elimination pathways, both metabolic and excretory, become saturated at high doses (Gehring *et al.* 1976, 1977, 1978). In the hands of the Dow research team, nonlinear-data-based compartmental models were ingen-

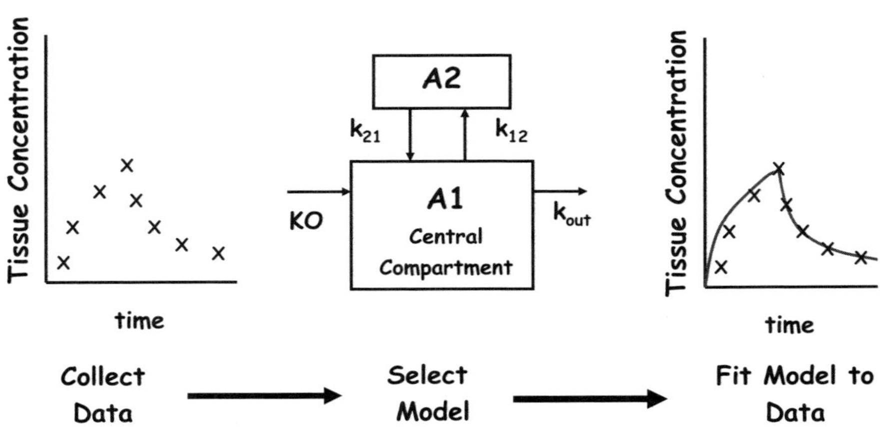

$$C_t = A\,e^{-k_a t} + B\,e^{-k_b t}$$

Figure 1.1 Schematic diagram depicting a method for developing data-based compartmental PK models.

iously applied to a series of compounds of toxicological and commercial importance including herbicides (Sauerhoff *et al.* 1976; 1977), solvents (McKenna *et al.* 1982), plastic monomers (McKenna *et al.* 1978a, 1978b), and hydrocarbons (Young *et al.* 1979; Ramsey *et al.* 1980). The final piece of technology needed to bring a full PBPK approach to studying factors that determine chemical disposition came with the rapid development of digital computation by the engineering community and the availability of these tools within the research laboratory at the Dow Chemical Company.

1.2.4 Digital Computation and PBPK Modeling

Scientists trained in chemical engineering and computational methods developed PBPK models for chemotherapeutic compounds—that is, chemicals used in cancer therapy (Bischoff *et al.* 1971). Many of these compounds are highly toxic and have therapeutic efficacy by being slightly more toxic to rapidly growing cells (the cancer cells) than to normal tissues. Initial successes with methotrexate (Bischoff *et al.* 1971) led to PBPK models for other compounds, including 5-fluorouracil (Collins *et al.* 1982) and cisplatin (Farris *et al.* 1988). These seminal contributions showed the ease with which realistic descriptions of physiology and relevant pathways of metabolism could be incorporated into PBPK models for chemical disposition and paved the way for more extensive use of PBPK modeling in toxicology and chemical risk assessment. These models took advantage of the increasing availability of digital computation on main frame computers for solving sets of MB-DEs.

Ramsey and Andersen (1984) applied a PBPK modeling approach to describe the disposition of styrene in rats and humans for a range of concentrations and for several routes of administration. One of these two scientists (J. C. Ramsey) was a member of the PK group developing nonlinear PK models for chemicals at Dow Chemical Company and solving these models with a modern software package for solving sets of MB-DEs by numerical integration. The other (M. E. Andersen) had worked in inhalation toxicology laboratories at the Wright-Patterson Air Force Base, OH, and developed a steady-state analysis of PBPK models for inhalation of metabolized vapors (Andersen 1981). This interinstitutional collaboration with styrene (Ramsey and Andersen 1984; Andersen *et al.* 1984) relied on advances from inhalation anesthesia, data-based compartmental modeling, pharmaceutics, chemical engineering, and digital computation, to create PBPK models that would support extrapolation across species, between exposure routes, and from high to low doses. Using scale-up methods common for engineering models (Dedrick 1973), the interspecies PBPK model for styrene (Fig. 1.2) was able to predict blood and exhaled air time-course curves for oral and intravenous dosing in the rat and for inhalation exposures in human volunteers. This ability to support extrapolation to untested (and sometimes untestable) conditions is an essential part of risk assessment and has made these PBPK models attractive tools in human health risk assessments of various kinds (Clewell and Andersen 1985; National Research Council 1987). In the styrene PBPK model, the liver was split off as a separate compartment (i.e., rather than embedded in a central compartment), metabolism in the liver was saturable (i.e., fol-

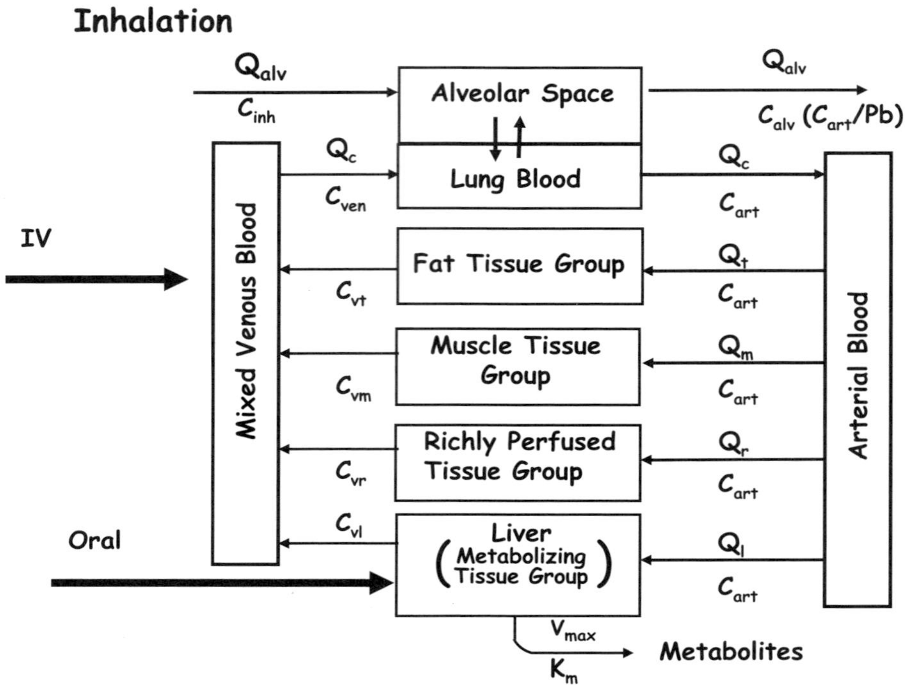

Figure 1.2 Schematic diagram of a multidose route PBPK model for a volatile compound.

lowed Michaelis–Menten kinetics), and styrene clearance from tissues was directly based on blood flow and metabolic characteristics of tissues.

1.3 EXPANSION OF PBPK MODEL APPLICATIONS

Even though the primary developments in PBPK approaches by the chemical engineering community were with pharmaceutical, primarily antineoplastic, compounds, the real expansion of the application of digital computation to create PBPK models of increasing complexity since the 1980s occurred when these methods were applied to environmental compounds and to chemical risk assessment. In pharmaceutical arenas, some of the inertia to developing PBPK models was due to the idea that extrapolations were unnecessary since PK data would eventually be developed in clinical studies. Some inappropriate "myths" which hampered the development of PBPK modeling in pharmaceutical industry and elsewhere have been discussed in a recent review on PBPK/PD modeling (Yang *et al.* 2004). This viewpoint neglects other attributes inherent in PBPK approaches. Among the opportunities offered by PBPK approaches are: (1) creating models from physiological, biochemical, and anatomical information, entirely separate from collection of detailed concentration time-course curves; (2) evaluating mechanisms by which biological processes govern disposition of a wide range of compounds by comparison of PK results with model predictions; (3) using chemicals as probes of the biological processes to gain more general information on the way chemical characteristics govern the importance of various transport pathways in the body; (4) applying the models in risk assessments for setting exposure standards; and (5) using annotation of a modeling database as a repository of information on toxicity and kinetics of specific compounds. Each of these is discussed in turn.

1.3.1. PBPK Models for Tissue Dosimetry from Secondary Data

The advent of biologically structured PBPK models had a dramatic influence on the nature of the experiments conducted to determine PK behavior and to estimate tissue dosimetry. In PBPK descriptions, time-course behavior is not an intrinsic property of the organism accessible only by direct experimentation. Instead, it is a composite behavior, governed by more fundamental physiological and biochemical processes. More importantly, these fundamental processes can be studied in simpler systems to obtain the necessary PBPK model parameters in experiments separate from collection of time-course concentration curves (Fig. 1.3). Based on these parameters and an appropriate model structure, tissue time-course behaviors can be predicted by computer simulation with PBPK models and compared to data as a test of model performance.

Volatile chemicals have provided a good test bed for examining this approach to PBPK modeling. The disposition of volatiles in the body is related to breathing rates, tissue volumes, tissue blood flow rates, tissue partition coefficients, and kinetic constants for metabolism of the chemical in particular tissues. Physiological factors important in developing an appropriate and useful PBPK model have been discussed

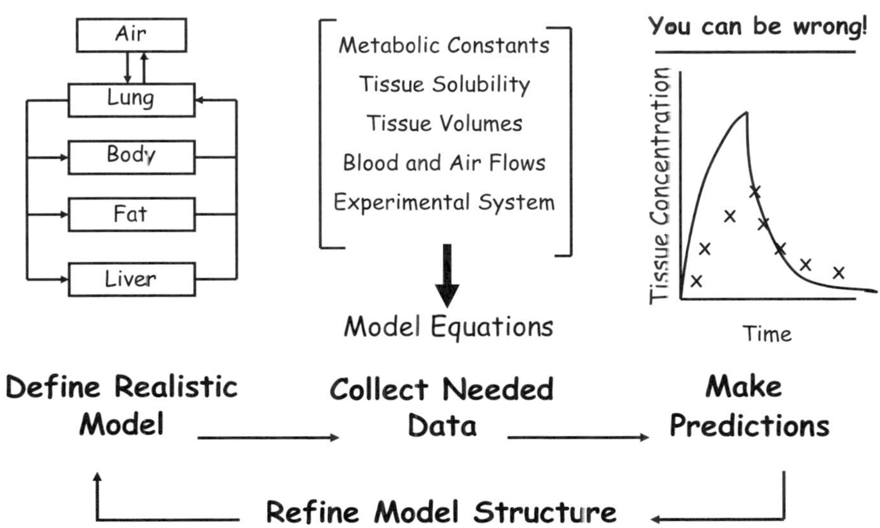

Figure 1.3 Schematic diagram depicting a method for developing PBPK models.

by Krishnan and Andersen (2001). Physiological parameters can be found in the biomedical literature, and they were recently compiled by Brown *et al.* (1997). Partition coefficients can be measured by equilibrating tissue homogenates in a vial with an atmosphere containing the test chemical (Sato and Nakajima 1979a; Gargas *et al.* 1989). Constants for metabolism (enzyme kinetic constants for saturable, first-order or second-order reactions) can be determined *in vitro* with tissue homogenates, microsomal preparations, liver slices, and so on, by supplementing these preparations with reactants to promote metabolic reactions (Sato and Nakajima 1979b; Hilderbrand *et al.* 1981; Kedderis *et al.* 1993). Another method for assessing metabolic parameters *in vivo* relies on closed chamber inhalation techniques. Here, a small numbers of live animals are placed in a closed chamber to measure the rate of loss of chemical at a variety of chamber concentrations (Hefner *et al.* 1975; Filser and Bolt 1979; Gargas *et al.* 1986). These *in vitro* and *in vivo* experiments can provide all the parameters necessary for constructing a PBPK model for the parent chemical; also, time-course behavior is now "predictable," based on results of these ancillary studies. Other approaches for developing predictive PBPK models include using structure–activity relationships (SARs) to estimate model parameters for classes of compounds (Parham *et al.* 1997; Parham and Portier 1998; Poulin and Krishnan 1996, 1999). While these approaches to parameterizing PBPK models by *in vitro*/simple *in vivo* studies are attractive for reducing the number of animals required for model development, some *in vivo* experimentation will nearly always be required to test the accuracy of the predicted behavior.

1.3.2 Biological Mechanisms Underlying Pharmacokinetic Behaviors

Predictions from PBPK models have a very useful property: They can be wrong. The ability to predict a particular outcome is a powerful tool for enhancing the information content of an experiment. In effect, PBPK models, based on proposed mechanisms of disposition, make predictions that become testable (Fig. 1.3). *Trans*-1,2-dichloroethylene (*t*DCE) provided a good example of a fairly spectacular, but enlightening, failure of a PBPK model. A simple PBPK model structure worked well in predicting the disappearance of a diverse group of volatile chemicals from a closed chamber (Gargas *et al.* 1986, 1990). This closed chamber study uses a small animal inhalation system with a recirculated atmosphere where carbon dioxide is removed by chemical adsorption and oxygen added back as it is utilized by respiration during the study. Chemical is added at time zero at various initial concentrations, and the diminution of chamber chemical is evaluated over time. The PBPK model for volatiles had time-invariant metabolic constants (maximum velocity and affinity constant, V_{max} and K_m, respectively) in liver and regarded the chamber atmosphere as another compartment. This model successfully described the behavior of many gases and vapors.

When applied to loss of *t*DCE from the chamber, the PBPK model was unable to fit the uptake curves (Gargas and Andersen 1988; Gargas *et al.* 1990). *t*DCE, or, more accurately, metabolites of *t*DCE, appeared to rapidly react with and inactivate the enzyme(s) responsible for *t*DCE metabolism (Lilly *et al.* 1999). A successful PBPK description for *t*DCE accounted for loss of *t*DCE metabolizing capacity over time, a complexity that was easily accommodated by virtue of the versatility of a physiological modeling framework. The model provided insights into (1) the mechanism of enzyme inactivation by metabolites, (2) the rate constants for inactivation by these metabolites, and (3) the number of molecules of *t*DCE metabolized for each mole of enzyme inactivated.

Another interesting example of the use of PBPK modeling in assessing the mechanism of distribution of a class of compounds is with 2,3,7,8-tetrachlorodibenzo-*p*-dioxin (TCDD) and related compounds (see Chapter 8). Over the past 25 years, PBPK models for the lipophilic compound TCDD have continuously evolved to include storage in fat, distribution via lipid transport pools in blood, induction of binding proteins throughout the liver and in specific regions of the liver, and alterations in clearance of endogenous hormones by induction of metabolizing enzymes in the liver. PBPK models for TCDD can then be extended to understand the pharmacokinetics of other polychlorinated dibenzo-*p*-dioxins. The unfolding story of TCDD illustrates the evolution of increasingly detailed PBPK models as new information becomes available for inclusion and exemplifies the natural extension of PBPK models to PBPD models as the biology underlying chemical disposition and response becomes increasingly understood.

1.3.3 Chemicals as Probes of Biological Processes

In the earliest PBPK models, tissues were well-mixed compartments and effluent blood emerging from the tissue was in equilibrium with the tissue contents. The distribution ratio between tissue and venous blood was described by a tissue : blood partition coefficient (P_t). Thus, the concentration of chemical in the venous blood exiting a tissue (C_{vt}) could be calculated as $C_{vt} = C_t/P_t$, where C_t is the concentration of chemical in the tissue (see Fig. 1.2). It was easy to write model equations using this assumption of venous equilibration, and many successful models include this assumption. However, there is considerable evidence that even low-molecular-weight chemicals, like methylene chloride (i.e., dichoromethane, DCM), do not clear as rapidly from tissues as predicted by a venous-equilibration model (Angelo *et al.* 1984). With styrene, some of the early rat PK studies included measurements of styrene concentrations in the liver, kidney, and blood over time during and after exposure in rats (Ramsey, unpublished observations, 1980). While these data were not included in the model formulation, they showed that tissue concentrations are more appropriately considered as composites of free chemical, partitioned chemical, and chemical bound in more slowly exchanging sites within the tissues, probably within specific lipid structures of cells. The ability to carefully study disposition of lipophilic compounds in these various compartments within a tissue *in vivo* is often hampered because highly lipophilic compounds are poorly metabolized and persistent in the body with long retention times. The rate of elimination is then determined by low metabolic clearance, and the rate of chemical moving between lipid stores is not limiting for the observed kinetics of these compounds. However, this generalization does not hold for chemicals that are both highly lipophilic and rapidly cleared from systemic circulation.

Cyclic siloxanes, used in some consumer products, have this constellation of properties. These volatile compounds have high hepatic, metabolic clearance and high clearance by exhalation, due to low blood:air partition coefficients. Total clearance from the body is very high with loss of free siloxane from blood occurring within minutes after cessation of an inhalation exposure in rats. With these compounds, movement between lipid storage sites becomes the dominant kinetic process observed at the end of the exposure period (Andersen *et al.* 2001; Dobrev *et al.* 2003). Studies of PK processes with cyclic siloxanes permitted the evaluation of lipid storage sites within specific tissues; they also uncovered kinetic evidence for lipid storage depots in blood that are not in communication with the free siloxanes circulating in blood. The PBPK models for these siloxanes incorporated new storage compartments that dominate the kinetics of this class of compounds, but which are also likely to be operative, though of relatively lesser importance in determining overall kinetic behavior, with all other lipophilic compounds. These siloxanes became a tool for understanding the role of lipophilicity in chemical transport and for unraveling pathways for lipid transport of chemicals in the body.

1.3.4 Risk Assessment Applications

The main reason for the rapid expansion of PBPK modeling over the past 10–15 years has undoubtedly been the contributions of the technology to dose-response assessment and extrapolations in chemical risk assessments. The first application of a PBPK model in a risk assessment was with DCM. A PBPK model for DCM (see Chapter 2) was first developed to explore causality between various measures of tissue dose (referred to as dose metrics) and carcinogenicity (Andersen *et al.* 1987). The PBPK model contained tissue clearance by oxidation and glutathione (GSH) conjugation in the liver and kidney, accounted for dosing by inhalation or drinking water, and allowed simulation of expected tissue dose metrics in mice and humans. With this PBPK model, it was possible to calculate expected tissue exposures to metabolites from the oxidative and conjugation pathways for different exposure conditions in the liver and lung for mice and humans.

For DCM the dose metrics chosen for analyzing tumor responses were integrated intensity of tissue exposure to reactive intermediates—that is, the rate of metabolism through a specific pathway per volume of tissue per time. The carcinogenic responses for DCM correlated well with the GSH pathway metabolism, but not with the oxidation pathway metabolism. The work with DCM was the first use of a PBPK model for low-dose and interspecies extrapolation based on tissue dose metrics. This extrapolation used human-specific parameters for tissue volumes, breathing rates, and distribution of enzymes involved in oxidation and conjugation in the model structure. Risk extrapolation assumed that mouse and human tissues would be equally responsive to equivalent tissue exposures to the reactive GSH pathway intermediates. This PBPK model has been cited and used in risk assessments by Health Canada (Health Canada 1993) and by the Occupational Safety and Health Administration (OSHA) in the United States (OSHA 1997). The risk assessment exercise with DCM established a pattern for application of a PBPK model in risk assessment that is still closely followed today (Table 1.1). In addition, the PBPK modeling spurred a variety of research to refine variability and uncertainty analysis

TABLE 1.1 Steps for Using PBPK Model Estimated Tissue Dose Metrics in Chemical Risk Assessment

1. Identify toxic effect(s) in animals and people and determine critical effect(s) for risk assessment.
2. Organize available data for mode(s) of action, metabolism, chemistry of compound/metabolites, and similar information for related compounds.
3. Describe potential mode(s) of action involved in critical effects.
4. Propose relationships between response and tissue dose, specifying the measure of tissue dose (i.e., the dose metric) associated with toxicity.
5. Develop an appropriate PBPK model to estimate the tissue dose metric for various routes of administration, at various doses, in test species and in humans.
6. Estimate tissue dose metrics during exposures in animals or people that produce toxicity.
7. Estimate risks in humans based on the tissue doses during human exposures, assuming a similar dose–response relationship in humans and rats based on the tissue dose metrics.

for PBPK models (Portier and Kaplan 1989; Clewell and Andersen 1996; Clewell 1995) and to conduct targeted research to confirm the association of toxicity with GSH-pathway metabolites. The refinement of the DCM PBPK model structure based on targeted mechanistic studies (schematized in Fig. 1.3) was particularly useful in establishing the GSH-pathway mode of action for DCM carcinogenicity and increasing confidence in application of this PBPK model in risk assessment.

Since the first proposal to use a PBPK model in risk assessment with DCM in 1987, the field of PBPK modeling in relation to risk modeling has grown steadily. In more recent years, there has been increasing acceptance of the use of these PBPK dosimetry models in a variety of risk assessments. The US EPA's Reference Concentration (RfC) documentation explicitly includes routine application of interspecies differences in dosimetry in assessing RfCs (US EPA 1994). The RfC documentation for vinyl chloride in the US EPA's Integrated Risk Information System (IRIS) specifically describes and uses a PBPK model for standard setting and dose route extrapolations (US EPA 2000a). A risk assessment with acrylic acid (Andersen *et al.* 2000) applies a PBPK model linked to computational fluid dynamic calculations of nasal airflow (Frederick *et al.* 1998) to derive an RfC. The Hazardous Air Pollutants (HAPS) Test Rule (Federal Register 1996) invited increased efficiency/efficacy in testing by providing possible substitution of certain oral toxicity tests instead of requiring new inhalation studies, for those instances where a validated PBPK model is available to conduct extrapolations across dose routes. Several case studies under the proposed cancer guidelines (US EPA 1996)—for example, those with formaldehyde (CIIT 1999), chloroform (International Life Sciences Institute 1977), and vinyl acetate (Bogdanffy *et al.* 1999)—base low-dose extrapolation on mode-of-action-specific target tissue doses calculated with PBPK models. In Chapter 8 of the US EPA's TCDD reassessment (US EPA 2000b), a variety of mechanistic protein induction models were evaluated to assess the predictions for low-dose risks for each of them. The discussion of these models includes a clear delineation of experiments that would be directly useful for establishing modes of action of the induction processes and for using the models to aid in risk assessment.

1.3.5 PBPK Models as Repository of Mechanistic Data on Distribution and Response

DCM caused lung tumors in the tracheobronchial regions of the lungs, with the tumors arising either from or near Clara cell-rich regions of the lung epithelium. The risk model for this compound equilibrated DCM in air with arterial blood and passed DCM onto tissues in the lungs via arterial blood flow. A variety of low-molecular-weight volatile compounds are known to cause lung tumors in mice and to cause damage to nasal tissues, especially to the olfactory epithelium. The initial PBPK models available for DCM or even more comprehensive models for styrene and its metabolites (Csanady *et al.* 1994), another compound that causes mouse lung tumors and rodent nasal tissue toxicity, were not developed to look at uptake of vapors directly from airway to the epithelial tissues. Airway equilibration of soluble, hydrophilic compounds restricts the uptake of chemical into the deeper lung and into systemic blood and PBPK models for equilibration with tissues have been developed for this phenomenon (Johanson 1991; Kumagai and Matsunaga 1995). PBPK models for nasal uptake of organic esters and acids with nasal tissue toxicity (Plowchalk *et al.* 1997; Bogdanffy *et al.* 2001; Frederick *et al.* 1998) have aided in risk assessments for these endpoints. These models also included airway transport to epithelial tissues with little delivery to these tissues from systemic tissues. What is the best model structure to use for relating nasal or lung responses of DCM or styrene?

Recently, more complete PBPK models for the airway/tissue interface in the lungs were applied to coordinate a risk assessment with styrene for mouse lung tumors and mouse and rat nasal tissue toxicity (Sarangapani *et al.* 2002). Another styrene model that includes airway equilibration and local metabolism in epithelial tissues in the lungs appeared recently (Csanady *et al.* 2003). This model utilized attributes of previous work on airflow modeling in the nasal airway and equilibration processes in the tracheobronchial region to relate toxic responses with metabolized dose of styrene in tissues. A wide range of research undergirds the assignment of toxic responses with tissue exposure to specific epoxide metabolites derived from oxidation of styrene by cytochrome P450 family enzyme complexes in epithelial tissues. The styrene PBPK model is an "evergreen" repository for quantitatively organizing information on the toxicology, dosimetry, and mode(s) of action with any compound; that is, the repository is continually updated as new information becomes available. The content of the repository includes information on (1) routes of administration in animal studies/human populations, (2) pathways of metabolism and interactions among groups of primary and secondary metabolites, (3) assignment of causality of responses with particular measures of dose (modes of action), and (4) definition in equation form of the accumulation and persistence of active toxic compounds at target sites. Model output supports risk assessment by calculating measures of dose for various situations in animals and exposed humans. It is possible to consider development of an annotated model with supporting information maintained in a linked database to a PBPK model structure that would serve drug development or safety assessments. We are not yet aware of any current software of this type, although similar applications for modeling signal transduction pathways are under development (www.bioinformaticsservices.com).

In developing and applying these newer models for styrene, discrepancies in dose–response relationships for inhalation toxicity in the lung, liver, and kidney became clarified for styrene and for other volatile compounds, such as chloroform, by elaborating the description of the respiratory tract. For relatively soluble compounds that have high systemic clearance, responses appear in pulmonary sites at lower exposure concentrations than in the liver and kidney because of the direct equilibration of airway tissues with inhaled gases and vapors. The liver and kidney only equilibrate with chemical in blood. This exercise also showed that earlier models with metabolism in the lung occurring only from compound delivered to pulmonary tissues by arterial blood were not realistic. Thus, the early model for DCM was not optimal for assessing pulmonary responses even though it had all the information available at the time for organizing the risk assessment. Ongoing research, both with DCM specifically and with broader sets of compounds affecting epithelial tissues in lung and nose, provided the impetus to create new model structures and new modeling approaches that included descriptions of airway equilibration, diffusion from air to epithelial tissues, and solubility and reaction directly in these tissues. We have to be prepared for these new ideas, new model structures, and the need to reanalyze datasets in the light of such new information.

Pharmacokinetics, as with studies of modes of action or metabolism, represents both (a) applications of mature model structures to specific compounds and (b) research applications to assess the adequacy of current model structures. Research needs for PK modeling will be repeatedly uncovered by data that do not behave as expected compared to models in current use, as occurred with blood concentrations of cyclic siloxanes after cessation of inhalation exposures, and when older model structures are applied to new questions, as with the comparison of nasal and liver toxicity of styrene. The continued development of PBPK methods in risk assessment must recognize that PBPK model building evolves in direct relation to the breadth of compounds and problems that have been evaluated. PBPK modeling has many applications, but at its core, it is very much a research endeavor to understand in detail the biological processes that control the concentrations of chemicals reaching target tissues in the body.

1.4 SUMMARY

In addition to the historical background for PBPK models, this overview on the use of PBPK dosimetry models in toxicology research and risk assessment has focused on several model applications: (a) *Exploratory evaluations* of proposed modes of action are possible by comparing biological responses in intact animals with various measures of dose—as done with evaluations of contributions from the two pathways of metabolism for DCM for carcinogenicity. (b) *Interpretive evaluations* occur in applying estimated dose metrics to assess acceptable exposure levels. In this use, the experimental exposures are first converted to internal dose metrics; the no-observed-adverse-effect levels (NOAELs) or benchmark doses from animal studies are then calculated in terms of these tissue dose metrics. Next, uncertainty factors are applied to target tissue dose metric and, finally, a human PBPK model is used to estimate the ambient exposure level that would give rise to this target tissue dose metric. The first time that this process was explicitly described was in a volume of *Drinking Water and Health* (NRC, 1987). (c) *Mechanistic evaluations* allow testing of the importance of various processes in controlling the tissue disposition and responses to chemicals. The discovery of suicide inhibition by *t*DCE or of the lipid storage compartments in blood presumed to exist from siloxane kinetics were noted because predictions from successful model structures failed to work in a new situation. The failure led to articulation of new ideas about determinants of kinetic behaviors, new model structures, and improvement of the fit between model and data. These three uses appear to be the dominant applications of PBPK modeling at the present time.

The following chapters attempt to put into some perspective the large number of articles dealing with applications of PBPK modeling in contemporary pharmacology, toxicology, and risk assessment. We found upwards of 1000 citations that were included in our review process. The organization of the chapters is on the types of compounds rather than on the nature of the problems attacked by use of these models. This initial chapter should provide some idea of the research, testing, and risk assessment questions that underlie the modeling for all the groups of compounds. We believe that the applications of PBPK modeling will remain diverse and will expand in both toxicology and pharmacology. For instance, PBPK modeling approaches for simple and complex mixtures are providing new opportunities to link groups of models for individual compounds together and to gain insights into the types of PK interactions possible when multiple compounds are present in test environments or in the real world. Despite the diversity of these various PBPK applications, it is important to note that they all, at their core, are investigations of the mechanisms by which chemicals move about in the body and reach target tissues in sufficient concentrations to elicit biological responses. As the mechanisms of distribution become clarified by quantitative evaluation of kinetic behavior with structured PBPK models, the mature models, containing our understanding of the full set of data on dosing, responses, and mode(s) of action, become central components of dose–response assessments for safety and risk assessments. The editors and authors hope that this volume provides a broad view of the current status of PBPK modeling activities and the spark to expand the applications of these approaches more broadly and especially to link these tools with PBPD models based on biological principles related to chemical actions in target tissues.

NOTATION

C_{vt}	the concentration of chemical in the venous blood exiting a tissue
C_t	the concentration of chemical in a tissue
DCM	dichloromethane
GSH	glutathione
HAPS	hazardous air pollutants
IRIS	Integrated Risk Information System
MB-DE	mass balance differential equation

NOAELS	no-observed-adverse-effect levels
PBPD	physiologically based pharmacodynamic
PBPK	physiologically based pharmacokinetic
PK	pharmacokinetic
P_t	tissue : blood partition coefficient
RfC	reference concentration
SAR	structure–activity relationship
TCDD	2,3,7,8-tetrachlorodibenzo-*p*-dioxin
***t*DCE**	*trans*-1,2-dichloroethylene

REFERENCES

Andersen, M. E. (1981). A physiologically based toxicokinetic description of the metabolism of inhaled gases and vapors—Analysis at steady state. *Toxicol. Appl. Pharmacol.* **60**, 509–526.

Andersen, M. E., Gargas, M. L., and Ramsey, J. C. (1984). Inhalation pharmacokinetics: Evaluating systemic extraction, total *in vivo* metabolism, and time course of enzyme induction for inhaled styrene in rats based on arterial blood : inhaled air concentration ratios. *Toxicol. Appl. Pharmacol.* **73**, 176–187.

Andersen, M. E., Clewell, H. J., III, Gargas, M. L., Smith, F. A., and Reitz, R. H. (1987). Physiologically-based pharmacokinetics and the risk assessment process for methylene chloride. *Toxicol. Appl. Pharmacol.* **87**, 185–205.

Andersen, M. E., Sarangapani, R., Gentry, P. R., Clewell, H. J., III, Covington, T. R., and Frederick, C. B. (2000). Application of a hybrid CFD-PBPK nasal dosimetry in an inhalation risk assessment: An example with acrylic acid. *Toxicol. Sci.* **57**, 312–325.

Andersen, M. E., Sarangapani, R., Reitz, R. H., Gallavan, R. H., Dobrev, I. D. and Plotzke, K. P. (2001). Physiological modeling reveals novel pharmacokinetic behavior for inhaled octamethylcyclotetrasiloxane in rats. *Toxicol. Sci.* **60**, 214–231.

Angelo, M. A., Bischoff, K. B., Pritchard, A. B., and Presser, M. A. (1984). A physiological model for the pharmacokinetics of methylene chloride in B6C3F1 mice following i.v. administration. *J. Pharmacokinet. Biopharm.* **12**, 413–436.

Bischoff, K. B., and Brown, R. H. (1966). Drug distribution in mammals. *Chem. Eng. Prog. Sym. Series* **62**, 33–45.

Bischoff, K. B., Dedrick, R. L., Zaharko, D. S., and Longstreth, J. A. (1971). Methotrexate pharmacokinetics. *J. Pharm. Sci.* **60**, 1128–1133.

Bogdanffy, M. S., Sarangapani, R., Plowchalk, D. R., Jarabek, A., and Andersen, M. E. (1999). A biologically-based risk assessment for vinyl acetate-induced cancer and non-cancer inhalation toxicity. *Toxicol. Sci.* **51**, 19–35.

Bogdanffy, M. S., Plowchalk, D. R., Sarangapani, R., Starr, T. B. Andersen, M. E. (2001). Mode-of-action-based dosimeters for interspecies extrapolation of vinyl acetate inhalation risk. *Inhal. Toxicol.* **13**, 377–396.

Brown, R. P., Delp, M. D., Lindstedt, S. L., Rhomberg, L. R., and Beliles, R. P. (1997). Physiological parameter values for physiologically based pharmacokinetic models. *Toxicol. Ind. Health* **13**, 407–484.

CIIT (1999). *Formaldehyde: Hazard characterization and dose-response assessment for carcinogenicity by the route of inhalation*, Chemical Industry Institute of Toxicology, Research Triangle Park, NC.

Clewell, H. J. (1995). The use of physiologically based pharmacokinetic modeling in risk assessment: a case study with methylene chloride. In Olin, S., Farland, W., Park, C., Rhomberg, L., Scheuplein, R., Starr, T., and Wilson, J. eds. Low-dose Extrapolation of cancer risk: Issues and Perspectives ILSI Press, Washington, DC.

Clewell, H. J., III, and Andersen, M. E. (1985). Risk assessment extrapolations and physiological modeling. *Toxicol. Ind. Health* **1**, 111–131.

Clewell, H. J., III, and Andersen, M. E. (1996). Use of physiologically based pharmacokinetic modeling to investigate individual versus population risk. *Toxicology*, **111**. 315–329.

Collins, J. M., Dedrick, R. L., Flessner, M. F., and Guarino, A. M. (1982). Concentration dependent disappearance of fluorouracil from peritoneal fluid in the rat: Experimental observations and distributed modeling. *J. Pharm. Sci.* **71**, 735–738.

Csanady, G. A., Mendrala, A. L., Nolan, R. J., and Filser, J. G. (1994). A physiologic pharmacokinetic model for styrene and 7,8-styrene oxide in mouse, rat, and man. *Arch. Toxicol.*, **68**, 143–157.

Csanady G. A., Kessler W., Hoffmann H. D., and Filser J. G. (2003). A toxicokinetic model for styrene and its metabolite styrene-7,8-oxide in mouse, rat and human with special emphasis on the lung. *Toxicol. Lett.* **138**, 75–102.

Dedrick, R. L. (1973). Animal scale-up. *J. Pharmacokinet. Biopharm.* **1**, 435–461.

Dobrev, I. D., Reddy, M. B., Plotzke, K. P., Varaprath, S., McNett, D. A., Durham, J., and Andersen, M. E. (2003). Closed chamber inhalation pharmacokinetic studies with hexamethyldisiloxane in the rat. *Inhal. Toxicol.* **15**, 589–617.

Farris, F. F., Dedrick, R. L., and King, F. G. (1988). Cisplatin pharmacokinetics: Applications of a physiological model. *Toxicol. Lett.* **43**, 117–137.

Federal Register (1996). Part II Environmental Protection Agency. 40 CFR 799. Proposed test rule for hazardous air pollutants; Proposed Rule. *Fed. Reg.*, June 26, 1996. 33177–33200.

Filser, J. G., and Bolt, H. M. (1979). Pharmacokinetics of halogenated ethylenes. *Archiv. Toxicol.* **42**, 123–136.

Fiserova-Bergerova, V. (1975). Mathematical modeling of inhalation exposure. *J. Combust. Toxicol.* **32**, 201–210.

Fiserova-Bergerova, V., and Holaday, D. A. (1979). Uptake and clearance of inhalation anesthetics in man. *Drug Metab. Rev.* **9**, 43–60.

Fiserova-Bergerova, V., Vlach, J., and Cassady, J. C. (1980). Predictable "individual differences" in uptake and excretion of gases and lipid soluble vapours simulation study. *Br. J. Ind. Med.* **37**, 42–49.

Frederick, C., Bush, M. L., Subramaniam, R. P., Black, K. A., Finch, L., Kimbell, J. S., Morgan, K. T., Subramaniam, R. P., Morris, J. B., and Ultman, J. S. (1998). Application of a hybrid computational fluid dynamics and physiologically based inhalation model for interspecies dosimetry extrapolation of acidic vapors in the upper airways. *Toxicol. Appl. Pharmacol.* **152**, 211–231.

Gargas, M. L., and Andersen, M. E. (1988). Physiologically based approaches for examining the pharmacokinetics of inhaled vapors. In: *Toxicology of the Lung*, D. E. Gardner, J. D. Crapo, and E. J. Massaro, eds., Raven Press, New York, pp. 449–476.

Gargas, M. L., Andersen, M. E., and Clewell, H. J., III (1986). A physiologically based simulation approach for determining metabolic constants from gas uptake data. *Toxicol. Appl. Pharmacol.* **86**, 341–352.

Gargas, M. L., Burgess, R. J., Voisard, D. E., Cason, G. H., and Andersen, M. E. (1989). Partition coefficients of low-molecular-weight volatile chemicals in various liquids and tissues. *Toxicol. Appl. Pharmacol.* **98**, 87–99.

Gargas, M. L., Clewell, H. J., III, and Andersen, M. E. (1990). Gas uptake inhalation techniques and the rates of metabolism of chloromethanes, chloroethanes, and chloroethylenes in the rat. *Inhal. Toxicol.* **2**, 285–309.

Gehring, P. J., Watanabe, P. G., and Blau, G. E. (1976). Pharmacokinetic studies in evaluation of the toxicological and environmental hazard of chemicals. In: *New Concepts in Safety Evaluation*, Advances in *Modern Toxicology*, Vol. 1, M. A. Mehlman, R. E. Shapiro, and H. Blumenthal, eds., Hemisphere Publishing Corp., New York, pp. 193–270.

Gehring, P. J., Watanabe, P. G., and Young, J. D. (1977). The relevance of dose-dependent pharmacokinetics in the assessment of carcinogenic hazard of chemicals. In: *Origins of Human Cancer, Book A: Incidence of Cancer in Humans*, Cold Spring Harbor Conferences on Cell Proliferation, Vol. 4, H. H. Hiatt, J. D. Watson, and J. A. Winsten, eds., Cold Spring Harbor Laboratory, Cold Spring Harbor, NY, pp. 187–203.

Gehring, P. J., Watanabe, P. G., and Park, C. N. (1978). Resolution of dose-response toxicity data for chemicals requiring metabolic activation: Example—vinyl chloride. *Toxicol. Appl. Pharmacol.* **44**, 581–591.

Gerlowski, L. E., and Jain, R. J. (1983). Physiologically based pharmacokinetic modeling: Principles and applications. *J Pharm. Sci.* **72**, 1103–1126.

Haggard, H. W. (1924a). The absorption, distribution, and elimination of ethyl ether. II. Analysis of the mechanism of the absorption and elimination of such a gas or vapor as ethyl ether. *J. Biol. Chem.* **59**, 753–770.

Haggard, H. W. (1924b). The absorption, distribution, and elimination of ethyl ether. III. The relation of the concentration of ether, or any similar volatile substance, in the central nervous system to the concentration in the arterial blood, and the buffer action of the body. *J. Biol. Chem.* **59**, 771–781.

Health Canada (1993). *Canadian Environmental Protection Act, Priority Substances List Assessment Report. Dichloromethane.* Canada Communications Group—Publishing, Ottawa, Canada, K1A 0S.

Hefner, R. E., Watanabe, P. G., and Gehring, P. J. (1975). Preliminary studies of the fate of inhaled vinyl chloride monomer in rats. *Ann. N.Y. Acad. Sci.*, **246**, 135–148.

Henderson, Y., and Haggard, H. W. (1943). *Noxions Gases and the Principles of Respiration Influencing their Action.* Second Edition, Reinhold Publishing Corporation. New York, New York.

Hilderbrand, R. L., Andersen, M. E., and Jenkins, L. J. (1981). Prediction of *in vivo* kinetic constants for metabolism of inhaled vapors from kinetic constants measured *in vitro. Fundam. Appl. Toxicol.* **1**, 403–409.

Himmelstein, K., and Lutz, R. J. (1979). A review of the application of physiologically based pharmacokinetic modeling. *J. Pharmacokinet. Biopharm.* **7**, 127–137.

International Life Sciences Institute (1977). *An Evaluation of EPA's Proposed Guidelines for Carcinogen Risk Assessment Using Chloroform and Dichloroacetic Acid as Case Studies: Report of an Expert Panel*, International Life Sciences Institute, Health and Environmental Sciences Institute, Washington, D.C.

Johanson G. (1991). Modeling of respiratory exchange of polar-solvents. *Ann. Occup. Hyg.* **35**, 323–339.

Kedderis, G. L., Carfagna, M. A., Held, S. D., Batra, R., Murphy, J. E., and Gargas, M. L. (1993). Kinetic analysis of furan biotransformation by F-344 rats *in vivo* and *in vitro. Toxicol. Appl. Pharmacol.* **123**, 274–282.

Kety, S. S. (1951). The theory and applications of the exchange of inert gas at the lungs. *Pharmacol. Rev.* **3**, 1–41.

Krishnan, K., and Andersen, M. E. (2001). Physiologically based pharmacokinetic modeling in toxicology. In: *Principles and Methods of Toxicology*, A. W. Hayes, ed., Taylor & Francis, Philadelphia, pp. 193–241.

Kumagai, S., and Matsunaga, I. (1995). Physiologically-based pharmacokinetic model for acetone. *Occup. Environ. Med.* **52**, 344–352.

Leung, H. W. (1991). Development and utilization of physiologically based pharmacokinetic models for toxicological applications. *J. Toxicol. Environ. Health* **32**, 247–267.

Lilly, P. D., Thorton-Manning, J., Gargas, M. L., Clewell, H. J., III, and Andersen, M. E. (1999). Kinetic characterization of CYP2E1 inhibition *in vivo* and *in vitro* by the chloroethylenes. *Arch. Toxicol.* **72**, 609–621.

Mapleson, W. W. (1963). An electric analog for uptake and exchange of inert gases and other agents. *J. Appl. Physiol.* **18**, 197–204.

McKenna, M. J., Zempel, J. A., Madrid, E. O., Braun, W. H., and Gehring, P. J. (1978a). Metabolism and pharmacokinetic profile of vinylidene chloride in rats following oral administration. *Toxicol. Appl. Pharmacol.* **45**, 821–835.

McKenna, M. J., Zempel, J. A., Madrid, E. O., and Gehring, P. J. (1978b). The pharmacokinetics of [^{14}C]vinylidene chloride in rats following inhalation exposure. *Toxicol. Appl. Pharmacol.* **45**, 599–610.

McKenna, M. J., Zempel, J. A., and Braun, W. H. (1982). The pharmacokinetics of inhaled methylene chloride in rats. *Toxicol. Appl. Pharmacol.* **65**, 1–10.

National Research Council (1986). *Drinking Water and Health*, Vol. 6, Chapter 6, National Academy Press, Washington, D.C., 1986.

National Research Council (1987). *Pharmacokinetics in Risk Assessment in Drinking Water and Health*, Vol. 8, National Academy Press, Washington, D.C., 487 pp.

Occupational Safety and Health Administration (OSHA) (1997). Occupational exposure to methylene chloride; final rule. *Fed. Reg.* **62**(7), 1493–1619.

Parham, F. M., Kohn, M. C., Matthews, H. B., DeRosa, C., and Portier, C. J. (1997). Using structural information to create physiologically based pharmacokinetic models for all polychlorinated biphenyls. *Toxicol. Appl. Pharmacol.* **144**, 340–347.

Parham, F. M., and Portier, C. J. (1998). Using structural information to create physiologically based pharmacokinetic models for all polychlorinated biphenyls. II. Rates of metabolism. *Toxicol. Appl. Pharmacol.* **151**, 110–116.

Plowchalk, D. R., Andersen, M. E., and Bogdanffy, M. S. (1997). Physiologically-based modeling of vinyl acetate uptake, metabolism and intracellular pH changes in the rat nasal cavity. *Toxicol. Appl. Pharmacol.* **142**, 386–400.

Portier, C. J., and Kaplan, N. L. (1989). Variability of safe dose estimates when using complicated model of the carcinogenic process. A case study: methylene chloride. *Fund. Appl. Toxicol.* **13**, 533–544.

Poulin, P., and Krishnan, K. (1996). Molecular structure-based prediction of the partition coefficients of organic chemicals for physiological pharmacokinetic models. *Toxicol. Methods* **6**, 117–137.

Poulin, P., and Krishnan, K. (1999). Molecular structure-based prediction of the toxicokinetics of inhaled vapors in humans. *Int. J. Toxicol.* **18**, 7–18.

Ramsey, J. C., and Andersen, M. E. (1984). A physiologically based description of the inhalation pharmacokinetics of styrene in rats and humans. *Toxicol. Appl. Pharmacol.* **73**, 159–175.

Ramsey, J. C., Young, J. D., Karbowski, R., Chenoweth, M. B., McCarty, L. P., and Braun, W. H. (1980). Pharmacokinetics of inhaled styrene in human volunteers. *Toxicol. Appl. Pharmacol.* **53**, 54–63.

Riggs, D. S. (1963). *The Mathematical Approach to Physiological Problems: A Critical Primer*. MIT Press, Cambridge, MA, 445 pp.

Sarangapani, R., Teeguarden, J. G., Cruzan, G., Clewell, H. J., and Andersen, M. E. (2002). Physiologically based pharmacokinetic modeling of styrene and styrene oxide respiratory-tract dosimetry in rodents and humans. *Inhal. Toxicol.* **14**, 789–834.

Sato, A., and Nakajima, T. (1979a). Partition coefficients of some aromatic hydrocarbons and ketones in water, blood and oil. *Br. J. Ind. Med.* **36**, 231–234.

Sato, A., and Nakajima, T. (1979b). A vial equilibration method to evaluate the drug metabolizing enzyme activity for volatile hydrocarbons. *Toxicol. Appl. Pharmacol.* **47**, 41–46.

Sauerhoff, M. W., Braun, W. H., Blau, G. E., and Gehring, P. J. (1976). The dose dependent pharmacokinetic profile of 2,4,5-trichlorophenoxy acetic acid following intravenous administration to rats. *Toxicol. Appl. Pharmacol.* **36**, 491–501 .

Sauerhoff, M. W., Braun, W. H., and LeBeau, J. E. (1977). Dose dependent pharmacokinetic profile of silvex following intravenous administration in rats. *J. Toxicol. Environ. Health* **2**, 605–618.

Teorell, T. (1937a). Kinetics of distribution of substances administered to the body. I. The extravascular modes of administration. *Arch. Int. Pharmacodyn.* **57**, 205–225.

Teorell, T. (1937b). Kinetics of distribution of substances administered to the body. II. The intravascular mode of administration. *Arch. Int. Pharmacodyn.* **57**, 226–240.

US Environmental Protection Agency (1994). *Methods for Derivation of Inhalation Reference Concentrations and Application of Inhalation Dosimetry*, EPA/600/8-90/066F, Office of Research and Development, Washington, D.C.

US Environmental Protection Agency (1996). *Proposed Guidelines for Carcinogen Risk Assessment*, EPA/600/P-92/003C, Office of Research and Development, Washington, D.C.

US Environmental Protection Agency (2000a). *Toxicological Review of Vinyl Chloride. Appendices A–D.* EPA/635R-00/004.

US Environmental Protection Agency (2000b). *Exposure and Human Health Reassessment of 2,3,7,8-Tetrachlorodibenzo-p-dioxin (TCDD) and Related Compounds. Part II. Health Assessment for 2,3,7,8-Tetrachlorodibenzo-p-dioxin (TCDD) and Related Compounds*, Chapter 8. Dose-Response Modeling for 2,3,7,8-TCDD. NCEA-1-0835. May 2000. Science Advisory Board Review Draft.

Yang, R. S. H., Andersen, M. E., Dennison, J. E., Ou, Y.C., Liao, K. H., and Reisfeld, B. (2004). Physiologically based pharmacokinetic and pharmacodynamic modeling. In: *Mouse Models of Cancer*, E. C. Holland, ed., John Wiley & Sons, New York, pp. 391–405.

Young, J. D., Ramsey, J. C., Blau, G. E., Karbowski, R. J., Nitschke, K. D., Slauter, R. W., and Braun, W. H. (1979). Pharmacokinetics of inhaled or intraperitoneally administered styrene in rats. *Toxicology and Occupational Medicine: Proceedings of the Tenth Inter-America Conference on Toxicology and Occupational Medicine*, Elsevier/North Holland, New York, pp. 297–310.

PART I

PBPK MODELING FOR VOLATILE ORGANIC COMPOUNDS

CHAPTER 2

HALOGENATED ALKANES

Micaela B. Reddy, Ivan D. Dobrev, and James E. Dennison

2.1 INTRODUCTION

In this chapter, we discuss PBPK models for many halogenated alkanes. This group of compounds includes volatile anesthetic gases, refrigerant gases, and halons. Additionally, many PBPK models have been developed for other halogenated methanes and ethanes. The purpose of this chapter is twofold. First, principles guiding the development of PBPK models for volatile gases are reviewed. Second, PBPK models for the various types of chemicals in this chapter are discussed to provide an understanding of the structure of fairly basic PBPK models. Because this chapter encompasses such a large area of research, it would be impossible to cover every contribution in detail. The tables summarizing the models are intended to be comprehensive, while the text focuses more narrowly on advances in experimental methods and on well-studied compounds and techniques.

2.2 PBPK MODEL DEVELOPMENT FOR VOLATILE ORGANICS

In PBPK model development, the body is treated as compartments representing specific physiological organs and/or tissues in parallel connected by blood flow.

Physiologically Based Pharmacokinetic Modeling, edited by Micaela B. Reddy, Raymond S. H. Yang, Harvey J. Clewell III, and Melvin E. Andersen.

Chemical can transfer from compartment to compartment by partitioning into and out of the blood. In model development, the simplest model possible treats every tissue as a single lumped compartment (see the decision tree structure in Fig. 2.1). While this approach is easy to implement, it does not represent the physiology well. One possible alternative to this simple treatment would be to split every organ and tissue out of this single lumped compartment for inclusion as an individual compartment with its own physiological, physicochemical, and biochemical properties. This approach represents the physiology well, but is complex and difficult to implement. A compromise between the two approaches results in a model structure that is physiologically accurate enough to be useful and yet simple enough to allow for physiologically meaningful parameterization.

2.2.1 Model Formulation

The level of detail included in PBPK models varies considerably. Some model structures are designed to contain all known factors that might influence kinetics. Frequently, these more detailed models are simplified by determining which of the model components can be eliminated without losing the ability to describe specific datasets. The other extreme of the spectrum is to develop model structures that are as simple as possible for any particular compound and route of exposure. In this latter approach, the Principle of Parsimony, also called Occam's razor, can be used to guide model development. According to this principle, the model structure should include no more complexity than is necessary to describe specific datasets and make predictions about specific exposures. Developing a PBPK model requires several steps that help determine an appropriate model structure based on this principle. First, the problem must be identified and defined so that a model structure can be chosen that will address the problem. Second, a literature search must be conducted. This literature search may result in information on model components essential for

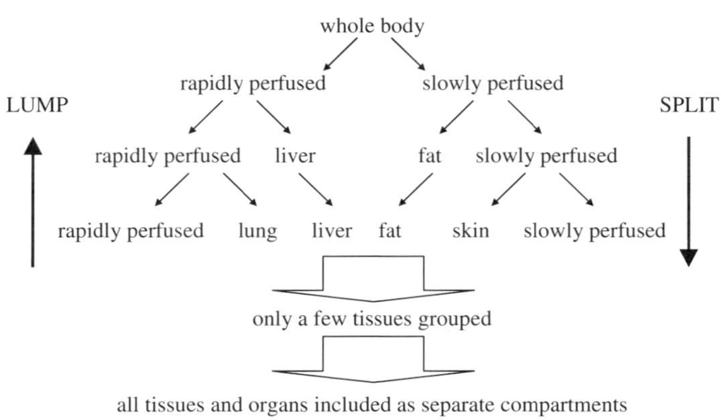

Figure 2.1 Schematic diagram of the method for lumping and splitting the body for a PBPK model.

appropriately describing pharmacokinetics. Third, a model structure, containing as few compartments as possible, must be developed. Reasons to split a tissue from a lumped compartment and treat it individually include:

1. The tissue is an exposure route.
2. The tissue is a target organ.
3. Biotransformation occurs within the tissue.
4. The tissue is the site of some other clearance process.
5. The tissue affects pharmacokinetics in some way.
6. Data on the tissue are available for comparison with simulation results.

The experimental apparatus included in a study may also be an important element of a model structure. For example, in a gas uptake study (Filser and Bolt 1979), the inhalation chamber must be included if the model is intended to simulate concentrations of chemical in the chamber air.

The complexity of the selected model structure must be consistent with the quality and breadth of the datasets available for model development. After determining the appropriate model structure, equations reflecting this model structure must be generated. A simple model structure is shown in Fig. 2.2. In this model, all tissues except the blood and liver are lumped into a single tissue compartment. The liver is included as the site of biotransformation. Equations for this simple PBPK model will be described in the following section.

2.2.2 Model Equations

For volatile organic compounds, the ability to describe inhalation exposures is an important consideration for modeling inhalation pharmacokinetics. By modeling the lungs as a well-mixed compartment with an average, one-directional airflow in the region of gas exchange (i.e., with air moving through the lungs with a constant flow rate equal to the alveolar ventilation rate, Q_P), and with rapid equilibration between lung air and blood in the lung alveoli, the concentration in the blood exiting the lungs, C_a, can be described as (Ramsey and Andersen 1984)

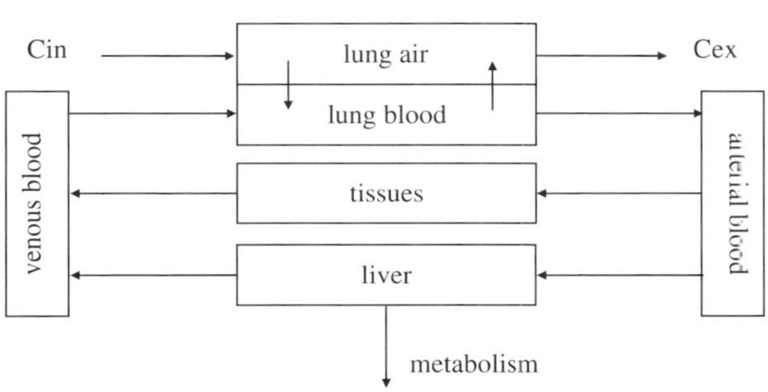

Figure 2.2 Schematic diagram of a simple PBPK model

$$C_a = \frac{Q_P \times C_{\text{in}} + Q_C \times C_{\text{BLV}}}{Q_C + Q_P / P_b} \tag{2.1}$$

where C_{BLV} is the chemical concentration in the venous blood compartment, P_b is the blood : air partition coefficient, Q_C is the cardiac output, and C_{in} is the inhaled concentration of chemical during the exposure and zero after the exposure ends. To determine the concentration of chemical in exhaled air, C_{ex}, the concentration of chemical in alveolar air, C_{alv} (i.e., C_a/P_b), must be adjusted for the concentration of chemical in the dead space of the lungs (i.e., the volume of the lungs where gas exchange does not occur) as follows:

$$C_{\text{ex}} = F_{\text{DS}} \times C_{\text{in}} + (1 - F_{\text{DS}}) \times C_{\text{alv}} \tag{2.2}$$

where F_{DS} is the fraction of dead space in the lungs, which is about 0.33 in rats and humans under typical physiological conditions.

In early PBPK models, constructed when computers were not very advanced and computational time required by the computer was a consideration, often only one blood compartment was included in the model structure. Today, computers are fast enough that the number of compartments does not noticeably increase the amount of time required by the computer for simulations, and separate compartments are typically included for the venous and arterial blood. The mass balance equation for the chemical concentration in the arterial blood compartment, C_{BLA}, is

$$V_{\text{BLA}} \frac{dC_{\text{BLA}}}{dt} = Q_C \times (C_a - C_{\text{BLA}}) \tag{2.3}$$

where t is time and V_{BLA} is the volume of the arterial blood compartment. The rate of mass transfer of chemical into tissue compartments is often limited by the blood flow to the tissues (i.e., uptake is perfusion limited), as shown in Eq. (2.4):

$$V_T \frac{dC_T}{dt} = Q_T \times \left(C_{\text{BLA}} - \frac{C_T}{P_T} \right) \tag{2.4}$$

where V_T is the volume of the tissue compartment, Q_T is the flow rate of blood to the tissue compartment, C_T is the concentration of chemical in the tissue compartment, and P_T is the tissue compartment : blood partition coefficient. As Eq. (2.4) is written, uptake of chemical by the tissue compartment is rapid enough that the rate of transfer into the tissue is limited by the rate at which chemical in blood reaches the compartment.

Sometimes, the mass transfer in some or all tissues can be limited by the rate of diffusion into the tissue, and the mass balance on the tissue compartment would become

$$V_T \frac{dC_T}{dt} = \text{PA}_T \times \left(C_{\text{BLA}} - \frac{C_T}{P_T} \right) \tag{2.5}$$

where PA_T is the permeability–area cross product for the tissue, which determines the rate at which chemical can enter the tissue compartment. In this case, the blood exiting the slowly perfused tissue compartment is not equilibrated with the tissue compartment (i.e., the concentration in blood leaving the tissue compartment, C_{EXT}, is not equal to C_T/P_T), as illustrated in Eq. (2.6):

$$C_{\mathrm{EXT}} = C_{\mathrm{BLA}} - \frac{\mathrm{PA}_T}{Q_T} \times \left(C_{\mathrm{BLA}} - \frac{C_T}{P_T} \right) \tag{2.6}$$

Often, the liver is treated as a blood-flow-limited compartment with Michaelis–Menten kinetics for metabolism (i.e., with a maximal reaction velocity $V_{\max}$ and Michaelis–Menten constant K_m), and the concentration of chemical in the liver, C_L, can be calculated using the following equation:

$$V_L \frac{dC_L}{dt} = Q_L \times \left(C_{\mathrm{BLA}} - \frac{C_L}{P_L} \right) - \frac{V_{\max}(C_L/P_L)}{K_m + (C_L/P_L)} \tag{2.7}$$

where V_L is the volume of the liver, Q_L is the blood flow rate to the liver, and P_L is the liver : blood partition coefficient. In Eq. (2.7), the rate of metabolism is related to C_L/P_L because this is the concentration of chemical in the venous return exiting the liver and is considered to be representative of the concentration of free chemical in the liver. Other terms for metabolic clearances following first-order and second-order metabolic processes have been included in some PBPK models. The blood flows from the organ compartments combine in a mixed venous return compartment:

$$V_{\mathrm{BLV}} \frac{dC_{\mathrm{BLV}}}{dt} = Q_L \times \frac{C_L}{P_L} + Q_T \times \frac{C_T}{P_T} - Q_C \times C_{\mathrm{BLV}} \tag{2.8}$$

where V_{BLV} is the volume of the venous blood compartment. For the case with diffusion-limited uptake in the tissue compartment, the term $Q_T \times C_T/P_T$ would be replaced by the term $Q_T \times C_{\mathrm{EXT}}$ in Eq. (2.8). At the beginning of the exposure, the concentration of chemical in all compartments is assumed to be zero (i.e., $C_{\mathrm{BLA}} = 0$, $C_T = 0$, $C_L = 0$, and $C_{\mathrm{BLV}} = 0$ at $t = 0$). Metabolism or binding can be added in any of the tissue compartments, including arterial or venous blood.

2.2.3 Model Parameterization

After developing the model equations, values for all parameters in the model must be specified. In general for PBPK model simulations, values for three types of parameters (i.e., physiological, physicochemical, and biochemical parameters) must be set to appropriate values. Physiological parameters include body weight (used to estimate volume), fractions of the body in each compartment cardiac output, blood flow rates to compartments, and the alveolar ventilation rate. This type of information has been compiled for many species by Brown *et al.* (1997).

Blood : air and tissue : air partition coefficients for volatile compounds, which can be used to calculate tissue : blood partition coefficients, have been experimentally determined for groups of compounds by several investigators. For example, blood : air, liver : air, muscle : air and fat : air partition coefficients for 55 chemicals in rat tissue were determined by Gargas *et al.* (1989). Biochemical parameters include metabolic rate constants and rate constants or dissociation constants for macromolecular binding. Biochemical parameters can be obtained by fitting a PBPK model to data from pharmacokinetic studies (e.g., time-course concentrations of chemical in blood and tissues) or from *in vitro* experiments. A detailed discussion of the physicochemical and biochemical parameters and the types of data available to determine appropriate values is found in Krishnan and Andersen (2001). Additionally, appropriate values for all three types of model parameters for many compounds can be found in previously published PBPK models.

2.2.4 Model Calculations

After model equations have been specified and values have been assigned to all model parameters, software packages designed to solve coupled ordinary differential equations—for example, acslXtreme® (AEgis Technologies Group), MatLab®, or Berkeley Madonna™ (Macey & Oster)—can be used to perform model simulations and calculations. PBPK models can be used for a number of useful applications. For example, workplace exposures to chemicals of interest in occupational health can be simulated, and target tissue doses resulting from specific exposures can be calculated. As more data and physiology are incorporated in model development, a model can be used for calculations and predictions with increasing confidence.

After a model has been developed, it can be used to examine the impact of variability and uncertainty on model output. Variability applies to parameters that are expected to be different for individual subjects (e.g., humans have a wide level of variability in body weight). In contrast, uncertainty refers to the situation where a parameter has an exact value, but the exact value is unknown and must be estimated (e.g., an individual person has an exact body weight, but this must be measured and there will be error associated with the measurement). The Monte Carlo technique (Portier and Kaplan 1989; Clewell and Andersen 1996) can be used to examine the impact of both effects. A PBPK model incorporating information on the range of variability in appropriate model parameters can be used to understand the variability in the response. While variability will remain the same, uncertainty can be reduced with additional experimentation. The Monte Carlo technique and/or a sensitivity analysis (Clewell *et al.* 1994; Reddy *et al.* 2003) can be used to determine which parameters have the greatest influence on model predictions, thereby indicating the key parameters that are the most important to know accurately and those that should be measured directly if possible.

2.3 EXPERIMENTAL METHODS DEMONSTRATED FOR GROUPS OF CHEMICALS

The introductory chapter traced the development of PBPK modeling approaches with anesthetics and other volatile gases from early conceptual work by Haggard and colleagues through more complete simulation models for metabolized gases and vapors of occupational interest by Fiserova-Bergerova and her colleagues. Steady-state analyses of the PBPK model for inhalation of metabolized vapors provided insight into the factors that determine the blood and tissue concentrations of the inhaled compound (Andersen 1981). Subsequent studies were completed to develop methods to assess specific parameters, including partition coefficients, and several approaches were evaluated for assessing kinetic constants for metabolism, such as *in vitro* studies with microsomal preparations (Hildebrand *et al.* 1981) and an *in vivo* method referred to as a gas uptake study (Hefner *et al.* 1975; Filser and Bolt 1979; Andersen *et al.* 1984).

Gas uptake studies were initially evaluated using compartmental pharmacokinetic models. Filser and Bolt (1979) used the compartmental modeling to assess kinetic constants for the series of halogenated ethylenes. In later work, results from gas uptake studies for a wide range of compounds were evaluated with a simple PBPK model. The inhalation kinetics of bromochloromethane, methyl chloroform, carbon tetrachloride, 1,1-dichloroethylene and diethyl ether were first studied in this fashion (Gargas *et al.* 1986a) and later chloromethane, chloroform, chloroethane, 1,1-dichloroethane, vinyl chloride and tetrachloroethylene were examined (Gargas *et al.* 1990). In these studies, time-course loss of the parent chemical is monitored in the chamber atmosphere, and the data are analyzed using a PBPK model. Metabolism parameters for saturable and/or first-order metabolism were included as terms in the liver compartment, and these constants were estimated by fitting model parameters to the data. These closed-chamber studies permitted rapid estimation of metabolic parameters in the intact animal by analysis with a PBPK model.

Another study on vapors with higher blood and tissue solubility used a different type of inhalation exposure regimen (i.e., a constant-concentration inhalation exposure). Here, the inhalation kinetics of 1,1,2-trichloroethane, 1,1,1,2-tetrachloroethane, 1,1,2,2-tetrachloroethane, pentachloroethane, hexachloroethane, and bromochloromethane were studied by exposing animals to a constant concentration in an exposure chamber for a set time. The rats were then moved to a chamber with clean air running through the chamber. Kinetic constants for metabolism were estimated by evaluating the time-course of accumulation of chemical in the chamber by exhalation (Gargas and Andersen 1989). Bromochloromethane inhalation pharmacokinetics in the rat had been previously studied. However, it was also included in this study to verify that metabolism parameters could be calculated by analyzing the time-course concentration data in the flow through chamber following a constant-concentration inhalation exposure with a PBPK model. A complication discovered in modeling these exhalation data was the confounding of the results by variable fur absorption and desorption in these whole-body exposures. 1,1,2-Trichloroethane and 1,1,1,2-tetrachloroethane did not accumulate appreciably on the rat fur during the exposure; significant amounts of 1,1,2,2-tetrachloroethane, pentachloroethane, and hexachloroethane did absorb onto rat fur. Rate constants for Michaelis–Menten biotransformation in the liver were successfully calculated for this group of halogenated ethanes when the fur desorption, independently evaluated by doing the study with a dead rat control group, was taken into consideration.

Methods similar to those used in these studies of rat inhalation kinetics have also been used to study human inhalation kinetics for groups of chemicals. Fiserova-Bergerova (1992) presented five-compartment and six-compartment human inhalation PBPK models for four volatile anesthetics (i.e., isoflurane, enflurane, halothane and methoxyflurane). The six-compartment model, which included the lungs, vessel-rich tissues, muscle and dermis tissues, inner adipose tissue and subcutaneous adipose tissue, performed better in simulating human pharmacokinetic data than the five-compartment model and was used to investigate the effects of body composition, metabolism, and the length of the exposure on the uptake and elimination of the four compounds. Model calculations revealed that elimination half-lives depended upon the size of the fat compartment (i.e., body composition) but not on the exposure duration. In contrast, the concentration of anesthetic agent in plasma was affected by the exposure duration but not by the size of the fat compartment. The use of two fat compartments did not affect simulated pharmacokinetics during the 24-hour period after the exposure, but affected pharmacokinetics at later times, indicating that clearance of anesthetic agent from the inner adipose tissue is the rate-limiting step for pulmonary clearance of anesthesia for times greater than 24 hours.

In a similar application of PBPK modeling to classes of chemicals, Vinegar *et al.* (1998) also examined the inhalation kinetics of halothane, isoflurane, and desflurane in humans using an inhalation PBPK model. The model contained a more detailed description of the respiratory tract in an effort to more accurately describe processes significantly affecting uptake during early times of the exposure (i.e., less than 5 minutes). This "breath by breath" model accounted for (1) the volumes of the dead space and pulmonary regions of the lung, (2) inhalation and exhalation (i.e., airflow was not treated as unidirectional), (3) the time delay resulting from air traveling through the dead space of the lung, and (4) the absorption of anesthetic by lung tissue in the pulmonary region, followed by equilibration of the tissue with lung blood. For model parameterization, both partition coefficients and metabolic constants were obtained from the literature. The PBPK models for short-term inhalation kinetics of halothane, isoflurane, and desflurane were each able to describe the end-alveolar air concentration for eight patients at times less than 5 minutes. This human PBPK model for anesthetic gases (Vinegar *et al.* 1998) was also used to simulate inhalation exposures to Halon 1211 for an actual case of accidental exposure. Later, this model structure was used in a study designed to determine safe exposure limits to prevent acute health effects from exposures to halon replacement chemicals. In this study (Vinegar *et al.* 2000), human PBPK models were developed for bromotrifluoromethane (Halon 1301), trifluoroiodomethane (CF_3I), pentafluoroethane (HFC-125), 1,1,1,3,3,3-hexafluoropropane (HFC-236fa), and 1,1,1,2,3,3,3-heptafluoropropane (HFC-227ea).

In a study that used gas uptake experiments to examine the pharmacokinetic interactions in a mixture of four trihalomethanes, da Silva *et al.* (1999) developed PBPK models for chloroform, bromoform, bromodichloromethane (BDCM), and dibromochloromethane (DBCM). As with the other studies in this section, tissue: air and blood: air partition coefficients were determined by the *in vitro* vial equilibration technique of Gargas *et al.* (1989) either as part of this study (bromoform) or from previously reported studies (chloroform, BDCM, DBCM). Using time-course data for blood concentrations following oral gavage doses of individual trihalomethanes, parameters for Michaelis–Menten kinetics and first-order GI uptake were estimated. In a later study (da Silva *et al.* 2000), the results of this first study were used to develop a PBPK model incorporating pharmacokinetic interactions demonstrated in rats administered a mixture of the four trihalomethanes. Gas uptake studies have been used to study inhalation pharmacokinetics of other mixtures (see Chapter 13).

These various studies demonstrated the reliability of PBPK models in predicting pharmacokinetics for a variety of compounds as experimental techniques were developed and perfected. Later, other investigators utilized the reliability of PBPK models, as demonstrated by many investigators, to assess human exposure to compounds such as dichloromethane, chloroform, carbon tetrachloride, and 1,1,1-trichloroethane (Yoshida 1993) and to calculate biological exposure indexes from occupational exposure limits for compounds such as methylene chloride (i.e., dibromomethane, DCM), chloroform, 1,2-dichloroethane and 1,1,1-trichloroethane (Leung 1992).

2.4 PBPK MODELS FOR HALOGENATED ALKANES

2.4.1 Anesthetic Gases

Inhalation PBPK models have been developed for five halogenated alkane volatile anesthetics (i.e., desflurane, enflurane, halothane, isoflurane, and methoxyflurane) and for two additional volatile anesthetics (i.e., nitrous oxide and diethyl ether). The majority of studies on these compounds have examined human pharmacokinetics, although PBPK models have been developed for halothane in the rat and dog and for isoflurane in the dog (Table 2.1). When administering anesthesia in a clinical setting, the goal is to manipulate the concentration of anesthesia inhaled so that the blood and expired concentrations remain constant (Fiserova-Bergerova 1992). This section will focus on a PBPK modeling approach developed by Lerou *et al.* (1991a), who developed a system model for closed-circuit inhalation anesthesia that incorporates a PBPK model to avoid the traditional assumption used in previous models that the concentration of anesthetic in the arterial blood remains constant in the calculation of anesthesia uptake.

This model, intended for clinical, teaching, and research applications, was initially developed for human inhalation exposures to nitrous oxide (N_2O), isoflurane, and enflurane (Lerou *et al.* 1991a). The model included 14 compartments including the kidney, brain, heart, liver, muscle, connective tissue, and adipose tissue interconnected by arterial and venous blood flows. The model also incorporated a description of the closed-circuit rebreathing device used to administer the anesthesia, in which it was assumed that a bolus of liquid anesthesia injected into the system vaporizes, mixes uniformly, and then behaves as an ideal gas. The injection of the bolus of anesthetic agent was simulated by the addition of an appropriate amount of vapor to the system during a 60-second period. The feedback-controlled administration of the anesthesia was simulated using a proportional-integral control scheme. The model, therefore, can be used to calculate the appropriate rate of administration of anesthesia to achieve a safe and effective alveolar concentration and the rate of uptake of anesthesia. Although simulation results were shown for N_2O, isoflurane, and enflurane, only N_2O simulation results were compared to experimental data. The authors concluded that the model uptake could simulate N_2O uptake, but additional validation was required. In another study, clinical validation for the system model

TABLE 2.1 Inhalation PBPK Models for Anesthetic Gases

Chemical	Species	Reference
Desflurane, 2-fluoro-2-(difluoromethoxy)-1,1,1-trifluoroethane	Human	Vinegar *et al.* (1998)
	Dog	Vinegar (2001)
Diethyl ether	Human	Munson *et al.* (1973)
Enflurane, 2-chloro-1-(difluoromethoxy)-1,1,2-trifluoroethane	Human	Lerou *et al.* (1991a)
	Human	Fiserova-Bergerova (1992)
	Human	Lerou *et al.* (1993)
	Human	Vermeulen *et al.* (2002)
Halothane, 2-bromo-2-chloro-1,1,1-trifluoroethane	Human	Ashman *et al.* (1970)
	Human	Smith *et al.* (1972)
	Human	Zwart *et al.* (1972)
	Human	Munson *et al.* (1973)
	Human	Fiserova-Bergerova (1992)
	Human	Vermeulen *et al.* (1995)
	Rat, human	Williams *et al.* (1996)
	Rat	Loizou *et al.* (1997)
	Human	Vinegar *et al.* (1998)
	Dog	Vinegar (2001)
Isoflurane, 2-chloro-2-(difluoromethoxy)-1,1,1-trifluoroethane	Human	Munson *et al.* (1973)
	Human	Lerou *et al.* (1991a)
	Human	Lerou *et al.* (1991b)
	Human	Fiserova-Bergerova (1992)
	Human	Vinegar *et al.* (1998)
	Dog	Vinegar (2001)
Methoxyflurane 2,2-dichloro-1,1-difluoro-1-methoxyethane	Human	Fiserova-Bergerova (1992)
Nitrous oxide	Human	Lerou *et al.* (1991a)

applied to isoflurane was performed (Lerou *et al.* 1991b). In this study, it was shown that a model incorporating no peripheral shunt (i.e., blood transferring directly from arterial blood into venous blood without passing through any compartments) performed better (i.e., more closely predicted patients' alveolar isoflurane concentrations) than a model incorporating a peripheral shunt with 16% of the Q_C.

In a later study testing the performance of their model (i.e., its ability to predict alveolar concentrations) for enflurane, the importance of the addition of nonpulmonary elimination (NPE) to the model was examined (Lerou *et al.* 1993). Metabolism was incorporated as the removal of a constant fraction of the anesthesia from the blood flow to the liver (i.e., as a concentration dependent linear clearance pathway). The model simulated enflurane kinetics more accurately when there was liver clearance. However, the authors pointed out that incorporating metabolism, while it improves the simulation results, may be an oversimplification that includes all forms of NPE (e.g., elimination in urine, sweat, and feces) in a single mechanism.

In a later study, the impact of NPE on the model simulation results for halothane was investigated, along with the importance of using age-adjusted partition coefficients (Vermeulen *et al.* 1995). For this study, metabolism was incorporated in the model using Michaelis–Menten kinetics. To incorporate the effects of age on partition coefficients, previously reported, experimentally determined values of the halothane blood : air partition coefficient and tissue : blood partition coefficients for the heart, brain, liver, muscle, and adipose tissue were used. Experimentally determined values for these partition coefficients were available for three to five different ages, and values for each patient's age had to be calculated by linear interpolation. By comparing simulations with and without NPE and age-adjusted partition coefficients incorporated into the model and comparing the simulation results to experimental data, it was demonstrated that the model performed better with the inclusion of NPE and age-adjusted partition coefficients. In the final article in this series (Vermeulen *et al.* 2002), experimentally determined cardiac output of individual patients and age-adjusted partition coefficients were used to assess the predictive performance of the model applied to enflurane. Although incorporating continuous cardiac output measurements did not improve model performance (i.e., the ability of the model to predict alveolar concentrations), the model did perform better with the inclusion of age-adjusted partition coefficients.

2.4.2 Chlorofluorocarbons (CFCs), Refrigerants, and Halons

Chlorofluorocarbons (CFCs), a class of halogenated alkanes commonly containing carbon, fluorine and chlorine, and hydrochlorofluorocarbons (HCFCs), which also contain hydrogen, have been used as refrigerants and in other applications (e.g., foam blowing, aerosols, and sterilants) over the past 50 years. Halons are another class of halogenated hydrocarbons, often containing bromine, that are typically used for fire and explosion protection. Desirable properties for these classes of compounds include low flammability, chemical stability, low environmental impact, compatibility with engineering materials, and minimal toxicity.

PBPK Models for Refrigerants and Related Compounds When the hazard presented by CFCs and halons to the stratospheric ozone was identified and recognized by the world community, old CFC refrigerants were scheduled to be slowly removed from commerce and new replacements had to be developed. The production of many CFCs (e.g., CFC-11, CFC-12, CFC-13) ceased in the United States 1995. Halons are generally no longer in use, although many countries still have halon fire suppression systems in aircrafts that use recycled chemical because no completely effective replacement for this application has been developed. One common measure of a chemical's ability to deplete the stratospheric ozone is called the ozone depletion potential (ODP). For CFC-11, Halon 1301, and HCFC-123, the ODP values are 1.0, 10.0, and 0.02, respectively. Some HCFCs and hydrofluorocarbons (HFCs) have been proposed as alternatives to CFCs. PBPK models have been developed for one chlorofluorocarbon, trichlorofluoromethane (CFC-11), and one halon, bromotrifluoromethane (Halon 1301) (Table 2.2). A PBPK model has also been developed for trifluoroiodomethane (CF_3I), which is a potential replacement for Halon 1301. Refrigerants and replacement refrigerants have been more extensively

TABLE 2.2 Inhalation PBPK Models for Refrigerants and Related Compounds

Chemical	ODP	Species	Reference
Bromotrifluoromethane, Halon 1301	10.00	Human	Vinegar and Jepson (1996)
		Human	Vinegar *et al.* (2000)
2,2-Dichloro-1,1,1-trifluoroethane, HCFC-123	0.02	Rat	Dodd *et al.* (1993)
		Rat	Vinegar *et al.* (1994)
		Rat	Loizou *et al.* (1994)
		Human	Williams *et al.* (1996)
		Human	Vinegar and Jepson (1996)
2-Chloro-1,1,1,2-tetrafluoroethane, HCFC-124	0.12	Rat, mouse, hamster	Loizou and Anders (1995)
1,1-Dichloro-1-fluoroethane, HCFC-141b	0.12	Rat	Loizou and Anders (1993)
Difluoromethane, HFC-32	0	Rat	Ellis *et al.* (1996)
1,1,1,2,3,3,3-Heptafluoropropane, HFC-227ea	0	Human	Vinegar and Jepson (1996)
		Human	Vinegar *et al.* (2000)
1,1,1,3,3,3-Hexafluoropropane, HFC-236fa	0	Human	Vinegar *et al.* (2000)
Pentafluoroethane, HFC-125	0	Human	Vinegar and Jepson (1996)
		Human	Vinegar *et al.* (2000)
Trichlorofluoromethane, CFC-11	1.0	Dog	Vinegar (2001)
Trifluoroiodomethane, CF_3I	0	Human	Vinegar *et al.* (1999)
		Human	Vinegar and Jepson (1996)
		Human	Vinegar *et al.* (2000)

studied as part of the effort to understand the potential health risks of these chemicals to support the evaluation process for potential replacements. It has been suggested that PBPK modeling can be a useful technique for understanding risks posed by CFC substitutes (Jarabek *et al.* 1994). In the following section, PBPK models for HCFC-123, for which the most PBPK models have been developed, will be reviewed.

HCFC-123 (2,2-Dichloro-1,1,1-trifluoroethane) The refrigerant HCFC-123, which can also be used as a foam blowing agent and a cleaning solvent, has been proposed as an alternative to Halon 1211. PBPK modeling has been used to assess the risks of this alternative refrigerant. The first PBPK model for HCFC inhalation exposures in the rat was developed by Dodd *et al.* (1993). This article described pharmacokinetic studies on five halon replacement candidates (i.e., HCFC-123, -124, and -142b, Halon 1211, and perfluorohexane) in the rat, but PBPK model simulations were only shown for HCFC-123. The model incorporated Michaelis–Menten metabolism in the liver, and the metabolism of HCFC-123 appeared to saturate at an exposure concentration of approximately 2000 ppm. However, metabolic clearance of these compounds tends to be lower than that observed for other low-molecular-weight volatiles.

In a study of HCFC-123 inhalation kinetics in male and female Sprague–Dawley rats, Loizou *et al.* (1994) developed another rat inhalation PBPK model for HCFC-123. The model, which included fat, lean tissue, richly perfused tissue, and liver compartments as well as a compartment for the inhalation chamber atmosphere, was adapted for male and female rats by adjusting appropriate physiological parameters to sex-specific values (e.g., cardiac output, alveolar ventilation, and body weight). The model also included the production of the primary oxidative metabolite of HCFC-123, trifluoroacetic acid (TFA), by estimating its rate of production in the liver as a fraction of the rate of total metabolism. Blood : air and tissue : blood partition coefficients were determined for both HCFC-123 and halothane, an analogous compound, *in vitro* using vial equilibration methods (Gargas *et al.* 1989). The experimentally determined partition coefficients for halothane were higher than values for HCFC-123 for blood and all tissues. The PBPK model accurately simulated TFA excretion in urine, and it indicated that TFA is the major metabolite of HCFC-123. However, the model failed to simulate the reduction in uptake of HCFC-123 observed in female rats at exposure levels of 2000 to 5000 ppm HCFC-123.

For the risk assessment of refrigerant replacements, PBPK models describing the disposition of these compounds in humans would be useful, but human pharmacokinetic data to use for PBPK model development are typically unavailable. To solve this problem, Jarabek *et al.* (1994) suggested an approach for using a structurally similar compound for which human data are available (i.e., halothane) to aid in model development. In this approach, PBPK models for halothane and HCFC-123 would be developed for the rat, and both models would be extrapolated to humans. The human HCFC-123 PBPK model could then be indirectly validated using human pharmacokinetic data for halothane, a volatile anesthetic gas for which human inhalation pharmacokinetic data were available. Vinegar *et al.* (1994) began this approach in a study with the goal of developing a PBPK model describing the fat and blood concentrations of HCFC-123 and TFA following HCFC-123 inhalation exposures of the Fischer 344 rat. Blood : air and tissue : blood partition coefficients were measured *in vitro* for HCFC-123. The model was similar to the earlier HCFC-123 model structure (Loizou *et al.* 1994) except that it included a gastrointestinal (GI) tissue compartment, the volume of the body fat compartment was set to age-appropriate values (i.e., it was higher for older rats), and the metabolism of HCFC-123 in the liver was modeled using Michaelis–Menten kinetics with substrate inhibition. Blood concentrations of rats exposed to high levels of HCFC-123 could only be simulated if substrate inhibition was included.

In the next step to developing a PBPK model for HCFC-123 in humans following the method outlined by Jarabek *et al.* (1994), Williams *et al.* (1996) developed PBPK models for halothane in rats and humans. Next, they used the rat HCFC PBPK model (Vinegar *et al.* 1994) and the human halothane PBPK model to guide the development of a human PBPK model for HCFC-123. An identical PBPK model structure was used to describe halothane and TFA pharmacokinetics in rats and humans and HCFC-123 pharmacokinetics in humans. This approach, used to develop the human HCFC-123 PBPK model, demonstrated the effectiveness with which "template model structures" for structurally similar compounds could aid in the development and validation of human PBPK models when human data are unavailable. The resulting human PBPK model for HCFC inhalation exposures, in the absence of human data, can be used to support risk assessments for HCFC-123. The human models of Williams *et al.* (1996) were later used by Vinegar and Jepson (1996) to guide the development of human inhalation models for Halon 1301, HFC-125, HFC-227ea, HCFC-123, and CF_3I.

2.4.3 Halogenated Alkanes

PBPK models have been developed for many halogenated methanes (Table 2.3) and halogenated ethanes (Table 2.4). DCM (Table 2.5), chloroform (Table 2.6), carbon tetrachloride (Table 2.7), and methyl chloroform (i.e., 1,1,1-trichloroethane) (Table 2.8) have been particularly well-studied. The following sections will focus on the two most-studied chemicals, DCM and chloroform.

Methylene Chloride, DCM DCM is widely used as a general all-purpose solvent in a variety of industrial applications, food processing, and agriculture. Long-term toxicity bioassays of DCM showed marked differences in the carcinogenic response of various species exposed via different routes. Dose-dependent increase in pulmonary and hepatocellular adenomas and carcinomas occurred in both male and female B6C3F1 mice exposed to 2000 and 4000 ppm DCM by inhalation, while a 2-year study with Syrian golden hamsters exposed to air concentrations of up to 3500 ppm failed to produce increased tumor incidence at any site (NTP 1986). Similarly, treatment related mammary tumors were observed in inhalation studies with Sprague Dawley and F344 rats, but there was no evidence of statistically significant increase in tumors during a drinking water study with B6C3F1 mice and

TABLE 2.3 PBPK Models for Halogenated Methanes[a]

Chemical	Species	Exposure route	Reference
Bromochloromethane, BCM	Rat	Inhalation	Gargas *et al.* (1986a)
Difluoromethane	Rat	Inhalation	Gargas *et al.* (1986b)
	Rat	Dermal	McDougal *et al.* (1986)[b]
	Rat	Dermal	Jepson and McDougal (1997)
	Rat	Dermal	Jepson and McDougal (1999)
Bromodichloromethane, BDCM	Rat	Inhalation, oral	Lilly *et al.* (1997)
	Rat	Oral	Lilly *et al.* (1998)
	Rat	Oral	da Silva *et al.* (1999)
Bromotrichloromethane	Rat	Inhalation	Thakore *et al.* (1991)
Bromoform	Rat	Oral	da Silva *et al.* (1999)
Chloromethane	Rat	Inhalation	Gargas *et al.* (1990)
Dibromochloromethane, DBCM	Rat	Oral	da Silva *et al.* (1999)
Dibromomethane, DBM	Rat	Dermal	McDougal *et al.* (1986)[b]
Fluorochloromethane	Rat	Inhalation	Gargas *et al.* (1986b)
	Rat	Dermal	Jepson and McDougal (1997)
	Rat	Dermal	Jepson and McDougal (1999)

[a] DCM, chloroform, and carbon tetrachloride are included in Tables 2.5, 2.6 and 2.7, respectively.

[b] This study examined the dermal absorption of chemical vapors in the rat.

TABLE 2.4 PBPK Models for Halogenated Ethanes[a]

Chemical	Species	Exposure route	Reference
Chloroethane	Rat	Inhalation	Gargas *et al.* (1990)
	Rat	Inhalation	DeJongh *et al.* (1998)
Hexachloroethane	Rat	Inhalation	Gargas and Andersen (1989)
	Rainbow trout	Gill	Nichols *et al.* (1991)
	Channel catfish	Gill	Nichols *et al.* (1993)
	Rainbow trout, channel catfish	Gill, dermal	Nichols *et al.* (1996)
	Lake trout	Gill	Lien *et al.* (2001)
1,2-Dibromoethane, EDB	Rat, human	Inhalation	Ploemen *et al.* (1997)
	Rat	Inhalation, oral, *i.v.*	Hissink *et al.* (2000)
	Human	Inhalation	Hissink *et al.* (2000)
1-Chloro-1,1-difluoroethane	Rat	Inhalation	Loizou *et al.* (1996)
1,1-Dichloro-1-fluoroethane	Rat	inhalation	Loizou *et al.* (1996)
1,1-Dichloroethane	Rat	Inhalation	Gargas *et al.* (1990)
1,2-Dichloroethane	Rat	Inhalation	Gargas *et al.* (1990)
	Human	Inhalation	Leung (1992)
	Rat	Inhalation	DeJongh *et al.* (1998)
Pentachloroethane	Rat	Inhalation	Gargas and Andersen (1989)
	Rainbow trout	Gill	Nichols *et al.* (1991)
	Channel catfish	Gill	Nichols *et al.* (1993)
	Rainbow trout, channel catfish	Gill, dermal	Nichols *et al.* (1996)
	Rat	Inhalation	DeJongh *et al.* (1998)
	Lake trout	Gill	Lien *et al.* (2001)
1,1,1,2-Tetrachloroethane	Rat	Inhalation	Gargas and Andersen (1989)
1,1,2,2-Tetrachloroethane	Rat	Inhalation	Gargas and Andersen (1989)
	Rainbow trout	Gill	Nichols *et al.* (1991)
	Channel catfish	Gill	Nichols *et al.* (1993)
	Rainbow trout, channel catfish	Gill, dermal	Nichols *et al.* (1996)
	Lake trout	Gill	Lien *et al.* (2001)
1,1,2-Trichloroethane	Rat	Inhalation	Gargas and Andersen (1989)
1,1,1-Trifluoroethane	Rat	Inhalation	Loizou *et al.* (1996)

[a] Methyl chloroform is included in Table 2.8.

F344 rats. These species and dose-route differences in response stimulated the effort to estimate the carcinogenic risk of DCM exposure to humans.

In general, tumor induction by DCM has been linked to biotransformation products. DCM is metabolized both *in vivo* and *in vitro* by two metabolic pathways: (1) cytosolic glutathione-S-transferase (GST), which requires the substrate glutathione (GSH) to produce carbon dioxide (CO_2) and other metabolites, and (2)

TABLE 2.5 PBPK Models for DCM

Species	Exposure route	Reference
Mouse	*I.v.*	Angelo *et al.* (1984)
Mouse, rat	*I.v.*, oral	Angelo and Pritchard (1984)
Rat	Dermal	McDougal *et al.* (1986)
Rat	Inhalation	Gargas *et al.* (1986b)
Mouse, rat, hamster, human	Inhalation, oral, *i.v.*	Andersen *et al.* (1987a)
Mouse, human	Inhalation, oral, *i.v.*	Andersen *et al.* (1987b)[a]
Rat, human	Inhalation	Andersen *et al.* (1987c)
Rat	Inhalation	Gargas *et al.* (1990)
Rat, human	Inhalation	Andersen *et al.* (1991)
Rat	Oral	Staats *et al.* (1991)
Human	Inhalation	Leung (1992)
Mouse	Inhalation	Andersen *et al.* (1994)[b]
Human	Inhalation, liver[c]	Yoshida (1993)
Mouse, hamster	Inhalation	Casanova *et al.* (1996)
Rat	Inhalation	DeJongh *et al.* (1998)
Mouse, human	Inhalation	El-Masri *et al.* (1999)
Human	Inhalation	Thrall *et al.* (2001)
Human	Inhalation	Jonsson *et al.* (2001)
Human	Inhalation	Jonsson and Johanson (2001)

[a] This paper is a condensed version of the Andersen *et al.* (1987a) paper.
[b] In this paper, model parameters (i.e., partition coefficients and metabolic parameters) were reported for both DCM and ^{2}H-dichloromethane, ^{2}H-DCM.
[c] In this model, it was assumed that all chemical ingested in drinking water and food entered the liver directly.

TABLE 2.6 PBPK Models for Chloroform

Species	Exposure route	Reference
Rat, mouse, human	Inhalation, oral, *i.p.*	Corley *et al.* (1990)
Rat, mouse, human	Inhalation, oral	Reitz *et al.* (1990)[a]
Rat	Inhalation	Gargas *et al.* (1990)
Rat	Oral	Staats *et al.* (1991)
Human	Inhalation	Leung (1992)
B6C3F1 mice	Inhalation	Gearhart *et al.* (1993)
Human	Dermal, inhalation	Chinery and Gleason (1993)
Human	Inhalation, liver[b]	Yoshida (1993)
Rat, mouse	Oral	Smith *et al.* (1995)
Human	Dermal	Roy *et al.* (1996b)
Rat	Oral	da Silva *et al.* (1999)
Human	Dermal	Levesque *et al.* (2000)
Human	Dermal	Corley *et al.* (2000)[c]
Rat	Inhalation	Evans *et al.* (2002)
Mouse	Inhalation	Tan *et al.* (2003)

[a] In this study, the model developed by Corley *et al.* (1990) was extended to describe the PD effect of cytotoxicity in the liver produced by chloroform metabolites.
[b] In this model, it was assumed that all chemical ingested in drinking water and food entered the liver directly.
[c] In this study, the human PBPK model developed by Corley *et al.* (1990) was extended to include the dermal exposure route.

TABLE 2.7 PBPK Models for Carbon Tetrachloride

Species	Exposure route	Reference
Rat	Inhalation	Gargas *et al.* (1986a)
Rat, human	Inhalation	Paustenbach *et al.* (1987)
Rat, monkey, human	Inhalation	Paustenbach *et al.* (1988)
Rat	Inhalation	Gargas *et al.* (1990)
Human	Inhalation	Leung (1992)
Human	Inhalation, liver[a]	Yoshida (1993)
Rat	Inhalation	Evans *et al.* (1994)
Rat	Inhalation	Evans and Simmons (1996)
Rat	*I.v.*	Thrall and Kenny (1996)
Rat	Oral	Semino *et al.* (1997)
Rat, mouse, hamster, human	Inhalation	Thrall *et al.* (2000b)

[a] In this model, it was assumed that all chemical ingested in drinking water and food entered the liver directly.

TABLE 2.8 PBPK Models for Methyl Chloroform

Species	Exposure route	Reference
Rat	Inhalation	Gargas *et al.* (1986a)
Rat, mouse, human	Inhalation, *i.v.*, oral	Reitz *et al.* (1988a)
Rat	Inhalation	DeJongh *et al.* (1998)
Human	Inhalation, liver[a]	Yoshida (1993)
Rat	Inhalation	Gargas *et al.* (1990)
Human	Inhalation	Leung (1992)
Human	Inhalation	Lapare *et al.* (1995)
Rat	Inhalation	Loizou *et al.* (1996)
Rat, human	Dermal	Thrall *et al.* (2000a)
Rat, human	Dermal	Poet *et al.* (2000)

[a] In this model, it was assumed that all chemical ingested in drinking water and food entered the liver directly.

cytochrome-P450-dependent mediated-oxidation (primarily CYP2E1), which produces carbon monoxide (CO), CO_2, and other products (Fig. 2.3). The GSH-mediated pathway is a low-affinity, high-capacity process, while oxidation is a high-affinity, low-capacity process. Each pathway generates reactive species that could be linked to the observed tumorogenicity.

Historically, the first modeling efforts with DCM were reported by Angelo *et al.* (1984) and Angelo and Pritchard (1984) (Table 2.5). The authors developed PBPK models describing the pharmacokinetics of DCM in mice and rats following *i.v.* and oral administration. The model incorporated separate compartments for arterial and venous blood, lungs, liver, kidneys, GI tract, and carcass. A fairly complex multicompartmental structure for GI uptake included four discrete compartments (upper/lower GI tissue and upper/lower GI lumen). To accommodate some observed discrepancies between tissue/blood concentrations in the course of

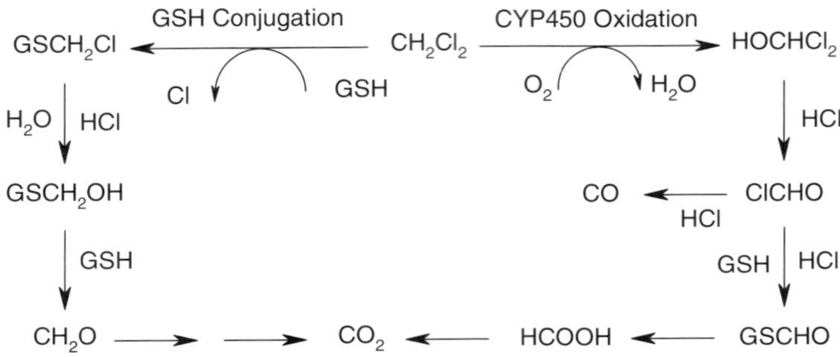

Figure 2.3 Reaction mechanisms for glutathione- and CYP450-mediated DCM metabolism. End products of glutathione (GSH) conjugation are formaldehyde (CH_2O) and carbon dioxide (CO_2), with the formation of chloromethyl glutathione ($GSCH_2Cl$) and hydroximethyl glutathione ($GSCH_2OH$) as possible intermediate products. Cytochrome P450 oxidation is initiated with the formation of dichloromethanol ($HOCHCl_2$) and monochloroformaldehyde (ClCHO), which ultimately leads to formic acid (HCOOH) and carbon monoxide (CO) as end products.

the experiment, deep tissue compartments were added to lungs, liver, kidney, and carcass. The tissue deep compartment was related to the lipid fraction of the specific tissue, and mass transfer between the deep and shallow compartments was governed by an intracompartmental partition rate. Metabolic clearance of DCM was by a single, saturable process taking place in the liver only, and a linear clearance term accounted for exhalation via lungs. The model was used to simulate DCM concentrations in liver for comparison with experimental data from an oral study in mice given a daily dose of 1000 mg/kg DCM in a corn oil vehicle. Additional simulations were performed to address differences in the uptake kinetics of DCM after oral dosing in water and corn oil vehicles.

To explore the causality between various internal dose metrics and carcinogenicity, Andersen *et al.* (1987a) developed an initial PBPK model for DCM in rats, mice, hamsters, and humans. The model included lung, liver, fat, and slowly and rapidly perfused tissues as separate well-mixed compartments, and metabolism took place in both the lung and the liver following the two separate pathways. Oxidation was described as a saturable process, while GSH conjugation was treated as a linear process described with a first-order rate constant. (Conjugation of DCM with GSH does not consume GSH, and no depletion of GSH occurs with this substrate.) For inhalation exposures, gas exchange occurred before DCM entered the lung tissue where it became available for metabolism. An alternate model in which metabolism and gas exchange occurred in a single compartment produced identical results. In drinking-water simulations, the daily dose of DCM was introduced into the liver compartment as a zero-order intake over the entire exposure period with complete absorption from the GI tract. The model was parameterized using literature values for physiological parameters, vial-equilibration estimates of partition coefficients, and closed chamber inhalation studies in rats, mice, and Syrian Hamster for calculation of whole-body kinetic constants for the two pathways of metabolism. For humans, *in vivo* estimates of biochemical constants for oxidative metabolism were available, while GST-mediated clearance was allometrically scaled from rodents. The model was validated in all animal species and humans using data from independent studies. With the PBPK model, the authors were able to calculate the expected tissue dosimetry of the two reactive metabolites for different dosing regimens in liver and in lung for mice and humans. Tumor responses from the animal studies were then related to the integrated intensity of tissue exposure to reactive intermediates—that is, the rate of metabolism through a specific pathway per volume of tissue per time.

From this analysis (Andersen *et al.* 1987a) it was concluded that the carcinogenic responses in tissues were closely correlated with the GST-mediated reactive metabolites or the concentration of DCM itself, but not with the oxidation metabolites. The model was also instrumental in the analysis of route-specific differences in carcinogenic response. The lack of carcinogenicity in drinking-water studies was related to slower intake and first-pass metabolism in the liver reducing tissue exposures to the GSH-pathway metabolites. Modeling-supported, low-dose and interspecies extrapolation of the tissue dose metrics revealed that the health risks associated with DCM exposures were expected to be much lower than previously estimated using default risk assessment procedures. The modeling exercise with DCM was the first use of PBPK modeling in risk assessment, and the template methodology developed in this study is still closely followed today. In addition, integration of PBPK modeling and targeted mechanistic studies has been particularly useful in establishing the GSH-pathway mode of action and increasing confidence in application of the PBPK model in risk assessment (Andersen 2003).

The concept of the dose metric—that is, the appropriate measure of the toxic form of chemical in a target tissue—has become a key idea in developing PBPK and PBPD models for responses to compounds. The metric with DCM was a surrogate measure of the intensity of exposure of tissue to a reactive metabolite, mg DCM metabolized by a pathway/gram tissue/time. These measures of tissue dose link through a series of biological steps to give rise to cancer. The dose metric also is closely tied to a mode of action, presumed to be that the reactive metabolites bind to DNA with the potential to initiate mutational events that can eventually produce tumors. These dual and interconnected concepts, dose metric and mode of action (Fig. 2.4), are the links bridging PBPK models (steps up to the dose metric) with PBPD models (from the initial mode of action to overt toxic responses in individuals).

One limitation of the initial use of the PBPK model for risk assessment with DCM was that site- and pathway-specific metabolism of DCM was apportioned between the lung and liver according to literature data on specific GST and CYP450 activities obtained from measurements with model substrates other than DCM. Since the CYP450 enzymes have higher activity for DCM than do GST enzymes, at low concentration most DCM metabolism occurs via oxidation. Human volunteer studies on DCM metabolism, typically performed at low exposure concentrations, were inadequate for estimating the rate of metabolism by the GST pathway, and kinetic constants for GST metabolism of DCM in humans were estimated by allometric scaling rather than from experimental data. Reitz *et al.* (1988b) updated the DCM

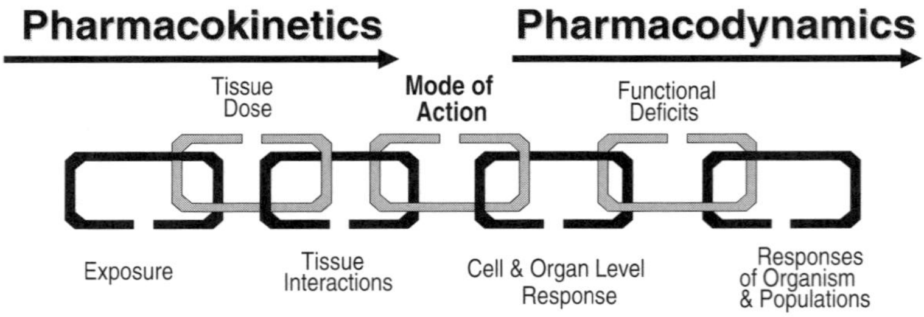

Figure 2.4 Exposure dose response. Mode of action is more explicitly defined as the linkage of the delivered dose and a biological consequence. Defined this way, we also achieve a better distinction between pharmacokinetics, steps coming before the mode of action linkage, and pharmacodynamics, steps that ensue from the mode of action.

PBPK model with *in vitro* measures of specific activities of GST- and CYP450-mediated DCM metabolism in lung and liver samples from mice, rats, hamsters, and humans. Mathematical optimization of these *in vitro* data was used to derive values of the metabolic rate constants $V_{\max}$ and K_m (i.e., the velocity at maximal substrate concentration and the Michaelis constant, respectively, for Michaelis–Menten enzyme kinetics) for GST and CYP450 metabolism in human and mouse liver; data from rat and hamster liver samples failed to produce satisfactory results.

The PBPK model for DCM was extended (Andersen *et al.* 1991) by studying in detail the kinetics of the oxidative DCM metabolism to CO and carboxyhemoglobin (COHb) in both rats and humans. Model structure and parameterization of the parent DCM metabolism portion of the model was very similar to the ones previously reported. Description of the CO and COHb kinetics was based on the Coburn–Forster–Kane equation adjusted for the metabolic production of CO, as described in an earlier PBPK model for dihalomethanes and their metabolites (Gargas *et al.* 1986b). Physiological and biochemical constants for the CO to COHb portion of the combined model were estimated from studies with male Fischer 344 rats exposed to 200 ppm CO for 2 hours and examining the time course of COHb after cessation of CO exposure. The human model was parameterized using allometric scaling of kinetic constants obtained for rats. Model performance was examined by comparing simulated with measured data from human inhalation studies with CO and DCM. The human PBPK model was found to provide a good representation of the observed kinetic behavior in all four human studies.

Effects of exercise and intersubject variability on DCM dose estimates were explored with a modified PBPK model (Dankovic and Bailer 1994). There were no changes to the basic model structure (Andersen *et al.* 1987a; Reitz *et al.* 1989). However, some physiological parameters, such as cardiac output, alveolar ventilation, and tissue blood perfusion, were adjusted from resting values to values consistent with a light work condition. To address interindividual variability, *in vivo* metabolic parameters were varied consistent with *in vitro* from the previous *in vitro* studies (Reitz *et al.* 1989). When compared with previous estimates, the modified model predicted increases in the GST pathway doses for the liver and lung ranging from 0 to 5.4 and 0 to 3.6 times those computed by the original model, respectively. The authors concluded that in some cases, occupational exposures could lead to several times higher individual doses of reactive GST metabolites than originally predicted.

PBPK modeling approaches also studied the mechanism of DCM oxidation in mice in gas uptake studies (Andersen *et al.* 1994). By monitoring kinetics of deuterated forms of DCM by gas uptake, *in vivo* rate constants were estimated for reaction of the proposed product of CYP450 oxidation of DCM, formylchloride. Somewhat unexpectedly, isotope effects were noted for K_m, but not for V_{max}. Two kinetic mechanisms were discussed to address the observed change in the apparent K_m after deuterium substitution: (1) rate-limiting product release from cytochrome P450 or a rate-limiting oxygen activation step followed by (2) a second-order reaction between the oxygen–enzyme complex and DCM. Both mechanisms emphasize the complex nature of the common pharmacokinetic constants V_{max} and K_m, and the significance of the C–H bond breaking for oxidation *in vivo* of DCM (Andersen *et al.* 1994). These studies utilized observations in the intact animal to assess aspects of the mechanism of catalysis with DCM.

A study on the effect of intrapopulation variability in physiological and biochemical model parameters on carcinogenic risk estimates has been reported (Portier and Kaplan 1989). Portier and Kaplan (1989) applied Monte Carlo methods to the methylene chloride model of Andersen *et al.* (1987) to investigate the effect of incorporating the variability of PBPK model parameters. Body weight, alveolar ventilation, partition coefficients, and metabolic constants were given truncated log-normal distributions, while some of the other physiological parameters such as organ weights and perfusion flows had normal distributions. For each parameter, the parameter means and standard deviations (where available) provided estimates for the mean and variance of their respective distribution. Using a one-hit model of carcinogenesis and resampling the bioassay data, they found that the distribution of virtually safe doses roughly spanned an order of magnitude for 20% model parameter variability and covered more than two orders of magnitude for 100% variability in some model parameters. The approach used by Portier and Kaplan (1989) provides a combined estimate of the uncertainty in the pharmacokinetic modeling and the uncertainty associated with the cancer bioassay. However, risk assessments typically use the 95% upper confidence limit on the cancer potency estimate to address uncertainty in the bioassay. A subsequent study in which only the pharmacokinetic parameters were varied reported pharmacokinetic variation (represented as the ratio of the 95th and 50th percentiles of the risk distribution) on the order of threefold (Clewell and Andersen 1996).

Subsequently, a more detailed study on the effects of GST polymorphism on the estimates of DCM to risk humans was reported (El-Masri *et al.* 1999). Among the GST gene superfamily, the theta form (GSTT1) has been linked to DCM metabolism to formaldehyde in mammals. GSTT1 is polymorphic in humans with frequencies of homozygous null genotype ranging from 10% to 60% in different ethnic and racial populations around the world. The effect of various GSTT1 genotype frequencies on the cancer risk estimates for DCM was estimated by application of Monte Carlo simulation methods in combination with PBPK modeling (El-Masri *et al.* 1999). This study used DNA-protein cross-links (DPX) caused by metabolism of DCM to formaldehyde as a better dose surrogate for cancer risk estimation. Previously, Casanova *et al.* (1996) reported use of a modified DCM model to study the relationship between the concentration of DCM and the formation of DPX in the mouse liver. The DPX formation was by a pseudo-first-order reaction of formaldehyde with DNA, and metabolic formaldehyde generation and elimination in the mouse liver was described with a separate submodel according to a previously developed PBPK model for formaldehyde disposition in the nasal mucosa (Casanova *et al.* 1991). The pharmacokinetic portion of the analysis was based on the basic DCM model, and calculations of the amount DPX formed in the liver were from the more complete model for formaldehyde reactivity (Casanova *et al.* 1996). DPX amounts were used as a dose surrogate, and Monte Carlo simulations were used to calculate distributions of risk estimates for a sample of 1000 runs, with each run representing a collection of biochemical and physiological parameters for a single person. The results demonstrated up to 30% higher average and median risk estimates when GSTT1 polymorphism was not included in the analysis.

Chloroform Chloroform (trichloromethane) is a colorless, nonflammable solvent used in many industrial or chemical operations as well as in the production of other chemicals. Significant indirect sources of chloroform release are reactions of chlorine with organic chemicals during the disinfection of drinking water and wastewaters. Chloroform has been detected in air, surface water, groundwater, and drinking water. Exposure to chloroform can occur through inhalation of indoor air, ingestion of tap water, and dermal absorption during bathing. Because people are exposed to chloroform by many different exposure pathways, PBPK models have been developed for oral, inhalation, dermal, and *i.p.* chloroform exposures (Table 2.6).

There is no strong evidence that chloroform possesses any significant genotoxic potential. Noncancer effects observed after repeated exposures of rodents to chloroform are limited to sustained cytotoxicity and persistent regenerative proliferation. Long-term bioassays with chloroform have resulted in liver tumors in mice and renal tumors in mice and rats. Chloroform has been shown to induce nasal lesions in rats and mice exposed by both inhalation and ingestion. There is a general agreement that chloroform carcinogenicity may be due to proliferative regenerative tissue response at cytotoxic concentrations.

Cytotoxicity is primarily related to the formation of reactive metabolites. Both oxidative and reductive pathways for chloroform metabolism have been identified, and both pathways are initiated with a cytochrome P450-dependent activation step (Fig. 2.5). The oxidative pathway generates CO_2 as well as phosgene and hydrochloric acid as the main reactive intermediates, while the reductive pathway generates a dichloromethylcarbene free radical. The balance between the oxidative and reductive pathways depends on species, tissue, dose, and partial oxygen pressure. Since under normal conditions reductive metabolism in liver and kidney is believed to be minor, chloroform toxicity has been traditionally attributed to the electrophilic metabolite phosgene.

Corley *et al.* (1990) developed the first comprehensive PBPK model for chloroform in mice, rats, and humans. Arterial and venous blood, lungs, liver, fat, kidney, slowly/rapidly perfused tissues, and the GI tract were described as separate, well-mixed compartments. Liver and kidney were considered the sites of saturable meta-

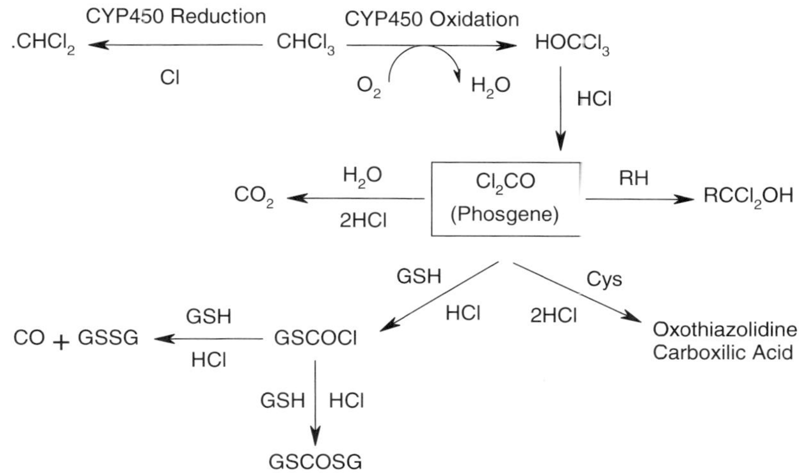

Figure 2.5 Reaction mechanisms for CYP450-mediated reductive and oxidative chloroform metabolism. Reductive metabolism leads to free dichloromethyl radical ($\cdot CHCl_2$) capable of forming covalent adducts with microsomal proteins. The chief oxidative product is trichloromethanol ($HOCCl_3$), which rapidly and spontaneously dehydrochlorinates to form phosgene. Phosgene reacts with the thiol group of glutathione (GSH), yielding *S*-chloro-carbonyl glutathione (GSCOCl), which in turn can interact further with glutathione to form either diglutathionyl dithiocarbonate (GSCOSG) or glutathione disulfide (GSSG, oxidized glutathione) and carbon monoxide (CO). Alternatively, phosgene can react with cellular proteins (RH) and cysteine (Cys), or decompose to carbon dioxide (CO_2). Adapted from USEPA (2001).

bolism, with the rate of kidney metabolism linked to the maximum velocity in the liver via a proportionality constant based on the relative tissue volumes. The metabolic constants were derived from *in vivo* gas-uptake studies with rodents and *in vitro* incubations with rodent and human microsomes. The model incorporated binding of a fraction of the total chloroform metabolites to cellular macromolecules as an indicator of the dose delivered, and it included corresponding terms for loss and resynthesis of the metabolizing enzyme. Macromolecular binding constants were estimated from in vivo binding data obtained following inhalation exposures to radiolabeled chloroform. The model was validated in all species by comparing simulated with experimentally observed organ-specific dose metrics. Model simulations of protein binding suggested a highest sensitivity of the mouse followed by rat and human, an observation consistent with the experimental data.

Subsequently, Reitz *et al.* (1990) modified the Corley model to include pharmacodynamic mechanisms associated with the experimentally observed cytotoxicity in the liver. Since the modeling exercise focused on the liver as a target organ for cytotoxicity, mechanistically the kidney compartment as described in the Corley model was included in the rapidly perfused tissue group. Cytotoxicity was linked to "cell killing" as a result of macromolecular cell binding, and it was described with a rate-independent parameter. In the pharmacodynamic portion of the model (i.e., concerning cytotoxicity), cells died when the rate of metabolism exceeded their ability to detoxify the metabolic products, and parameters were derived from quantitative histopathology and cell replication experiments. In addition, the probability of cell death was characterized as a normal distribution function depending on both the rate of metabolism and the time of exposure. After exploring two relevant dose measures for liver—that is, the daily average of total macromolecular binding and cytotoxicity—the latter was chosen as the dose surrogate best reflecting carcinogenic response. Two computational approaches were used to derive safety factors for chloroform exposure and compared with the earlier predictions (Corley *et al.* 1990). The model was also used to explore (a) the significance of the dosing route and vehicle for the overall rate of metabolism *in vivo* and (b) its relation to hepatotoxicity.

The values of some of the physiological and biochemical parameters required for the PBPK models vary with body temperature. Using temperature–telemetry devices, significant decreases in the body temperature of $B6C3F_1$ mice were found in mice exposed to high chloroform concentrations (Gearhart *et al.* 1993). To account for the possible effect of this temperature difference on pharmacokinetics, *in vitro* activities of CYP450 were investigated at temperatures ranging from 24°C to 40°C, and showed a linear decrease in metabolic activity with decreasing temperature. Similar *in vitro* studies on tissue partitioning revealed lower tissue/blood partition coefficients with decreasing temperature. By modifying tissue/blood partition coefficients and metabolism according to body temperature, Gearhart *et al.* (1993) reproduced gas uptake data without the need of enzyme loss inclusion, as suggested in other studies (Corley *et al.* 1990). For high exposure concentrations, a first-order metabolic rate constant was incorporated into the model to account for additional pathways of chloroform biotransformation.

Chinery and Gleason (1993) developed a human PBPK model that included dermal exposure to chloroform in people while showering. Modifications included a multicompartment skin submodel and a shower model. The physiological description of the skin included a dilute surface-water solution compartment, a stratum corneum layer as the site of diffusion-limited absorption, and a viable epidermis compartment in dynamic equilibrium with the skin blood. The shower model was represented as a single well-mixed compartment with airflow in and out for simulating the mass exchange of chloroform from the water phase to the air. Mass transfer to and between each skin compartment was described with linear terms, while the shower model used nonlinear constants. Exhaled breath concentrations of volunteers exposed to chloroform were used to calibrate the model. The authors used the model to explore the impact of various skin model parameters on the ratio of inhaled to dermally absorbed dose.

The relative contribution of different exposure routes to target tissue dosimetry was studied by Blancato and Chiu (1994), who also included dermal exposure route from water while bathing or swimming. Using macromolecular binding as a dose surrogate, a simulated 10 minutes of dermal exposure (shower) to chloroform accounted for about 25% and 53% of the total dose in the liver and kidney, respectively. In contrast, drinking-water exposures were predicted to contribute only 7% to the kidney dose. This difference was attributed to a significant first-pass effect during oral intake of chloroform. Smith *et al.* (1995) also worked with the basic chloroform model for rats and mice to examine various estimates of internal dose as predictors of carcinogenicity in rodent bioassays. Rate-dependent dosimetry such as the rate of metabolism and fraction of hepatocytes killed per day were found to be in good agreement with the liver bioassay data. In contrast, none of the above dose measures correlated well with the kidney tumor data, and best results were obtained with the administered dose scaled to body surface area.

Georgopoulos *et al.* (1994) and Roy *et al.* (1996a,b) used chloroform as an example of the application of PBPK modeling to reconstruct short-term exposure concentrations and tissue dosimetry resulting from multiroute and multimedia exposures to volatile organic compounds. Here, dermal absorption was modeled with two different approaches: one identical to Chinery and Gleason (1993) using different skin permeability values, and the second one implementing Fick's second law for diffusive transport (i.e., treating the skin as a membrane instead of a compartment). The change of chloroform concentration in the stratum corneum with the distance from the skin surface was described with partial differential equations. For both models, the ability to predict single and combined inhalation and dermal exposures was evaluated using available data on ambient and exhaled breath concentration from shower and swimming pool studies. Both models predicted exposure concentrations ranging by factor of two from the measured values. The differences were attributed to the incomplete resolution of the internal dose between the different routes of exposure (Roy *et al.* 1996a).

More recently, Corley *et al.* (2000) used PBPK modeling to investigate the temperature dependence of dermal absorption of chloroform following bath water exposures. Continuous real-time breath analysis was used as a more accurate technique to study the kinetics of chloroform in human subjects exposed skin-only via bath water. Model structure was the same as previously reported (Corley *et al.* 1990), with the addition of a separate perfusion-limited skin compartment describing the dermal uptake of chloroform as a function of blood flow and diffusion according to the Fick's law. The model was used to study the relative contributions of oral, dermal, and inhalations exposures of humans to the total body burdens of chloroform for various exposure scenarios. Considering relevant dermal and drinking-water exposures, the relative contribution of dermal route was estimated between 1% and 28% of the total dose, depending on the water temperature.

Levesque *et al.* (2000) adapted the Corley *et al.* (1990) and Roy *et al.* (1996a) PBPK models to study temporal variations in internal chloroform dosimetry in resting and exercising swimmers. Skin exposure was described with a single permeability constant derived in experiments with swimmers exposed only via the dermal route. The average macromolecular binding of active chloroform metabolites in the liver and kidney was used as a dose surrogate to estimate the cancer risk associated with chloroform exposure for competitive and leisure swimmers. The model estimated up to a two times higher internal exposure in competitive swimmers. Subsequently, the same dose measures were used to calculate the cancer risk associated with exposure to chloroform during typical household activities, such as taking a bath or shower or washing clothes (Levesque *et al.* 2002). The model, essentially the same as reported earlier, included the dermal, inhalation, and drinking water exposure routes. Average and worst-case exposure scenarios were simulated for all exposure routes. For average multiroute exposure scenarios, 5- and 10-fold increases were calculated in kidney and liver dosimetry, respectively. The worst-case scenario resulted in further 3.5-fold increase in both dose metrics. Nevertheless, liver concentrations remained about 6000 times lower than the lowest exposure level associated with no increase in tumor incidence in animal studies, thus providing a considerable safety margin during household exposures to chloroform.

A panel of scientists assembled by the International Life Sciences Institute developed a risk assessment for chloroform, following the revised US EPA cancer guidelines (International Life Sciences Institute 1997). Dose–response assessment in this risk evaluation was based on using an interspecies PBPK model from dose metrics in liver and kidney, the target organs for cancer in rodents. The dose metrics considered for the risk assessments were both a daily average (metabolized dose/gram tissue/day) and peak (metabolized dose/gram tissue) dose. The latter metric appears to be a better marker for toxicity and subsequent carcinogenicity.

In a study designed to clarify the inhaled concentrations and exposure durations leading to cytolethality and regenerative cell proliferation in the liver, Constan *et al.* (2002) studied chloroform-induced regenerative cell proliferation in the B6C3F1 mouse liver. PBPK modeling was used to relate dose to anticipated response in both the mouse and human. The chloroform model of Corley *et al.* (1990) has been adapted by Tan *et al.* (2003) to describe chloroform inhalation exposures in female B6C3F1 mice (the original model was for male mice) and extended to include the PD effects of chloroform (i.e., cytolethality and regenerative cellular proliferation) observed by Constan *et al.* (2002). The pharmacodynamic model included the production of damage at a rate proportional to the metabolism rate, repair of damage as a saturable process, stimulation of the rate of cell death from damage, and cell division rate stimulation (Fig. 2.6). This model, which can be used to relate the tissue dose of various dose metrics to cytolethality and regenerative cell proliferation, can

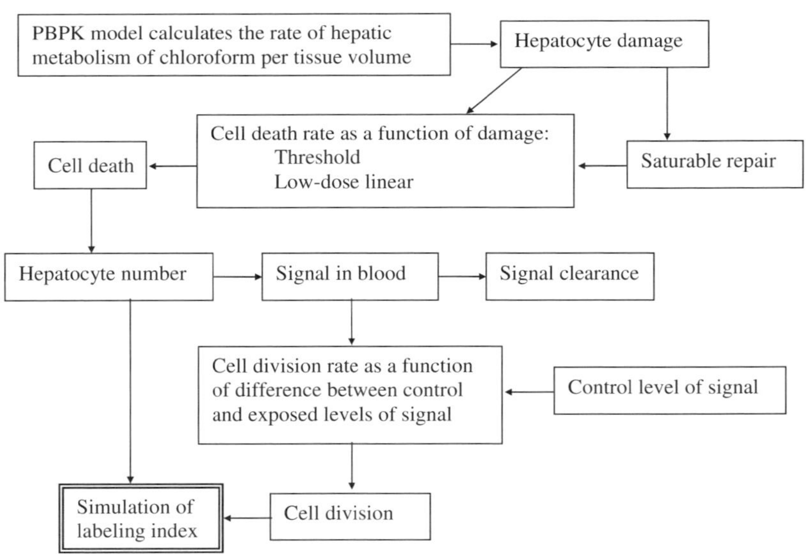

Figure 2.6 Pharmacodynamic model structure of Tan *et al.* (2003). Adapted with permission from the Society of Toxicology, Copyright 2003.

be used to test the hypothesis that stimulated cell proliferation stimulated by chloroform cytolethality, is directly related to tumor response.

2.5 SUMMARY

The halogenated alkanes represent a watershed group of compounds in discussing the expansion of PBPK modeling as an indispensable tool in toxicological research, human exposure analysis, and dose–response assessments. After the slow march of technology development required to support PBPK modeling, the initial application in toxicology/risk assessment using DCM displayed the advantages of these approaches as tools to allow extrapolations across route, dose, and species and provided a clear example of the potential for their broader application. Experimental methods were quickly refined to measure partition coefficients, blood flows, and metabolic constants, allowing more rapid construction of PBPK models for a much larger numbers of compounds. Model analysis tools—for example, Monte Carlo and sensitivity analysis—were refined and routinely applied in evaluating model behaviors and supporting risk assessments for populations of individuals rather than for an idealized member of a human population. As computational ability and experimental methods improved, the utility of PBPK models became readily apparent. PBPK models have been demonstrated useful in determining the amount of anesthesia to administer to a patient and in understanding the health risks posed by replacement refrigerant gases.

The applications of model analysis to assist in designing experiments to derive sensitive parameters with more confidence and to focus on questions of dose metrics led to an appreciation of the iterative flow of information between laboratory research and quantitative modeling of the data. This attribute is evident in the mechanistic studies with DCM that followed the assertion, based on modeling results, that the glutathione pathway was more likely than the CYP450 pathway to be involved in cancer causation with this compound. Routinely, models for parent compounds were extended to include important metabolites that might also play a role in toxicity or be important in refining estimates of rates of production of reactive intermediates in metabolis pathways. This chapter focused primarily on the models for DCM and chloroform, but the wider applications of PBPK modeling with the halogenated alkanes, emphasized by the groups of articles noted in the various tables, indicates the important place these compounds hold in the history and current refinement of PBPK modeling in environmental health.

NOTATION

BCM bromochloromethane
BDCM bromodichloromethane
C_a the concentration in the blood exiting the lungs
C_{alv} concentration of chemical in alveolar air
C_{BLA} chemical concentration in the arterial blood compartment

C_{BLV}	chemical concentration in the venous blood compartment
^{14}C-DBM	^{14}C-dibromomethane
C_{ex}	concentration of chemical in exhaled air
C_{EXT}	concentration in blood leaving the tissue compartment for diffusion-limited uptake
CFC	chlorofluorocarbon
CF_3I	trifluoroiodomethane
$CHCl_3$	chloroform
CH_2O	formaldehyde
C_{in}	inhaled concentration of chemical
C_L	concentration of chemical in the liver compartment
CO	carbon monoxide
CO_2	carbon dioxide
COHb	carboxyhemoglobin
C_T	concentration of chemical in the tissue compartment
CYP450	cytochrome p450
DBCM	dibromochloromethane
DBM	methylene chloride (i.e., dibromomethane)
DPX	DNA-protein cross-links
EDB	1,2-dibromoethane
F_{DS}	fraction of dead space in the lungs
GI	gastrointestinal
GSH	glutathione
GST	glutathione-S-transferase
HCFC	hydrochlorofluorocarbon
HFC	hydrofluorocarbon
i.p.	intraperitoneal
i.v.	intravenous
K_m	the Michaelis–Menten constant
MFO	mixed-function oxidase
N_2O	nitrous oxide
NPE	nonpulmonary elimination
ODP	ozone depletion potential
PA_T	permeability-area cross product for diffusion-limited uptake
P_b	blood : air partition coefficient
PD	pharmacodynamic
P_L	liver : blood partition coefficient
P_T	tissue compartment : blood partition coefficient
Q_C	cardiac output
Q_L	blood flow rate to the liver
Q_P	alveolar ventilation rate
Q_T	flow rate of blood to the tissue compartment,
t	time
TFA	trifluoroacetic acid
V_{BLA}	volume of the arterial blood compartment
V_{BLV}	volume of the venous blood compartment

V_L	volume of the liver
V_{max}	the reaction velocity for metabolism at maximal substrate concentration
V_T	volume of the tissue compartment

REFERENCES

Andersen, M. E. (1981). A physiologically based toxicokinetic description of the metabolism of inhaled gases and vapors—Analysis at steady state. *Toxicol. Appl. Pharmacol.* **60**, 509–526.

Andersen, M. E. (2003). Toxicokinetic modeling and its applications in chemical risk assessment. *Toxicol. Lett.* **138**, 9–27.

Andersen, M. E., Gargas, M. L., and Ramsey, J. C. (1984). Inhalation pharmacokinetics: Evaluating systemic extraction, total *in vivo* metabolism, and time course of enzyme induction for inhaled styrene in rats based on arterial blood : inhaled air concentration ratios. *Toxicol. Appl. Pharmacol.* **73**, 176–187.

Andersen, M. E., Clewell, H. J., Gargas, M. L., Smith, F. A., and Reitz, R. H. (1987a). Physiologically based pharmacokinetics and the risk assessment process for methylene chloride. *Toxicol. Appl. Pharmacol.* **87**, 185–205.

Andersen, M. E., Clewell, H. J., and Reitz, R. H. (1987b). Physiologically based pharmacokinetics and methylene chloride cancer risk assessment considerations. In: *Human Risk Assessment. The Role of Animal Selection and Extrapolation*, Ed. M. V. Roloff, Taylor and Francis Ltd., London, pp. 129–139.

Andersen, M. E., MacNaughton, M. G., Clewell, H. J., and Paustenbach, D. J. (1987c). Adjusting exposure limits for long and short exposure periods using a physiologically based pharmacokinetic model. *Am. Ind. Hyg. Assoc. J.* **48**, 335–342.

Andersen, M. E., Clewell, H. J., Gargas, M. L., MacNaughton, M. G., Reitz, R. H., Nolan, R. J., and McKenna, M. J. (1991). Physiologically based pharmacokinetic modeling with dichloromethane, its metabolite, carbon-monoxide, and blood carboxyhemoglobin in rats and humans. *Toxicol. Appl. Pharmacol.* **108**, 14–27.

Andersen, M. E., Clewell, H. J., Mahle, D. A., and Gearhart, J. M. (1994). Gas uptake studies of deuterium isotope effects on dichloromethane metabolism in female B6C3F1 mice *in vivo*. *Toxicol. Appl. Pharmacol.* **128**, 158–165.

Angelo, M. J., and Pritchard, A. B. (1984). Simulations of methylene chloride pharmacokinetics using a physiologically based model. *Regul. Toxicol. Pharmacol.* **4**, 329–339.

Angelo, M. J., Bischoff, K. B., Pritchard, A. B., and Presser, M. A. (1984). A physiological model for the pharmacokinetics of methylene chloride in $B_6C_3F_1$ mice following i.v. administrations. *J. Pharmacokinet. Biopharm.* **12**, 413–436.

Ashman, M. N., Blesser, W. B., and Epstein, R. M. (1970). A nonlinear model for the uptake and distribution of halothane in man. *Anesthesiology* **33**, 419–429.

Blancato, J. N., and Chiu, N. (1994). Use of pharmacokinetic models to estimate internal doses from exposure. In: *Drinking Water Contaminants and Health: Integration of Exposure Assessment, Toxicology, and Risk Assessment*, R. Wang, ed., Marcel Dekker, New York, pp. 217–239.

Brown, R. P., Delp, M. D., Lindstedt, S. L., Rhomberg, L. R., and Beliles, R. P. (1997). Physiological parameter values for physiologically based pharmacokinetic models. *Toxicol. Ind. Health* **13**, 407–484.

Casanova, M., Morgan, K. T., Steinhagen, W. H., Everitt, J. I., Popp, J. A., and Heck, H. d'A. (1991). Covalent binding of inhaled formaldehyde to DNA in the respiratory tract of rhesus monkeys: Pharmacokinetics, rat-to-monkey interspecies scaling, and extrapolation to man. *Fundam. Appl. Toxicol.* **17**, 409–428.

Casanova, M., Conolly, R. B., and Heck, H. D. (1996). DNA-protein cross-links (DPX) and cell proliferation in B6C3F$_1$ mice but not Syrian golden hamsters exposed to dichloromethane: Pharmacokinetics and risk assessment with DPX as dosimeter. *Fundam. Appl. Toxicol.* **31**, 103–116.

Chinery, R. L., and Gleason, A. K. (1993). A compartmental model for the prediction of breath concentration and absorbed dose of chloroform after exposure while showering. *Risk Anal.* **13**, 51–62.

Clewell, H. J., and Andersen, M. E. (1996). Use of physiologically based pharmacokinetic modeling to investigate individual versus population risk. *Toxicology* **111**, 315–329.

Clewell, H. J. III, Lee, T., and Carpenter, R. L. (1994). Sensitivity of physiologically based pharmacokinetic models to variation in model parameters: Methylene chloride. *Risk Anal.* **14,** 521–531.

Constan, A. A., Wong, B. A., Everitt, J. I., and Butterworth, B. E. (2002). Chloroform inhalation exposure conditions necessary to initiate liver toxicity in female B6C3F1 mice. *Toxicol. Sci.* **66**, 201–208.

Corley, R. A., Mendrala, A. L., Smith, F. A., Staats, D. A., Gargas, M. L., Conolly, R. B., Andersen, M. E., and Reitz, R. H. (1990). Development of a physiologically based pharmacokinetic model for chloroform. *Toxicol. Appl. Pharmacol.* **103**, 512–527.

Corley, R. A., Gordon, S. M., and Wallace, L. A. (2000). Physiologically based pharmacokinetic modeling of the temperature-dependent dermal absorption of chloroform by humans following bath water exposures. *Toxicol. Sci.* **53**, 13–23.

da Silva, M. L., Charest-Tardif, G., Krishnan, K., and Tardif, R. (1999). Influence of oral administration of a quaternary mixture of trihalomethanes on their blood kinetics in the rat. *Toxicol. Lett.* **106**, 49–57.

da Silva, M. L., Charest-Tardif, G., Krishnan, K., and Tardif, R. (2000). Evaluation of the pharmacokinetic interactions between orally administered trihalomethanes in the rat. *J. Toxicol. Env. Health Pt. A*, **60**, 343–353.

Dankovic, D. A., and Bailer, A. J. (1994). The impact of exercise and intersubject variability on dose estimates for dichloromethane derived from a physiologically-based pharmacokinetic model. *Fundam. Appl. Toxicol.* **22**, 20–25.

DeJongh, J., Verhaar, H. J. M., and Hermens, J. L. M. (1998). Role of kinetics in acute lethality of non-reactive volatile organic compounds (VOCs). *Toxicol. Sci.* **45**, 26–32.

Dodd, D. E., Brashear, W. T., and Vinegar, A. (1993). Metabolism and pharmacokinetics of selected halon replacement candidates. *Toxicol. Lett.* **68**, 37–47.

Ellis, M. K., Trebilcock, R., Naylor, J. L., Tseung, K., Collins, M. A., Hext, P. M., and Green, T. (1996). The inhalation toxicology, genetic toxicology, and metabolism of difluoromethane in the rat. *Fundam. Appl. Toxicol.* **31**, 243–251.

El-Masri, H. A., Bell, D. A., and Portier, C. J. (1999). Effects of glutathione transferase theta polymorphism on the risk estimates of dichloromethane to humans. *Toxicol. Appl. Pharmacol.* **158**, 221–230.

Evans, M. V., Crank, W. D., Yang, H. M., and Simmons, J. E. (1994). Applications of sensitivity analysis to a physiologically based pharmacokinetic model for carbon tetrachloride in rats. *Toxicol. Appl. Pharmacol.* **128**, 36–44.

Evans, M. V., and Simmons, J. E. (1996). Physiologically based pharmacokinetic estimated metabolic constants and hepatotoxicity of carbon tetrachloride after methanol pretreatment in rats. *Toxicol. Appl. Pharmacol.* **140**, 245–253.

Evans, M. V., Boyes, W. K., Simmons, J. E., Litton, D. K., and Easterling, M. R. (2002). A comparison of Haber's rule at different ages using a physiologically based pharmacokinetic (PBPK) model for chloroform in rats. *Toxicology* **176**, 11–23.

Filser, J. G., and Bolt, H. M. (1979). Pharmacokinetics of halogenated ethylenes. *Archiv. Toxicol.* **42**, 123–136.

Fiserova-Bergerova, V. (1992). Inhalation anesthesia using physiologically based pharmacokinetic models. *Drug Metab. Rev.* **24**, 531–557.

Gargas, M. L., and Andersen, M. E. (1989). Determining kinetic constants of chlorinated ethane metabolism in the rat from rates of exhalation. *Toxicol. Appl. Pharmacol.* **99**, 344–353.

Gargas, M. L., Andersen, M. E., and Clewell, H. J. (1986a). A physiologically based simulation approach for determining metabolic constants from gas uptake data. *Toxicol. Appl. Pharmacol.* **86**, 341–352.

Gargas, M. L., Clewell, H. J., III, and Andersen, M. E. (1986b). Metabolism of inhaled dihalomethanes *in vivo*: Differentiation of kinetic constants for two independent pathways. *Toxicol. Appl. Pharmacol.* **82**, 211–223.

Gargas, M. L., Burgess, R. J., Voisard, D. E., Cason, G. H., and Andersen, M. E. (1989). Partition coefficients of low-molecular-weight volatile chemicals in various liquids and tissues. *Toxicol. Appl. Pharmacol.* **98**, 87–99.

Gargas, M. L., Clewell, H. J., and Andersen, M. E. (1990). Gas uptake inhalation techniques and the rates of metabolism of chloromethanes, chloroethanes, and chloroethylenes in the rat. *Inhal. Toxicol.* **2**, 295–319.

Gearhart, J. M., Seckel, C., and Vinegar, A. (1993). *In vivo* metabolism of chloroform in B6C3F1 mice determined by the method of gas uptake: The effects of body temperature on tissue partition coefficients and metabolism. *Toxicol. Appl. Pharmacol.* **119**, 258–266.

Georgopoulos, P. G., Roy, A., and Gallo, M. A. (1994). Reconstruction of short-term multiroute exposure to volatile organic compounds using physiologically-based pharmacokinetic models. *J. Exp. Anal. Environ. Epidemiol.* **4**, 309–328.

Hefner, R. E., Watanabe, P. G., and Gehring, P. J. (1975). Preliminary studies of the fate of inhaled vinyl chloride monomer in rats. *Ann. N.Y. Acad. Sci.* **246**, 135–148.

Hilderbrand, R. L., Andersen, M. E., and Jenkins, L. J. (1981). Prediction of *in vivo* kinetic constants for metabolism of inhaled vapors from kinetic constants measured *in vitro*. *Fundam. Appl. Toxicol.* **1**, 403–409.

Hissink, A. M., Wormhoudt, L. W., Sherratt, P. J., Hayes, J. D., Commandeur, J. N. M., Vermeulen, N. P. E., and van Bladeren, P. J. (2000). A physiologically-based pharmacokinetic (PB-PK) model for ethylene dibromide: relevance of extrahepatic metabolism. *Food Chem. Toxicol.* **38**, 707–716.

International Life Sciences Institute, Health and Environmental Sciences Institute. (1997). *An Evaluation of EPA's Proposed Guidelines for Carcinogen Risk Assessment Using Chloroform and Dichloroacetic Acid as Case Studies: Report of an Expert Panel*. Washington, D.C.

Jarabek, A. M., Fisher, J. W., Rubenstein, R., Lipscomb, J. C., Williams, R. J., Vinegar, A., and McDougal, J. N. (1994). Mechanistic insights aid the search for CFC substitutes: Risk assessment of HCFC-123 as an example. *Risk Anal.* **14**, 231–250.

Jepson, G. W., and McDougal, J. N. (1997). Physiologically based modeling of nonsteady state dermal absorption of halogenated methanes from an aqueous solution. *Toxicol. Appl. Pharmacol.* **144**, 315–324.

Jepson, G. W., and McDougal, J. N. (1999). Predicting vehicle effects on the dermal absorption of halogenated methanes using physiologically based modeling. *Toxicol. Sci.* **48**, 180–188.

Jonsson, F., Bois, F. Y., and Johanson, G. (2001). Physiologically based pharmacokinetic modeling of inhalation exposure of humans to dichloromethane during moderate to heavy exercise. *Toxicol. Sci.* **59**, 209–218.

Jonsson, F., and Johanson, G. (2001). A Bayesian analysis of the influence of GSTT1 polymorphism on the cancer risk estimate for dichloromethane. *Toxicol. Appl. Pharmacol.* **174**, 99–112.

Krishnan, K., and Andersen, M. E. (2001). Physiologically based pharmacokinetic modeling in toxicology. In: *Principles and Methods of Toxicology*, A. W. Hayes, ed., Taylor & Francis, Philadelphia, pp. 193–241.

Lapare, S., Tardif, R., and Brodeur, J. (1995). Effect of various exposure scenarios on the biological monitoring of organic solvents in alveolar air. 2. 1,1,1-Trichloroethane and trichloroethylene. *Int. Arch. Occup. Environ. Health* **67**, 375–394.

Lerou, J. G. C., Dirksen, R., Beneken Kolmer, H. H., and Booij, L. H. D. J. (1991a). A system model for closed-circuit inhalation anesthesia. I. Computer study. *Anesthesiology* **75**, 345–355.

Lerou, J. G. C., Dirksen, R., Beneken Kolmer, H. H., Booij, L. H. D. J., and Borm, G. F. (1991b). A system model for closed-circuit inhalation anesthesia. II. Clinical validation. *Anesthesiology* **75**, 230–237.

Lerou, J. G. C., Vermeulen, P. M., Dirksen, R., Booij, L. H. D. J., and Borm, G. F. (1993). The predictive performance of a system model for enflurane closed-circuit inhalation anesthesia. *Anesthesiology* **79**, 932–942.

Leung, H. W. (1992). Use of physiologically based pharmacokinetic models to establish biological exposure indexes. *Am. Ind. Hyg. Assoc. J.* **53**, 369–374.

Levesque, B., Ayotte, P., Tardif, R., Charest-Tardif, G., Dewailly, E., Prud'Homme, D., Gingras, G., Allaire, S., and Lavoie, R. (2000). Evaluation of the health risk associated with exposure to chloroform in indoor swimming pools. *J. Toxicol. Env. Health Pt. A* **61**, 225–243.

Levesque, B., Ayotte, P., Tardif, R., Ferron, L., Gingras, S., Schlouch, E., Gingras, G., Levallois, P., and Dewailly, E. (2002). Cancer risk associated with household exposure to chloroform. *J. Toxicol. Env. Health Pt. A* **65**, 489–502.

Lien, G. J., McKim, J. M., Hoffman, A. D., and Jenson, C. T. (2001). A physiologically based toxicokinetic model for lake trout (*Salvelinus namaycush*). *Aqua. Toxicol.* **51**, 335–350.

Lilly, P. D., Andersen, M. E., Ross, T. M., and Pegram, R. A. (1997). Physiologically based estimation of *in vivo* rates of bromodichloromethane metabolism. *Toxicology* **124**, 141–152.

Lilly, P. D., Andersen, M. E., Ross, T. M., and Pegram, R. A. (1998). A physiologically based pharmacokinetic description of the oral uptake, tissue dosimetry, and rates of metabolism of bromodichloromethane in the male rat. *Toxicol. Appl. Pharmacol.* **150**, 205–217.

Loizou, G. D., and Anders, M. W. (1993). Gas-uptake pharmacokinetics and biotransformation of 1,1-dichloro-1-fluoroethane (HCFC-141b). *Drug Metab. Dispos.* **21**, 634–639.

Loizou, G. D., and Anders, M. W. (1995). Gas-uptake pharmacokinetics and metabolism of 2-chloro-1,1,1,2-tetrafluoroethane (HCFC-124) in the rat, mouse, and hamster. *Drug Metab. Dispos.* **23**, 875–880.

Loizou, G. D., Urban, G., Dekant, W., and Anders, M. W. (1994). Gas-uptake pharmacokinetics of 2,2-dichloro-1,1,1-trifluoroethane (HCFC-123). *Drug Metab. Dispos.* **22**, 511–517.

Loizou, G. D., Eldirdiri, N. I., and King, L. J. (1996). Physiologically based pharmacokinetics of uptake by inhalation of a series of 1,1,1-trihaloethanes: Correlation with various physicochemical parameters. *Inhal. Toxicol.* **8**, 1–19.

Loizou, G. D., Tran, C. L., and Anders, M. W. (1997). Physiologically based pharmacokinetic analysis of the concentration-dependent metabolism of halothane. *Xenobiotica* **27**, 87–99.

McDougal, J. N., Jepson, G. W., Clewell, H. J., MacNaughton, M. G., and Andersen, M. E. (1986). A physiological pharmacokinetic model for dermal absorption of vapors in the rat. *Toxicol. Appl. Pharmacol.* **85**, 286–294.

Munson, E. S., Eger, E. I., II, and Bowers, D. L. (1973). Effects of anesthetic-depressed ventilation and cardiac output on anesthetic uptake: A computer nonlinear simulation. *Anesthesiology* **38**, 251–259.

Nichols, J. W., McKim, J. M., Lien, G. J., Hoffman, A. D., and Bertelsen, S. L. (1991). Physiologically based toxicokinetic modeling of three waterborne chloroethanes in rainbow-trout (*Oncorhynchus mykiss*). *Toxicol. Appl. Pharmacol.* **110**, 374–389.

Nichols, J. W., McKim, J. M., Lien, G. J., Hoffman, A. D., Bertelsen, S. L., and Gallinat, C. A. (1993). Physiologically-based toxicokinetic modeling of three waterborne chloroethanes in channel catfish, *Ictalurus punctatus*. *Aquat. Toxicol.* **27**, 83–112.

Nichols, J. W., McKim, J. M., Lien, G. J., Hoffman, A. D., Bertelsen, S. L., and Elonen, C. M. (1996). A physiologically based toxicokinetic model for dermal absorption of organic chemicals by fish. *Fundam. Appl. Toxicol.* **31**, 229–242.

NTP (National Toxicology Program). (1986). *Toxicology and Carcinogenesis Studies of Dichloromethane (Methylene Chloride) (CAS No. 75-09-2) in F344/N Rats and B6C3F1 Mice (Inhalation Studies).* U.S. Department of Health and Human Services. Technical Report No. 306. MH Publication No. 86-2562.

Paustenbach, D. J., Clewell, H. J., Gargas, M. L., and Andersen, M. E. (1987). Development of a physiologically based pharmacokinetic model for multiday inhalation of carbon tetrachloride. In: *Drinking Water and Health*, Vol. 8, National Academy of Science, Washington, D.C., pp. 310–324.

Paustenbach, D. J., Clewell, H. J., Gargas, M. L., and Andersen, M. E. (1988). A physiologically based pharmacokinetic model for inhaled carbon tetrachloride. *Toxicol. Appl. Pharmacol.* **96**, 191–211.

Ploemen, J., Wormhoudt, L. W., Haenen, G., Oudshoorn, M. J., Commandeur, J. N. M., Vermeulen, N. P. E., DeWaziers, I., Beaune, P. H., Watabe, T., and VanBladeren, P. J. (1997). The use of human *in vitro* metabolic parameters to explore the risk assessment of hazardous compounds: The case of ethylene dibromide. *Toxicol. Appl. Pharmacol.* **143**, 56–69.

Poet, T. S., Thrall, K. D., Corley, R. A., Hui, X. Y., Edwards, J. A., Weitz, K. K., Maibach, H. I., and Wester, R. C. (2000). Utility of real time breath analysis and physiologically based pharmacokinetic modeling to determine the percutaneous absorption of methyl chloroform in rats and humans. *Toxicol. Sci.* **54**, 42–51.

Portier, C. J., and Kaplan, N. L. (1989). Variability of safe dose estimates when using complicated models of the carcinogenic process. A case study: Methylene chloride. *Fundam. Appl. Toxicol.* **13**, 533–544.

Ramsey, J. C., and Andersen, M. E. (1984). A physiologically based description of the inhalation pharmacokinetics of styrene in rats and humans. *Toxicol. Appl. Pharmacol.* **73**, 159–175.

Reddy, M. B., Andersen, M. E., Morrow, P. E., Dobrev, I. D., Varaprath, S., Plotzke, K. P., and Utell, M. J. (2003). Physiological modeling of inhalation kinetics of octamethylcyclotetrasiloxane (D4) in humans during rest and exercise. *Toxicol. Sci.* **72**, 3–18.

Reitz, R. H., McDougal, J. N., Himmelstein, M. W., Nolan, R. J., and Schumann, A. M. (1988a). Physiologically based pharmacokinetic modeling with methylchloroform—Implications for interspecies, high-dose low-dose, and dose route extrapolations. *Toxicol. Appl. Pharmacol.* **95**, 185–199.

Reitz, R. H., Mendrala, A. L., Park, C. N., Andersen, M. E., and Guengerich, F. P. (1988b). Incorporation of *in vitro* enzyme data into the physiologically-based pharmacokinetic (PB-PK) model for methylene chloride: Implications for risk assessment. *Toxicol. Lett.* **43** 97–116.

Reitz, R. H., Mendrala, A. L., and Guengerich, F. P. (1989). *In vitro* metabolism of methylene chloride in human and animal tissues—Use in physiologically based pharmacokinetic models. *Toxicol. Appl. Pharmacol.* **97**, 230–246.

Reitz, R. H., Mendrala, A. L., Corley, R. A., Quast, J. F., Gargas, M. L., Andersen, M. E., Staats, D. A., and Conolly, R. B. (1990). Estimating the risk of liver-cancer associated with human exposures to chloroform using physiologically based pharmacokinetic modeling. *Toxicol. Appl. Pharmacol.* **105**, 443–459.

Roy, A., Weisel, C. P., Ballo, M. A., and Georgopoulos, P. G. (1996a). Studies of multiroute exposure/dose reconstruction using physiologically based pharmacokinetic models. *Toxicol. Ind. Health* **12**, 153–163.

Roy, A., Weisel, C. P., Gallo, M., and Georgopoulos, P. (1996b). Studies of multiroute exposure/dose reconstruction using physiologically based pharmacokinetic models. *J. Clean Technol. Environ. Toxicol. Occup. Med.* **5**, 285–295.

Semino, G., Lilly, P., and Andersen, M. E. (1997). A pharmacokinetic model describing pulsatile uptake of orally-administered carbon tetrachloride. *Toxicology* **117**, 25–33.

Smith, A. E., Gray, G. M., and Evans, J. S. (1995). Ability of predicted internal dose measures to reconcile tumor bioassay data for chloroform. *Regul. Toxicol. Pharmacol.* **21** 339–351.

Smith, N. T., Zwart, A., and Beneken, J. E. W. (1972). Interaction between the circulatory effects and the uptake and distribution of halothane: Use of a multiple model. *Anesthesiology* **37**, 47–58.

Staats, D. A., Fisher, J. W., and Connolly, R. B. (1991). Gastrointestinal absorption of xenobiotics in physiologically based pharmacokinetic models—A 2-compartment description. *Drug Metab. Dispos.* **19**, 144–148.

Tan, Y. M., Butterworth, B. E., Gargas, M. L., and Conolly, R. B. (2003). Biologically motivated computational modeling of chloroform cytolethality and regenerative cellular proliferation. *Toxicol. Sci.* **75**, 192–200.

Thakore, K. N., Gargas, M. L., Andersen, M. E., and Mehendale, H. M. (1991). PB-PK derived metabolic constants, hepatotoxicity, and lethality of BrCCl3 in rats pretreated with chlordecone, phenobarbital, or mirex. *Toxicol. Appl. Pharmacol.* **109**, 514–528.

Thrall, K. D., and Kenny, D. V. (1996). Evaluation of a carbon tetrachloride physiologically based pharmacokinetic model using real-time breath-analysis monitoring of the rat. *Inhal. Toxicol.* **8**, 251–261.

Thrall, K. D., Poet, T. S., Corley, R. A., Tanojo, H., Edwards, J. A., Weitz, K. K., Hui, X., Maibach, H. I., and Wester, R. C. (2000a). A real-time *in-vivo* method for studying the percutaneous absorption of volatile chemicals. *Int. J. Occup. Environ. Health* **6**, 96–103.

Thrall, K. D., Vucelick, M. E., Gies, R. A., Zangar, R. C., Weitz, K. K., Poet, T. S., Springer, D. L., Grant, D. M., and Benson, J. M. (2000b). Comparative metabolism of carbon tetrachloride in rats, mice, and hamsters using gas uptake and PBPK modeling. *J. Toxicol. Env. Health Pt. A* **60**, 531–548.

Thrall, K. D., Callahan, P. J., Weitz, K. K., Edwards, J. A., Brinkman, M. C., and Kenny, D. V. (2001). Design and evaluation of a breath-analysis system for biological monitoring of volatile compounds. *Am. Ind. Hyg. Assoc. J.* **62**, 28–35.

US EPA. (2001). *Toxicological Review of Chloroform In Support of Summary Information on the Integrated Risk Information System (IRIS)*, U.S. Environmental Protection Agency, Washington, D.C. (at http://www.epa.gov/iris).

Vermeulen, P. M., Lerou, J. G. C., Dirksen, R., Booij, L. H. D. J., and Borm, G. F. (1995). A system model for halothane closed-circuit anesthesia. *Anesthesiology* **83**, 515–527.

Vermeulen, P. M., Kalkman, C. J., Dirksen, R., Knape, J. T. A., Moons, K. G. M., and Borm, G. F. (2002). Predictive performance of a physiological model for enflurane closed-circuit anaesthesia: Effects of continuous cardiac output measurements and age-related solubility data. *Br. J. Anaesth.* **88**, 38–45.

Vinegar, A. (2001). PBPK modeling of canine inhalation exposures to halogenated hydrocarbons. *Toxicol. Sci.* **60**, 20–27.

Vinegar, A., and Jepson, G. W. (1996). Cardiac sensitization thresholds of halon replacement chemicals predicted in humans by physiologically-based pharmacokinetic modeling. *Risk Anal.* **16**, 571–579.

Vinegar, A., Williams, R. J., Fisher, J. W., and McDougal, J. N. (1994). Dose-dependent metabolism of 2,2-dichloro-1,1,1-trifluoroethane—A physiologically-based pharmacokinetic model in the male Fischer-344 rat. *Toxicol. Appl. Pharmacol.* **129**, 103–113.

Vinegar, A., Jepson, G. W., and Overton, J. H. (1998). PBPK modeling of short-term (0 to 5 min) human inhalation exposures to halogenated hydrocarbons. *Inhal. Toxicol.* **10**, 411–429.

Vinegar, A., Jepson, G. W., Hammann, S. J., Harper, G., Dierdorf, D. S., and Overton, J. H. (1999). Simulated blood levels of CF3I in personnel exposed during its release from an F-15 jet engine nacelle and during intentional inhalation. *Am. Ind. Hyg. Assoc. J.* **60**, 403–408.

Vinegar, A., Jepson, G. W., Cisneros, M., Rubenstein, R., and Brock, W. J. (2000). Setting safe acute exposure limits for halon replacement chemicals using physiologically based pharmacokinetic modeling. *Inhal. Toxicol.* **12**, 751–763.

Williams, R. J., Vinegar, A., McDougal, J. N., Jarabek, A. M., and Fisher, J. W. (1996). Rat to human extrapolation of HCFC-123 kinetics deduced from halothane kinetics: A corollary approach to physiologically based pharmacokinetic modeling. *Fundam. Appl. Toxicol.* **30**, 55–66.

Yoshida, K. (1993). Preliminary exposure assessment of volatile chlorinated hydrocarbons in Japan. *Chemosphere* **27**, 621–630.

Zwart, A., Smith, N. T., and Beneken, J. E. W. (1972). Multiple model approach to uptake and distribution of halothane: The use of an analog computer. *Comp. Biomed. Res.* **5**, 228–238.

CHAPTER 3

HALOGENATED ALKENES

Dong-Soon Bae, Melvin E. Andersen, and Harvey J. Clewell III

3.1 INTRODUCTION

Halogenated alkenes are metabolized in mammals primarily by cytochrome P450 family enzymes and by conjugation with glutathione, GSH (Henschler 1985; Mansuy 1985). The double bonds in halogenated alkenes are oxidized to reactive intermediates, frequently epoxides, that are highly reactive and can covalently bind to nucleic acids (Henschler and Bonse 1977). Glutathione conjugates formed from trichloroethylene and perchloroethylene are also broken down by normal renal processing of glutathione conjugates to metabolites that are toxic and mutagenic. For all members of this group of chemicals, metabolites are more toxic than parent chemicals. Knowledge of the rates of metabolic activation for halogenated alkenes and formation or depletion of intermediate reactive products by GSH conjugation becomes important for describing the disposition and elimination of the parent chemical and has to be considered when building PBPK models for evaluating risks associated with exposures to these compounds. With respect to human health risks, both carcinogenicity and target organ toxicity (primarily liver) are of some concern. The International Agency for Research on Cancer (IARC) has labeled vinyl chloride as carcinogenic to humans, vinyl fluoride and trichloroethylene as probably carcinogenic to humans, and perchloroethylene and β-chloroprene as possibly carcinogenic to humans. PBPK models for several of these halogenated ethylenes have been developed to support human health risk assessments for cancer endpoints using dose metrics related to rates of production of reactive metabolites or tissue concentrations

Physiologically Based Pharmacokinetic Modeling, edited by Micaela B. Reddy, Raymond S. H. Yang, Harvey J. Clewell III, and Melvin E. Andersen.

of more stable metabolites, such as trichloroacetic acid in the case of trichloroethylene and perchloroethylene.

3.2 THE CHLOROETHYLENES: BACKGROUND

This chapter reviews PBPK models developed for several haloethylenes: vinyl chloride (1-chloroethylene, VC), vinyl fluoride (VF), vinylidene chloride (1,1-dichloroethene, VDC), *cis*-1,2-dichloroethylene (*c*DCE), *trans*-1,2-dichloroethylene (*t*DCE), trichloroethylene (TCE), and tetrachloroethylene (perchloroethylene, PERC) (Tables 3.1 and 3.2). Models for a halogenated propene, allyl chloride (3-chloro-1-propene, AC), and two halogenated butenes, β-chloroprene (2-chloro-1,3-butadiene, CD) and hexachlorobutadiene (HCB), are also discussed. Several PBPK models that focused on halogenated alkenes in mixtures (Table 3.3) are discussed in Chapter 13.

The restricted group of two-carbon chlorinated ethylenes (i.e., VC, VDC, *c*DCE, *t*DCE, TCE, and PERC) has represented an attractive class of compounds for developing methods for determining partition coefficients (Gargas *et al.* 1989) and assessing rates of metabolism in intact animals (Filser and Bolt 1979; Gargas *et al.* 1986, 1990). The characteristics of metabolism and toxicity of these compounds are now fairly well-established in rats (Table 3.1; Fig. 3.2). CYP450-mediated metabolism through oxidation of the double bonds leads to oxiranes (see VC in Table 3.2) or in some cases to direct production of downstream oxidative products such as chloral (trichloroacetaldehyde) with TCE (Fig. 3.2). In those cases where oxiranes are formed, these intermediates may either react with tissue components, rearrange by chloride-ion shifts, or be hydrolyzed by water addition. The oxirane of VC, for instance, could rearrange to chloroacetaldehyde or could add water to form 2-hydroxy-acetaldehyde. Rearrangement of oxiranes with VDC, *c*DCE, *t*DCE, and PERC would produce highly reactive acylchlorides. Water addition to the oxiranes from *c*DCE and *t*DCE should produce a highly reactive compound, dialdehyde, that may play a role in the suicide inhibition noted with metabolism of these compounds (Lilly *et al.* 1998).

PBPK models were first used to study the kinetic constants for these oxidation reactions using analysis of gas uptake studies. All of the chlorinated ethylenes with the exception of PERC were found to have relatively high affinities for oxidation, with dissociation constants estimated in the range of several micromoles. The maximum rates of metabolism (here expressed as micromoles per hour for a 1-kg rat) depended on the substitution patterns of the chlorines. The three substrates with no more than one-chlorine per carbon (i.e., VC, *c*DCE, and *t*DCE) have similar values of V_{max}. VDC and TCE, with one carbon having two chlorine atoms attached, have V_{max} values that are almost double those for VC, *c*DCE, and *t*DCE. With PERC, it was difficult to accurately establish kinetic constants of metabolism from gas uptake studies alone because the chamber loss was much less than observed with the other chlorinated ethylenes. This compound has a lower V_{max} and considerably higher K_m than the other members of the class. Toxic responses to these chloroethylenes depend on the nature of the metabolites produced under particular exposure

TABLE 3.1 PBPK Models for Halogenated Alkenes[a]

Chemical (abbreviation)	Structure	Species	Reference
Vinyl chloride (VC)	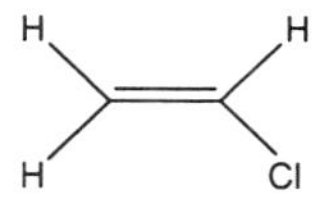	Rat, mouse, human	Chen and Blancato (1989)
		Rat	Gargas *et al.* (1990)
		Rat, mouse, human	Reitz *et al.* (1996a)
		Rat, mouse, human	Clewell *et al.* (2001)
Vinyl fluoride (VF)	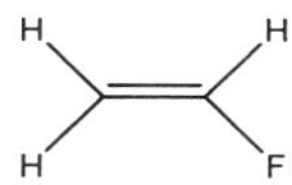	Rat, mouse	Cantoreggi and Keller (1997)
Vinylidene chloride (VDC)	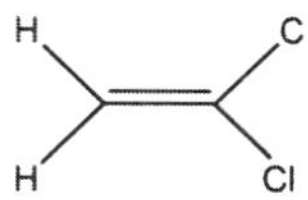	Rat	Gargas *et al.* (1986)
		Rat	D'Souza and Andersen (1988)
		Rat	Gargas *et al.* (1990)
		Human	Leung (1992)
cis-1,2-Dichloroethylene (*c*DCE)	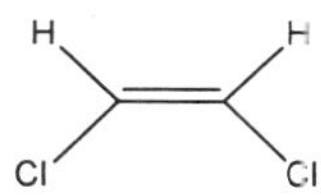	Rat	Gargas *et al.* (1990)
		Rat	Lilly *et al.* (1998)
trans-1,2-Dichloroethylene (*t*DCE)	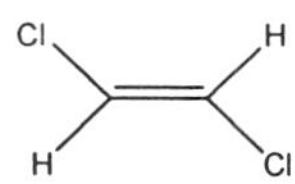	Rat	Gargas *et al.* (1990)
		Rat	Lilly *et al.* (1998)
Tetrachloroethylene (PERC)	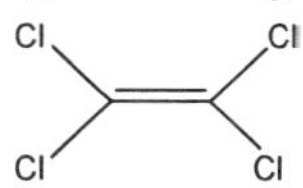	Human	Guberan and Fernandez (1974)
		Human	Droz *et al.* (1989b)
		Rat, human	Koizumi (1989)
		Human	Bois *et al.* (1990)
		Rat	Gargas *et al.* (1990)
		Human	Leung (1992)
		Human	Rao and Brown (1993)
		B6C3F1 mouse	Gearhart *et al.* (1993)
		Rat, dog	Dallas *et al.* (1994a)
		Rat	Dallas *et al.* (1994b)
		Rat, mouse, human	Reitz *et al.* (1996b)
		Human	Jang and Droz (1997)
		Rat	DeJongh *et al.* (1998)
		Human	Loizou (2001)
		Rat, human	Poet *et al.* (2002)
Allyl chloride (AC)	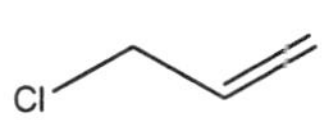	Rat	Clewell and Andersen (1987)
β-Chloroprene (CD)	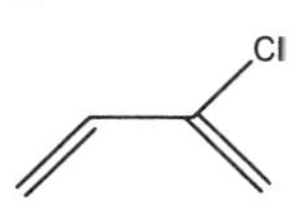	Fisher and Wistar rat, mouse, hamster, human	Himmelstein *et al.* (2004a)
Hexachlorobutadiene (HCB)	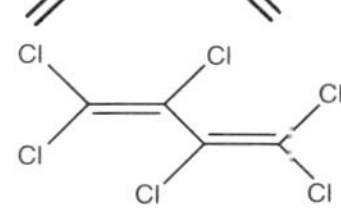	Rat, human	Green *et al.* (2003)

[a] Trichloroethylene (TCE) models are listed in Table 3.4.

TABLE 3.2 Metabolism Parameters and Some Biological Responses with Chloroethylenes in Rodents

Compound	Metabolite	Products	V_{max} (μmol/hr)	K_m (μM)	Glutathione reactions	Responses
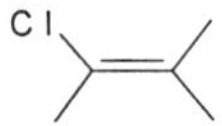 (VC)	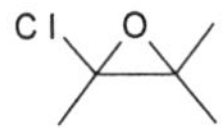	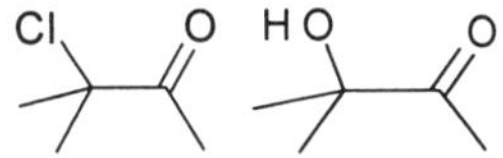	48[a]	1.6	Depletion by metabolites	Potent multispecies carcinogen
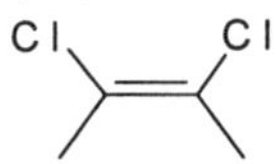 (cDCE)	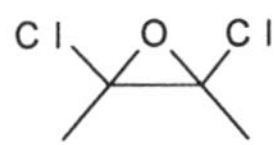	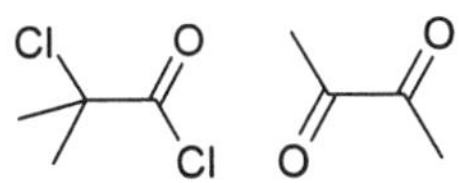	47[b]	1.96	Not known	Suicide enzyme inhibition (k_d = 2.7)[d]
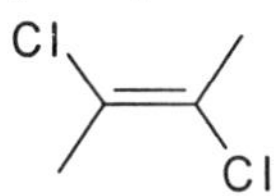 (tDCE)	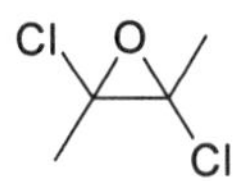	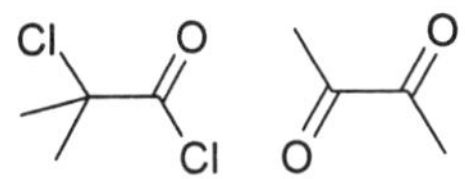	44[b]	0.82	Not known	Suicide enzyme inhibition (k_d = 496)[d]
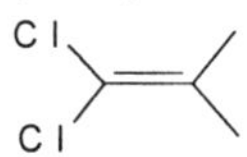 (DCE)	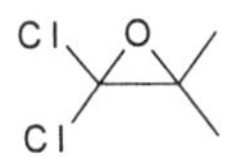	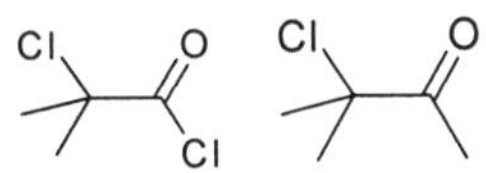	78[a]	1.03	Depletion by metabolites	Potent hepatotoxicant
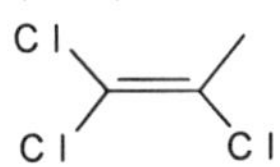 (TCE)	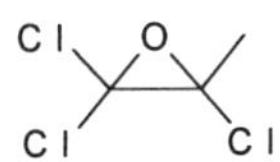	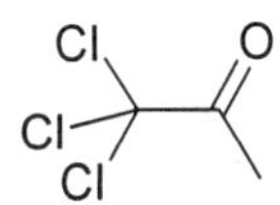 Chloral	83.6[a]	1.90	Direct TCE conjugation	Weakly carcinogenic in liver, lung and kidney
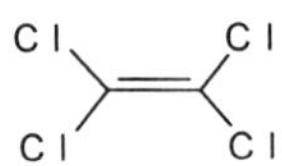 (PERC)	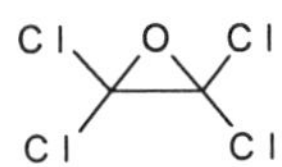	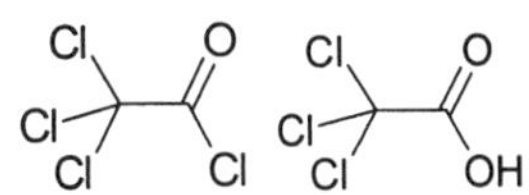	5.6[c]	34.6	Direct PERC conjugation	Weakly carcinogenic in mouse liver

[a] Gargas *et al.* (1990).
[b] Lilly *et al.* (1998).
[c] Dobrev *et al.* (2002).
[d] These k_d values, from Lilly *et al.* (1998), are complexly related to enzyme loss and should be regarded as relative values between the two isomers.

TABLE 3.3 PBPK Models for Halogenated Alkenes in Mixtures

Chemical mixture	Species	Reference
TCE, VC	Rat	Barton *et al.* (1995)
TCE, VDC	Rat	Andersen *et al.* (1987)
	Rat	ElMasri *et al.* (1996)
TCE, ethanol	Human	Sato *et al.* (1991a)
TCE, toluene	Rat	Thrall and Poet (2000)
TCE, PERC, methyl chloroform	Rat	Dobrev *et al.* (2001)
	Human	Dobrev *et al.* (2002)

situations (a qualitative consideration) and the amounts metabolized in the exposure situation (a quantitative consideration). The PBPK models for these compounds may track parent compound in various tissue compartments, rates of metabolism to reactive metabolites per tissue volume, concentrations of stable metabolites, or any combination of these measures of dose to tissues.

3.3 REVIEW OF PBPK MODELS

3.3.1 Vinyl Chloride (VC)

VC is an important industrial chemical because of the wide variety of end-use products from VC polymers and its low cost as a feedstock for polymer production (Cowfer and Magistro 1985). It is also produced from biodegradation of TCE by bacteria in the soil. Thus, past spills of TCE may lead to exposures of the public to VC in drinking water or via other environmental contamination. VC is carcinogenic in both humans and experimental animals (US EPA 1986a). At high exposure concentrations in rats, VC causes depletion of hepatic GSH. The current risk assessment for VC carcinogenicity, published by the US EPA (US EPA, 2000a), quantitatively incorporates pharmacokinetic information for VC into the human risk calculations using a PBPK model. The more recent PBPK modeling development for VC has been geared toward providing support for the risk assessment process.

The inhalation pharmacokinetics of VC in the rat was studied using closed-chamber techniques and vial-equilibration (Gargas *et al.* 1989, 1990). The constants for metabolism and partition coefficients from this earlier work were the inputs for a four-compartment PBPK model used by Chen and Blancato (1989) to correlate metabolized dose in animals with incidence of liver hemangiosarcoma and hepatocellular carcinoma in rats. Metabolism occurred only in the liver compartment. The risk estimates from the animal data using metabolized dose were consistent with human cancer incidences in exposed worker populations.

Next, Reitz *et al.* (1996a) developed a PBPK model for VC exposures in rats, mice, and humans, also intended to incorporate biological detail into the VC risk assessment. This model had a compartmental structure similar to that used in the earlier model, and it calculated dose as VC metabolize/volume liver/day. Strategies were developed to evaluate metabolic parameters in mice and humans from previous studies with low-molecular-weight hydrocarbon such as methylene chloride and chloroform. The estimated metabolic parameters were checked by evaluating gas uptake studies in mice and from human exposure studies. Good agreement was observed in assessing liver tumor incidence in mice and rats when compared on the basis of daily average rate of metabolism per unit tissue volume.

A more extensive PBPK model, initially discussed by Clewell *et al.* (1995) and later presented in detail (Clewell *et al.* 2001), described the uptake, distribution, and metabolism of VC in the mouse, rat, hamster, and human following inhalation or oral exposure. Four tissue compartments were described: a richly perfused tissue compartment that includes all of the organs except the liver; a slowly perfused tissue compartment that includes the muscle and skin tissues; a fat compartment; and a liver compartment. Metabolism of VC occurs only in the liver and is modeled by two saturable pathways: (1) a high-affinity, low-capacity pathway representing the cytochrome P450 2E1 enzyme (CYP2E1) and (2) a low-affinity, high-capacity pathway representing metabolism by other P450 isozymes such as CYP2C11/6 and CYP1A1/2. The parameters for the two oxidative pathways in the mouse, rat, hamster, and human were estimated by fitting the model to data from closed-chamber inhalation exposures for each of the species and strains of interest. In the rat, additional data on total metabolism and glutathione depletion were also used in the estimation of the parameters for the two pathways. Only the high-affinity pathway was included in the human description.

Two reactive products of VC metabolism by the P450 pathway, chloroethylene oxide and chloroacetaldehyde, can be metabolized further, leading to CO_2, or can react with GSH or other biological molecules (including DNA in the case of the oxirane). To account for GSH depletion, a description of GSH kinetics was also included in the model, using a description previously developed for a PBPK model of VDC, a related compound that also has the potential to deplete tissue glutathione (D'Souza and Andersen 1988). This VC PBPK model was used to predict the total production of reactive metabolites from VC per volume of liver, both in the animal bioassays and in human exposure scenarios. These measures of internal exposure were used in a dose–response model to predict the risk associated with lifetime exposure to VC in air or drinking water.

Extensive pharmacokinetic sensitivity and Monte Carlo uncertainty/variability analyses were also performed on this model (Clewell *et al.* 2001). Eight parameters showed a significant impact on model predictions based on the dose metric (i.e., the total amount of metabolism divided by the volume of the liver): They were body weight, alveolar ventilation, cardiac output, liver blood flow, liver volume, blood/air partition coefficient, the capacity and affinity for metabolism by CYP2E1, and, in the case of oral gavage, the rate of oral uptake. To consider parameter interactions and uncertainty, a Monte Carlo analysis was performed to estimate the combined impact of uncertainty regarding the values of all the parameters. Pharmacokinetic uncertainty/variability was determined to be a relatively small contributor to the overall uncertainty in a risk assessment for VC since the 95th percentiles of the distributions of risk estimates were within a factor of two of the means.

Animal-based risk estimates for human inhalation exposure to VC using total metabolism estimates from the PBPK model were consistent with risk estimates based on human epidemiological data (Clewell *et al.* 2001). This article is a good example of the contribution of PBPK modeling in cancer risk assessment. A version of this model was applied by the EPA in their risk assessment for VC (US EPA 2000a).

3.3.2 Vinyl Fluoride (VF)

VF is a gas used in the polymer industry to produce polyvinyl fluoride and other fluoropolymers. Analogous to other monohalogenated ethylenes such as VC and vinyl bromide, VF is a carcinogen in rats and mice (Bogdanffy 1995). The principal neoplastic lesions induced in both male and female rats were hepatic hemangiosarcomas and hepatocellular carcinomas. Male and female mice also had high incidences of hepatic hemangiosarcomas, and bronchioalveolar and mammary gland adenocarcinomas were also noted (Bogdanffy 1995). Mice were about threefold more sensitive than rats toward tumor formation. A PBPK model describing the *in vivo* pharmacokinetics of VF in rats and mice was constructed to allow for a better understanding of the difference in sensitivity between rats and mice to tumor induction by exposure to VF (Cantoreggi and Keller 1997). The usual four-compartment PBPK model for inhalation of volatile chemicals was adapted for VF. The model was fit to the time-course closed-chamber air concentrations to estimate metabolism parameters. It was concluded that the greater metabolic capacity of mice (V_{maxc} = 0.3 mg/hr-kg for whole body, V_{max} = 3.5 nmol/hr-mg protein for microsomes) compared to that of rats (V_{maxc} = 0.1 mg/hr-kg for whole body, V_{max} = 1.1 nmol/hr-mg protein for microsomes) for VF contributed to the greater incidence of tumor formation in mice. (The mouse would have a higher tissue dose metric for dose metabolized per volume liver per time.) A model with low capacity and high apparent affinity ($K_m \cong 0.001$ mg/L) best described the uptake curves for rats, whereas mice showed higher capacity but lower apparent affinity ($K_m \cong 0.02$ mg/L). To establish a rational basis for animal to human extrapolation in risk assessment, VF pharmacokinetics were also investigated *in vitro* in rodent and human microsomes. It was suggested that the good correlation obtained between *in vivo* and *in vitro* metabolism estimates in rodents should provide increased confidence in the use of human *in vivo* predictions based on *in vitro* data (Cantoreggi and Keller 1997).

The low K_m values estimated for the liver metabolism of VF, as well as of other compounds similar to the haloethylenes, may be misleading. When compounds have high affinities (low K_m values), liver clearance *in vivo* at low concentrations becomes limited by liver blood flow rather than constants for metabolism. When fitting the metabolic parameters, uptake profiles can be improved by lowering K_m or by increasing liver blood flow. The correlation between these parameters makes it difficult to unambiguously determine K_m from uptake studies alone (see Gargas *et al.* 1990). The K_m values estimated for *c*DCE and *t*DCE, in the next section, are less dependent on blood flow and more dependent on mechanisms of suicide inhibition with these compounds. The K_m values estimated for these compounds are more likely to be valid estimates.

3.3.3 *cis*-1,2-Dichloroethylene (*c*DCE) and *trans*-1,2-Dichloroethylene (*t*DCE)

Initial studies of the closed-chamber gas uptake kinetics of these compounds, with a compartmental analysis of the uptake curves (Filser and Bolt 1979), indicated that these two compounds were not very well metabolized. Subsequently, gas uptake studies were examined with a PBPK model for rats using appropriate values of tissue partition coefficients and physiological parameters, with a single saturable pathway for metabolism in the liver. For these compounds, it was impossible to generate a good fit of the PBPK model to all exposure concentrations with a single value of $V_{\max}$ and K_m (Gargas *et al.* 1990). When an attempt was made to fit the high-concentration data preferentially (Fig. 3.1A) from a closed-chamber gas uptake experiment, the model could not accurately describe chamber concentrations for any of the other exposure levels. The chamber concentrations rapidly decreased at early times, and then they decreased less rapidly than could be accounted for by the standard physiological model. The same phenomenon was observed for *c*DCE. These observations, obtained from analysis with a standard PBPK model using time-invariant metabolic constants, suggested that the rate constant of metabolism decreased with time at higher exposure concentrations. At the same time, the model's consistent underprediction of metabolic clearance for the two lower concentrations indicated that this time-dependent decrease was less severe at lower concentrations. Together, these observations were consistent with loss of enzyme due to *c*DCE and *t*DCE exposure—that is suicide inhibition with resynthesis (Gargas *et al.* 1990).

Lilly *et al.* (1998) used PBPK modeling of *in vivo* studies coupled with *in vitro* experimentation to explore the possible mechanisms for the decrease in metabolic activity during the course of exposures to *c*DCE and *t*DCE. Four possible mechanisms of interaction with the CYP2E1 enzyme were examined for *c*DCE and *t*DCE: (1) A short-lived metabolite reacts with the enzyme–substrate complex, (2) the metabolite reacts with the total enzyme, (3) the metabolite reacts with free enzyme, and (4) enzyme is inactivated by a bound intermediate. To test these hypotheses, mathematical descriptions for each mechanism were incorporated into PBPK models and then the model with enzyme loss was compared against the *in vivo* experimental data. The best fits were obtained for the model in which short-lived reactive metabolites produced during *c*DCE and *t*DCE metabolism reacted with the enzyme–substrate complex—that is mechanism (1)—to inactivate metabolizing enzyme in an irreversible manner. In addition, it was also necessary to include enzyme resynthesis in the model to obtain an accurate representation of all experimental data (Fig. 3.1B). *In vitro* results with microsomal incubations confirmed that *c*DCE and *t*DCE inhibited the CYP2E1 enzyme.

The analysis of the *in vivo* gas uptake data with a PBPK model incorporating descriptions of different mechanistic hypotheses allowed the identification of the most likely mechanism of inhibition. Both *in vivo* and *in vitro* results demonstrated that CYP2E1 was more potently inhibited by *t*DCE metabolites than by *c*DCE metabolites (Lilly *et al.* 1998). This difference is noted in Table 3.2 by the values for k_d with these two compounds: 496 versus 2.7 for *t*DCE and *c*DCE, respectively. The estimated $V_{\max}$ terms for VC, *c*DCE, and *t*DCE, when suicide inhibition was

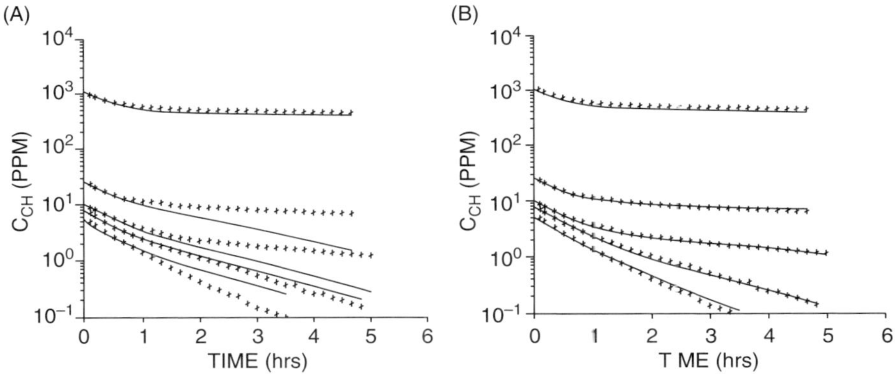

Figure 3.1 Closed-chamber gas uptake behavior for three rats exposed to *t*DCE at initial chamber concentrations of 5, 7.25, 10.5, 25, and 1125 ppm. The symbols are experimental data and the curves are model simulations. (**A**) The simulated curves were the best-fit of the model parameters that could be produced using the PBPK gas uptake that included only saturable metabolism with time invariant metabolic parameters. (**B**) For the same data, the simulated curves generated using the PBPK model incorporating suicide inhibition–resynthesis better described the data. The figure is from Gargas *et al.* (1990).

included for the latter two, turn out to be remarkably similar. Suicide inhibition was suggested to result from formation of dialdehyde by water addition to the epoxide intermediates of these two compounds (Table 3.2). The sensitivity coefficients for loss of compound from the chamber at low concentrations are highly sensitive to the K_m for metabolism because the mechanism of inhibition, involving rate of production and bound enzyme, leads to a square dependence on K_m. The estimates of K_m for these two compounds are perhaps the most reliable of any of the fitted K_m values for this group of compounds from *in vivo* studies.

It is noteworthy that the investigation of the mechanism of suicide inhibition by *t*DCE and *c*DCE was motivated by the failure of the initial PBPK model. The fits that could be achieved between simulated and experimental results were constrained by the physiological structure of the PBPK model. When simulations cannot adequately simulate experimental results, new hypotheses are required to account for the failure of the original model structure (Gargas *et al.* 1990). The suicide inhibition by *c*TCE and *t*DCE, overlooked in earlier compartmental analyses with these compounds (Filser and Bolt 1979), became obvious with use of the structured PBPK model.

3.3.4 Vinylidene Chloride (VDC)

VDC, an air and drinking-water contaminant (EPA 1982; US EPA 1986b), is a very potent liver toxicant. The metabolism of VDC has been extensively studied (Henschler and Bonse 1977; Leibman and Ortiz 1977; Filser and Bolt 1979). VDC is metabolized by CYP450 enzymes to reactive metabolites that are responsible for its toxicity. Since these metabolites are detoxified by GSH, liver GSH status is an important factor in the expression of VDC toxicity.

Metabolism of VDC was initially represented as a single saturable process in a PBPK model for closed-chamber inhalation exposures of rats (Gargas *et al.* 1986, 1990). D'Souza and Andersen (1988) developed a more extensive PBPK model for VDC in the rat based on oxidative metabolism of VDC and subsequent GSH detoxification of its metabolite under a variety of experimental conditions. This model successfully predicted concentrations of VDC in the blood, tissue, and exhaled air and also predicted depletion/resynthesis of liver GSH levels as a function of exposure concentrations, dose, and route of administration. The amount of VDC metabolized was sensitive to the rate of absorption due to the low blood : air partition coefficient of VDC and due to its saturable metabolism. The terminal half-life of VDC in the blood is not representative of metabolism rates but of redistribution of VDC from fat. Although the PBPK model for VDC was used to interpret the metabolism and kinetic behavior of VDC in the rodent, it was not used for interspecies extrapolation of animal data to predict human kinetics. The VDC model of D'Souza and Andersen (1988) was later used as the basis for a human inhalation PBPK model for VDC, albeit without any direct validation with human data (Leung 1992).

3.3.5 Trichloroethylene (TCE)

Because of its widespread use as a degreaser in the 1960s and 1970s, TCE is now a widespread environmental contaminant in the United States. TCE is readily taken up into systemic circulation by oral and inhalation routes of exposure. It is rapidly metabolized by the hepatic P450 system and, to a much lesser degree, by direct conjugation with GSH (Fig. 3.2). The principal stable metabolites of TCE are trichloroethanol (TCOH) and trichloroacetic acid (TCA). Other minor metabolites include dichloroacetic acid (DCA), oxalic acid, and *N*-(hydroxyacetyl)aminoethanol. Two of the oxidative metabolites of TCE, TCA and DCA, have been suggested to be responsible for the hepatocellular tumor formation in mice observed with TCE exposure (Herren-Freund *et al.* 1987). The small proportion of TCE that is metabolized by enzymatic conjugation with GSH in the liver is further metabolized in the kidney to the cysteine conjugate, 1,2-dichlorovinylcysteine (DCVC). Reactive products of further metabolic processing of DCVC in the kidney are toxic and mutagenic. The kidney toxicity of TCE and carcinogenic responses in this tissue are believed to be due to these metabolites. Given that its volume of use, its history of environmental release, and its potential for human exposure are all high, it is not surprising that quite a number of PBPK models have been developed for TCE (Table 3.4).

PBPK Models Focusing on Parent Compound A number of PBPK models developed for TCE include pharmacokinetic descriptions for parent compound, but not for metabolite (Table 3.4). These models have been used successfully for predicting TCE concentrations in the blood and tissues, for calculating the uptake and elimination profiles of TCE during inhalation exposures (Dallas *et al.* 1991), for investigating the impact of different exposure conditions (e.g., exposure concentra-

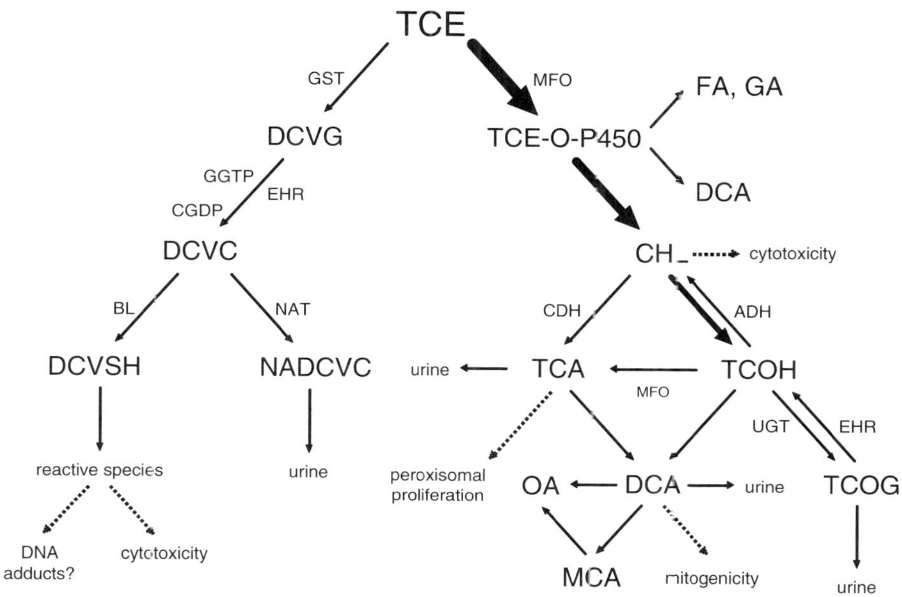

Figure 3.2 Metabolism of TCE. Abbreviations not given in text: (*right pathway*) CDH, chloral dehydrogenase (aldehyde oxidase); EHR, enterohepatic recirculation; FA, formic acid; GA, glyoxylic acid; OA, oxalic acid; TCE-O-P450, oxygenated TCE-Cytochrome P450 transition state complex; TCOG, TCOH glucuronide; UGT, UDP glucuronosyl transferase; (*left pathway*) BL, cysteine conjugate β-lyase; CGDP, cysteinyl-glycine dipeptidase; DCVG, dichlorovinyl glutathione; DCVSH, dichlorovinyl mercaptan; GGTP, γ-glutamyl transpeptidase; NADCVC, *N*-acetyl dichlorovinylcysteine; NAT, *N*-acetyl transferase.

tion and physical workload), and of interindividual variations in physiology (e.g., body build and liver function) on the kinetics of TCE by combining statistical simulation techniques with PBPK modeling (Droz 1989a, 1989b). Parent chemical PBPK models with estimates of total metabolism have also been employed to calculate the total metabolized dose in support of a cancer risk assessment for TCE (Bogen 1988; Koizumi 1989), to use occupational exposure limits to calculate biological exposure indexes (Leung 1992), and to assess human exposures resulting from TCE levels in ambient air and food (Yoshida 1993).

Several studies have focused on understanding the pharmacokinetics of TCE. A competitive metabolic inhibition model was developed to examine uptake of inhaled mixtures of TCE and VDC in rats (Andersen *et al.* 1987). These studies were best described by incorporating competitive inhibition in the liver equations for the two. By inhibiting VDC metabolism, TCE actually reduced the hepatotoxicity of inhaled VDC. The PBPK model of Dallas *et al.* (Dallas *et al.* 1991) accurately predicted the time courses of TCE concentrations in the blood and exhaled breath of rats during and following continuous inhalation exposure to 50 to 500 ppm TCE. This model included a separate lung tissue compartment and a mass transfer coef-

TABLE 3.4 PBPK Models for Trichloroethylene, TCE

Model type	Species	Reference
Models for parent compound	Human	Bogen (1988)
	Rat, human	Koizumi (1989)
	Human	Droz *et al.* (1989b)
	Rat	Staats *et al.* (1991)
	Rat	Gargas (1990)
	Rat	Gargas *et al.* (1990)
	Rat	Dallas *et al.* (1991)
	Human	Leung (1992)
	Human	Yoshida (1993)
	Rat, human	Poet *et al.* (2000)
	Long–Evans rat	Albanese *et al.* (2002)
	Long–Evans rat	Simmons *et al.* (2002)
Models for parent compound and metabolites	Human	Fernandez (1977)
	Human	Sato *et al.* (1991b)
	Human	Sato (1991)
	Rat	Fisher *et al.* (1989)[a]
	Rat	Fisher *et al.* (1990)[a]
	Rat, mouse	Fisher *et al.* (1991)
	Human	Allen and Fisher (1993)
	B6C3F1 mouse	Abbas and Fisher (1997)
	Human	Lapare *et al.* (1995)
	B6C3F$_1$ mouse	Greenberg *et al.* (1999)
	Rat, mouse, human	Clewell *et al.* (2000)

[a] This article involves perinatal transfer of TCE and is discussed in Chapter 12.

ficient controlling the bidirectional rate of transfer of chemical between the lung and alveolar air.

The effect of the fat compartment description on the simulation of TCE pharmacokinetics has recently been investigated. In this study, three different descriptions of adipose tissue were compared (Albanese *et al.* 2002; Banks and Potter 2002). The two common descriptions, a perfusion-limited model and a diffusion-limited model, could be implemented with ordinary differential equations alone. A third description was developed that included partial differential equations to simulate axial-dispersion in the adipose tissue, in order to capture the physiological heterogeneity of fat tissue, including widely varying fat cell sizes, lipid distribution, and blood flow properties. The authors conclude that the more complex description may be well-suited to predict the experimental behavior of TCE in adipose tissue using parameter estimation techniques. However, no comparison of the predictions of the three descriptions with experimental data was actually performed.

Although most TCE PBPK models have focused on inhalation exposures, other exposure routes have been studied. A PBPK model developed by Staats *et al.* (1991) described the gastrointestinal (GI) absorption of TCE along with methylene chloride, chloroform, and dichloroethane after oral gavage in a water or corn oil vehicle. It was necessary to use a two-compartment representation of the GI tract to account for time courses of absorption in these rats. In another study, a PBPK modeling approach was used to interpret data from a study of the dermal absorption of TCE in rats and humans from water or soil vehicles (Poet *et al.* 2000). After the dermal dose was applied, the amount of TCE exhaled was determined, and these data were analyzed using a PBPK model to determine the rate and extent of dermal absorption.

Recently, a strain-specific PBPK model for TCE in the male Long–Evans rat was developed (Simmons *et al.* 2002). This model included the brain as a separate physiological compartment to aid in evaluation of neurotoxicity data. Other features of the model were (1) incorporation of strain-specific tissue volumes; (2) use of a more quantitative measure of model fit to guide selection of input parameter values when strain-specific information was not available; (3) analysis of the impact of model input parameters on tissue concentration of TCE across multiple tissues, external exposure concentrations, and times; and (4) use of the model to aid in estimating a measure of internal dose that correlates with several measures of neurotoxicity.

Sato *et al.* (1977) formulated a PBPK model for respiratory exposure of humans to TCE. The model included three compartments, with intercompartment exchange of TCE governed solely by intertissue diffusion. Metabolic and respiratory excretion was assumed to occur in the richly perfused tissue compartment. A more complete PBPK model, which accurately predicted respiratory elimination of TCE and cumulative urinary excretion of TCE metabolites in humans, was constructed by Fernandez *et al.* (1977). This model included the three compartments of Sato *et al.* (1977), as well as a liver compartment with blood-flow-limited metabolism and a lung compartment for respiratory absorption and elimination of TCE. In a series of publications, Sato and co-workers described the use of a PBPK model of the kinetics of TCE and the urinary excretion of total metabolites in humans to (a) evaluate the impact of changes in physiological factors (e.g., sex, body size, and body fat content; Sato *et al.* 1991b) and environmental factors (Sato 1991) on the kinetics of TCE in the human and (b) predict the effects of interactions with ethanol consumption on TCE kinetics (Sato *et al.* 1991a).

These parent chemical, total metabolism models are useful for understanding and interpreting TCE pharmacokinetics. However, they cannot be used for predicting tissue exposure to specific metabolites that are considered more relevant to TCE carcinogenicity in mice and rats. The next section reviews PBPK models that include descriptions of metabolites.

Models of Both TCE and Its Metabolites PBPK models have been developed to simulate the kinetics of TCE and its principal metabolite, TCA, in the pregnant rat following inhalation and ingestion of TCE from drinking water or gavage dosing (Fisher *et al.* 1989). Additional compartments (e.g., mammary tissue, placenta, and fetus) were incorporated into the model, and selected physiological parameters changed during pregnancy. This model provided a good representation of TCE and TCA levels measured experimentally in maternal and fetal blood at a limited number of times post-exposure. A PBPK models has also been developed to describe the transfer of TCE from the lactating rat (exposed to TCE by inhalation or in drinking water) to the rat pup through breast milk (Fisher *et al.* 1990). PBPK models involving the perinatal transfer of xenobiotics, including those for TCE (Fisher *et al.* 1989, 1990), are reviewed in Chapter 12.

A PBPK model for the kinetics of TCE and its principal metabolite, TCA, in the mouse (Fisher *et al.* 1991) has also been developed. These rodent models, together with a similar model of TCE and TCA in the human (Allen and Fisher 1993), served as the basis for a PBPK-based risk assessment for TCE liver carcinogenicity based on either total metabolism of TCE or AUC for TCA in blood (Fisher and Allen 1993). These models provided the first successful cross-species pharmacokinetics description for a metabolite of TCE.

Lapare *et al.* (1995) explored the influence of different exposure conditions (e.g., exposure duration or work load) on the elimination of TCE in alveolar air and excretion of TCA and TCOH in the urine in human volunteers. Four exposure scenarios differing in exposure concentration, exposure duration, or various concentration and daily work load were adequately simulated by the optimized models. For exposure situations likely to be encountered in the workplace, PBPK modeling appeared to be a useful tool both for predicting alveolar air concentrations under a given set of exposure conditions and for developing strategies for biological monitoring of exposure to TCE.

A more comprehensive PBPK model (Clewell *et al.* 2000), building on the work of Fisher and Allen (Fisher *et al.* 1991; Allen and Fisher 1993), incorporated descriptions of several additional metabolites: TCOH (including enterohepatic recirculation of its glucuronide), DCA (in the plasma), chloral (in the lung), and DCVC (in the kidney). On the basis of sensitivity and uncertainty analyses performed with the model, it was suggested that the model could provide reasonably accurate and precise estimates of dose metrics for TCE and its major metabolites, TCA and TCOH, in both experimental animals and humans, but that dose estimates for other metabolites (DCA, chloral, and DCVC) were more uncertain. The application of this PBPK model in a mode of action directed risk assessment for TCE has been discussed in detail (Clewell and Andersen 2004).

Fisher and colleagues have continued to elaborate and refine their PBPK models for TCE, focusing on the metabolites of interest for liver carcinogenicity. Published models include (1) a model of the kinetics of TCE, CHL, TCA, DCA, and TCOH in the B6C3F1 mouse based on data from corn oil gavage exposures (Abbas and Fisher 1997), (2) a model of TCE, TCA, and TCOH in the human based on data from controlled human inhalation exposures (Fisher *et al.* 1998), (3) a model of TCE, TCA, and TCOH kinetics in the rat that considers enterohepatic recirculation of TCA and TCOH following oral or intravenous exposure to TCE (Stenner *et al.* 1998), and (4) a model of inhaled TCE and its oxidative metabolites in the B6C3F1 mouse (Greenberg *et al.* 1999). Together, these models provide a capability for estimating dose metrics in the mouse, rat, and human in support of a risk assessment for TCE liver carcinogenicity. A potential advantage of these mouse PBPK models is that their calibration includes data on TCA concentrations in the liver. However, since there was no human data on liver concentrations, the human model could not be similarly calibrated. Therefore, the relationship of liver and blood TCA dosimetry must still be inferred from data on binding of TCA (Clewell and Andersen 2004).

3.3.6 Tetrachloroethylene (PERC)

PERC is widely used dry cleaning and as a metal degreasing solvent. It is a hazardous air pollutant, a common contaminant at Superfund waste sites, and a surface-water and groundwater pollutant. The rodent carcinogenicity of PERC in the liver is considered to result from the generation of reactive intermediates during its metabolism (Clewell *et al.* 2004). As in the case of TCE, the pathways for PERC bioactivation and detoxification are complex and several of the enzymes involved exhibit sex- and species-dependent differences. Consequently, it has been suggested that information about species differences in multiple pathways would be necessary to extrapolate results from experimental animals to humans with certainty (Lash and Parker 2001). However, the metabolism of PERC has not been as extensively studied as it has for TCE; as a result, the PBPK models for PERC have been limited to descriptions of the parent chemical and its major metabolite, TCA. Although the oxidative rate of metabolism for PERC is much lower than for TCE (Table 3.2), most of PERC oxidation by CYP450 enzyme produces TCA; the stoichiometric yield of TCA from TCE metabolism is much lower.

PBPK models for PERC have been developed by a number of investigators (Table 3.1). With few exceptions, the PBPK models for PERC share the simple four-compartment structure (liver, fat, rapidly perfused tissues, and slowly perfused tissues) and steady-state description of lung equilibration used by Ramsey and Andersen (1984) with styrene. Only one of the published models (Gearhart *et al.* 1993) provides a description of the kinetics of TCA, the major metabolite of PERC. None of the models provide a description of the glutathione conjugation metabolic pathway that has been implicated in the kidney lesions produced by PERC. Therefore it is not possible to estimate dose metrics for kidney toxicity as has been done in the case of TCE. For the most part, the differences between the models reflect the different data used by the authors in their development.

Several pharmacokinetic modeling studies have been performed to characterize the kinetics and metabolism of PERC in the mouse, rat, and human (Chen and Blancato 1987; Ward *et al.* 1988; Gearhart *et al.* 1993; Reitz *et al.* 1996b). All of the studies used a single saturable Michaelis–Menten description for the metabolism of PERC. However, while the models of Chen and Blancato (1987) and Reitz *et al.* (1996b) also included only saturable metabolism in the rodent, the Ward *et al.* (1988) and Gearhart *et al.* (1993) rodent models included both a saturable and a linear component in the equation for metabolism. In Ward *et al.* (1988) it was assumed that the saturable pathway represented oxidative P450 metabolism and that the linear pathway represented conjugation with glutathione, whereas Gearhart *et al.* (1993) assumed that both components represented oxidation by P450 enzymes. Ward *et al.* (1988) did not attempt to justify the pathway assignments, but Gearhart *et al.* (1993) reported that the assumption that both components were oxidative was necessary to provide a consistent description of both the closed-chamber PERC clearance data and the data on TCA concentrations following oral gavage with PERC.

The Gearhart *et al.* (1993) model, which was only for the mouse, also differs from the other models in that it includes two fat compartments with different perfusion ratios. More significantly, the Gearhart *et al.* (1993) model described the kinetics of the principal metabolite, TCA, using a single-compartment model with first-order urinary excretion and assuming that the amount of TCA produced represents 60% of the total amount of PERC metabolized. The kinetic parameter values for TCA were taken from a description of the same metabolite in a model of TCE (Fisher and Allen 1993).

Development of PBPK models of PERC has also been conducted in the rat and dog. These parent chemical models focused on partition coefficients and tissue distribution (Dallas *et al.* 1994a), tissue concentration–time profile data (Dallas *et al.* 1994b), prediction of systemic uptake and respiratory elimination (Dallas *et al.* 1994c), and prediction of differences due to species, dose, and exposure route (Dallas *et al.* 1995). The conclusions of these studies were as follows: (1) The model accurately simulated percentage uptake and cumulative uptake of PERC over time; (2) predicted tissue levels were in close agreement with the levels measured over time in seven tissues and in blood; (3) the PBPK model was shown to have utility in predicting the impact of metabolism and exhalation on pharmacokinetics of PERC in different species of widely differing size (rat and dog); and (4) differences in the kinetics of PERC across species, route of administration, and high-to-low dose can be predicted with reasonable success in rats and dogs using a PBPK model.

Several additional PBPK models of PERC have been developed for special applications. For example, models have been used to understand the effects of variability in exposure conditions an in physiology on inhalation kinetics (Droz 1989a, 1989b) and to calculate biological exposure indexes for occupational exposures (Leung 1992). Another study used PBPK models for PERC and TCE to present an approach for using data to predict human toxicity (Koizumi 1989). Loizou (2001) developed a PBPK model for PERC in the human to investigate workplace exposures. Using the PERC metabolism and partitioning parameters from Gearhart *et al.* (1993), he was able to successfully simulate published experimental data on exhaled PERC concentrations following both inhalation and dermal exposures, as well as on the time course of blood concentrations and urinary excretion of TCA following inhalation of PERC. The model was then used to analyze occupational exposure data from dry-cleaning operations. A PBPK model for PERC (Rao and Brown 1993) was also used to predict the concentration of PERC in the brain following inhalation and dermal exposures due to bathing and showering use of the water from a private well contaminated with PERC. In another study, data on the exhalation of PERC following dermal exposures of rats and humans to PERC from a soil matrix has been analyzed with a PBPK model to determine the rate and extent of dermal absorption during the exposures (Poet *et al.* 2002).

Two similar Monte Carlo investigations have been conducted to determine the relative importance of the various parameters in a PBPK model for PERC. One study by Bois *et al.* (1990) reported that kinetic parameters defining the metabolic rate were the most important model parameters when considering the precision and sensitivity of PBPK models for mice, rats, and humans (Bois *et al.* 1990). In a second study, Gearhart *et al.* (1993) exercised their PBPK model for PERC to examine the effects of PBPK model parameter variability on PBPK model predictions in a risk assessment. The authors concluded that in comparison with the uncertainty regarding mode of action and choice of dose surrogates, parameter uncertainty is not a significant source of variability in the use of this model for risk assessment.

The large number of PBPK models for PERC in the literature stimulated several studies comparing the output of different PBPK models. In the first of these, seven previously published human and rodent PBPK parent chemical models for PERC were compared. The choice of the dataset for the calibration of metabolic parameters was determined to be the key factor differentiating the models, and this choice led to significant variability in the prediction of low-dose metabolism rates for humans (Hattis *et al.* 1990). In a later study (Hattis *et al.* 1993), 10 different human PBPK parent chemical models for PERC were compared with several types of experimental data (i.e., data on the amount absorbed and the concentrations of PERC in alveolar air and venous blood following inhalation exposures from seven human studies). All the models deviated from actual observations to some extent, and it was recommended that more physiological detail in the model structure (e.g., including heterogeneity in the fat compartment or the movement of TCE between the fat and muscle compartments) could improve the accuracy of model predictions.

A more recent study (Clewell *et al.* 2004) performed a critical evaluation of the alternative PBPK models of PERC in the mouse and human to assess their usefulness for conducting a risk assessment for PERC based on its liver carcinogenicity, which would be based on the amount of PERC metabolized per volume liver. All of the models (Chen and Blancato 1987; Ward *et al.* 1988; Bois *et al.* 1990, 1996; Gearhart *et al.* 1993; Reitz *et al.* 1996b) provided reasonably accurate simulations of some of the pharmacokinetic data available for PERC in mice or humans, and could therefore be considered, to some extent, to be validated. However, while similar predictions of metabolism were obtained with the alternative models in the mouse, predictions of metabolism in the human with different models varied considerably. This species difference in the variation of the PBPK model estimates of metabolism appeared to stem from the different kinds of data used to identify the metabolism parameter values in mice and humans. While the mouse models made use of data that were highly informative regarding metabolism, including radiolabel disposition, metabolite excretion, or closed chamber clearance studies, many of the human models relied on parent chemical kinetic data that did not directly reflect metabolism.

The alternative human models were then evaluated by Clewell *et al.* (2005) using data, not used in the development of any of the models, on the urinary excretion of TCA for human subjects exposed to relatively low exposure concentrations of PERC. The model of Gearhart *et al.* (1993), which was the only model to include a description of TCA kinetics, provided the closest predictions (within a factor of two) of the urinary excretion observed in these low-concentration exposures. Other models overestimated metabolite excretion in this study by 5- to 15-fold. During the course of this evaluation, a systematic discrepancy between model predictions and experimental data for the time-course of the urinary excretion of TCA suggested a contribution from TCA formed by metabolism of PERC in the kidney and excreted directly into the urine. A modification of the model of Gearhart *et al.* (1993) to include metabolism of PERC to TCA in the kidney at 10% of the capacity of the liver, with direct excretion of the TCA formed in the kidney into the urine, markedly improved agreement with the experimental time-course data, without altering predictions of liver metabolism (Clewell *et al.* 2005). This case study with PERC demonstrated the danger of relying on parent chemical kinetic data to validate a model that will be used for the prediction of metabolism.

3.3.7 Allyl Chloride (AC)

The results of a PBPK model for AC were presented by Clewell and Andersen (1987). AC is metabolized by CYP450-mediated oxidation to metabolites that react with and consume GSH. It is also directly conjugated with GSH, as were TCE and PERC. Both the oxidative and the conjugative GSH pathways appeared to be dose dependent (Fig. 3.3A). These data points were obtained in the gas-uptake studies, and the smooth curves in Fig. 3.3A are the best-fit curves, assuming a saturable pathway and a first-order pathway with rate constants that are independent of exposure concentration. The model predicted curves did not fit the overall data points; the prediction at high concentrations was lower than the data points, and at intermediate concentrations the prediction was uniformly higher than the data. This dis-

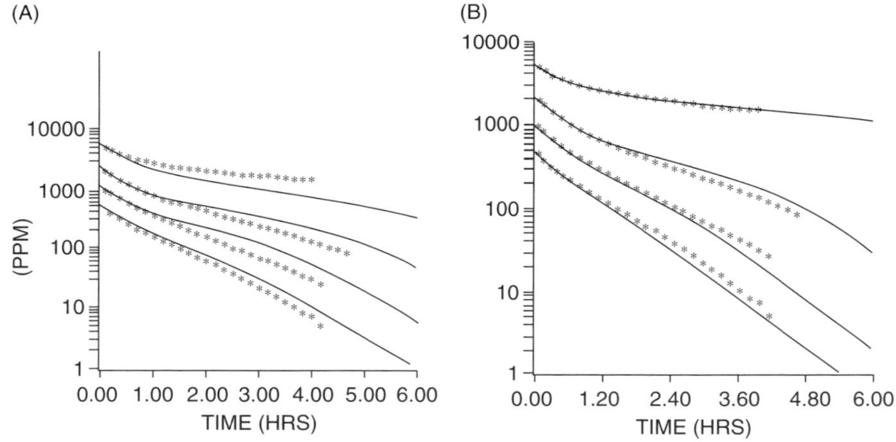

Figure 3.3 Cofactor depletion in gas-uptake experiments with AC. Concentration of AC in a closed, recirculated chamber containing three Fischer 344 rats. Initial chamber concentrations were 500, 1000, 2000, and 5000 ppm. The symbols are the experimental data and the curves are model simulations. (**A**) The curves represent the best-fit of the allyl chloride model with time-invariant rate constants. (**B**) For the same data, the curves show the predictions of the expanded model which not only included depletion of GSH by reaction with AC but also provided for the regulation of GSH biosynthesis. The figure is from Clewell and Andersen (1987).

crepancy is reminiscent of the challenges in fitting a simple PBPK model to *t*DCE (Fig. 3.1) but has a different basis.

To build a model for examining the biological basis of the kinetic behavior of AC, it was necessary to include time-dependence of hepatic GSH concentrations (Clewell and Andersen 1987). The basic GSH depletion/resynthesis model had a zero-order production of GSH and a first-order consumption rate that was increased by reaction of the GSH with AC directly and with metabolites formed from AC oxidation. In the refined PBPK model used in Fig. 3.3B, GSH resynthesis was regulated by controlling the concentration of the rate-limiting enzyme for GSH biosynthesis. The production of this enzyme was inversely related to the instantaneous GSH concentration. This description, coupling the loss of AC from the chamber and depletion of the GSH concentration in the liver, provided much-improved agreement between the data and the predicted behavior (Fig. 3.3B) and provided insight as to the biological mechanisms that may be important in controlling the rate of AC metabolism.

3.3.8 β-Chloroprene (CD)

A PBPK model has been developed for CD, which is used to produce polychloroprene rubber, to analyze the relationship between dose of CD and lung tumor response in several species (Himmelstein *et al.* 2004a). PBPK models were developed for CD in the mouse, Fischer and Wistar rats, hamster, and human. Metabolism was included in the lung and liver with metabolic parameters derived from *in vitro* experiments using lung and liver tissues from each species of animal (Himmelstein *et al.* 2004b). Additionally, tissue:air partition coefficients were experimentally measured *in vitro* in blood, lung, liver, muscle, fat and kidney samples from each species for the calculation of appropriate tissue:blood partition coefficients. This modeling work was used to estimate the external exposure level that would result in a comparable internal dose in humans to the benchmark internal dose derived from the rodent lung tumor data.

3.3.9 Hexachlorobutadiene, HCB

HCB has been reported to be toxic to the rat kidney; the toxicity is believed to be a consequence of the metabolism of HCB by glutathione conjugation in the liver and the beta-lyase pathway in the kidney. This pathway follows a similar sequence as the reactions of TCE with GSH and processing of the conjugate (Fig. 3.2). To support prediction of human toxicity, the hepatic conjugation of HCB with glutathione, the metabolism of the cysteine conjugate by renal beta-lyases and *N*-acetyltransferases, and the metabolism of the *N*-acetylcysteine conjugate by renal acylases were compared *in vitro* in rat and human tissues (Green *et al.* 2003). Rates for each metabolic step were lower in humans than in rats: 5-fold for glutathione conjugation, 3-fold for beta-lyase, and 3.5-fold for *N*-acetyltransferase. Acylase activity could not be detected in human kidney cytosol. A PBPK model incorporating these data predicted that beta-lyase metabolism in humans would be more than an order of magnitude lower than that in rats.

3.4 SUMMARY

PBPK modeling with haloethylenes has progressed substantially from the efforts of 20 to 25 years ago to simply establish kinetic constant for metabolism in the living animals. Complexities of metabolism, including dose-dependent metabolism, cofactor depletion, and suicide inhibition occurring with some of these compounds, were identified and evaluated quantitatively through use of structured PBPK models. The many early models that estimated concentrations of parent chemical and tissue-specific rates of metabolism were sequentially enlarged to account for tissue exposures to a variety of metabolites, best exemplified by the continuing elaboration of PBPK models for TCE and its multiple active metabolites, including TCA, DCA, and DCVC. Many of the PBPK models for these haloethylenes, especially with VC, VF, TCE, PERC, CD, and HCB, have been used for interspecies extrapolation of metabolism and dosimetry and for conducting human health risk assessments. In a recent human PBPK model for PERC, detailed time-course curves for renal elimination of TCA indicated systematic deviations from expectations for a simple description of TCA elimination from a single, central compartment, indicating direct excretion of TCA formed in kidney to urine. These haloethylene PBPK models show the power of a structured model to create hypotheses for chemical disposition that are tested by experiment and the concept, with TCE, that these models can serve as repositories for organizing information on modes of action, tissue dosimetry, and risk assessment.

NOTATION

AC	allyl chloride
CD	β-chloroprene
***c*DCE**	*cis*-1,2-dichloroethylene
CHL	chloral
CYP	cytochrome P450
DCA	dichloroacetic acid
DCVC	1,2-dichlorovinylcysteine
GI	gastrointestinal
GSH	glutathione
HCB	hexachlorobutadiene
IARC	International Agency for Research on Cancer
LMS	linearized multistage
PERC	tetrachloroethylene, perchloroethylene
TCA	trichloroacetic acid
TCE	trichloroethylene
TCOH	trichloroethanol
***t*DCE**	*trans*-1,2-dichloroethylene
VC	vinyl chloride, 1-chloroethylene
VDC	vinylidene chloride, 1,1-dichloroethylene
VF	vinyl fluoride
VOC	volatile organic compound

REFERENCES

Abbas, R., and Fisher, J. W. (1997). A physiologically based pharmacokinetic model for trichloroethylene and its metabolites, chloral hydrate, trichloroacetate, dichloroacetate, trichloroethanol, and trichloroethanol glucuronide in B6C3F1 mice. *Toxicol. Appl. Pharmacol.* **147**, 15–30.

Albanese, R. A., Banks, H. T., Evans, M. V., and Potter, L. K. (2002). Physiologically based pharmacokinetic models for the transport of trichloroethylene in adipose tissue. *Bull. Math. Biol.* **64**, 97–131.

Allen, B. C., and Fisher, J. W. (1993). Pharmacokinetic modeling of trichloroethylene and trichloroacetic acid in humans. *Risk Anal.* **13**, 71–86.

Andersen, M. E., Gargas, M. L., Clewell, H. J. D., and Severyn, K. M. (1987). Quantitative evaluation of the metabolic interactions between trichloroethylene and 1,1-dichloroethylene *in vivo* using gas uptake methods. *Toxicol. Appl. Pharmacol.* **89**, 149–157.

Banks, H. T., and Potter, L. K. (2002). Model predictions and comparisons for three toxicokinetic models for the systemic transport of trichloroethylene. *Math. Comput. Model.* **35**, 1007–1032.

Barton, H. A., Creech, J. R., Godin, C. S., Randall, G. M., and Seckel, C. S. (1995). Chloroethylene mixtures—pharmacokinetic modeling and *in-vitro* metabolism of vinyl-chloride, trichloroethylene, and *trans*-1,2- dichloroethylene in rat. *Toxicol. Appl. Pharmacol.* **130**, 237–247.

Bogdanffy, M. S., Makovec, G. T., and Frame, S. R. (1995). Inhalation oncogenicity bioassay in rats and mice with vinyl fluoride. *Fundam. Appl. Toxicol.* **26**, 223–238.

Bogen, K. T. (1988). Pharmacokinetics for regulatory risk analysis: The case of trichloroethylene. *Regul. Toxicol. Pharmacol.* **8**, 447–466.

Bois, F. Y., Zeise, L., and Tozer, T. N. (1990). Precision and sensitivity of pharmacokinetic models for cancer risk assessment: tetrachloroethylene in mice, rats, and humans. *Toxicol. Appl. Pharmacol.* **102**, 300–315.

Bois, F. Y., Gelman, A., Jiang, J., Maszle, D. R., Zeise, L., and Alexeef, G. (1996). Population toxicokinetics of tetrachloroethylene. *Arch. Toxicol.* **70**, 347–355.

Cantoreggi, S., and Keller, D. A. (1997). Pharmacokinetics and metabolism of vinyl fluoride *in vivo* and *in vitro*. *Toxicol. Appl. Pharmacol.* **143**, 130–139.

Chen, C., and Blancato, J. (1987). Role of pharmacokinetic modeling in risk assessment—Perchloroethylene (PCE) as an example. In: *Pharmacokinetics and Risk Assessment, Drinking Water and Health*, Vol. 8, National Academy Press, Washington, D.C., pp. 369–390

Chen, C. W., and Blancato, J. N. (1989). Incorporation of biological information in cancer risk assessment: Example—vinyl chloride. *Cell Biol. Toxicol.* **4**, 417–444.

Clewell, H. J., III, and Andersen, M. E. (1987). Dose, species, and route extrapolation using physiologically based pharmacokinetic models. In: *Drinking Water and Health, Pharmacokinetics in Risk Assessment*, National Academy Press, Washington, D.C., pp. 159–182.

Clewell, H.J., III, and Andersen, M. E. (2004). Applying mode-of-action and pharmacokinetic considerations in contemporary cancer risk assessments: An example with trichloroethylene. *Crit. Rev. Toxicol.* **34**, 1–61.

Clewell, H. J., Gentry, P. R., Kester, J. E., and Andersen, M. E. (2005). Evaluation of physiologically based pharmacokinetic models in risk assessment: an example with perchloroethylene. *Crit. Rev. Toxicol.* in press.

Clewell, H. J., Gentry, P. R., Gearhart, J. M., Allen, B. C., and Andersen, M. E. (1995). Considering pharmacokinetic and mechanistic information in cancer risk assessments for environmental contaminants—Examples with vinyl-chloride and trichloroethylene. *Chemosphere* **31**, 2561–2578.

Clewell, H. J., Gentry, P. R., Covington, T. R., and Gearhart, J. M. (2000). Development of a physiologically based pharmacokinetic model of trichloroethylene and its metabolites for use in risk assessment. *Environ. Health Perspect.* **108**, 283–305.

Clewell, H. J., Gentry, P. R., Gearhart, J. M., Allen, B. C., and Andersen, M. E. (2001). Comparison of cancer risk estimates for vinyl chloride using animal and human data with a PBPK model. *Sci. Total Environ.* **274**, 37–66.

Cowfer, J. A., and Magistro, A. J. (1985). Vinyl polymers: Vinyl chloride. In: *Kirk–Othmer Concise Encyclopedia of Chemical Technology*, John Wiley & Sons, New York, pp. 1229–1233.

Dallas, C. E., Gallo, J. M., Ramanathan, R., Muralidhara, S., and Bruckner, J. V. (1991). Physiological pharmacokinetic modeling of inhaled trichloroethylene in rats. *Toxicol. Appl. Pharmacol.* **110**, 303–314.

Dallas, C. E., Chen, X. M., Muralidhara, S., Varkonyi, P., Tackett, R. L., and Bruckner, J. V. (1994a). Use of tissue disposition data from rats and dogs to determine species-differences in input parameters for a physiological model for perchloroethylene. *Environ. Res.* **67**, 54–67.

Dallas, C. E., Chen, X. M., Muralidhara, S., Varkonyi, P., Tackett, R. L., and Bruckner, J. V. (1995). Physiologically-based pharmacokinetic model useful in prediction of the influence of species, dose, and exposure route on perchloroethylene pharmacokinetics. *J. Toxicol. Environ. Health* **44**, 301–317.

Dallas, C. E., Chen, X. M., Obarr, K., Muralidhara, S., Varkonyi, P., and Bruckner, J. V. (1994b). Development of a physiologically-based pharmacokinetic model for perchloroethylene using tissue concentration-time data. *Toxicol. Appl. Pharmacol.* **128**, 50–59.

Dallas, C. E., Muralidhara, S., Chen, X. M., Ramanathan, R., Varkonyi, P., Gallo, J. M., and Bruckner, J. V. (1994c). Use of a physiologically-based model to predict systemic uptake and respiratory elimination of perchloroethylene. *Toxicol. Appl. Pharmacol.* **128**, 60–68.

DeJongh, J., Verhaar, H. J. M., and Hermens, J. L. M. (1998). Role of kinetics in acute lethality of non-reactive volatile organic compounds (VOCs). *Toxicol. Sci.* **45**, 26–32.

Dobrev, I. D., Andersen, M. E., and Yang, R. S. H. (2001). Assessing interaction thresholds for trichloroethylene in combination with tetrachloroethylene and 1,1,1-trichloroethane using gas uptake studies and PBPK modeling. *Arch. Toxicol.* **75**, 134–144.

Dobrev, I. D., Andersen, M. E., and Yang, R. S. H. (2002). *In silico* toxicology: Simulating interaction thresholds for human exposure to mixtures of trichloroethylene, tetrachloroethylene, and 1,1,1-trichloroethane. *Environ. Health Perspect.* **110**, 1031–1039.

Droz, P. O., Wu, M. M., and Cumberland, W. G. (1989a). Variability in biological monitoring of solvent exposure. II. Application of a population physiological model. *Br. J. Ind. Med.* **46**, 547–558.

Droz, P. O., Wu, M. M., Cumberland, W. G., and Berode, M. (1989b). Variability in biological monitoring of solvent exposure. I. Development of a population physiological model. *Br. J. Ind. Med.* **46**, 447–460.

D'Souza, R. W., and Andersen, M. E. (1988). Physiologically based pharmacokinetic model for vinylidene chloride. *Toxicol. Appl. Pharmacol.* **95**, 230–240.

ElMasri, H. A., Tessari, J. D., and Yang, R. S. H. (1996). Exploration of an interaction threshold for the joint toxicity of trichloroethylene and 1,1-dichloroethylene: Utilization of a PBPK model. *Arch. Toxicol.* **70**, 527–539.

EPA (1982). *1,1-Dichloroethylene. Occurrence in Drinking Water, Food and Air*. Office of Drinking Water, U.S. Environmental Protection Agency, Washington, D.C.

Fernandez, J. G., Droz, P. O., Humbert, B. E., and Caperos, J. R. (1977). Trichloroethylene exposure. Simulation of uptake, excretion, and metabolism using a mathematical model. *Br. J. Ind. Med.* **34**, 43–55.

Filser, J. G., and Bolt, H. M. (1979). Pharmacokinetics of halogenated ethylenes in rats. *Arch. Toxicol.* **42**, 123–136.

Fisher, J. W., and Allen, B. C. (1993). Evaluating the risk of liver cancer in humans exposed to trichloroethylene using physiological models. *Risk Anal.* **13**, 87–95.

Fisher, J. W., Whittaker, T. A., Taylor, D. H., Clewell, H. J., and Andersen, M. E. (1989). Physiologically based pharmacokinetic modeling of the pregnant rat—A multiroute exposure model for trichloroethylene and its metabolite, trichloroacetic-acid. *Toxicol. Appl. Pharmacol.* **99**, 395–414.

Fisher, J. W., Whittaker, T. A., Taylor, D. H., Clewell, H. J., and Andersen, M. E. (1990). Physiologically based pharmacokinetic modeling of the lactating rat and nursing pup—A multiroute exposure model for trichloroethylene and its metabolite, trichloroacetic-acid. *Toxicol. Appl. Pharmacol.* **102**, 497–513.

Fisher, J. W., Gargas, M. L., Allen, B. C., and Andersen, M. E. (1991). Physiologically based pharmacokinetic modeling with trichloroethylene and its metabolite, trichloroacetic-acid, in the rat and mouse. *Toxicol. Appl. Pharmacol.* **109**, 183–195.

Fisher, J. W., Mahle, D. A., and Abbas, R. (1998). A human physiologically based pharmacokinetic model for trichloroethylene and its metabolites, trichloroacetic acid and free trichloroethanol. *Toxicol. Appl. Pharmacol.* **152**, 339–359.

Gargas, M. L. (1990). An exhaled breath chamber system for assessing rates of metabolism and rates of gastrointestinal absorption with volatile compounds. *J. Am. Coll. Toxicol.* **9**, 447–453.

Gargas, M. L., Andersen, M. E., and Clewell, H. J. (1986). A physiologically based simulation approach for determining metabolic constants from gas uptake data. *Toxicol. Appl. Pharmacol.* **86**, 341–352.

Gargas, M. L., Burgess, R. J., Voisard, D. E., Cason, G. H., and Andersen, M. E. (1989). Partition coefficients of low molecular weight volatile chemicals in various liquids and tissues. *Toxicol. Appl. Pharmacol.* **98**, 81–99.

Gargas, M. L., Clewell, H. J., and Andersen, M. E. (1990). Gas uptake inhalation techniques and the rates of metabolism of chloromethanes, chloroethanes, and chloroethylenes in the rat. *Inhal. Toxicol.* **2**, 295–319.

Gearhart, J. M., Mahle, D. A., Greene, R. J., Seckel, C. S., Flemming, C. D., Fisher, J. W., and Clewell, H. J. (1993). Variability of physiologically-based pharmacokinetic (PBPK) model parameters and their effects on PBPK model predictions in a risk assessment for perchloroethylene (PCE). *Toxicol. Lett.* **68**, 131–144.

Green, T., Lee, R., Farrar, D., and Hill, J. (2003). Assessing the health risks following environmental exposure to hexachlorobutadiene. *Toxicol. Lett.* **138**, 63–73.

Greenberg, M. S., Burton, G. A., and Fisher, J. W. (1999). Physiologically based pharmacokinetic modeling of inhaled trichloroethylene and its oxidative metabolites in B6C3F(1) mice. *Toxicol. Appl. Pharmacol.* **154**, 264–278.

Guberan, E., and Fernandez, J. (1974). Control of industrial exposure to tetrachloroethylene by measuring alveolar air concentrations: Theoretical approach using a mathematical model. *Br. J. Ind. Med.* **31**, 159–167.

Hattis, D., White, P., Marmorstein, L., and Koch, P. (1990). Uncertainties in pharmacokinetic modeling for perchloroethylene. I. Comparison of model structure, parameters, and predictions for low-dose metabolism rates for models derived by different authors. *Risk Anal.* **10**, 449–458.

Hattis, D., White, P., and Koch, P. (1993). Uncertainties in pharmacokinetic modeling for perchloroethylene. II. Comparison of model predictions with data for a variety of different parameters. *Risk Anal.* **13**, 599–610.

Henschler, D. (1985). Halogenated alkenes and alkynes. In: *Bioactivation of Foreign Compounds* M. W. Anders, ed., Academic Press, New York, pp. 317–347.

Henschler, D., and Bonse, G. (1977). Metabolic activation of chlorinated ethylenes: Dependence of mutagenic effect on electrophilic reactivity of the metabolically formed epoxides. *Arch. Toxicol.* **39**, 7–12.

Herren-Freund, S. L., Pereira, M. A., Khoury, M. D., and Olson, G. (1987). The carcinogenicity of trichloroethylene and its metabolites, trichloroacetic acid and dichloroacetic acid, in mouse liver. *Toxicol. Appl. Pharmacol.* **90**, 183–189.

Himmelstein, M. W., Carpenter, S. C., Evans, M. V., Hinderliter, P. M., and Kenyon, E. M. (2004a). Kinetic modeling of beta-chloroprene metabolism. II. The application of physiologically based modeling for cancer dose response analysis. *Toxicol. Sci.* **79**, 28–37.

Himmelstein, M. W., Carpenter, S. C., and Hinderliter, P. M. (2004b). Kinetic modeling of beta-chloroprene metabolism. I. *In vitro* rates in liver and lung tissue fractions from mice, rats, hamsters, and humans. *Toxicol. Sci.* **79**, 18–27.

Jang, J. Y., and Droz, P. O. (1997). Ethnic differences in biological monitoring of several organic solvents. 2. A simulation study with a physiologically based pharmacokinetic model. *Int. Arch. Occup. Environ. Health* **70**, 41–50.

Koizumi, A. (1989). Potential of physiologically based pharmacokinetics to amalgamate kinetic data of trichloroethylene and tetrachloroethylene obtained in rats and man. *Br. J. Ind. Med.* **46**, 239–249.

Lapare, S., Tardif, R., and Brodeur, J. (1995). Effect of various exposure scenarios on the biological monitoring of organic-solvents in alveolar air. 2. 1,1,1- trichloroethane and trichloroethylene. *Int. Arch. Occup. Environ. Health* **67**, 375–394.

Lash, L. H., and Parker, J. C. (2001). Hepatic and renal toxicities associated with perchloroethylene. *Pharmacol. Rev.* **53**, 177–208.

Leibman, K. C., and Ortiz, E. (1977). Metabolism of halogenated ethylenes. *Environ. Health Perspect.* **21**, 91–98.

Leung, H. W. (1992). Use of physiologically based pharmacokinetic models to establish biological exposure indexes. *Am. Ind. Hyg. Assoc. J.* **53**, 369–374.

Lilly, P. D., Thornton-Manning, J. R., Gargas, M. L., Clewell, H. J., and Andersen, M. E. (1998). Kinetic characteristics of CYP2E1 inhibition *in vivo* and *in vitro* by the chloroethylenes. *Arch. Toxicol.* **72**, 609–621.

Loizou, G. D. (2001). The application of physiologically based pharmacokinetic modeling in the analysis of occupational exposure to perchloroethylene. *Toxicol. Lett.* **124**, 59–69.

Mansuy, D., and Battioni, P. (1985). Particular ability of cytochrome P-450 to form reactive intermediates and metabolites. In: *Drug Metabolism, Molecular Approaches and Pharmacological Implications*, G. Seist, ed., Pergamon Press, New York, pp. 195–203.

Poet, T. S., Corley, R. A., Thrall, K. D., Edwards, J. A., Tanojo, H., Weitz, K. K., Hui, X., Maibach, H. I., and Wester, R. C. (2000). Assessment of the percutaneous absorption of trichloroethylene in rats and humans using MS/MS real-time breath analysis and physiologically based pharmacokinetic modeling. *Toxicol. Sci.* **56**, 61–72.

Poet, T. S., Weitz, K. K., Gies, R. A., Edwards, J. A., Thrall, K. D., Corley, R. A., Tanojo, H., Hui, X., Maibach, H. I., and Wester, R. C. (2002). PBPK modeling of the percutaneous absorption of perchloroethylene from a soil matrix in rats and humans. *Toxicol. Sci.* **67**, 17–31.

Ramsey, J. C., and Andersen, M. E. (1984). A physiologically based description of the inhalation pharmacokinetics of styrene in rats and humans. *Toxicol. Appl. Pharmacol.* **73**, 159–175.

Rao, H. V., and Brown, D. R. (1993). A physiologically based pharmacokinetic assessment of tetrachloroethylene in groundwater for a bathing and showering determination. *Risk Anal.* **13**, 37–49.

Reitz, R. H., Gargas, M. L., Andersen, M. E., Provan, W. M., and Green, T. L. (1996a). Predicting cancer risk from vinyl chloride exposure with a physiologically based pharmacokinetic model. *Toxicol. Appl. Pharmacol.* **137**, 253–267.

Reitz, R. H., Gargas, M. L., Mendrala, A. L., and Schumann, A. M. (1996b). *In vivo* and *in vitro* studies of perchloroethylene metabolism for physiologically based pharmacokinetic modeling in rats, mice, and humans. *Toxicol. Appl. Pharmacol.* **136**, 289–306.

Sato, A. (1991). The effect of environmental factors on the pharmacokinetic behaviour of organic solvent vapours. *Ann. Occup. Hyg.* **35**, 525–541.

Sato, A., Endoh, K., Kaneko, T., and Johanson, G. (1991a). Effects of consumption of ethanol on the biological monitoring of exposure to organic-solvent vapors—A simulation study with trichloroethylene. *Br. J. Ind. Med.* **48**, 548–556.

Sato, A., Endoh, K., Kaneko, T., and Johanson, G. (1991b). A simulation study of physiological factors affecting pharmacokinetic behaviour of organic vapours. *Br. J. Ind. Med.* **48**, 342–347.

Sato, A., Nakajima, T., Fujiwara, Y., and Murayama, N. (1977). A pharmacokinetic model to study the excretion of trichloroethylene and its metabolites after an inhalation exposure. *Br. J. Ind. Med.* **34**, 56–63.

Simmons, J. E., Boyes, W. K., Bushnell, P. J., Raymer, J. H., Limsakun, T., McDonald, A., Sey, Y. M., and Evans, M. V. (2002). A physiologically based pharmacokinetic model for trichloroethylene in the male long-evans rat. *Toxicol. Sci.* **69**, 3–15.

Staats, D. A., Fisher, J. W., and Connolly, R. B. (1991). Gastrointestinal absorption of xenobiotics in physiologically based pharmacokinetic models—A 2-compartment description. *Drug Metab. Dispos.* **19**, 144–148.

Stenner, R. D., Merdink, J. L., Fisher, J. W., and Bull, R. (1998). Physiologically-based pharmacokinetic model for trichloroethylene considering enterohepatic recirculation of major metabolites. *Risk Anal.* **18**(3), 261–269.

Thrall, K. D., and Poet, T. S. (2000). Determination of biokinetic interactions in chemical mixtures using real-time breath analysis and physiologically based pharmacokinetic modeling. *J. Toxicol. Env. Health Pt. A* **59**, 653–670.

US EPA (1986a). Guidelines for carcinogen risk assessment. *Federal Register*, pp. 33992–43003.

US EPA (1986b). *Health Assessment Document for Vinylidene Chloride*. Final Report. US EPA, Washington, D.C.

US EPA (2000a). Toxicological review of vinyl chloride. Appendices A-D. EPA/635R-00/004.

Ward, R. C., Travis, C. C., Hetrick, D. M., Andersen, M. E., and Gargas, M. L. (1988). Pharmacokinetics of tetrachloroethylene. *Toxicol. Appl. Pharmacol.* **93**, 108–117.

Yoshida, K. (1993). Preliminary exposure assessment of volatile chlorinated hydrocarbons in Japan. *Chemosphere* **27**, 621–630.

CHAPTER 4

ALKENE AND AROMATIC COMPOUNDS

James E. Dennison

4.1 INTRODUCTION

Alkenes and aromatic compounds tend to share similar pathways of metabolism as well as similar modes of toxicity. Many aromatic and unsubstituted alkene compounds are metabolized first by cytochrome P450 (CYP)-mediated oxidation of a double bond to an epoxide, followed by hydration of the epoxide to diols and conjugation of the diol by sulfation or glucuronidation. Sometimes, initial oxidation of a double bond is followed by subsequent oxidation of a second double bond, if present. Most of the chemicals discussed in this section are carcinogenic in animals, at least in some species at high doses, an effect usually thought to be associated with reactivity of epoxide intermediates. Many of the chemicals also cause noncarcinogenic effects by other modes of action. These similarities have led to several common approaches in the strategies for developing pharmacokinetic (PK) descriptions of the chemicals' metabolism and the types of PBPK models that have been developed. Nevertheless, there are still many differences in both the pharmacokinetics and pharmacodynamics of these chemicals, and some of these differences are reflected in the models.

A few chemicals for which the major route of metabolism is not through an epoxide are also included in this section due to structural similarities. Also, some

Physiologically Based Pharmacokinetic Modeling, edited by Micaela B. Reddy, Raymond S. H. Yang, Harvey J. Clewell III, and Melvin E. Andersen.

hydrocarbons that do form epoxides are not included here. For some of these chemicals, PBPK models have yet to be developed. Others have been included in another section (e.g., halogenated ethylenes in Chapter 3) because of additional chemical or toxicological similarities to the principal chemicals in other sections. The major efforts in the area, and thus the organization of this section, include PBPK models for the chemicals benzene, styrene, butadiene, isoprene, ethylene and ethylene oxide, naphthalene and other polycyclic aromatic hydrocarbons (PAHs), certain halobenzenes, and some miscellaneous compounds. Some chemicals that have received less attention in the literature (e.g., hydroquinone and propylene) are grouped with structurally related chemicals in this chapter.

Chronic animal bioassays, mutagenicity assays, DNA binding studies, human toxicity studies, and other assays with a number of the chemicals discussed here have frequently yielded equivocal or variable results. Typically, these studies have found some evidence of carcinogenic effects, but the effects may vary based on dose, species, gender, tissue, or other factors. There is also little, if any, consensus for most of the chemicals as to which metabolite or combination thereof is the proximal carcinogenic metabolite(s).

These difficulties with classical toxicological tests create many opportunities for PBPK modeling to be used to organize and integrate other information. Thus, models for these chemical have been developed as an approach to deciphering the basis for species and gender differences in toxicity, suggesting modes of action and hypotheses to be tested, presenting a means for assessing target tissue dose in humans for potential toxic intermediates, and serving as a platform for human health risk assessment, among other uses. Alkenes and aromatic compounds that are common in industry or the environment do not represent a vast number of chemicals *per se*, but have been relatively well studied using PBPK approaches, with almost 100 published models in the class. In this chapter, models for benzene (Table 4.2), styrene (Table 4.3), butadiene (Table 4.4), and isoprene (Table 4.5) are reviewed. Other chemicals covered in this chapter include ethylene/ethylene oxide (Table 4.6), PAHs (Table 4.7), halobenzenes (Table 4.8), and some additional related chemicals (e.g., acrylamide, Table 4.9). Alkyl benzenes are included in Chapter 6 because they generally do not form epoxides.

A review of the models for these chemicals provides an interesting historical picture. For the best-studied chemicals such as benzene and styrene, the earliest models were typically very simple, but were followed by more complex models based on more detailed descriptions of their metabolism. A variety of model structures and strategies to incorporate aspects of the underlying biology and pharmacokinetics have been used throughout these PBPK models. Some of these strategies have been applied to more than one chemical in this section, reflecting biological commonalities pertaining to epoxide chemistry. For example, some form of the "privileged access" concept (Johanson and Filser 1993; Kohn and Melnick 2000) that addresses enhanced hydrolysis of epoxide intermediates has been used with butadiene, styrene, and ethylene. Prior to describing specific modeling approaches, we will review some of the common metabolic motifs and approaches for simulating the biological fate of these chemicals.

4.2 PK AND PHARMACODYNAMIC PROPERTIES IMPORTANT IN PBPK MODEL DEVELOPMENT FOR AROMATIC AND ALKENE COMPOUNDS

4.2.1 Metabolism and Mode of Action

The generalized mode of action for epoxide-mediated carcinogenesis involves (1) formation of the reactive epoxide intermediate, (2) binding to DNA, and (3) mutagenesis. As Fig. 4.1 shows, the first significant step in biotransformation of aromatic and alkene hydrocarbons discussed in this chapter is the formation of the epoxide intermediate. For these lower-molecular-weight compounds, it is generally believed that the insertion of molecular oxygen is mediated predominately by the CYP 2E1 isoenzyme in the liver, although metabolism by some other CYP isoforms may also occur. Also, some investigators have suggested that extrahepatic metabolism, especially in epithelial cells of the upper and lower respiratory tract and kidney, may be quantitatively important for metabolism of some chemicals by CYP 2E1 or other CYPs. Subsequent biotransformation consists of several possible steps, with the

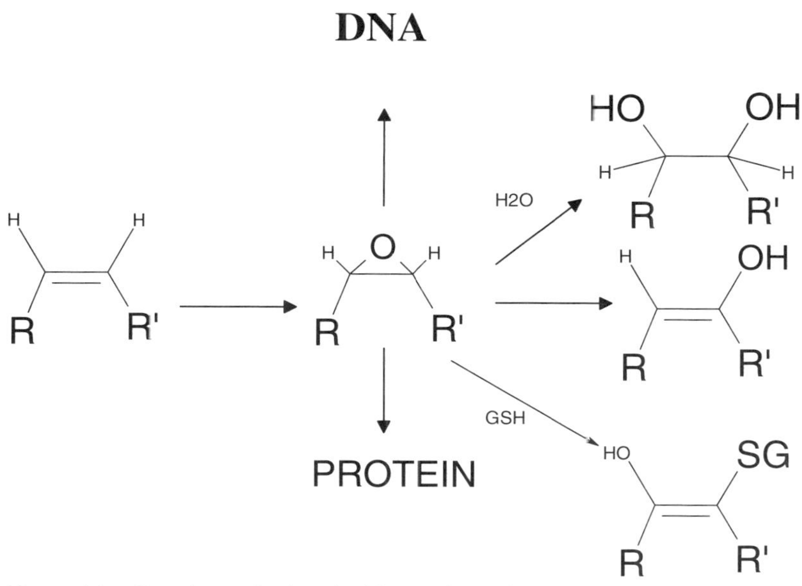

Figure 4.1 Generic mechanism for biotransformation and mode of action of epoxide-mediated carcinogenesis in unsubstituted aromatic and alkene hydrocarbons. The generalized mode of biotransformation of aromatic compounds involves a CYP-mediated addition of molecular oxygen to the double bond, forming an epoxide. The epoxide can bind to DNA or cellular protein or be further biotransformed. Usually, further biotransformation involves GST-mediated conjugation with GSH, nonenzymatic rearrangement to the corresponding alcohol, or hydrolysis via epoxide hydratase to form a diol.

order and significance (based on relative rates) varying from chemical to chemical (and species, gender, or other factors). These subsequent steps may include:

1. Rearrangement of the epoxide to an alcohol
2. Nonenzymatic hydrolysis to the diol
3. Hydrolysis of the epoxide by epoxide hydratase to diols
4. Conjugation of the epoxide with glutathione
5. A second epoxidation by CYP (if other double bonds are present)
6. Conjugation of the alcohols as glucuronide or sulfate esters

After formation of the epoxides or corresponding alcohols, the chemical may be conjugated or excreted directly. Conjugation is common with glutathione (GSH) via enzymatic processes (sometimes nonenzymatically), but enzymatic conjugation with sulfate or glucuronide also occurs. The epoxide intermediates formed from these chemicals are relatively long-lived *in vivo* and can usually be measured in blood. The various models in this section treat metabolism in varying ways.

4.2.2 Model Structures

This biotransformation paradigm has been incorporated into PBPK models with varying levels of complexity, driven perhaps by the purpose of the model, the availability of data, and/or the hypothesized mode(s) of action. These models could be described as "simple," "more complex," and "full" models (Table 4.1). While full models may provide more information on internal doses of certain metabolites, simple models are often quite useful and indeed can be better or more appropriate under various circumstances. Full models, on the other hand, are never "complete" either. Simple models allow for absorption of the parent chemical and clearance of the chemical. Thus, tissue levels are usually only computed for the parent compound. "Clearance" refers to the removal of a parent chemical or a metabolite that results in reduction of concentration but no identification of the downstream metabolite. However, a chemical can be cleared by excretion, and the amount in the excreta (e.g., urine, exhaled air, feces) may be determined. In the more complex formulations, the rates of formation of the epoxide, and second generation metabolites are

TABLE 4.1 Model Structures and Resulting Dose Metrics Available

Model type	Metabolism	Typical dose metrics
Simple	Metabolic clearance of parent	AUC of parent compound or rate of metabolism
More complex	Parent to epoxide, epoxide to alcohol or diol, conjugates	Above, plus epoxides, alcohols, or conjugate concentrations in tissues or excreta, binding
Full	All relevant pathways	Above plus additional metabolites, tissues, or excreted metabolites

usually included in the model. Thus, tissue concentrations of metabolites are determined, as well as the rate of clearance of the second-generation metabolites to excreta. Full models would typically include all significant metabolites and metabolic pathways, as well as time courses of metabolites in excreta. Typical dose metrics or biomarkers for these different model structures are indicated in Table 4.1. The dose metrics can be determined as peak or average chemical concentrations, areas under the curve (AUC), and so on.

4.2.3 PK Differences

The chemicals in this section vary tremendously in lipophilicity, vapor pressure, and other properties that affect their distribution in the animal. Consequently, the models vary widely in sensitivity to some parameters such as the volume of fat tissue. The compounds physically range from the gases such as ethylene or propylene to liquids such as benzene and *o*-dichlorobenzene, to solids such as the PAHs. Many of the chemicals are sufficiently large and lipophilic that they are thought to exhibit diffusionally constrained transport to tissues. Some models have considered this property (for example, Sweeney *et al.* 1997; Sarangapani *et al.* 2002a). A few chemicals are poorly metabolized and lipophilic, and they accumulate in fat. With these more persistent compounds, dosing studies have been performed over several weeks. Rates of change in body mass and the relative volume of each tissue have to be considered in these cases (for example, Freeman *et al.* 1989).

4.2.4 Extrahepatic Metabolism and Transport of Metabolites

Some models reviewed in this chapter incorporate metabolism in lung or other extrahepatic tissues. Many of the epoxides produced from unsubstituted aromatic or alkene hydrocarbons are relatively stable. Measurable levels of epoxides can usually be found in blood or other tissues after treatment with the parent chemical. Thus, some models have been designed to account for distribution of epoxides outside of the liver, as opposed to more reactive species which are often regarded as never leaving their originating tissues, such as dichloromethane. The inclusion of partitioning of the metabolites, extrahepatic metabolism, and nonmetabolic target tissues is thus a feature of some of the models.

4.2.5 GSH Conjugation

Due to the prevalence of GSH conjugation, cellular GSH has been an important issue in some of the work described in this section (Csanady *et al.* 1994; Johanson and Filser 1993). Cellular levels of GSH can be a determinant of conjugation rates (when metabolism is sufficiently fast) and a marker of toxicity. Reactions with GSH have been treated as first order or second order, and many models consider depletion by explicitly including GSH concentrations, necessitating a GSH synthesis model (Kohn and Melnick 1996). Some models have been calibrated or validated against cellular GSH levels (Krishnan *et al.* 1992).

4.2.6 Endogenous Production

Ethylene and isoprene are endogenously produced in mammals at levels that are very likely significant in the context of low level xenobiotic exposure (Their and Bolt 2000; Filser *et al.* 1996). Thus, tissue dose consists of both endogenous and exogenous alkenes. To explore this issue, some models estimate the tissue doses that arise from endogenous production as compared to external exposures (Filser *et al.* 1996; Csanady *et al.* 2000). Ethylene and ethylene oxide (EtO) are interesting in that both parent chemical and metabolite are produced and used commercially and both are also produced endogenously. While EtO is sufficiently reactive to be used as a commercial disinfectant, it is stable in the body for sufficient time to distribute to tissues. Therefore, tissue exposure to EtO depends on endogenous ethylene metabolized to EtO, inhaled ethylene metabolized to EtO, and EtO inhaled directly (Csanady *et al.* 2000). Another article described the potential risk associated with inhalation of ambient styrene oxide in occupational settings (Tornero-Velez and Rappaport 2001).

4.2.7 Reactivity with DNA and Protein

Covalent binding of metabolites is a significant PK phenomenon for many of the chemicals in this chapter. Obviously, binding to DNA is an important issue for genotoxic metabolites. Of additional significance is reversible or nonreversible binding to other macromolecules, as these processes serve as a means for (1) sequestration and subsequent release of metabolites or (2) clearance, respectively. Many of the "full" models described below include equations to address reactions with macromolecules and use protein or DNA binding (i.e., adduct) datasets for model calibration or validation (e.g., Krishnan *et al.* 1992; Csanady *et al.* 2003).

4.2.8 Inhibition of Second Oxidative Steps

Metabolites of many of the chemicals in this section can also undergo oxidation, usually catalyzed by the same enzyme systems responsible for the first oxidative reaction. Because both parent compound and metabolites are common substrates for oxidation by the same enzyme, they are likely to exhibit inhibitory interactions under appropriate conditions. The decision to include mutual inhibition in PBPK models depends on factors such as the relative rates of oxidation, the degree of inhibition, and the identity of the proximal carcinogen. While these inhibitory interactions may not be significant in many cases, initial efforts to describe the pharmacokinetics of parent-metabolite inhibition have been reported in some of these models (Cole *et al.* 2001), as first described for the sequential oxidation of *n*-hexane to 2,5-hexanedione, the putative neurotoxicant (see Chapter 6).

4.2.9 Variability and PK Differences

Several investigators have addressed questions regarding variability and differences between species, genders, ethnicity, and genotype. Typically, differences in human pharmacokinetics have been explored with relatively simple models. Interesting results have come from animal models with regard to species and gender differences.

General efforts to address parameter variability have been explored especially with benzene models but with some other chemicals as well. A description of modeling efforts such as Monte Carlo simulation and Bayesian analysis is outside the scope of this section, but the existence of these models will be briefly mentioned.

4.2.10 Subcompartments in PBPK Models

Many of the models also described the pharmacokinetics of the chemical using tissues with subcompartments. While not unique to alkene and aromatic hydrocarbons, some interesting approaches have been used. Subcompartments have been described for lung, liver, blood, fat, muscle, GI tract, and nose (e.g., Csanady *et al.* 2003; Sarangapani *et al.* 2002b; Evelo *et al.* 1993). Oral absorption components of these models have generally been relatively simple; however, some models have been more detailed in this respect.

4.2.11 "Privileged Access" of Epoxide Hydratase to Epoxide Substrates

Some investigators have noted that initially formed epoxides are metabolized by epoxide hydratase at a rate that is significantly higher than expected. Different approaches have been taken to account for this in some of the PBPK models of aromatic and alkene compounds. One approach visualizes the epoxide as initially bound to the endoplasmic reticular membrane and not immediately available for conjugation (Kohn and Melnich 2000, 2001). In an earlier approach, the epoxide is formed in a formal subcompartment within the liver and must first transfer into the outer liver compartment prior to conjugation (Csanady *et al.* 1995; Johanson and Filser 1993, 1996). These concepts have been applied to styrene, 1,3-butadiene, and ethylene.

In this brief introduction, we have described some of the common elements of the PBPK models for alkenes and aromatic hydrocarbons. In the remainder of the section, we describe the structures of many of the models by chemical group. Not every model could be described; in general, the first model of a given structure is described, and similar models may be mentioned only briefly. Due to the large number of PBPK models reviewed in this section, the discussion focuses on model structure and the strategies used to accommodate toxic responses. Discussion of model validation, results of the modeling efforts, and conclusions drawn by the authors is necessarily limited.

4.3 REVIEW OF AROMATIC AND ALKENE PBPK MODELS

4.3.1 Benzene—A Known Human Carcinogen with an Uncertain Mode of Action

Benzene is toxic to the hematopoietic system and is a known human leukemogen. Numerous theories exist for benzene's mode of action as a carcinogen, but no one of them has been widely accepted (Smith 1996). The bone marrow target is unusual compared to other, more common sites for chemical carcinogens (e.g., liver and kidney). In general, benzene exposure in animals reproduces toxicity to blood forming tissues, but produces different spectra of tumors in animals than those seen in humans.

Benzene closely follows the general aromatic biotransformation paradigm, with consecutive oxidative steps, epoxide rearrangement, epoxide hydrolysis, formation of diols, conjugation with GSH, glucuronide, or sulfate, and excretion in the urine. A ring-opening metabolite is also significant (muconaldehyde and its oxidation product, muconic acid). In bone marrow, the aromatic diols (hydroquinone and catechol) can undergo redox cycling to corresponding benzoquinones accompanied by reactive oxygen release. Proposed carcinogenic metabolites include muconaldehyde, hydroquinone/*p*-benzoquinone, catechol/*o*-benzoquinone, trihydroxybenzene, and phenol. Some of the dose metrics proposed include the AUC in blood or bone marrow for these various metabolites, total benzene metabolism to capture all of these (but also include the nontoxic GSH conjugate), or total benzene metabolism minus GSH conjugation.

Benzene models have a short but rich history. After the first models were developed in 1989, 34 additional models have been published, although several of them are similar or identical in structure (Table 4.2). Several authors have explored benzene pharmacokinetics with relatively simple models. An early model published by Travis in 1990 for mice, rats, and humans included extrahepatic metabolism in bone marrow and assumed that 80% of the metabolite would be excreted as phenol in urine (Travis *et al.* 1990). Numerous datasets were used to build and validate the model, including data from four different dosing routes and benzene levels in blood, bone marrow, and exhaled air, as well as the amount of benzene metabolized. The same model was later verified with data from human exposures in dry cleaning and during or after cigarette smoking (Travis *et al.* 1991). For the smoking scenario, a step function was used to estimate the uptake of benzene from episodic exposures.

This research group used a similar model in two subsequent papers. In the first, the benzene model was used to simulate benzene, toluene, and benzene/toluene mixture pharmacokinetics in the rat by incorporating enzyme inhibition parameters (Purcell *et al.* 1990). A second model simulated benzene pharmacokinetics when the rats were treated with benzene alone or with benzene in gasoline (Travis *et al.* 1992). The pharmacokinetics of benzene in gasoline were described by altering the V_{max} and K_m values for benzene metabolism from those for the single chemical. A similar mixture model has been published for benzene in two-, three-, four-, or five-component mixtures (Haddad *et al.* 1999, 2000), with inhibition being treated using conventional, biochemical representations for these interactions. Interestingly, Purcell *et al.* (1990) and Haddad *et al.* (1999) concluded that the mechanism of enzyme inhibition between benzene and toluene were different (noncompetitive vs. competitive, respectively). A similar model was used to evaluate the ability of metabolic parameter values measured *in vitro* to adequately describe uptake and excretion data collected *in vivo* (deJongh and Blaauboer 1996).

The largest body of work on benzene PBPK modeling has been published by a number of authors working at the Chemical Industry Institute of Toxicology (CIIT) and other institutions using models of the more complex or full variety. The general

TABLE 4.2 PBPK Models for Benzene

Reference	Type	Distinguishing feature
Travis *et al.* (1990)	Simple	Metabolism in bone marrow, phenol in urine
Travis *et al.* (1991)	Simple	Modeled benzene uptake from cigarettes
Purcell *et al.* (1990)	Simple	Mixture of benzene and toluene, metabolic inhibition
Travis *et al.* (1992)	Simple	Benzene alone or in gasoline
Haddad *et al.* (1999)	Simple	Four component mixture with competitive inhibition
Haddad *et al.* (2000)	Simple	Five component mixture with competitive inhibition
DeJongh and Blaauboer (1996)	Simple	Used *in vitro* data for metabolism
Medinsky *et al.* (1989c)	MC	Several metabolic pathways modeled in rats and mice
Medinsky *et al.* (1989b)	MC	Calculated dose metrics
Medinsky *et al.* (1989d)	MC	Human scale-up
Sun *et al.* (1990)	MC	Hemoglobin binding
McMahon *et al.* (1994)	MC	Urinary clearance and young vs. old mice emphasized
Seaton *et al.* (1995)	MC	Simple metabolism in two compartments
Kenyon *et al.* (1996)	MC	Analyzed gender differences in mice
Kenyon and Medinsky (1995)	MC	Three-stage liver with different enzymes present
Cole *et al.* (2001)	Full	Metabolic inhibition, statistical approach
Lindstrom *et al.* (1997)	MC	Included benzene oxide data in blood
Hilderbrand *et al.* (1981)	Simple	Steady-state analysis of benzene pharmacokinetics
Brown *et al.* (1998)	Simple	Differences between genders in humans
Sinclair *et al.* (1999)	Simple	Metabolism in bone marrow, dermal absorption
Bois and Paxman (1992)	MC	Monte Carlo analysis
Bois *et al.* (1991a)	MC	Monte Carlo analysis
Spear and Bois (1994)	MC	Monte Carlo analysis
Spear *et al.* (1991)	MC	Monte Carlo analysis
Thomas *et al.* (1996)	MC	Monte Carlo analysis
Woodruff and Bois (1993)	MC	Monte Carlo analysis
Woodruff *et al.* (1992)	MC	Monte Carlo analysis
Bois *et al.* (1996)	MC	Bayesian analysis
Droz *et al.* (1989a)	MC	Related variability analyses
Droz *et al.* (1989b)	MC	Related variability analyses
Krewski *et al.* (1995)	MC	Related variability analyses
Corley *et al.* (2000)	MC	Hydroquinone

[a] MC, more complex model.

intent of this work appears to be the development of PBPK models to determine tissue levels of putative carcinogenic metabolites, to explore hypotheses for interspecies sensitivity to benzene, and to support interspecies extrapolation for human risk assessment applications. The first in this series of models included a basic physiological structure for rats and mice with a more detailed description of benzene metabolism than found in most benzene models (Medinsky *et al.* 1989c). The metabolic description included (a) saturable metabolism of benzene to benzene oxide/phenol and (b) saturable metabolism of benzene oxide/phenol to phenol conjugates, mercapturic acid species, muconic species, and hydroquinone conjugates. The sum of metabolites in each pathway was used for model calibration. Similar models were subsequently used to calculate dose metrics and to scale up to human (Medinsky *et al.* 1989b, 1989d).

Subsequent experiments by this group indicated that benzene metabolites bind covalently with hemoglobin. Using the radiolabeled data from these experiments, the group extended the Medinsky *et al.* (1989c) model to incorporate hemoglobin binding by unidentified metabolites (Sun *et al.* 1990). Binding had both a linear and saturable component, suggesting to the authors that more than one process was involved. Later, the original mouse model (without hemoglobin binding) was extended with rates for renal clearance of metabolites and used to assess differences in metabolism and clearance between young and old mice (McMahon *et al.* 1994). Again, the metabolites phenol, hydroquinone, muconic acid, and mercapturic acid were grouped with other conjugated forms (e.g., all hydroquinones) and compared to similarly grouped urinary marker data. Age-specific physiological, chemical, and metabolic parameters were used and indicated differences between the young and old mice in urinary elimination and in the rate of formation of hydroquinone conjugates.

The next model (Seaton *et al.* 1995) modified metabolism to include four pathways: benzene oxidation to phenol, phenol oxidation to hydroquinone, phenol sulfation, and hydroquinone glucuronidation (other metabolites were not included). The model was calibrated using the *in vitro* data for linearized metabolism. The physiological compartmental model included only blood and liver compartments and was used to fit parameters at steady-state concentrations. Inter- and intraspecies variability was assessed based on the model fits to the *in vitro* data.

In addition to age, interindividual, and interspecies variation, the CIIT group has investigated gender differences in benzene metabolism (Kenyon *et al.* 1996). The PBPK model used in the work on gender differences was a simplified version of their earlier one for inhalation exposures, without secondary metabolism. Clearance of phenol and hydroquinone after intravenous administration of the metabolites was also analyzed using classical PK methods.

One issue affecting the pharmacokinetics of benzene and probably some other chemicals is the heterogeneous distribution of certain enzymes in the liver acinus. A number of reports have indicated that CYP 2E1 is mainly located in the centrolobular part of the acinus, "downstream" from the preponderance of sulfotransferases, that are mainly located in the periportal region. Glucuronosyltransferases also appear to be enriched in the centrolobular region. The implications of this distribution have been discussed in the literature, but the topic has received only limited quantitative attention. Yet, this heterogeneous distribution of enzymes may explain the observation that phenol does not appear to be carcinogenic even though the putative carcinogenic metabolites arise from downstream metabolites formed from phenol. The hypothesis is that phenol becomes conjugated with sulfate in a first pass through the periportal region of the liver, yielding very low levels of phenol available for hydroquinone production in the centrilobular region. In contrast, benzene bypasses the sulfotransferase and may be oxidized sequentially to phenol and on to hydroquinone in the centrolobular region. A model was developed to simulate the distributed enzyme paradigm, and the results support the hypothesis regarding the localization of the enzymes and the consequent effect on circulating levels of activated versus detoxified metabolites (Kenyon and Medinsky 1995). The model included a yet more detailed description of metabolism and was validated with data for various metabolites in blood, liver, and urine.

Another feature of this model was an initial effort to address another issue, that of potential inhibition of the second oxidative step by parent chemical and *vice versa*. This issue was addressed in the most extensive published model structure for benzene (Cole *et al.* 2001). This model included (a) the structures previously suggested for metabolite binding in tissues, (b) the formation of benzene oxide and other metabolites previously described, and (c) the product of tertiary oxidation, trihydroxybenzene. Metabolic inhibition of each of the oxidative steps was built into the model. The model also estimated a confidence interval around the best estimate of the parameter value for fitted metabolic parameters. While this is frequently done in other ways with population-based models using Monte Carlo or other statistical methods, this feature is not typically found in deterministic models.

In these earlier models, benzene oxide (the first metabolite of benzene biotransformation, the epoxide of benzene) was transient. Benzene oxide would then undergo rearrangement or further biotransformation before it leaves the liver. Thus, the reactions of benzene to phenol could be modeled as a single step. Another group measured benzene oxide in blood. This group modified the Medinsky *et al.* (1989c) model to simulate benzene oxide, assuming that benzene oxide measured in blood was formed in the liver (Lindstrom *et al.* 1997). This model-based analysis estimated that about 4% of metabolized benzene became available in blood as benzene oxide.

Other benzene models have explored some related PK issues. An early study assessed benzene kinetics using a steady-state PK model (Hilderbrand *et al.* 1981), but did not include differential equations for the pharmacokinetics of the chemical in physiological compartments. In another study, differences in male and female human pharmacokinetics were determined using a simple benzene model (Brown *et al.* 1998). Another human model was used to simulate phenol in urine and benzene in exhaled breath (Sinclair *et al.* 1999). This model included metabolism in liver and bone marrow. Dermal uptake was incorporated in the model by using a normal skin compartment separate from a "skin entry" compartment. The model subsequently compared the simulations with controlled human exposure data as well as substantial datasets previously collected in the workplace (Sherwood and Sinclair 1999). Unfortunately, some details were not clear. The authors stated that the model equations and parameter values would be published later; efforts to locate a subsequent publication were unsuccessful.

Numerous benzene models have been developed for use in variability and sensitivity analysis. These topics are outside the scope of this section, so the models will be mentioned only by citation. These models have used a variety of techniques, including standard Monte Carlo variability propagation (Bois and Paxman 1992; Bois *et al.* 1991a; Spear and Bois 1994; Spear *et al.* 1991; Thomas *et al.* 1996; Woodruff and Bois 1993; Woodruff *et al.* 1992), Bayesian methods, and/or Markov Chain Monte Carlo (Bois *et al.* 1996) and related methods (Droz *et al.* 1989a, 1989b; Krewski *et al.* 1995). A comparison of three existing models (Bois *et al.* 1991b) and analyses with existing models (Cox 1996; Cox and Ricci 1992; Weisel *et al.* 1996) have also been performed.

A detailed model for hydroquinone as a single chemical has been developed (Corley *et al.* 2000), incorporating several tissues, metabolism in and uptake from the GI tract, irreversible and reversible binding to proteins, conjugation of hydroquinone to glucuronide and sulfate, formation of benzoquinone in liver, and conjugation of benzoquinone with glutathione. A number of parameters were measured *in vitro* from existing or new data, and the description of the model was explicit regarding the approaches used to build the model and to test it, with data in two species of rats that had differing responses and in humans.

The suite of benzene models in the literature are varied in purpose and design and had considerable variation in the description of metabolism. This variability in structure could be attributable to the lack of consensus over the metabolite(s) responsible for carcinogenicity. Many of the models were used for hypothesis testing. For example, models developed by Medinsky and Kenyon *et al.* were used to quantify the AUC for various metabolites for rats versus mice, males versus female, and so on; the model developed by Lindstrom *et al.* (1997) quantified the amount of benzene oxide leaving the liver. While the current suite of benzene PBPK models may not rise to the level where they can be used for a biologically based risk assessment at this point, they are making a contribution toward that ultimate goal.

4.3.2 Styrene—Early PBPK Models

Oxidation of benzene occurs across a double bond in the ring, destroying the aromatic character of the compound. This double bond is stronger than a typical alkene bond and is attacked in benzene because there is no other possible site of oxidation. When alkyl side chains are present, they are preferentially oxidized by C–H bond oxidation (toluene). With a vinyl side chain (styrene), epoxidation of the side chain is also highly favored over aromatic bond-breaking.

Styrene was the subject of some of the earliest PBPK models developed. Though there is equivocal evidence for styrene's carcinogenicity, the chemical has been of continued interest in the PK field. Perhaps this may be due to the fact that, unlike many other carcinogenic monomers, styrene is often used in small-scale manufacturing processes where exposures may not be as well-controlled as in large-scale plastics manufacturing. Consequently, the potential for human exposure may be significant. Noncarcinogenic endpoints, particularly acute effects, have also motivated the development of several styrene models. A simplified illustration of the predominant metabolic pathways for styrene is illustrated in Fig. 4.2. While such figures will not be provided for all chemicals in this section due to space limitations, the one for styrene provides a sense of the complexity of metabolism. Twenty-one styrene PBPK models have been published; many of them contain new and unusual elements (Table 4.3). Interestingly, although hippuric acid is a significant urinary metabolite of styrene (Ecetoc 1992), it has not been included in any of the styrene models. Moreover, styrene was the subject of the first PBPK models for xenobiotics published in the peer-reviewed literature. We shall start by describing these early, simple models.

Droz and Guillemin (1983) were motivated by a need to estimate appropriate biomarker levels for styrene exposure. Using what has become almost a standard

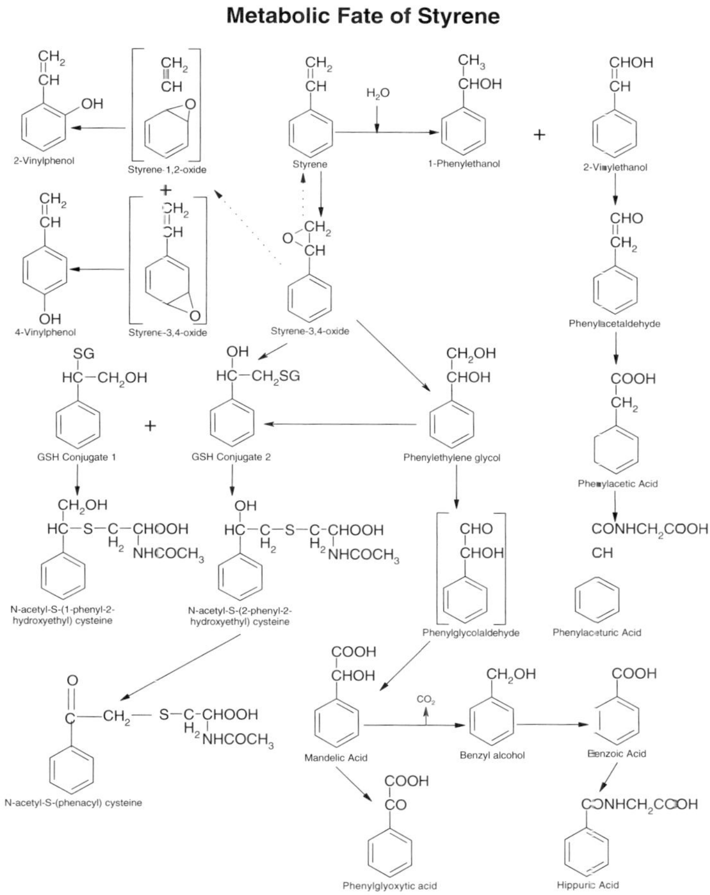

Figure 4.2 Simplified styrene metabolism. Adapted from Ecetoc (1992).

for simple models, a five-compartment model was constructed (liver, slowly perfused, rapidly perfused, fat, and lung), with perfusion-limited uptake, lung:alveolar air and blood:tissue equilibrium, and metabolism restricted to the liver. In this model, all metabolism was linear and included two reactions:

$$\text{Styrene} \rightarrow \text{Mandelic acid (MA)} \rightarrow \text{Phenylglyoxylic acid (PGA)}$$

Characteristically, the first pathway is a composite of four biotransformation steps. The Droz *et al.* model simulated human data for exhaled styrene and for MA and PGA in urine. Four levels of activity were included (rest, light work, medium work, and hard work) with corresponding physiological parameter values. Metabolic constants were determined from halftime and excretion data. Droz *et al.* used a commonly applied technique for the water-soluble metabolites MA and PGA. After formation, these metabolites are considered to reside in body water, and they are therefore modeled by a one-compartment model (a volume of distribution) rather

TABLE 4.3 PBPK Models for Styrene

Reference	Type	Distinguishing feature
Andersen and Ramsey (1983)	Simple	Enzyme induction during chronic exposure in rats
Andersen *et al.* (1984)	Simple	Enzyme induction during chronic exposure in rats
Droz and Guillemin (1983)	MC	Modeled two metabolites in urine, exercise
Ramsey and Andersen (1984)	Simple	Human scale-up
Paterson and McKay (1986)	Simple	Fugacity approach
Perbellini *et al.* (1988)	MC	Human model for two metabolites, five-day exposures
Wang *et al.* (1996)	MC	Estimated metabolic rates using classical PK methods
Leavens and Bond (1996)	MC	Estimated inhibition in mixture with butadiene
Pierce *et al.* (1998)	Simple	Styrene in blood and brain
Johanson and Naslund (1988)	Simple	Two muscle compartments (working and resting), done in spreadsheet
Lof and Johanson (1993)	MC	Styrene in blood and mandelic acid in urine
Jonsson and Johanson (2002)	MC	Bayesian analysis, washin–washout effect modeled
Hetrick *et al.* (1991)	Simple	Sensitivity and variability
Jang *et al.* (1999)	Simple	Sensitivity and variability
Jang and Droz (1997)	Simple	Ethnic differences in PKs
Filser *et al.* (1993)	MC	Butadiene and styrene models with interactions
Csanady *et al.* (1994)	MC	Intracellular first-pass hydrolysis, several dosing routes
Tornero-Velez and Rappaport (2001)	MC	Modeled inhalation of styrene and styrene oxide
Sarangapani *et al.* (2002a)	MC	Series of layered lung compartments with metabolism
Sarangapani *et al.* (2002b)	MC	Three-compartment lung model for steady-state calculations
Csanady *et al.* (2003)	MC	Two-compartment lung model, washin–washout, adducts

than in the typical five compartments used for parent chemicals. One feature of this model that has not been carried forward by other investigators is the use of partial pressure as a concentration unit in tissues. Although they are mathematically equivalent, after adjustment of units, the use of traditional concentration values has been preferred by other researchers.

A similar model was published at about the same time that implemented the concept of interspecies scaling (Ramsey and Andersen 1984). A single, saturable metabolic pathway was included for clearance of parent compound; however, no metabolites were described. The model was developed using rat inhalation data and then allometrically scaled to humans (including scaling of Vmax using allometric scaling methods that are often still used) and validated with human data for styrene in blood and exhaled air.

Examining data for some longer exposure periods in rats, these authors concluded that styrene was capable of inducing the enzyme(s) responsible for its metabolism during an acute exposure. The model described above was expanded to address the issue of enzyme induction (Andersen and Ramsey 1983; Andersen *et al.* 1984). Based on available PK data, enzyme induction was described using a lag time (induction was described at a basal level until the lag time was reached), and it was then increased by a first-order rate until a maximum value was reached. Different estimates of the lag time, first-order rate constant, and maximum velocity of metabolism were obtained for different exposure levels as well as for animals that were chemically induced (phenobarbital or repeated exposure) or inhibited (pyrazole). This model also addressed oral routes of exposure.

The next model to be published was essentially a translation of the Ramsey and Andersen (1984) model (Paterson and Mackay 1986). In the Paterson approach, the model equations were expressed in the mathematically equivalent chemical engineering parameter "fugacity," which has been more extensively used in modeling chemical fate and transport in the environment. Fugacity is a thermodynamic parameter closely related to chemical activity. Writing a model in terms of fugacity offers the benefit that model output reflects the status of equilibrium between compartments. In other words, fugacity calculations denote which direction chemical flux is going between two compartments. Similar computations can easily be performed in PBPK models using concentration and partition coefficients. Fugacity-based models have been popular in environmental modeling but did not find continued use in PBPK.

Another human model emphasized accumulation of chemical and metabolites over repeated exposures (Perbellini *et al.* 1988). This eight-compartment model included two livers: one for chemical flux from blood and another for metabolism. Metabolites were distributed to body water and excreted in urine. Rest and light work activities were simulated for a five-day workweek and included nonsaturable metabolism for Styrene → MA → PGA.

Another relatively simple human styrene model included just three compartments: fat, other tissues, and liver (Wang *et al.* 1996). By assuming that tissues were in thermodynamic equilibrium during final portions of the decay curves, the authors estimated hepatic clearances using classical PK methods and individual subject's height and weight. This analysis suggested to the authors that metabolism of styrene was not significantly faster in the subjects who had previous styrene exposure compared to naive subjects. Another simple model (Leavens and Bond 1996) considered metabolism of both styrene and butadiene and the mixture of the two. Two possible oxidative pathways were considered in the model (i.e., a second CYP for initial oxidation of styrene parallel to oxidation by CYP 2E1). Optimization of the necessary kinetic parameters including the inhibitory constant was used to confirm the hypothesis that more than one CYP was significantly involved in styrene metabolism.

Pierce *et al.* (1998) developed a styrene PBPK model for the purposes of predicting styrene in the blood and brain. The model was validated with prior data and was used to predict blood–brain levels of styrene after controlled human exposures to varying workplace levels of styrene. Venous blood drawn from the antecubital vein was considered to be a mix of 15% flow from fat and 85% flow from muscle tissues. Dose was assumed to result from 65% uptake of inhaled styrene. The model was used to determine the blood level and thus the inhaled concentration at which neurotoxicity has been seen in published studies where exposure concentrations were reported.

Three human styrene models have been developed by Johanson and co-workers. The first simple model had a different construct for muscle (Johanson and Naslund 1988). The model incorporated two skin/muscle compartments representing resting and working muscle, both distinct from the slowly perfused compartment. Working muscle was that part of the individual's muscle mass that received increased perfusion during exercise. This model was also used to demonstrate how PBPK models could be computed in spreadsheet programs on personal computers. Another human styrene PBPK model (Lof and Johanson 1993) was a four-compartment model that incorporated only the styrene to MA metabolic step. Individual human data from controlled exposure studies were modeled using styrene in blood and MA in urine as dose metrics.

In a more recent and complex model, Bayesian analysis was used to work with human variability issues with styrene (Jonsson and Johanson 2002). Several compartments as well as working and resting muscle and two types of fat were included. Metabolites were not modeled. The authors also included the concept of washin–washout of parent chemicals in the upper non-exchanging airways. The washin–washout concept is the suggestion that some chemicals may be retained in the upper airways by diffusing into mucous or deeper layers, reducing the effective concentration at the alveoli. During exhalation, chemical absorbed in the mucus or deeper layers could diffuse back into expired air. To deal with the issue, Jonsson *et al.* applied correction factors to both inhalation and exhalation rates, using different factors for work and resting states. A series of datasets after inhalation exposures, including styrene in subcutaneous fat, were used to validate the model. Antecubital venous blood was modeled as a mixture of resting muscle and shunted arterial blood.

Other styrene models have also been used to evaluate sensitivity and variability issues (Hetrick *et al.* 1991; Jang *et al.* 1999). The latter model was used to investigate some differences in styrene pharmacokinetics between Caucasians and Asians (Jang and Droz 1997). Using a seven-compartment model that included MA and PGA, Jang *et al.* ran the model using actual body weights for each study subject. Differences in styrene kinetics were found between the two groups but could be simulated by changing parameters in the model.

Two PBPK models have also been published by Filser and co-workers. An early mixture model for styrene and butadiene including competitive inhibition was constructed using gas uptake data (Filser *et al.* 1993). In brief, gas uptake studies involve measuring the rate of uptake of a chemical into a rodent, determined by measuring the decline in chemical concentration in the atmosphere of a sealed chamber with test animals inside. The decline reflects absorption, distribution, and metabolism. Rates of metabolism and other PK model parameters can be determined using a PBPK model and gas uptake data, or the data can be used to validate an existing model. The gas uptake system has been described in a series of papers and has been used with one-compartment models (whole body) (Filser 1992; Filser *et al.* 1995) or with PBPK models (Csanady *et al.* 1994; Gargas *et al.* 1986, 1990, 1995).

The second styrene model in this group was a much more detailed styrene model (Csanady *et al.* 1994). This model included four routes of exposure to simulate the administration of either styrene or styrene oxide (SO) in mice, rats, and humans. Metabolism was represented as styrene oxidized to SO and SO metabolized to either the diol or GSH conjugate. GSH conjugation was modeled with a kinetic equation that addressed the ping-pong reaction mechanism involving the binding of cofactors and substrates and intervening release of products. Acid hydrolysis was incorporated in the stomach.

Perhaps most interesting was the inclusion of kinetic enhancement for hydrolysis of the epoxide. This paradigm was previously developed by the same group for butadiene metabolism and was used in butadiene and ethylene PBPK models (Csanady *et al.* 1994, 1996; Johanson and Filser 1993, 1996). The idea that hydrolysis was kinetically enhanced arose from the observation that hydrolysis rates measured *in vitro* did not provide an adequate fit to *in vivo* data. Filser *et al.* suggested that epoxide hydratase (EH) *in vivo* was located in a more proximal position to the CYPs than were cytoplasmic glutathione-*S*-transferase (GST) enzymes, resulting in faster hydrolysis. Others have even suggested that EH and the CYP may actually form a metabolic complex in the endoplasmic reticulum (ER). This hypothesis has been referred to as the "privileged access" concept (Kohn and Melnick 2000) with respect to other alkenes. The Filser *et al.* model addressed this matter by lowering the K_m value for the EH-mediated reaction until a suitable fit to the data was achieved. The model was built and verified using data from three species and four exposure routes generated in different labs.

The Csanady model was later adapted by another group to determine the ratio of the body burden of SO from direct SO inhalation versus that produced via metabolism of inhaled styrene (Tornero-Velez and Rappaport 2001). Workers exposed to styrene also inhale a small proportion of SO that derives from atmospheric oxidation of the airborne styrene. The authors concluded that since inhaled styrene undergoes a significant degree of intrahepatic first-pass hydrolysis after the epoxide is formed, inhalation of small proportions of SO actually contributed more to systemic levels of SO than inhalation of styrene itself. These same comparisons are made in some of the isoprene and ethylene/ethylene oxide models discussed below.

Many PBPK models are developed to assess chemical concentrations in blood and target organs such as liver and kidney. Recent interest in respiratory tract toxicity has stimulated the development of more complex descriptions of the lung to allow descriptions of local metabolism and toxicity. In particular, several models that describe the pharmacokinetics of reactive vapors are discussed in Chapter 5. Similar models were also developed for styrene.

In laboratory rats and mice, inhalation of styrene causes epithelial cell toxicity. Mice develop lesions in the deep lung and have increased incidences of lung tumors. Sarangapani *et al.* (2002a) developed a PBPK model for mice, rats, and humans that described the lung as a series of four compartments with reduced airflow in the deeper two. Also, styrene and/or SO entered both epithelial and submucosal layers of each lung compartment by diffusion, but was metabolized in the epithelial layer as well as in the liver. Kinetic enhancement of hydrolysis of SO was included by assuming that SO was produced in a microsomal subcompartment and could move into a cytosolic subcompartment by diffusion. GSH production, degradation, and depletion were also included.

This model application explained the tendency of nasal and lung tissues to show toxicity at levels lower than those in the liver, based on local concentrations of SO. Also, due to metabolic differences between species, it explained why mice developed tumors at lower exposure levels and suggested that humans are at lower risk as well. Finally, a description of stereospecific metabolism of SO enantiomers was also consistent with the tumor data when expected differences in rodent metabolism were taken into account.

A similar model for styrene (and chloroform) was also produced (Sarangapani *et al.* 2002b). This model simplified the lung into three compartments—the nasal cavity, conducting airways, and pulmonary compartments—whereas the model described above also included a transitional airway compartment. In both models, gas exchange occurred in the pulmonary compartment, but uptake by diffusion occurred in the more proximal compartments. The three-compartment model (Sarangapani *et al.* 2002b) was used to verify a steady-state calculation approach. For styrene and chloroform, the steady-state calculations indicated that local concentrations of metabolites can be higher in lung tissues than in the liver or elsewhere in the body.

Another styrene PBPK model was developed with an enhanced description of the lung (Csanady *et al.* 2003). In this model, the lung was divided into two compartments, the conducting zone and the alveolar zone. In the conducting zone, absorption and metabolism of styrene was permitted, as in the Sarangapani *et al.* (2002a,b) models; however, ventilation was two-way and "washin–washout" (see discussion below) was also permitted. The model was designed for inhalation and oral routes of exposure in mice, rats, and humans. The final difference between this version and the models developed by Sarangapani *et al.* was that it also simulated hemoglobin and DNA (lymphocytes) adducts.

In this section, PBPK models for styrene have been described from some of the earliest PBPK models to some of the most recent and complex models. The series of models reflect the general growth and development of PBPK modeling as a whole. Early models were very simple in structure. Additional features were added over time; as more experience was obtained with modeling, more insight was gained into a number of physiological processes, when better types of additional physiological and PK data became available to support model development and when models could then be used to assess risks for more demanding endpoints.

4.3.3 1,3-Butadiene

The PK behavior of 1,3-butadiene (BD) has many features that are analogous to the vinyl side-chain oxidation of styrene. Metabolism pathways include epoxidation, hydrolysis of the epoxide, and conjugation with glutathione. However, several properties of the chemical cause variations on the general metabolism scheme presented in Fig. 4.1. With BD, a second oxidation step occurs, forming diepoxides and related compounds. Also, due to the existence of multiple epoxide species (Fig. 4.3), the identity of the proximal carcinogen is not well established. As a whole, PBPK

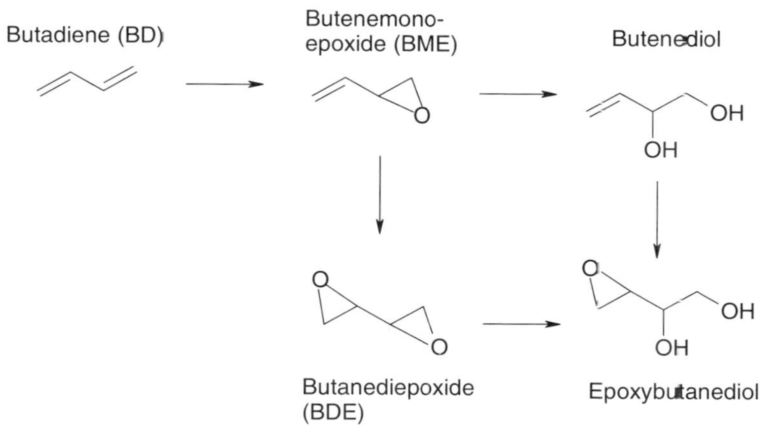

Figure 4.3 Simplified metabolic scheme for butadiene. Double bonds on butadiene may be oxidized to epoxides, primarily by CYP2E1 in the liver. Epoxides are hydrolyzed by epoxide hydratase. Epoxybutanediol can be formed through two pathways. BME and BDE can be conjugated with GSH. Alcohols may be further conjugated with sulfate or glucuronides.

models for BD (Table 4.4) have therefore been more metabolically complex than other models for alkenes.

An early PBPK model for BD in rats and mice was developed by Hattis and co-workers (Hattis and Wasson 1987). In this model, Hattis combined the liver and richly perfused tissues into one metabolically active compartment. Other modelers have generally split the liver and richly perfused tissues, based on the potential for flow limited metabolism. Muscle and fat compartments were also included. This model was used to calculate the AUC for the first epoxide [butenemonoepoxide (BME)] for use as a dose metric. The risk assessment also performed using this dose metric serves as an early example of a biologically based risk assessment.

Another simple model for BD was used to estimate AUC of BD in lung as a dose metric (Hallenbeck 1992). This dose metric was then used as a dose surrogate in conjunction with rodent cancer bioassay study data, in a comparison with the administered dose, to develop a biologically based dose–response curve. Limited model details and references were provided, but the model may have been otherwise similar to the one developed by Hattis *et al.* (1987).

A different conceptual approach for the airways was implemented in another BD model (Evelo *et al.* 1993). This model included typical compartments, except that the lung was subdivided into alveolar and bronchial regions. Metabolism was included in both lung subcompartments (and liver), but gas exchange only occurred in the alveolar lung.

A later version of a simple BD model was produced to analyze pretreatment data (Jackson *et al.* 2000). Using the model, the authors simulated gas uptake data after mice were given either no pretreatment, 1,2-*trans*-dichloroethylene to inhibit CYP 2E1, or 1-aminobenzotriazole to inhibit all cytochrome P450s. The results sug-

TABLE 4.4 PBPK Models for Butadiene

Reference	Type	Distinguishing feature
Hattis and Wasson (1987)	Simple	AUC of first epoxide for risk assessment
Hallenbeck (1992)	Simple	Area under the curve for BD in lung
Evelo *et al.* (1993)	Simple	Two compartment lung model with metabolism
Jackson *et al.* (2000)	Simple	Rats given CYP inhibitors for metabolic rate estimation
Johanson and Filser (1993)	MC	Intrahepatic compartment for first pass hydrolysis
Johanson and Filser (1996)	MC	Similar to 1993 model scaled up to humans
Filser *et al.* (1993)	MC	Butadiene and styrene models with interactions
Csanady *et al.* (1996)	MC	Similar to Johanson and Filser (1993, 1996) models with protein binding
Kohn and Melnick (1993)	MC	Several metabolic pathways varied with species
Kohn and Melnick (1996)	MC	Added complexity to the Kohn and Melnick (1993) model
Bond *et al.* (1994)	Simple	Single oxidative step
Medinsky *et al.* (1994)	MC	Added complexity to the Bond *et al.* (1994) model
Sweeney *et al.* (1996a)	MC	GSH submodel, some diffusional processes
Sweeney *et al.* (1997)	MC	Added complexity to Sweeney *et al.* (1996a) model
Sweeney *et al.* (2001)	MC	Added complexity to Sweeney *et al.* (1997) model
Kohn and Melnick (2000)	MC	Privileged access concept for free and bound epoxide
Kohn and Melnick (2001)	MC	Privileged access concept for free and bound epoxide
Bois *et al.* (1999)	Simple	Two-compartment nonphysiological model
Smith *et al.* (2001)	MC	Human model for analysis of variability issues

gested that CYP isoforms other than CYP 2E1 may play a significant role in BD metabolism. The frequency of micronuclei in harvested erythrocytes was also correlated with the amount of butadiene metabolized determined by the model under different experimental regimens.

Butadiene serves as a prime example of interspecies differences in carcinogenicity between rats and mice. Both Sprague–Dawley rats and B6C3F1 mice develop tumors at multiple sites after chronic butadiene treatment, but mice are much more sensitive, developing tumors at 6 ppm versus 1000 or 8000 ppm for rats. This difference in sensitivity is more exaggerated than for most chemicals, and it has stimulated numerous efforts to determine the causes. Thus, most of the models for this chemical were developed for both species, and a number of them were scaled up to human.

The next series of models investigated these issues by including simple metabolism of BD, typically including metabolism of butadiene to BME and clearance of BME via GSH conjugation and EH hydrolysis. Some also computed but didn't track butadiene diepoxide (BDE) formation. Johanson *et al.* developed models for BD pharmacokinetics in mice, rats, and humans. In a model for rats and mice (Johanson and Filser 1993), the authors modeled BD and BME gas uptake data and liver GST data using a model that combined the slowly and rapidly perfused compartments. Metabolism was represented by oxidation of BD and hydrolysis or GSH conjugation of the epoxide (using ping-pong kinetics). The metabolite BME was distributed to the same tissues as the parent and could be exhaled. Hydrolysis of BME by epoxide hydratase (hydrolase) was treated in a novel manner, described below. A scaled-up human model used a similar model structure (Johanson and Filser 1996). The Filser *et al.* model previously discussed in the styrene section was also used for BD pharmacokinetics (Filser *et al.* 1993). A follow-up model to these (Csanady *et al.* 1996) also included covalent binding with hemoglobin.

The first of several models developed by Kohn and co-workers included this simple metabolism structure (Kohn and Melnick 1993). In this model, Kohn argued for the use of a lower liver flow rate than found in most models, a choice that can affect total clearance of highly extracted chemicals. Metabolism was included in the liver, lung, and richly perfused tissues based on *in vitro* data. Some of the metabolic pathways or kinetic expressions varied by tissue and/or species, such as the GST mechanism. Oxidation of BME to BDE was included for mice only, and competitive inhibition by BD was included. Using this model, Kohn suggested that species differences in body burden of BME were influenced more by physiological differences than by metabolic differences between species. The follow-up model (Kohn and Melnick 1996) was produced to incorporate more available data. This model included tissue blood, alveolar space and GI tract compartments, a GSH depletion/resynthesis model, and included conjugation of BME with GSH in blood as well as metabolizing tissues.

Another simple model (Bond *et al.* 1994) was not described in detail, but appears to include only the first oxidative step. This model was also used to perform a "what-if" simulation of butadiene and styrene or benzene mixture exposures, using competitive inhibition as the mode of interaction. A follow-up model added the hydrolysis and GSH conjugation steps (Medinsky *et al.* 1994). These investigators indicated that lung metabolism was necessary to permit the model to fit the gas uptake data used in the simulation.

The next model from the same group of investigators introduced additional refinements (Sweeney *et al.* 1996a). Only briefly described in the publication, the model apparently included a GSH model that included diurnal variation in GSH, nonenzymatic clearance of BME, and diffusion-limited transport of BME. Also, to account for ongoing biotransformation that occurs between the time animals are sacrificed and when the tissues are frozen, the model permitted metabolism to continue for the time estimated for this process.

Subsequent models developed in the late 1990s and thereafter expanded the description of metabolism and added other refinements. Sweeney *et al.* (1997) expanded their earlier work by including BDE explicitly, including BDE clearance by conjugation with GSH, hydration by EH, and by nonenzymatic processes. The nonenzymatic rates constants were set equal to rates of nonspecific loss of chemical from vial equilibrium studies after GST and EH were inhibited. *In vitro* metabolic rate constants were used; and where metabolism was higher *in vivo*, additional metabolism was assumed to be linear. Initial metabolism of BD to another set of metabolites (aldehydes) was assumed to clear a fractional amount of BD. Diffusion limited flux of the epoxides to and from the liver was also included. GSH conjugation was modeled using ping-pong kinetics, and a second isoform of CYP-mediated oxidation was included. Following this, another model refinement was published

(Sweeney *et al.* 2001). A simplified GSH resynthesis model was included, and body weight change during treatment was simulated. Inhibition of BME oxidation to the diepoxide by BD was included. The model, scaled to humans, was designed to allow estimation of the diepoxide as a dose metric.

A similar metabolic description, without nonenzymatic clearance of epoxides, was developed (Kohn and Melnick 2000). The GSH resynthesis model was modified assuming that γ-glutamyl synthetase activity was the rate-limiting step in GSH resynthesis with end-product inhibition. Finally, rather than using a washin–washout model as proposed by Filser, Johanson, and co-workers, this model used an empirical extraction ratio parameter that controlled the amount of chemical that could partition into the lung.

Several investigators have noted that the rate of hydration of BME formed in the liver by BD oxidation cannot be simulated without using unreasonable affinity constants. The general hypothesis is that the epoxide derived from a precursor (BME from BD, in this case) has a rate of hydrolysis *in vivo* that is significantly higher than predicted from microsomal incubations of the epoxide. The rationale for this observation is that the two enzymes—the oxidative CYP P450 and EH—are in close proximity, and the local concentration of the product of oxidation is much higher than the average concentration throughout the tissue.

Different approaches have been used in models to account for this possibility. In the Kohn *et al.* models (Kohn and Melnick 2000, 2001), the phenomenon has been called "privileged access." These authors proposed that after BD is oxidized to BME, the BME is initially bound to the ER membrane and then released via a first-order process. Free BME can be conjugated with GSH or hydrolyzed by EH, but bound BME could only be hydrolyzed by EH. Also, hydrolysis of bound BME had a lower K_m than hydrolysis of free BME. These enzymatic differences constitute a "privileged access" to EH relative to GST.

This bound and free privileged access concept was retained in a subsequent Kohn *et al.* model and applied to all three epoxides—that is, BME, BDE, and 1,2-butanediol-monoepoxide (Kohn and Melnick 2001). A unique feature of this model was its revision to the GSH submodel based on cysteine utilization by γ-glutamyl synthase where cysteine is synthesized by zero-order processes. Comparing with data, cysteine depletion paralleled GSH depletion. An analysis of the goodness of fit of three BD models with different structures for tissue blood was also published (Kohn 1997).

In the Johanson *et al.* models (Csanady *et al.* 1996; Johanson and Filser 1993, 1996), the privileged access concept was addressed not by assuming bound and free forms of the substrate, but by formulating an intrahepatic compartment within the liver. Reaction with EH is within the intrahepatic compartment where BME is formed from BD; reaction with GSH is in the liver compartment. Transfer of BME to the mixed liver compartment from the intrahepatic compartment was modeled as a flow clearance. Finally, the K_m value for hydrolysis was optimized at a value about one-fifth of the rate estimated from *in vitro* studies. BD models developed by others have not included explicit expressions for this hypothesized mechanism. In styrene models, Csanady *et al.* used an approach similar to the intrahepatic compartment approach used for BD (Csanady *et al.* 1994). With ethylene, one modeling approach used was to reduce the liver concentration of the epoxide by a species specific amount (Csanady *et al.* 2000).

Bayesian analysis was used in the context of a two-compartment nonphysiological model (Bois *et al.* 1999). This model was used to determine the most efficient experimental design for a subsequent human volunteer study (Smith *et al.* 2001). The Smith *et al.* (2001) model included slowly perfused and fat compartments and simple metabolism of BD to BME in the rapidly perfused compartment. Many parameters were measured in the subjects, and the remaining parameters were fitted to the model with Markov Chain Monte Carlo approaches with posteriors generated for each subject. The model output was extensively analyzed for differences between gender, ethnicity, age, and genotype.

4.3.4 Isoprene

Another monomer used in the synthetic plastics industry, isoprene, is produced endogenously in the body at relatively high levels. It is also found in the environment as a bioeffluent from flora. Four different models have addressed the pharmacokinetics of the compound and its potentially mutagenic metabolites (Table 4.5). In the first model, clearance of inhaled isoprene was simulated using saturable metabolism (Filser *et al.* 1996). Ten percent of metabolism was extrahepatic as total metabolism was greater than the rate of delivery of chemical to the liver. The rate of endogenous production of isoprene in human was determined by monitoring exhaled isoprene in a closed system and was included in the PBPK model as a zero-order process. A follow-up article using a similar model (Csanady and Filser 2001) provided additional model validation. This subsequent model was used to compute the AUC for isoprene in blood for exposed and nonexposed humans. The authors suggested a workplace exposure limit of 10 ppm (8 hours) based on model findings that this exposure level would result in an isoprene blood AUC about four times that resulting from endogenous production. Another model was also developed and used to calculate tissue dose metrics for a chronic bioassay (Melnick and Kohn 2000; NTP 1999).

A more complex model was later developed with a more detailed description of the metabolism of isoprene in mice (Bogaards *et al.* 2001). This model included metabolism of isoprene to either of its two monoepoxides, metabolism of both monoepoxides to the diepoxide, hydrolysis of all epoxides, and GSH conjugation of the monoepoxides. All metabolic steps were included in the lung and liver, and

TABLE 4.5 PBPK Models for Isoprene

Reference	Type	Distinguishing feature
Filser *et al.* (1996)	MC	Determined rate of endogenous production
Csanady and Filser (2001)	MC	Follow-up to Filser *et al.* (1996)
Melnick and Kohn (2000)	MC	Calculated tissue dose metrics for chronic bioassay
Bogaards *et al.* (2001)	Full	Calculated amounts of numerous metabolites in several tissues

epoxide metabolites were distributed to lung, liver, and body compartments. After including metabolic rate data determined *in vitro* for each step, the model was validated with a small dataset of blood isoprene levels in the mouse. The model was also scaled up to rats and humans.

4.3.5 Ethylene, Propylene, and Their Oxides

Metabolism of the smaller alkenes is strikingly similar to the generic scheme for metabolism of olefins. Both ethylene and propylene are oxidized to epoxides that are largely cleared by GSH conjugation and hydration, with significant binding to protein and/or DNA. Models have been constructed for ethylene oxide (EtO) and ethylene with EtO as a metabolite (Table 4.6). These models have included a large number of tissue compartments, including testes, a target organ for EtO. The first model for EtO was fairly simple but included binding to hemoglobin and DNA adducts and GSH depletion (Krishnan *et al.* 1992). Model development and validation used data from different sources, including GSH depletion in three tissues (including testes), levels of four distinct hemoglobin adducts, DNA adducts, closed chamber data, and a GSH conjugate in urine after inhalation or other dosing regimens. A similar model was later published (Fennell and Brown 2001) that included diffusion-limited uptake in the testes and nonenzymatic hydrolysis as well as GSH conjugation and enzymatic hydrolysis. Nonphysiological PK models (one- or two-compartment) have also been published (Filser and Bolt 1984; Filser *et al.* 1992; Selinski *et al.* 2000).

Csanady *et al.* (2000) developed a model for both ethylene and EtO. Endogenous production of ethylene was modeled as a zero-order process. A washin–washout effect was addressed by adding an upper respiratory tract compartment that did not exchange chemical, and it serves to reduce the amount of chemical that reaches the alveolar region by a fractional amount that was chemical- and species-specific. Csanady *et al.* noted that this approach was preferred instead of reducing pulmonary ventilation because a reduction in pulmonary ventilation affected exhaled chemical simulations. Clearance of EtO by GST and EH was combined as a linear process. In analogy to the privileged access concept with the BD models, the authors found that only part of the EtO produced from ethylene became systemically available. In their EtO model, the authors used a fractional availability parameter to reduce the EtO concentration in liver. This was only applied to rats and not to humans and mice. The model was developed and validated with a rich dataset. This

TABLE 4.6 PBPK Models for Ethylene, Propylene, and Their Oxides

Reference	Type	Distinguishing feature
Krishnan *et al.* (1992)	MC	EtO binding to DNA and protein and GSH depletion
Fennell and Brown (2001)	MC	Diffusion-limited uptake in testes
Csanady *et al.* (2000)	MC	Ethylene and its oxide included in model, endogenous ethylene production, washin–washout
Filser *et al.* (2000)	MC	Propylene model, three species, washin–washout

model structure was later used with propylene in three species (Filser *et al.* 2000). The washin–washout effect for propylene required a 40% reduction in inhaled propylene.

4.3.6 Naphthalene and Other PAHs

PBPK models for relatively few PAHs have constructed to date (Table 4.7). Examples include naphthalene, pyrene, benzo(*a*)pyrene (BaP), and nitropyrene. Considering the prevalence of these compounds in the environment and their known carcinogenic potential, it is surprising that PBPK modeling work on this class of chemicals has been fairly limited. Reasons for this may be that: (1) PAH exposure often involves exposure to mixtures of PAHs and other chemicals; (2) exposures are typically to low levels of PAHs, thus exacerbating the difficulties associated with low-level extrapolation from high-level models; (3) at least some of the airborne chemical present is adsorbed to surfaces of particles, complicating the assumption of equilibrium between ambient concentrations and the concentration in lung blood; and (4) dermal exposure is also an important route of exposure.

While all PAHs probably can be metabolized to epoxides, carcinogenicity is frequently associated with "bay-region" diol epoxide intermediates (Parkinson, 2001). However, of the chemicals discussed in this section, only BaP has a bay region. The PBPK model descriptions of PAH metabolism have usually been quite simple. Only the naphthalene models include more than metabolism of the parent chemical to the first epoxide. Species differences were addressed with the naphthalene models discussed below, with mice again being much more sensitive than rats. Nevertheless, all of the models in this section on PAHs include some interesting features.

The first PBPK model for a PAH was for the carcinogen nitropyrene (Medinsky *et al.* 1989a) in rats. Due to the experimental exposure system that was used to generate modeling data, this model was constructed with a novel concept for routes of exposure. The authors estimated that 5% of the total dose was inhaled

TABLE 4.7 PBPK Models for Naphthalene and Other PAHs

Reference	Chemical	Type	Distinguishing feature
Medinsky *et al.* (1989a)	Nitropyrene	Simple	Model includes ingestion and inhalation
Roth and Vinegar (1990)	BaP	Simple	Model includes metabolism in liver/lung and macromolecule binding
Moir (1999)	BaP	Simple	Diffusional limitation included
Haddad *et al.* (1998)	Pyrene	Simple	Diffusional limitation and binding included for
Law *et al.* (1991)	Pyrene	Simple	Trout model
Sweeney *et al.* (1996b)	Naphthalene	MC	Model with stereochemistry
Quick and Shuler (1999)	Naphthalene	MC	Model with GSH depletion, binding
Willems *et al.* (2001)	Naphthalene	MC	Added diffusion limitations to Quick and Shuler (1999) model

and deposited in the upper respiratory tract. The model took this into account by having the deposited fraction ingested and absorbed systemically from the gut. Another 25% of the dose was expected to be deposited in the lung and instantly absorbed. Another 25% of the dose was thought to be adsorbed onto animal fur, ingested during preening, and absorbed from the gut in a manner similar to the chemical in the upper respiratory tract. Nitropyrene metabolites were excreted mainly in feces but also in urine in a first-order manner. Binding to macromolecules and clearance of bound label was also included.

Two models for BaP have been published. The earlier one (Roth and Vinegar 1990) was a simple model including (a) metabolism in liver and lung and (b) binding in three tissues. The model was used with data for BaP tissue concentrations in naive and 3-methylcholanthrene-induced animals. Results of the model indicated that induction of metabolizing enzymes increased the amount of chemical that was metabolized in the lung relative to the liver. Diffusional limitation was included in another simple BaP model (Moir 1999) by empirically assuming that a fraction of the normal blood flow went to tissues from a tissue blood compartment. The lack of fit for the liver data suggested to the author that protein binding needed to be included in some tissues.

A simple PAH model was also constructed for pyrene. In one pyrene model (Haddad *et al.* 1998), diffusion-limited uptake to tissues and binding in liver and lung were included. The latter property was incorporated by using "apparent partition coefficients." Another pyrene PBPK model was based on pharmacokinetics in trout (Law *et al.* 1991).

Naphthalene models have been developed by a group at Cornell University. Both models addressed metabolism according to the schematic shown in Fig. 4.4. One basic theme in these models is the idea of a compound being metabolized to a reactive intermediate in one tissue, circulating to another tissue, and binding to macromolecules in the second tissue. The first PBPK model for naphthalene pharmacokinetics in rodents was published in 1996 (Sweeney *et al.* 1996b). This model incorporated oxidation of naphthalene into two enantiomers of naphthalene oxide as well as the other metabolic steps (Fig. 4.4). Binding of naphthalene oxide to tissue macromolecules was modeled as an irreversible process in liver and lung. Bioavailability from oral dosing studies was assumed to be 70–100%, and absorption was first order. In the second model (Quick and Shuler 1999), the authors refined the model and used it to simulate data on GSH depletion, epoxide binding to protein, naphthalene in blood, and mercapturates in urine. The model was used to support a hypothesis regarding differential toxicity in mice and rats, similar to some of the benzene models discussed above. Mice have a higher maximal velocity for oxidation on a weight-adjusted basis, leading to greater GSH depletion. GSH depletion resulted in higher naphthalene oxide (NO) levels capable of reacting with macromolecules, including DNA.

Both of these naphthalene models relied on kinetic parameter estimates from *in vitro* measurements or from their cell culture analog experimental system (Ghanem and Shuler 2000; Sweeney *et al.* 1995). The cell culture analog is a novel system in which cultures of immortalized lung and liver cells were connected by recirculating medium to create an artificial "physiological" system. The data from

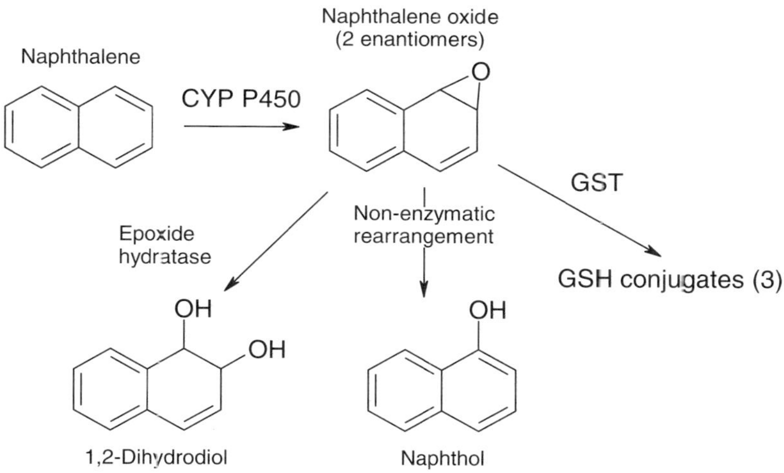

Figure 4.4 Initial steps of naphthalene biotransformation. Some PBPK models for naphthalene have taken stereospecificity into account.

these experiments were analyzed with a kinetic model referred to as a PBPK model to develop estimates of various biochemical parameters.

The most recent model (Willems *et al.* 2001) further explored the species-specific toxicity issue. This model used metabolic structures similar to that in Sweeney *et al.* (1996b) and Quick and Schuler (1999) and some of the same rate constants. However, the newer effort also included diffusion-limited distribution in tissues. The authors concluded that naphthalene levels in the female mouse lung and in rat lungs did not correlate with tumor formation rates. Lung levels of naphthalene oxide did show an association with tumor rates in these species and genders, but other metabolites or modes of action may also play a part in explaining susceptibility.

4.3.7 Halobenzenes

The halobenzenes are a diverse group of chemicals with a limited number of published models (Table 4.8). By our count, there are well over 100 possible halobenzenes if limited to fluoro, chloro, and bromo species. Eight models are available: chlorobenzene, some dichlorobenzenes, hexachlorobenzene bromobenzene, and chloropentafluorobenzene (CPFB). Nevertheless, the chemicals in this group have differing characteristics that carry through to their modes of action. They were selected for modeling based upon interest from a risk assessment standpoint.

A simple model has been developed for chlorobenzene in humans (Kumagai and Matsunaga 1995). The model assumed that a constant fraction of the chlorobenzene was metabolized to chlorocatechol by linear kinetics and other metabolites were ignored. The model also included urinary excretion of chlorocatechol as a linear process for model validation. The purpose of the model was to allow computation of some interesting dose metrics, including (a) one-hour time-weighted average

TABLE 4.8 PBPK Models for Halobenzenes

Reference	Chemical	Type	Distinguishing feature
Kumagai and Matsunaga (1995)	Chlorobenzene	Simple	Model includes metabolites in urine
Hissink *et al.* (1996)	DCB	Simple	Models includes enzyme induction
Hissink *et al.* (1997)	DCB	MC	Model includes more complex metabolism
Morris *et al.* (1993)	Bromobenzene	MC	Perfused nose bromobenzene model, two flow pathways
Freeman *et al.* (1989)	HCB	Simple	1000 hour simulations included tissue growth
Roth *et al.* (1993)	HCB	MC	Detailed GI tract and bile flows
Vinegar *et al.* (1990)	CPFB	Simple	Model incorporated ventilation data
Kinkead *et al.* (1990)	CPFB	Simple	Added repeated dose data to Vinegar *et al.* (1990) model
Crank and Vinegar (1992)	CPFB	Simple	Added mixing in lung to Vinegar *et al.* (1990) model
Clewell and Jarnot (1994)	CPFB	Simple	Alveolar compartment

blood chlorobenzene concentrations and (b) the amount of chlorocatechol excreted to urine during the last two and last four hours of the exposure period as well as during the first two hours of the next day. Variable exposures were also simulated to assess the effect on the dose metrics. The authors concluded that it was acceptable to consider average daily exposure as a basis for controlling chronic toxicity, but that one-hour levels would be important in terms of acute toxicity.

A model for 1,2-dichlorobenzene (DCB) and 1,4-DCB was developed for rats and briefly described (Hissink *et al.* 1996). V_{max} and K_m values for both isomers were estimated from the model and for 1,4-dichlorobenzene with naive and induced rats. 1,2-Dichlorobenzene was cleared more rapidly than 1,4-dichlorobenzene, although the maximum concentration of the epoxide was lower. V_{max} for 1,4-dichlorobenzene at 250 mg/kg was increased by inducers; however, the reported V_{max} for 250 mg/kg was less than the V_{max} at 50 mg/kg, raising the possibility of optimization issues. A subsequent model focusing on 1,2-DCB in rats and humans was described in more detail (Hissink *et al.* 1997). This model again incorporated formation of the epoxide and further metabolism to the GSH conjugate and conversion to phenol. It assumed that a constant fraction of the epoxide was bound to protein, conjugated to GSH, and metabolized to dichlorophenol. GSH depletion and regeneration was included and compared with data on liver GSH.

One model has been developed for brominated benzenes (Morris *et al.* 1993) and other chemicals described as "nonreactive." This model has similarities to models developed for reactive vapors absorbed in the nose (Chapter 5). The bromobenzene model is essentially a perfused nose model, with no other tissue groups included. Even though the whole organism was not included in the model's structure, it could be regarded as a PBPK model because it included absorption, distribution, metabolism, and clearance of chemicals in a physiologically based compartmental model structure. The unique feature of the model was its modeling of two distinct airflow pathways: the dorsal medial pathway that contacted both respiratory and olfactory epithelium and the lateral ventral pathway that contacted only respiratory mucosa. In addition, tissues were modeled as parallel compartments stacked upon each other, similar to early dermal PBPK models. The tissue compartments included an anterior mucus layer, two to four epithelial compartments (depending on depth of tissue), and two submucosal compartments. Equilibrium between air concentrations and the mucus layer was included. The chemical diffused between compartments was metabolized in the epithelium and submucosa and was cleared from the submucosa by blood. This model structure was applied to a variety of compounds in this article. Importantly, this innovative approach to assessing parameters important for nasal uptake spurred a much larger body of work to refine and extend these models, eventually leading to the use of this type of nasal uptake model for risk assessments for deriving reference concentrations with various gases and vapors, as further described in Chapter 5.

A set of two interesting but different models have been developed for hexachlorobenzene (HCB). Pertinent features of HCB are its high lipophilicity and slow metabolism. Indeed, in these models, no metabolism was included even for long timecourse simulations. Both models also included a large number of tissue compartments. The first model (Freeman *et al.* 1989) was run for up to 1000 hours. Because of this long time course, tissue weight parameters were included as time-varying parameters rather than as a single time-invariant value. Partition coefficients were estimated from *in vivo* data at steady state. The authors determined that the decline in HCB concentration in fat in the post-dosing period was principally caused by growth.

A related research group published a follow-up model for HCB (Roth *et al.* 1993) with several new features. The gastrointestinal tract was subdivided into three compartments with fecal flow through them. Flows for chyme, bile, lymph, and urine were also included. The lymph flow was included based on the concept that uptake of chemical from the GI tract was associated with absorption of fatty acids into lymph. Also, due to the long duration of time simulated by the models, the authors incorporated loss of chemical from the gut into the gut lumen as a consequence of sloughing of mucosal cells containing HCB. The model was verified with HCB levels in liver, fat, urine, blood, bile, and feces.

Models for chloropentafluorobenzene (CPFB) have been developed to assist in performing safety assessments with respect to intentional human exposures. CPFB was considered by the U.S. military for use in chemical warfare training. During an exercise, soldiers would be exposed to CPFB. After removing respirators, their breath would be monitored with direct reading sensors for exhaled CPFB to determine if the respirators were effective. For this reason, models for CPFB focused on exhaled breath dose metrics. One interesting feature of CPFB models is the use of measured data for alveolar ventilation as opposed to average values from literature. CPFB is rapidly cleared through exhalation post-exposure. Since the rate and concentration of CPFB exhalation were very dependent on alveolar ventilation, time-dependent measured values for this parameter have been used in CPFB models. CPFB models have been developed for primates as well as rats and humans.

The first model implemented this concept for exhaled air experiments in rats (Vinegar *et al.* 1990). A subsequent rat model (Kinkead *et al.* 1990) used the same structure but provided additional validation from 21-day exposures that included blood sampling for the parent chemical. The third model in the series scaled the rat model up to monkeys (Crank and Vinegar 1992). This model included a more detailed description of ambient and retained air mixing in the lung. Also, the data used for validation in monkeys were collected from anesthetized monkeys using a breathing apparatus. As this apparatus absorbed some of the chemical and influenced exhalation processes, it was explicitly included as a submodel. Fundamentally, such an apparatus model is analogous to the use of chamber loss rates in a closed chamber system.

The fourth and most complete model covered four species: mice, rats, primates, and humans (Clewell and Jarnot 1994). An alveolar compartment was added to improve the biological realism of the model. In this compartment, inspired air mixes with the alveolar air that is in equilibrium with blood. A series of calculated dose metrics were used to determine the human exposure level that would produce a level of the dose metric (inclusive of uncertainty factor adjustments) equivalent to that experienced by animals at the NOAEL. Monte Carlo simulations were also used to assess the impact of variability on model outcomes. As with HCB, the dose metrics for CPFB toxicity were all associated with the parent chemical rather than metabolites.

4.3.8 Miscellaneous Related Compounds

In this section, models for acrylamide, acrylonitrile (ACN), 2,4-toluene diamine, and 4-aminobiphenyl are reviewed (Table 4.9). To date, there are three versions of a model for ACN and a single model for acrylamide. A quasi-PBPK model of aminobiphenyl has also been developed as well as a model for 2,4-toluene diamine.

ACN is metabolized in a similar manner to the aromatic and unsubstituted alkenes already discussed. After epoxidation, it can be conjugated with glutathione. However, hydrolysis by EH was not observed, and the parent chemical can also be conjugated with GSH. Therefore, the metabolic scheme for ACN in the first of three

TABLE 4.9 PBPK Models for Other Compounds Reviewed in This Chapter

Reference	Chemical	Type	Distinguishing feature
Gargas *et al.* (1995)	ACN	MC	Epoxidation and conjugation of ACN and metabolites
Kedderis *et al.* (1996)	ACN	MC	Added stomach compartment to Gargas *et al.* (1995) ACN model
Sweeney *et al.* (2003)	ACN	MC	Human scaled-up ACN model
Kirman *et al.* (2003)	Acrylamide	MC	Model with several metabolic steps
DeJongh *et al.* (1999)	Acrylamide	MC	Model in rats
Luu *et al.* (1998)	2,4-Toluenediamine	MC	Model for exposure from breast implants
Kadlubar *et al.* (1991)	4-Aminobiphenyl	MC	Model for dogs, urothelium compartment

rat ACN models (Gargas *et al.* 1995) included epoxidation and conjugation of both the parent and its metabolite cyanoethylene oxide (CEO). The model also included first-order reaction of ACN and CEO with both hemoglobin and other blood proteins. In a follow-up model (Kedderis *et al.* 1996), the authors added a stomach compartment and nonenzymatic reaction of ACN with GSH. In the early model, GSH conjugation was first order, but this was modified to a second-order rate expression in the subsequent model. These models were subsequently used in risk assessment for ACN (Kirman *et al.* 2000). A human model for ACN was subsequently developed (Sweeney *et al.* 2003). This model indicated that blood and brain concentrations of ACN and CEO were similar in rats and humans after inhalation exposure, but that rats had higher blood ACN levels after exposure in drinking water.

An acrylamide model for rats (Kirman *et al.* 2003) examined the pharmacokinetics of acrylamide and its epoxide, glycidamide. As with ACN, both parent and metabolite can be conjugated with GSH. However, glycidamide can also be hydrolyzed by EH. The model included arterial and venous blood compartments, liver, lung, and the rest of the body as additional compartments. A submodel also partitioned the blood compartments into a serum and a cellular blood compartment. The GSH submodel included depletion and regeneration. Binding of acrylamide and glycidamide were included via second-order rate equations for hemoglobin and first-order equations for binding to liver, blood, or tissue macromolecules. Acrylamide is an important food contaminant, found in many fried foods and many snack foods. The mode of action for carcinogenicity is unknown. This PBPK model was developed to integrate and organize new information on metabolism and toxicity as they become available and to correlate measures of target tissue dose with responses. A simpler PBPK model for acrylamide in rats was published earlier (DeJongh *et al.* 1999). In this model, acrylamide was metabolized by saturable metabolism and cleared by other processes using a linear function. The model was used to support risk assessment for subchronic and acute toxicity.

A PBPK model for 2,4-toluene diamine (TDA) in rats and humans was developed to support a risk assessment for breast implants (Luu *et al.* 1998). In this model, TDA was assumed to be 32% metabolized to 4-acetylamino 2-aminotoluene (AAT), and it was assumed that this was the toxic metabolite; 10% of TDA was assumed bound. AAT was subjected to metabolism and excretion.

A model for 4-aminobiphenyl was developed for the dog (Kadlubar *et al.* 1991). This model is not a typical PBPK model, yet has similar features in many ways and includes some different physiological processes in its description. 4-Aminobiphenyl (ABP) is a bladder carcinogen. The mode of action that this model explored occurs when the ABP is *N*-hydroxylated and subsequently glucuronidated in the liver, excreted into urine, deconjugated, and transformed into a reactive species capable of binding to urothelial DNA and causing genetic damage. The purpose of the model was to help investigate the potential effect of infrequent voiding and lowering of urinary pH to increase the urinary concentration of unconjugated *N*-hydroxy-4-ABP. The model thus included only the tissues of interest—namely, body and urothelium—as well as urine. Several metabolic steps were modeled; of particular interest is the nonenzymatic hydrolysis of the glucuronide conjugate in the urine. Hemoglobin binding was a significant process in the model, and the model addressed resorption of *N*-hydroxy-4-ABP and ABP into urothelium. Beyond this description, model equations and parameter values were not provided. The model indicated that tissue dose to *N*-hydroxy-4-ABP and adduct formation depended primarily on voiding frequency and less on urine pH.

4.4 SUMMARY

PBPK models for alkenes and aromatic compounds form a diverse and extensive group. Many of the models have been extended to risk assessment applications, in particular to support interspecies extrapolations. Primarily, these efforts have focused on epoxides and DNA adducts, reflecting the fact that nearly of the compounds are considered to be genotoxic carcinogens. A number of very interesting attributes can be found in this group of models. These include relatively robust descriptions of metabolism, numerous Phase II metabolic pathways (conjugation), involvement of nonenzymatic reactions, descriptions of biotransformation by epoxide hydratase, sequential oxidations of dienes, simulation of intermediates (epoxides) in blood, GSH synthesis and depletion, and binding to macromolecules. While the group as a whole more often than not used similar PBPK model structures, numerous variations were created based on various aspects of the underlying biology. Some of these strategies can probably be applied to compounds in other classes when the need arises. While considerable insight into the pharmacokinetics (and sometimes pharmacodynamics) of these compounds has been gained through this large body of work, the group should serve as a foundation for new generations of models that are of an increasingly robust characteristic.

NOTATION

AAT	4-acetylamino 2-aminotoluene
ABP	4-aminobiphenyl
ACN	acrylonitrile
AUC	area under the curve
BaP	benzo(*a*)pyrene
BD	1,3-butadiene
BDE	butadiene diepoxide
BME	butenemonoepoxide
CEO	cyanoethylene oxide
CIIT	chemical Industry Institute of Technology
CPFB	chloropentafluorobenzene
CYP	cytochrome P450
DCB	dichlorobenzene
EH	epoxide hydratase
ER	endoplasmic reticulum
EtO	ethylene oxide
GI	gastrointestinal

GSH	glutathione
GST	glutathione-*S*-transferase
HCB	hexachlorobenzene
K_m	affinity constant
MA	mandelic acid
MC	more complex (model)
NO	naphthalene oxide
PAHs	polycyclic aromatic hydrocarbons
PGA	phenylglyoxylic acid
PK	pharmacokinetic
SO	styrene oxide
TDA	2,4-toluene diamine
V_{max}	maximal rate of metabolism

REFERENCES

Andersen, M. E., and Ramsey, J. C. (1983). A physiologically-based description of the inhalation pharmacokinetics of styrene in rats and humans. In: *Developments in the Science and Practice of Toxicology*, A. W. Hayes, R. C. Schnell, and T. S. Miya, eds., Elsevier, Rockville, MD, pp. 415–418.

Andersen, M. E., Gargas, M. L., and Ramsey, J. C. (1984). Inhalation pharmacokinetics: evaluating systemic extraction, total *in vivo* metabolism, and the time course of enzyme induction for inhaled styrene in rats based on arterial blood : inhaled air concentration ratios. *Toxicol. Appl Pharmacol.* **73**, 176–187.

Bogaards, J. J., Freidig, A. P., and van Bladeren, P. J. (2001). Prediction of isoprene diepoxide levels *in vivo* in mouse, rat and man using enzyme kinetic data *in vitro* and physiologically-based pharmacokinetic modelling. *Chem. Biol. Interact.* **138**, 247–265.

Bois, F. Y., and Paxman, D. G. (1992). An analysis of exposure rate effects for benzene using a physiologically based pharmacokinetic model. *Regul. Toxicol. Pharmacol.* **15**, 122–136.

Bois, F. Y., Smith, M. T., and Spear, R. C. (1991a). Mechanisms of benzene carcinogenesis: Application of a physiological model of benzene pharmacokinetics and metabolism. *Toxicol. Lett.* **56**, 283–298.

Bois, F. Y., Woodruff, T. J., and Spear, R. C. (1991b). Comparison of 3 physiologically based pharmacokinetic models of benzene disposition. *Toxicol. Appl. Pharmacol.* **110**, 79–88.

Bois, F. Y., Jackson, E. T., Pekari, K., and Smith, M. T. (1996). Population toxicokinetics of benzene. *Environ. Health Perspect.* **104**(Suppl 6), 1405–1411.

Bois, F. Y., Smith, T. J., Gelman, A., Chang, H. Y., and Smith, A. E. (1999). Optimal design for a study of butadiene toxicokinetics in humans. *Toxicol. Sci.* **49**, 213–224.

Bond, J. A., Csanady, G. A., Gargas, M. L., Guengerich, F. P., Leavens, T., Medinsky, M. A., and Recio, L. (1994). 1,3-Butadiene—Linking metabolism, dosimetry, and mutation–induction. *Environ. Health Perspect.* **102**, 87–94.

Brown, E. A., Shelley, M. L., and Fisher, J. W. (1998). A pharmacokinetic study of occupational and environmental benzene exposure with regard to gender. *Risk Anal.* **18**, 205–213.

Clewell, H. J., 3rd, and Jarnot, B. M. (1994). Incorporation of pharmacokinetics in noncancer risk assessment: Example with chloropentafluorobenzene. *Risk Anal.* **14**, 265–276.

Cole, C. E., Tran, H. T., and Schlosser, P. M. (2001). Physiologically based pharmacokinetic modeling of benzene metabolism in mice through extrapolation from *in vitro* to *in vivo*. *J. Toxicol. Env. Health Pt. A* **62**, 439–465.

Corley, R. A., English, J. C., Hill, T. S., Fiorica, L. A., and Morgott, D. A. (2000). Development of a physiologically based pharmacokinetic model for hydroquinone. *Toxicol. Appl. Pharmacol.* **165**, 163–174.

Cox, L. A. (1996). Reassessing benzene risks using internal doses and Monte-Carlo uncertainty analysis. *Environ. Health Perspect.* **104**, 1413–1429.

Cox, L. A., Jr., and Ricci, P. F. (1992). Reassessing benzene cancer risks using internal doses. *Risk Anal.* **12**, 401–410.

Crank, W. D., and Vinegar, A. (1992). A physiologically based pharmacokinetic model for chloropentafluorobenzene in primates to be used in the evaluation of protective equipment against toxic gases. *Toxicol. Ind. Health* **8**, 21–35.

Csanady, G. A., and Filser, J. G. (2001). Toxicokinetics of inhaled and endogenous isoprene in mice, rats, and humans. *Chem. Biol. Interact.* **135–136**, 679–685.

Csanady, G. A., Mendrala, A. L., Nolan, R. J., and Filser, J. G. (1994). A physiological pharmacokinetic model for styrene and styrene-7,8-oxide in mouse, rat and man. *Arch. Toxicol.* **68**, 143–157.

Csanady, G. A., Kreuzer, P. E., Baur, C., and Filser, J. G. (1996). A physiological toxicokinetic model for 1,3-butadiene in rodents and man: blood concentrations of 1,3-butadiene, its metabolically formed epoxides, and of haemoglobin adducts—Relevance of glutathione depletion. *Toxicology* **113**, 300–305.

Csanady, G. A., Denk, B., Putz, C., Kreuzer, P. E., Kessler, W., Baur, C., Gargas, M. L., and Filser, J. G. (2000). A physiological toxicokinetic model for exogenous and endogenous ethylene and ethylene oxide in rat, mouse, and human: formation of 2-hydroxyethyl adducts with hemoglobin and DNA. *Toxicol. Appl. Pharmacol.* **165**, 1–26.

Csanady, G. A., Kessler, W., Hoffmann, H. D., and Filser, J. G. (2003). A toxicokinetic model for styrene and its metabolite styrene-7,8-oxide in mouse, rat and human with special emphasis on the lung. *Toxicol. Lett.* **138**, 75–102.

DeJongh, J., and Blaauboer, B. J. (1996). *In vitro*-based and *in vivo*-based simulations of benzene uptake and metabolism in rats. *ATLA-Altern. Lab. Anim.* **24**, 179–190.

DeJongh, J., Nordin-Andersson, M., Ploeger, B. A., and Forsby, A. (1999). Estimation of systemic toxicity of acrylamide by integration of in vitro toxicity data with kinetic simulations. *Toxicol. Appl. Pharmacol.* **158**, 261–268.

Droz, P. O., and Guillemin, M. P. (1983). Human styrene exposure. V. Development of a model for biological monitoring. *Int. Arch. Occup. Environ. Health* **53**, 19–36.

Droz, P. O., Wu, M. M., and Cumberland, W. G. (1989a). Variability in biological monitoring of organic solvent exposure. II. Application of a population physiological model. *Br. J. Ind. Med.* **46**, 547–558.

Droz, P. O., Wu, M. M., Cumberland, W. G., and Berode, M. (1989b). Variability in biological monitoring of solvent exposure. I. Development of a population physiological model. *Br. J. Ind. Med.* **46**, 447–460.

Ecetoc (1992). *Styrene Toxicology Investigations on the Potential for Carcinogenicity*, European Centre for Ecotoxicology and Toxicology of Chemicals, Technical Report No. 52, November 1992, pp. 1–61.

Evelo, C. T. A., Oostendorp, J. G. M., Tenberge, W. F., and Borm, P. J. A. (1993). Physiologically-based toxicokinetic modeling of 1,3-butadiene lung metabolism in mice becomes more important at low doses. *Environ. Health Perspect.* **101**, 496–502.

Fennell, T. R., and Brown, C. D. (2001). A physiologically based pharmacokinetic model for ethylene oxide in mouse, rat, and human. *Toxicol. Appl. Pharmacol.* **173**, 161–175.

Filser, J. G. (1992). The closed chamber technique—Uptake, endogenous production, excretion, steady-state kinetics and rates of metabolism of gases and vapors. *Arch. Toxicol.* **66**, 1–10.

Filser, J. G., and Bolt, H. M. (1984). Inhalation pharmacokinetics based on gas uptake studies. VI. Comparative evaluation of ethylene oxide and butadiene monoxide as exhaled reactive metabolites of ethylene and 1,3-butadiene in rats. *Arch. Toxicol.* **55**, 219–223.

Filser, J. G., Denk, B., Tornqvist, M., Kessler, W., and Ehrenberg, L. (1992). Pharmacokinetics of ethylene in man; body burden with ethylene oxide and hydroxyethylation of hemoglobin due to endogenous and environmental ethylene. *Arch. Toxicol.* **66**, 157–163.

Filser, J. G., Johanson, G., Kessler, W., Kreuzer, P. E., Stei, P., Baur, C., and Csanady, G. A. (1993). *A Pharmacokinetic Model to Describe Toxicokinetic Interactions Between 1,3-Butadiene and Styrene in Rats: Predictions for Human Exposure*, IARC Scientific Publications, Lyon, Publication 127, pp. 65–78.

Filser, J. G., Csanady, G. A., Kreuzer, P. E., and Kessler, W. (1995). Toxicokinetic models for volatile industrial chemicals and reactive metabolites. *Toxicol. Lett.* **82–83**, 357–366.

Filser, J. G., Csanady, G. A., Denk, B., Hartmann, M., Kauffmann, A., Kessler, W., Kreuzer, P. E., Putz, C., Shen, J. H., and Stei, P. (1996). Toxicokinetics of isoprene in rodents and humans. *Toxicology* **113**, 278–287.

Filser, J. G., Schmidbauer, R., Rampf, F., Baur, C. M., Putz, C., and Csanady, G. A. (2000). Toxicokinetics of inhaled propylene in mouse, rat, and human. *Toxicol. Appl. Pharmacol.* **169**, 40–51.

Freeman, R. A., Rozman, K. K., and Wilson, A. G. (1989). Physiological pharmacokinetic model of hexachlorobenzene in the rat. *Health Phys.* **57**, 139–147.

Gargas, M. L., Andersen, M. E., and Clewell, H. J. D. (1986). A physiologically based simulation approach for determining metabolic constants from gas uptake data. *Toxicol. Appl. Pharmacol.* **86**, 341–352.

Gargas, M. L., Clewell, H. J., and Andersen, M. E. (1990). Gas uptake inhalation techniques and the rates of metabolism of chloromethanes, chloroethanes, and chloroethylenes in the rat. *Inhal. Toxicol.* **2**, 295–319.

Gargas, M. L., Andersen, M. E., Teo, S. K., Batra, R., Fennell, T. R., and Kedderis, G. L. (1995). A physiologically based dosimetry description of acrylonitrile and cyanoethylene oxide in the rat. *Toxicol. Appl. Pharmacol.* **134**, 185–194.

Ghanem, A., and Shuler, M. L. (2000). Combining cell culture analogue reactor designs and PBPK models to probe mechanisms of naphthalene toxicity. *Biotechnol. Prog.* **16**, 334–345.

Haddad, S., Charest-Tardif, G., Tardif, R., and Krishnan, K. (2000). Validation of a physiological modeling framework for simulating the toxicokinetics of chemicals in mixtures. *Toxicol. Appl. Pharmacol.* **167**, 199–209.

Haddad, S., Tardif, R., Charest-Tardif, G., and Krishnan, K. (1999). Physiological modeling of the toxicokinetic interactions in a quaternary mixture of aromatic hydrocarbons. *Toxicol. Appl. Pharmacol.* **161**, 249–257.

Haddad, S., Withey, J., Lapare, S., Law, F., and Krishnan, K. (1998). Physiologically-based pharmacokinetic modeling of pyrene in the rat. *Environ. Toxicol. Pharmacol.* **5**, 245–255.

Hallenbeck, W. H. (1992). Cancer risk assessment for the inhalation of 1,3-butadiene using physiologically based pharmacokinetic modeling. *Bull. Environ. Contam. Toxicol.* **49**, 66–70.

Hattis, D., and Wasson, J. (1987). *Pharmacokinetic/Mechanism-Based Analysis of the Carcinogenic Risk of Butadiene.* NTIS Report #PB88-202817. November 1987. Center for Technology, Policy and Industrial Development, Massachusetts Institute of Technology, Cambridge, MA. CTPID, 1987.

Hetrick, D. M., Jarabek, A. M., and Travis, C. C. (1991). Sensitivity analysis for physiologically based pharmacokinetic models. *J. Pharmacokinet. Biopharm.* **19**, 1–20.

Hilderbrand, R. L., Andersen, M. E., and Jenkins, L. J. (1981). Prediction of *in vivo* kinetic constants for metabolism of inhaled vapors from kinetic constants measured *in vitro*. *Fund. Appl. Toxicol.* **1**, 403–409.

Hissink, E., van Ommen, B., Bogaards, J. J., and van Bladeren, P. J. (1996). Hepatic epoxide concentrations during biotransformation of 1,2- and 1,4-dichlorobenzene. The use of *in vitro* and *in vivo* metabolism, kinetics and PB-PK modeling. *Adv. Exp. Med. Biol.* **387**, 129–133.

Hissink, A. M., van Ommen, B., Kruse, J., and van Bladeren, P. J. (1997). A physiologically based pharmacokinetic (PB-PK) model for 1,2-dichlorobenzene linked to two possible parameters of toxicity. *Toxicol. Appl. Pharmacol.* **145**, 301–310.

Jackson, T. E., Lilly, P. D., Recio, L., Schlosser, P. M., and Medinsky, M. A. (2000). Inhibition of cytochrome P450 2E1 decreases, but does not eliminate, genotoxicity mediated by 1,3-butadiene. *Toxicol. Sci.* **55**, 266–273.

Jang, J. Y., and Droz, P. O. (1997). Ethnic differences in biological monitoring of several organic solvents. 2. A simulation study with a physiologically based pharmacokinetic model. *Int. Arch. Occup. Environ. Health* **70**, 41–50.

Jang, J. Y., Droz, P. O., and Chung, H. K. (1999). Uncertainties in physiologically based pharmacokinetic models caused by several input parameters. *Int. Arch. Occup. Environ. Health* **72**, 247–254.

Johanson, G., and Filser, J. G. (1993). A physiologically based pharmacokinetic model for butadiene and its metabolite butadiene monoxide in rat and mouse and its significance for risk extrapolation. *Arch. Toxicol.* **67**, 151–163.

Johanson, G., and Filser, J. G. (1996). PBPK model for butadiene metabolism to epoxides: Quantitative species differences in metabolism. *Toxicology* **113**, 40–47.

Johanson, G., and Naslund, P. H. (1988). Spreadsheet programming—A new approach in physiologically based modeling of solvent toxicokinetics. *Toxicol. Lett.* **41**, 115–127.

Jonsson, F., and Johanson, G. (2002). Physiologically based modeling of the inhalation kinetics of styrene in humans using a bayesian population approach. *Toxicol. Appl. Pharmacol.* **179**, 35–49.

Kadlubar, F. F., Dooley, K. L., Teitel, C. H., Roberts, D. W., Benson, R. W., Butler, M. A., Bailey, J. R., Young, J. F., Skipper, P. W., and Tannenbaum, S. R. (1991). Frequency of urination and its effects on metabolism, pharmacokinetics, blood hemoglobin adduct formation, and liver and urinary bladder DNA adduct levels in beagle dogs given the carcinogen 4-aminobiphenyl. *Cancer Res.* **51**, 4371–4377.

Kedderis, G. L., Teo, S. K., Batra, R., Held, S. D., and Gargas, M. L. (1996). Refinement and verification of the physiologically based dosimetry description for acrylonitrile in rats. *Toxicol. Appl. Pharmacol.* **140**, 422–435.

Kenyon, E. M. and Medinsky, M. A. (1995). Incorporation of heterogeneous enzyme distribution into a physiological model for benzene, phenol, and hydroquinone. Abstract 263. *Toxicologist*, **15**(1).

Kenyon, E. M., Kraichely, R. E., Hudson, K. T., and Medinsky, M. A. (1996). Differences in rates of benzene metabolism correlate with observed genotoxicity. *Toxicol. Appl. Pharmacol.* **136**, 49–56.

Kinkead, E. R., Wall, H. G., Hixson, C. J., Tice, R. R., Kutzman, R. S., and Vinegar, A. (1990). Chloropentafluorobenzene—Short-term inhalation toxicity, genotoxicity and physiologically-based pharmacokinetic model development. *Toxicol. Ind. Health* **6**, 533–550.

Kirman, C. R., Hays, S. M., Kedderis, G. L., Gargas, M. L., and Strother, D. E. (2000). Improving cancer dose-response characterization by using physiologically based pharmacokinetic modeling: an analysis of pooled data for acrylonitrile-induced brain tumors to assess cancer potency in the rat. *Risk Anal.* **20**, 135–151.

Kirman, C. R., Gargas, M. L., Deskin, R., Tonner-Navarro, L., and Andersen, M. E. (2003). A physiologically based pharmacokinetic model for acrylamide and its metabolite, glycidamide, in the rat. *J. Toxicol. Environ. Health Pt. A* **66**, 253–274.

Kohn, M. C. (1997). The importance of anatomical realism for validation of physiological models of disposition of inhaled toxicants. *Toxicol. Appl. Pharmacol.* **147**, 448–458.

Kohn, M. C., and Melnick, R. L. (1993). Species-differences in the production and clearance of 1,3-butadiene metabolites—A mechanistic model indicates predominantly physiological, not biochemical, control. *Carcinogenesis* **14**, 619–628.

Kohn, M. C., and Melnick, R. L. (1996). Effects of the structure of a toxicokinetic model of butadiene inhalation exposure on computed production of carcinogenic intermediates. *Toxicology* **113**, 31–39.

Kohn, M. C., and Melnick, R. L. (2000). The privileged access model of 1,3-butadiene disposition. *Environ. Health Perspect.* **108**, 911–917.

Kohn, M. C., and Melnick, R. L. (2001). Physiological modeling of butadiene disposition in mice and rats. *Chem. Biol. Interact.* **135–136**, 285–301.

Krewski, D., Wang, Y., Bartlett, S., and Krishnan, K. (1995). Uncertainty, variability, and sensitivity analysis in physiological pharmacokinetic models. *J. Biopharm. Stat.* **5**, 245–271.

Krishnan, K., Gargas, M. L., Fennell, T. R., and Andersen, M. E. (1992). A physiologically based description of ethylene-oxide dosimetry in the rat. *Toxicol. Ind. Health* **8**, 121–140.

Kumagai, S., and Matsunaga, I. (1995). Effect of variation of exposure to airborne chlorobenzene on internal exposure and concentrations of urinary metabolite. *Occup. Environ. Med.* **52**, 65–70.

Law, F. C., Abedini, S., and Kennedy, C. J. (1991). A biologically based toxicokinetic model for pyrene in rainbow trout. *Toxicol. Appl. Pharmacol.* **110**, 390–402.

Leavens, T. L., and Bond, J. A. (1996). Pharmacokinetic model describing the disposition of butadiene and styrene in mice. *Toxicology* **113**, 310–313.

Lindstrom, A. B., Yeowell-O'Connell, K., Waidyanatha, S., Golding, B. T., Tornero-Velez, R., and Rappaport, S. M. (1997). Measurement of benzene oxide in the blood of rats following administration of benzene. *Carcinogenesis* **18**, 1637–1641.

Lof, A., and Johanson, G. (1993). *Dose-Dependent Kinetics of Inhaled Styrene in Man*, IARC Scientific Publications, Lyon, Vol. 127, pp. 89–99.

Luu, H. M., Hutter, J. C., and Bushar, H. F. (1998). A physiologically based pharmacokinetic model for 2,4-toluenediamine leached from polyurethane foam-covered breast implants. *Environ. Health Perspect.* **106**, 393–400.

McMahon, T. F., Medinsky, M. A., and Birnbaum, L. S. (1994). Age-related changes in benzene disposition in male C57bl/6n mice described by a physiologically-based pharmacokinetic model. *Toxicol. Lett.* **74**, 241–253.

Medinsky, M. A., Bond, J. A., Hunsberger, S., and Griffith, W. C., Jr. (1989a). A physiologically based model of 1-nitropyrene metabolism after inhalation or ingestion. *Health Phys.* **57**, 149–155.

Medinsky, M. A., Sabourin, P. J., Henderson, R. F., Lucier, G., and Birnbaum, L. S. (1989b). Differences in the pathways for metabolism of benzene in rats and mice simulated by a physiological model. *Environ. Health Perspect.* **82**, 43–49.

Medinsky, M. A., Sabourin, P. J., Lucier, G., Birnbaum, L. S., and Henderson, R. F. (1989c). A physiological model for simulation of benzene metabolism by rats and mice. *Toxicol. Appl. Pharmacol.* **99**, 193–206.

Medinsky, M. A., Sabourin, P. J., Lucier, G., Birnbaum, L. S., and Henderson, R. F. (1989d). A toxikinetic model for simulation of benzene metabolism. *Exp. Pathol.* **37**, 150–154.

Medinsky, M. A., Leavens, T. L., Csanady, G. A., Gargas, M. L., and Bond, J. A. (1994). *In vivo* metabolism of butadiene by mice and rats: A comparison of physiological model predictions and experimental data. *Carcinogenesis* **15**, 1329–1340.

Melnick, R. L., and Kohn, M. C. (2000). Dose-response analyses of experimental cancer data. *Drug Metab. Rev.* **32**, 193–209.

Moir, D. (1999). Physiological modeling of benzo(a)pyrene pharmacokinetics in the rat. *Toxic. Assess. Altern.* 79–95.

Morris, J. B., Hassett, D. N., and Blanchard, K. T. (1993). A physiologically based pharmacokinetic model for nasal uptake and metabolism of nonreactive vapors. *Toxicol. Appl. Pharmacol.* **123**, 120–129.

NTP (1999). *Toxicology and carcinogenicity studies of isoprene (CAS No. 78-79-5) in F344/N rats (inhalation studies), Technical Report NTP-TR-486.* NIH Publication No. 97-3976, National Institutes of Health, Research Triangle Park, NC, pp. 1–176.

Parkinson, A. (2001). Biotransformation of Xenobiotics. In: *Casarett and Doull's Toxicology: The Basic Science of Poisons*, 5th ed., C. D. Klaassen, ed.; McGraw-Hill, New York p. 199.

Paterson, S., and Mackay, D. (1986). A pharmacokinetic model of styrene inhalation with the fugacity approach. *Toxicol. Appl. Pharmacol.* **82**, 444–453.

Perbellini, L., Mozzo, P., Turri, P. V., Zedde, A., and Brugnone, F. (1988). Biological exposure index of styrene suggested by a physiologico-mathematical model. *Int. Arch. Occup. Environ. Health* **60**, 187–193.

Pierce, C. H., Becker, C. E., Tozer, T. N., Owen, D. J., and So, Y. (1998). Modeling the acute neurotoxicity of styrene. *J. Occup. Env. Med.* **40**, 230–240.

Purcell, K. J., Cason, G. H., Gargas, M. L., Andersen, M. E., and Travis, C. C. (1990). *In vivo* metabolic interactions of benzene and toluene. *Toxicol. Lett.* **52**, 141–152.

Quick, D. J., and Shuler, M. L. (1999). Use of *in vitro* data for construction of a physiologically based pharmacokinetic model for naphthalene in rats and mice to probe species differences. *Biotechnol. Prog.* **15**, 540–555.

Ramsey, J. C., and Andersen, M. E. (1984). A physiologically based description of the inhalation pharmacokinetics of styrene in rats and humans. *Toxicol. Appl. Pharmacol.* **73**, 159–175.

Roth, R. A., and Vinegar, A. (1990). Action by the lungs on circulating xenobiotic agents, with a case-study of physiologically based pharmacokinetic modeling of benzo(*a*)pyrene disposition. *Pharmacol. Ther.* **48**, 143–155.

Roth, W. L., Freeman, R. A., and Wilson, A. G. (1993). A physiologically based model for gastrointestinal absorption and excretion of chemicals carried by lipids. *Risk Anal.* **13**, 531–543.

Sarangapani, R., Teeguarden, J. G., Cruzan, G., Clewell, H. J., and Andersen, M. E. (2002a). Physiologically based pharmacokinetic modeling of styrene and styrene oxide respiratory-tract dosimetry in rodents and humans. *Inhal. Toxicol.* **14**, 789–834.

Sarangapani, R., Clewell, H. J., Cruzan, G., and Andersen, M. E. (2002b). Comparing respiratory-tract and hepatic exposure–dose relationships for metabolized inhaled vapors: A pharmacokinetic analysis. *Inhal. Toxicol.* **14**, 835–854.

Seaton, M. J., Schlosser, P., and Medinsky, M. A. (1995). *In vitro* conjugation of benzene metabolites by human liver: Potential influence of interindividual variability on benzene toxicity. *Carcinogenesis* **16**, 1519–1527.

Selinski, S., Golka, K., Bolt, H. M., and Urfer, W. (2000). Estimation of toxicokinetic parameters in population models for inhalation studies with ethylene. *Environmetrics* **11**, 479–495.

Sherwood, R. J., and Sinclair, G. C. (1999). New PBPK model applied to old occupational exposure to benzene. *Am. Ind. Hyg. Assoc. J.* **60**, 259–265.

Sinclair, G. C., Gray, C. N., and Sherwood, R. J. (1999). Structure and validation of a pharmacokinetic model for benzene. *Am. Ind. Hyg. Assoc. J.* **60**, 249–258.

Smith, M. T. (1996). The mechanism of benzene-induced leukemia: a hypothesis and speculations on the causes of leukemia. *Environ. Health Perspect.* **104**(Suppl 6), 1219–1225.

Smith, T. J., Lin, Y. S., Mezzetti, M., Bois, F. Y., Kelsey, K., and Ibrahim, J. (2001). Genetic and dietary factors affecting human metabolism of 1,3-butadiene. *Chem. Biol. Interact.* **135–136**, 407–428.

Spear, R. C., and Bois, F. Y. (1994). Parameter variability and the interpretation of physiologically based pharmacokinetic modeling results. *Environ. Health Perspect.* **102**(Suppl 11), 61–66.

Spear, R. C., Bois, F. Y., Woodruff, T., Auslander, D., Parker, J., and Selvin, S. (1991). Modeling benzene pharmacokinetics across three sets of animal data: parametric sensitivity and risk implications. *Risk Anal.* **11**, 641–654.

Sun, J. D., Medinsky, M. A., Birnbaum, L. S., Lucier, G., and Henderson, R. F. (1990). Benzene hemoglobin adducts in mice and rats: Characterization of formation and physiological modeling. *Fund. Appl. Toxicol.* **15**, 468–475.

Sweeney, L. M., Shuler, M. L., Babish, J. G., and Ghanem, A. (1995). A cell-culture analog of rodent physiology—Application to naphthalene toxicology. *Toxicol. Vitro* **9**, 307–316.

Sweeney, L. M., Himmelstein, M. W., Schlosser, P. M., and Medinsky, M. A. (1996a). Physiologically based pharmacokinetic modeling of blood and tissue epoxide measurements for butadiene. *Toxicology* **113**, 318–321.

Sweeney, L. M., Shuler, M. L., Quick, D. J., and Babish, J. G. (1996b). A preliminary physiologically based pharmacokinetic model for naphthalene and naphthalene oxide in mice and rats. *Ann. Biomed. Eng.* **24**, 305–320.

Sweeney, L. M., Schlosser, P. M., Medinsky, M. A., and Bond, J. A. (1997). Physiologically based pharmacokinetic modeling of 1,3-butadiene, 1,2-epoxy-3-butene, and 1,2:3,4-diepoxybutane toxicokinetics in mice and rats. *Carcinogenesis* **18**, 611–625.

Sweeney, L. M., Himmelstein, M. W., and Gargas, M. L. (2001). Development of a preliminary physiologically based toxicokinetic (PBTK) model for 1,3-butadiene risk assessment. *Chem. Biol. Interact.* **135–136**, 303–322.

Sweeney, L. M., Gargas, M. L., Strother, D. E., and Kedderis, G. L. (2003). Physiologically based pharmacokinetic model parameter estimation and sensitivity and variability analyses for acrylonitrile disposition in humans. *Toxicol. Sci.* **71**, 27–40.

Their, R., and Bolt, H. F. (2000). Carcinogenicity and genotoxicity of ethylene oxide: New aspects and recent advances. *Crit. Rev. Toxicol.* **30**, 595–608.

Thomas, R. S., Bigelow, P. L., Keefe, T. J., and Yang, R. S. H. (1996). Variability in biological exposure indices using physiologically based pharmacokinetic modeling and Monte Carlo simulation. *Am. Ind. Hyg. Assoc. J.* **57**, 23–32.

Tornero-Velez, R., and Rappaport, S. M. (2001). Physiological modeling of the relative contributions of styrene-7,8-oxide derived from direct inhalation and from styrene metabolism to the systemic dose in humans. *Toxicol. Sci.* **64**, 151–161.

Travis, C. C., Craig, P. H., and Bowers, J. C. (1991). Characterization of human exposure to ambient levels of benzene using pulmonary "wash-out" data. *Atmos. Environ.* **25A**, 1643–1647.

Travis, C. C., Fox, M. T., Simmons, W. M., and Lyon, B. F. (1992). Coexposure to gasoline vapor decreases benzene metabolism in Fischer-344 rats. *Toxicol. Lett.* **62**, 231–240.

Travis, C. C., Quillen, J. L., and Arms, A. D. (1990). Pharmacokinetics of benzene. *Toxicol. Appl. Pharmacol.* **102**, 400–420.

Vinegar, A., Winsett, D. W., Andersen, M. E., and Conolly, R. B. (1990). Use of a physiologically based pharmacokinetic model and computer simulation for retrospective assessment of exposure to volatile toxicants. *Inhal. Toxicol.* **2**, 119–128.

Wang, Y., Kupper, L. L., Lof, A., and Rappaport, S. M. (1996). Comparison of average estimated metabolic rates for styrene in previously exposed and unexposed groups with pharmacokinetic modelling. *Occup. Environ. Med.* **53**, 601–605.

Weisel, C., Yu, R., Roy, A., and Georgopoulos, P. (1996). Biomarkers of environmental benzene exposure. *Environ. Health Perspect.* **104**, 1141–1146.

Willems, B. A., Melnick, R. L., Kohn, M. C., and Portier, C. J. (2001). A physiologically based pharmacokinetic model for inhalation and intravenous administration of naphthalene in rats and mice. *Toxicol. Appl. Pharmacol.* **176**, 81–91.

Woodruff, T. J., and Bois, F. Y. (1993). Optimization issues in physiological toxicokinetic modeling—a case-study with benzene. *Toxicol. Lett.* **69**, 181–196.

Woodruff, T. J., Bois, F. Y., Auslander, D., and Spear, R. C. (1992). Structure and parameterization of pharmacokinetic models: Their impact on model predictions. *Risk Anal.* **12**, 189–201.

CHAPTER 5

REACTIVE VAPORS IN THE NASAL CAVITY

James E. Dennison and Melvin E. Andersen

5.1 INTRODUCTION

5.1.1 Nasal Effects and Risk Assessment

Several PBPK models have been developed for chemicals that are of concern due to toxicity to tissues lining the nasal passage. With some of these chemicals, various nasal epithelial responses in animals, including toxicity to respiratory and olfactory epithelial tissues and cancer, have served as the basis for risk assessments. In the absence of compound-specific information on nasal tissue dosimetry, exposure limits have been established for these types of compounds using defaults, such as the US EPA Reference Concentration (RfC) methodology (US EPA 1994) to extrapolate delivered dose from rats to humans. These default approaches may ignore some features of the biological processes that determine toxicity, such as differences in regional uptake of chemical into nasal tissue, bioactivation of chemicals within tissues, and differing modes of irritant action of the compounds in the epithelial

Physiologically Based Pharmacokinetic Modeling, edited by Micaela B. Reddy, Raymond S. H. Yang, Harvey J. Clewell III, and Melvin E. Andersen.

tissues. Differences in these various processes between rat and human noses may be substantial, and they need to be taken into account to assess the risks posed by these chemicals in humans. Thus, PBPK models for these inhaled compounds need to be based on the biological processes that are connected to suspected modes of action in the nose in order to have a biologically based platform for risk assessment extrapolation.

The history of the development of these nasal PBPK models is similar to that for many others types of chemicals, starting with simpler models and moving toward more complex structures. Here we focus on specific PBPK models developed for vinyl acetate (VA), ethyl acrylate (EA), acrylic acid (AA), methyl methacrylate (MMA), formaldehyde (HCHO), epichlorohydrin (EPI), and hydrogen sulfide (H_2S). Some modeling efforts for respiratory tract uptake of compounds are not included in this chapter. A model for ozone described tissue uptake throughout the deeper regions of the respiratory tract focusing more on fluxes to tissues along the walls of the tract rather than tissue phase processes (Miller *et al.* 1985; Overton *et al.* 1987). In addition, other chemicals such as styrene have effects on nasal epithelial tissues. PBPK modeling efforts with styrene include airway uptake of compound and metabolism in nasal tissues. Models for styrene are discussed in Chapter 4.

5.1.2 General Models for Nasal Uptake

A simple flow-through tube model for nasal uptake of gases and vapors was first described over 30 years ago (Aharonson *et al.* 1974). This model included flow rates through the nose and net mass transfer from the air phase into the tissue. The description was similar to parallel tube models for liver clearance that were developed in the same period (Pang and Rowland 1977). The net mass transfer from air to epithelial tissues for any particular chemical depends on air-phase mass transfer, tissue diffusion, tissue metabolism/reactivity, and tissue blood flow that can carry off absorbed compounds in the venous effluent blood from the nasal tissues. These models predict a concentration gradient along the longitudinal direction of the tube. This initial modeling focused on uptake of soluble vapors, including acetone and sulfur dioxide, and technically is not regarded as a PBPK model for the air spaces or the tissue phases.

Nonetheless, all subsequent PBPK models for nasal uptake of gases and vapors have had to account for four specific characteristics that are explicitly or implicitly contained in this simpler model. First, there is the presence of an air phase from which compounds are removed. Second, the air stream passes over tissue surfaces where uptake along the tissue reduces concentrations that can enter subsequent regions. Third, the chemicals have to pass between phases, air to liquid, to gain access to the sites of action. Fourth, the tissues are not as well perfused with blood as many internal organs and tissues, requiring consideration of diffusion from the air–mucus interface into deeper tissues and the bloodstream. In the tube model these last two considerations are all subsumed in the net mass transfer coefficient estimated for uptake along the tube. Over the past decade, PBPK models developed for vapor uptake in the nose have wrestled with approaches to consider these various tissue characteristics.

5.1.2.1 *Air Phase* The initial PBPK models for nasal uptake (Morris *et al.* 1993; Plowchalk *et al.* 1997) lacked an explicitly defined air phase. In these descriptions the entire mass of chemical in the air equilibrated with the mucus layer. Anatomical studies and quantitative fluid flow calculations, based on computational fluid dynamic (CFD) modeling, subsequently provided the basic information needed to describe the volume of nasal air spaces (meatuses) and the gas flow rates between these nasal regions (Kimbell *et al.* 1993). In many of the more recent PBPK models for nasal uptake, computational results from CFD models have been used to ascribe volumes of air spaces and input and efflux airflow rates to specific regions within the nose.

5.1.2.2 *Specific Nasal Regions* The Aharonson *et al.* (1974) tube model had a single compartment with one airflow stream. In most of the other PBPK models, the ability to account for gradients along the nose was determined by the number of nasal epithelial regions included in the descriptions. Most models break the flow into a dorsal and ventral pathway, especially when attempting to assess responses of the olfactory epithelium (Frederick *et al.* 1998). The dorsal air stream passes over a region of respiratory epithelium (described as a single compartment) and then on to an olfactory region. Each of these air-phase regions is treated as well-mixed, so the gradient is artificially achieved with single concentration values over the dorsal respiratory epithelium and a single concentration over the olfactory epithelium. Some models only evaluate uptake for unidirectional flow. Others include bidirectional flow assessing uptake over a more realistic breathing pattern. The numbers of nasal air regions in the models vary considerably. The model with the largest number of separate regions was developed for HCHO where a CFD model was used to account for the mass fluxes to tissues and tissue regions with similar mass fluxes were lumped together in "bins" (Kimbell *et al.* 2001a). The flux in each bin was then used as an input into the tissue compartments in each of the binned areas.

5.1.2.3 *Air-Phase Mass Transfer Coefficients* A difficult challenge in creating the initial PBPK models for nasal uptake was estimating the mass transfer coefficient from air to the tissue. The uptake under any particular exposure situation depends on various air-phase and tissue-phase factors. However, the air-phase mass transfer coefficient is one of the most important, especially for highly extracted compounds. Morris *et al.* (1993) and Plowchalk *et al.* (1997) implicitly assumed an air-phase mass transfer that is very large with respect to airflow rates through the nose. Frederick *et al.* (1998) were the first to estimate the air-phase mass transfer coefficient for each region from information provided by CFD model calculations with a CFD boundary condition of zero concentration at the wall. This boundary provides maximal uptake from each compartment. For a well-mixed compartment, the maximum extraction in the compartment is the air-phase mass transfer coefficient times surface area (MTC × SA) divided by the sum of this term and the airflow (Q_{air}) through the compartment.

$$\text{Extraction maximum} = (\text{MTC} \times \text{SA}) \big/ ((\text{MTC} \times \text{SA}) + Q_{air}) \qquad (5.1)$$

The models for AA, EA, and MMA and the later work with VA used this approach to estimate the air-phase mass transfer coefficient.

5.1.2.4 Interfacial Mass Transfer Coefficient Chemical is transferred from the lumen into the tissue by dissolving in mucus and transporting deeper into tissue by diffusion. In the simplest representation of the interfacial mass transfer coefficient (IMTC), for an otherwise inert tissue phase, the net mass transfer coefficient contains terms for resistance to mass flux given by the air-phase diffusion, the tissue diffusion coefficient, and the tissue: gas-phase partition coefficient (Henry's law constant). This representation, however, requires adjustment when there is reactivity of the compound in the tissue boundary (Bush *et al.* 1998).

The models for AA, VA [except for the Plowchalk *et al.* (1997) description], and EA use the simple description of the IMTC. With MMA, the input to the mucus/uppermost tissue layer in the epithelial tissues was provided by the air-phase mass transfer coefficient with tissue clearance by metabolism and diffusion in the uppermost layer. In this approach the IMTC is not provided as an input to the PBPK model. Instead, it is a consequence of metabolism, ionization, and diffusion characteristics that are included in the description of the air phase and the tissue phase. The IMTC becomes an especially important issue for compounds like VA or styrene which have moderate tissue solubility and high uptake due to clearance by metabolizing enzymes in the mucus layer or in tissue layers adjacent to the air phase in some regions of the nasal epithelium. In the approach used with MMA, the first tissue compartment in contact with air becomes the boundary which determines the IMTC.

5.1.2.5 Tissue Diffusion To account for diffusion through a significant depth of tissue before chemical becomes available for equilibration in venous blood, the tissues are subdivided into a variable number of compartments as the chemical passes deeper into the nasal epithelium. The Morris *et al.* (1993) and Plowchalk *et al.* (1997) models used 10 μm slices of tissues that were treated as separate well-mixed compartments. By dividing into a number of slices, the series of well-mixed compartments approximates diffusion through the tissues. With separations of 10 μm, the intercompartmental mass transfer can still be described using the tissue phase diffusion coefficient for the compound. It is also possible to place enzymatic activities into the various subcompartments based on known distributions of enzymes throughout the nose, as was done for several enzyme activities involved in metabolizing VA or its metabolites (Plowchalk *et al.* 1997).

A review of many of these issues and modeling considerations, including the issues of physiology, cell and tissue differences, enzyme localization, and mass transfer coefficients, has been published (Bogdanffy and Sarangapani 2003).

5.2 NO AIR-PHASE MODELS

5.2.1 The "Perfused Nose" Model

An early PBPK model for nasal absorption and metabolism in the F344 rat was developed by Morris *et al.* (1993). This model, based on earlier formulations that did not incorporate metabolism in nasal tissues (Gerde and Dahl 1991), set forth a general paradigm for the nasal tissue PBPK model structures. While equations that describe movement of chemical from the airway lumen into tissues varied in later models, the structural paradigm for tissue compartments in this model has persisted with various modifications into most subsequent models for nasal uptake of gases and vapors.

Morris *et al.* (1993) described the nose as consisting of separate respiratory and olfactory tissues in two distinct airflow pathways (Fig. 5.1). One pathway contained the dorsal/medial airflow in which the inspired air first contacted a section of respiratory mucosa and then contacted a section of olfactory mucosa. In the other pathway, the lateral/ventral pathway, only respiratory mucosa was present. Total ventilation was divided between these two airflows. These divisions were a simplification, but were based on anatomical considerations.

Each section of mucosa was further divided into a series of stacked compartments that represented mucus, epithelial tissues, and submucosal tissues, with multiple compartments of epithelium and/or submucosa, the exact number of which depended on the mucosa section. Of these, only the submucosal tissue was perfused. Metabolic activity in nasal tissue varies with cell type, so the activity of various enzymes, including cytochrome P450 (CYP P450), carboxylesterase, and alcohol dehydrogenase, was assigned to submucosal or epithelial compartments, depending on the tissues (olfactory or respiratory), in accordance with earlier studies. In some later uses of this model structure, metabolism was also included in the mucus/uppermost tissue layer.

Thus, the model included three sections of mucosa, each of which was comprised of some number of vertically-stacked compartments. Each slice of tissue in

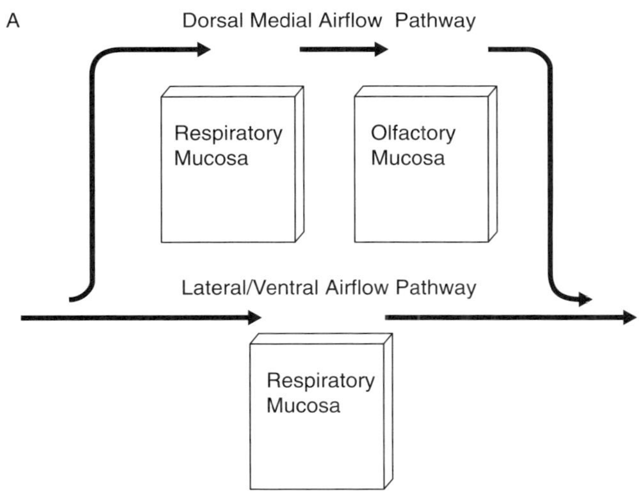

Figure 5.1 The two-airflow pathway model of the rat nose in Morris *et al.* (1993). The dorsal medial pathway had 8% of the total flow passing over respiratory and then olfactory mucosa. The lateral ventral pathway represented a composite of several pathways, and it had the remaining flow passing over respiratory mucosa only. Reproduced with permission from *Toxicology and Applied Pharmacology*.

each section was a compartment just like a "liver compartment" in a classical model, as the chemical mass balance was maintained for each slice, and chemical flux from proximal (closer to airway lumen) or distal (closer to systemic circulation) slices occurred. During exposure, chemicals were first absorbed into the mucus compartment according to the mucus:air partition coefficient. They then diffused into lower compartments where metabolism could occur. Upon reaching the submucosa, they could partition into blood and be cleared from the tissues. As this model did not include other compartments such as liver, it was essentially a "perfused nose" PBPK model.

The important feature of the Morris *et al.* (1993) model is that the chemical uptake from the air phase is treated as airflow-limited (i.e., net mass transfer from air to tissue is assumed to be much greater than diffusion from air to tissues). The Morris *et al.* (1993) model was calibrated and validated using data that described the fractional uptake of chemical in the upper respiratory tract. Fractional uptake was determined by measuring the concentration differences between the chemical entering and leaving the nasal passage through a cannula. Datasets of this absorption for xylene, ethyl acetate, isoamyl alcohol, acetone, and bromobenzene were selected as representatives of relatively nonmetabolized chemicals or chemicals that are metabolized by CYP P450s, esterases, and alcohol dehydrogenase. An excellent correlation between predicted and measured uptake was obtained. These calibration datasets are limited to total extraction over the entire nose rather than regional information of concentration gradients in the nose.

Another PBPK model similar to the model described by Morris *et al.* (1993) has been published (Overton 1990), although this model included the whole respiratory tract instead of focusing on the nose. In each of three regions—the upper respiratory tract, tracheobronchial region, and the pulmonary region—there were three compartments. One compartment was for the air space in the corresponding region, and there were two tissue compartments in parallel, both with perfusion. An interesting feature of the model was that the air space compartments were not of fixed size; they expanded or contracted with cyclic breathing cycles modeled as square waves. The model was developed with uptake data for styrene and was also used to model HCHO uptake (Overton 2001).

5.2.2 Vinyl Acetate

VA has been investigated with PBPK models to conduct a risk assessment for carcinogenic responses seen in the rat nose. Based on histopathology and other evidence, the tumors were observed at concentrations that caused toxicity in the nasal epithelium. In particular, regenerative proliferation after cytotoxicity was noted and has been ascribed to cellular acidification during metabolism of inhaled VA. VA is metabolized by carboxylesterases in nasal epithelial tissues to form acetaldehyde (AAld) and then by aldehyde dehydrogenase to acetic acid. While AAld is weakly clastogenic, acetic acid is primarily associated with cytotoxicity due to intracellular acidification, toxicity, and altered cell proliferation. Therefore, the general objective of PBPK models was to describe uptake of VA in different tissue types, the ensuing concentrations of VA, AAld, and acetic acid in tissues, and presumed pH changes.

Models for VA drew from the original model structure for the nasal compartment developed by Morris *et al.* (1993). The first PBPK model in a series of VA models (Plowchalk *et al.* 1997) had the same airflow patterns as Morris *et al.* (1993). The mucosal tissue compartments were also similar to those in Morris *et al.* (1993). The respiratory epithelium included a single mucus layer and three epithelial cell, one basal cell, and two submucosal cell compartments. The olfactory tissue contained the same stacked compartments, except that there were four epithelial tissue compartments instead of three (Fig. 5.2). Perfusion was included in the submucosal tissues.

Histochemical studies showed that enzyme activity was not uniform throughout different tissue types in the nasal compartment. Based on these studies, carboxylesterase activity determined *in vitro* was divided between mucus, epithelium, and basal cell compartments in the respiratory mucosa and between all compartments in olfactory mucosa. Aldehyde dehydrogenase activity was included in epithelium and basal cells of the respiratory mucosa and in basal cells and submucosa in olfactory epithelium. Acetyl-CoA synthetase activity, which served to clear acetic acid from the tissues, was distributed throughout all tissues.

Cellular acidification was included to determine the concentration of H^+ in each compartment. Cytoplasmic buffering capacity was estimated, and Na^+/H^+ antiporter activity was estimated and included to yield an estimate of intracellular pH. Transport of VA, AAld, and acetic acid in tissues was modeled using diffusional approaches.

The model was validated against whole-nose uptake data for VA uptake and for appearance of AAld. Similar to the uptake data used by Morris *et al.* (1993), the concentration of chemical leaving the upper respiratory tract represented the amount of VA depleted due to tissue uptake and the amount of AAld diffusing out of mucus

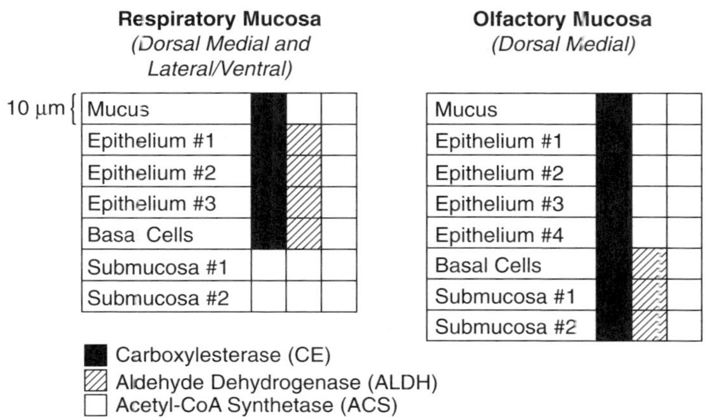

Figure 5.2 Conceptual representation of tissue stacks in the respiratory and olfactory mucosa in Plowchalk *et al.* (1997). Histochemical experiments were performed to assay enzyme activity for assignment to mucus, epithelium, basal cell, or submucosal cells, as indicated by shading. Reproduced with permission from *Toxicology and Applied Pharmacology*.

after metabolism in mucus or in deeper tissue compartments. The model was calibrated by comparison with steady-state nasal uptake studies at exposures from 76 to 2070 ppm. The final model reasonably simulated the exiting concentrations of VA and AAld at all levels, with the exception of AAld at the highest concentration. Treatment with a carboxylesterase inhibitor decreased VA uptake by tissues, but suggested that inhibition was not complete. The model predicted that pH would decrease from 7.4 to about 6.3 in respiratory epithelium and to 6.7 in olfactory epithelium at an inhaled concentration of 600 ppm VA. As with models for other reactive vapors, calculations with the model for VA suggested that respiratory epithelium experience high chemical fluxes despite lower indications of toxicity.

5.3 CREATING THE AIR-PHASE COMPARTMENTS

5.3.1 Computational Fluid Dynamics

CFD is a mathematical modeling method used for calculating patterns of airflow and their velocities for varying geometries and pressures. In CFD modeling for nasal airflow, the interior air spaces in the nose are divided into an appropriate number of boxes with defined boundaries. Using physical properties of the fluid (air or liquid), the velocity of the fluid can be computed at the limits of each of the boxes. This description provides an anatomically realistic definition of all airflows within the nasal passage.

The first CFD model of the nose used nasal casts of F344 rats to determine the morphology of the airways (Kimbell *et al.* 1993). CFD simulations were conducted and verified with experimental data on fluid flow velocities in different regions of the nose. The model was then used to estimate the flux of HCHO to different regions of the nasal mucosa—that is, how much chemical was delivered to the air:mucus interface in different regions. A good correspondence was found between the predicted delivery of chemical to the surfaces and locations of tumors.

5.3.2 Estimating the Air-Phase Mass Transfer Coefficient

The first model to couple the CFD model of the airway with a PBPK model of the nose was designed for three "nonreactive" chemicals (Bush *et al.* 1998). In this model, the nose was described in a manner similar to that in the model by Morris *et al.* (1993). Four airflows pathways were included. Along each of these airflow paths there were four serial sections of respiratory or olfactory mucosa, each of which contained stacks of mucus, epithelial, and submucosal compartments. Validation of the model was performed against the nasal uptake data for the acetone, AA, and isoamyl alcohol (with enzyme inhibition) studies (Morris *et al.* 1993).

In the model for the three nonreactive vapors (Bush *et al.* 1998), equations were developed that included delivery of chemical to a tissue compartment with uptake to tissue given by a calculated IMTC from air to mucus. The composite mass transfer coefficient for AA included terms to account for tissue ionization of AA that decreases resistance to tissue diffusion. This model was not designed to account for metabolism in the tissue compartments.

The CFD model for the rat has been extensively used in PBPK modeling for HCHO and with other chemicals, as described below. It was also extended for applications with nonhuman primate and human nasal passages (Kimbell and Subramaniam 2001). The rat nasal mold was also used to predict deposition of particulates in the nose (Kelly *et al.* 2001).

5.3.3 Estimating Air-Phase Mass Transfer Coefficients—Acrylic Acid

In a subsequent model (Frederick *et al.* 1998), the air-phase mass transfer coefficient for various regions was estimated from Eq. (5.1). The CFD model was first used to estimate the maximum regional extraction. With the flow through the region known from the CFD model, the air-phase mass transfer coefficient was easily calculated. The estimates of the air-phase mass transfer coefficients were included into a hybrid PBPK-CFD model for AA that considered (a) tissue ionization of AA and (b) diffusion of both ionic forms of the chemical. Model validation was performed by adjusting the model parameters to account for the uptake of the slowly metabolized nonionizable compounds studied by Morris *et al.* (1993). The flow-dependent unidirectional uptake (fractional nasal deposition) was determined and compared with experimental results for acetone, bromobenzene, isoamyl alcohol, xylene, and AA (Fig. 5.3).

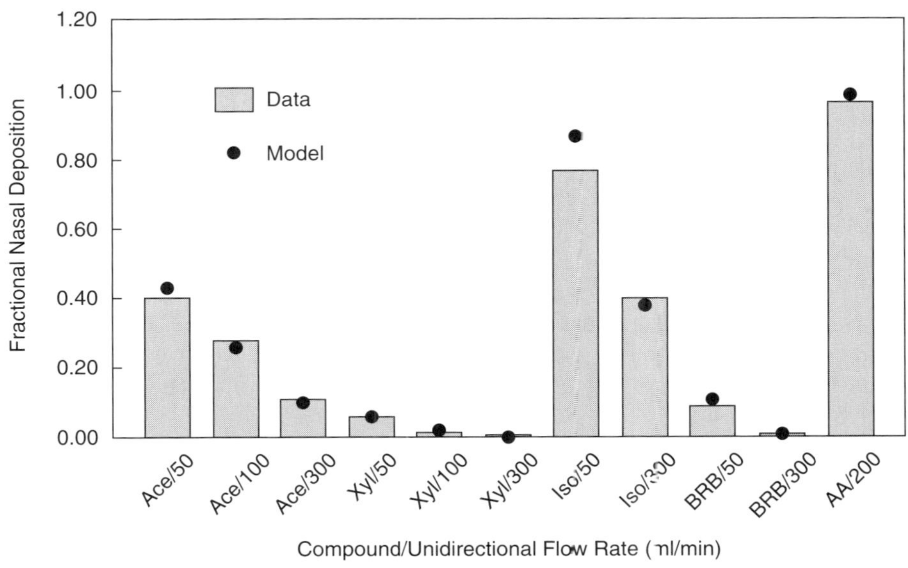

Figure 5.3 Nasal deposition data (shaded columns) and predictions by the model of Frederick *et al.* (1998) (solid circles). Inspiratory flow rates (ml/min) are listed after compound abbreviation. Ace: Acetone, a nonmetabolized vapor. Xyl: Xylene, a CYP P450 substrate. Iso: Iso-amyl alcohol, an alcohol dehydrogenase substrate. BRB: Bromobenzene, a CYP P450 substrate. AA: Acrylic acid. Reproduced with permission from *Toxicology and Applied Pharmacology*.

A more complete PBPK model for AA uptake and distribution in the nose was described in a separate article (Frederick *et al.* 2001). This model was used to evaluate the influence of cyclic versus unidirectional airflow and to calculate an RfC based on tissue dosimetry of AA (Andersen *et al.* 2000).

5.4 OTHER MODELS FOR VAPORS AFFECTING NASAL TISSUES

The following section describes the development of PBPK models that were primarily designed to understand the factors that control tissue uptake of vapors throughout the nose. Many of these models were then used to support various risk assessment calculations. We have not discussed major conclusions from the applications of these models to support risk assessments, emphasizing model structure and the choices of dose metrics rather than specific recommendations arising from use of the models.

5.4.1 Vinyl Acetate

To estimate the rate constants for metabolism of VA by carboxylesterase and of AAld by aldehyde dehydrogenase, an "*in vitro* headspace PBPK" model was developed (Bogdanffy *et al.* 1998). Tissue sections from human and rat olfactory and respiratory mucosa were placed in headspace vials. Headspace concentrations of VA (disappearance) and AAld (appearance) were analyzed with a PBPK model that included only the air and tissue compartments (six stacked tissues similar to those used in the *in vivo* models). Values of the maximum velocity of metabolism in rats and humans were similar in olfactory tissues but dissimilar in respiratory epithelium.

A refined PBPK model for VA was developed by Bogdanffy *et al.* (1999a) and scaled up to predict VA uptake and nasal metabolism in humans. Additional olfactory and/or respiratory epithelium compartments were included (depending on species) to allow for clearance in upstream regions. Air-phase resistance to transfer from air to mucus was also included. The model made use of the kinetic data obtained in the "*in vitro*" headspace PBPK model, and uptake experiments were performed to obtain refined uptake data for VA and AAld in the nose. The model was then used (1) to perform risk assessment calculations using a variety of dose metrics and (2) to examine possibility of CYP P450-mediated metabolism (Bogdanffy *et al.* 1999b).

5.4.2 Ethyl Acrylate and Its Metabolite, Acrylic Acid

EA is a reactive monomer used in production of several polymers. During chronic bioassays, low systemic toxicity was found, although irritation at the site of application was observed. One study did find tumors in the rodent forestomach. An early PBPK model for EA was developed to assess the uptake and distribution of orally administered EA and concomitant effects such as glutathione (GSH) depletion (Frederick *et al.* 1992). While this model did not formally include inhalation exposures, later models for EA as a reactive vapor were based in part of this original model structure.

The Frederick *et al.* (1992) EA PBPK model had a large number of physiological compartments (Fig. 5.4), including skin, kidney, muscle, liver, lung, fat, other perfused tissues, and a series of compartments representing the gastrointestinal tract. Metabolic clearance was represented by saturable metabolism via carboxylesterase and second-order reaction with GSH in each compartment. A GSH resynthesis model was included that also permitted induction of GSH synthesis. Binding to reactive proteins was a second order process. The liver was divided into three serial compartments. To model data from oral gavage studies with EA in corn oil, the model assumed that chemical would be transferred from an oily phase to the aqueous phase

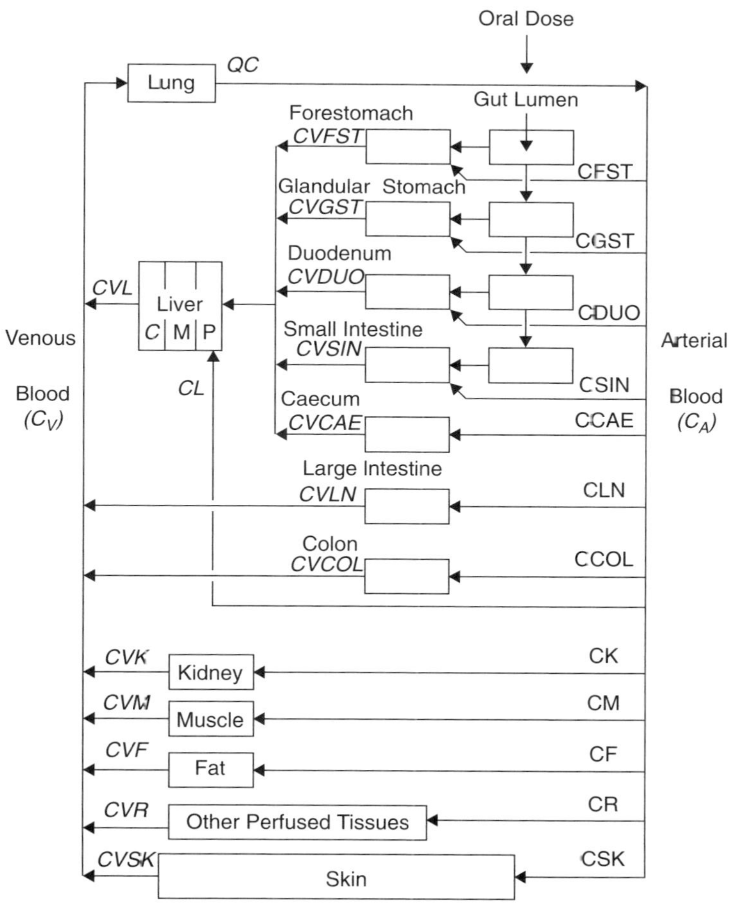

Figure 5.4 Structure of a PBPK model for EA developed by Frederick *et al.* (1992) for oral gavage exposure. The model included GSH induction, resynthesis, and depletion for use as a dose metric. Reproduced with permission from *Toxicology and Applied Pharmacology*.

and absorbed by the gut wall by first order processes. Further metabolism of AA to carbon dioxide was also included. Uptake into the forestomach was described with a permeation coefficient from the gut contents, similar to approaches used in some dermal absorption models. Only 10% of the administered dose was considered available for absorption in the forestomach.

The model primarily used rate constants developed *in vitro* or in separate experiments. EA is so rapidly metabolized that time course data for EA disappearance or formation of AA, its major metabolite, were difficult to collect. The model was instead validated against data for GSH depletion, carbon dioxide production, and protein binding. An excellent correspondence was obtained between the model predictions and measured GSH in forestomach (Fig. 5.5). The authors then developed a dose metric related to GSH depletion in this tissue. As GSH varied by a factor of two normally, the dose metric used was area under the curve (AUC) of GSH depletion by over 50%. This dose metric was plotted against chronic bioassay data for irritation, hyperplasia, and carcinomas (Fig. 5.6). The generally good correspondence between AUC (<50%) GSH and the outcomes suggested that the carcinogenic response was due to toxicity and altered cell proliferative responses, much as responses in the nasal cavity with VA appear to be associated with cytotoxicity.

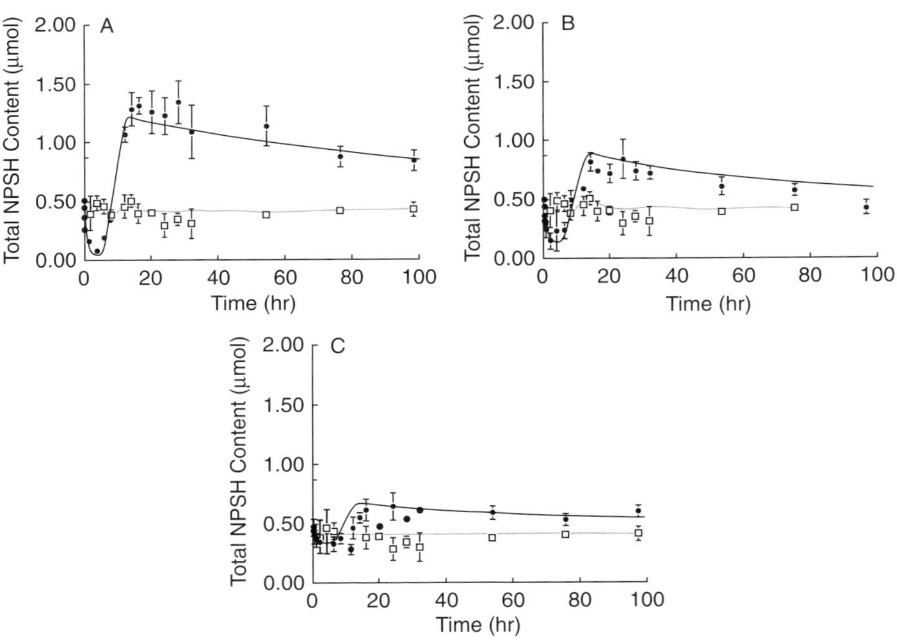

Figure 5.5 GSH, measured as total nonprotein sulfhydryl (NPSH) content in the forestomach of rats after EA dosing by corn oil gavage. Curves are the PBPK model output of Frederick *et al.* (1992), and symbols are experimental data for control (open squares) and treated rats (closed circles) exposed to 200, 50, and 10 mg/kg (panels A–C, respectively). The model adequately reproduces the very complex behavior of GSH in these experiments.

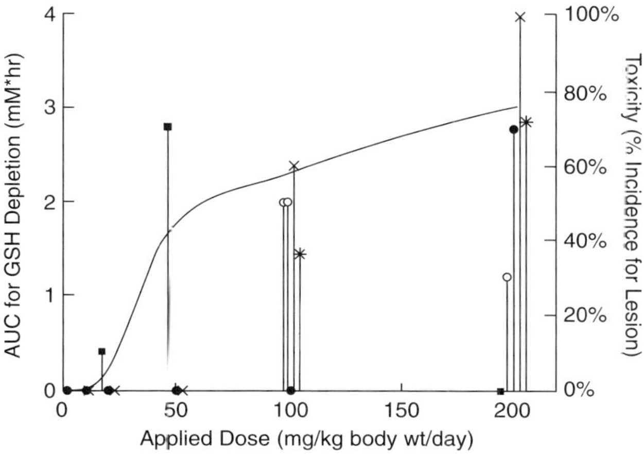

Figure 5.6 Meta-analysis of several toxic endpoints measured in two bioassays and applied dose and internal dose for EA in Frederick *et al.* (1992). Applied dose (mg/kg/day) on the bottom axis is shown with incidence data on right axis (mild hyperplasia (squares), moderate hyperplasia (diamonds), marked hyperplasia (circles), inflammation or ulceration (crosses), and papillomas or carcinomas (asterisks). Model-predicted AUC for 50% GSH depletion is shown in the curve for each exposure level. The GSH dose metric reasonably correlates with toxicity.

A CFD-PBPK model was later developed for EA for nasal toxicity risk assessment (Frederick *et al.* 2002). This model coupled the CFD-PBPK model structure previously used for other chemicals (Bush *et al.* 1998), the EA metabolic and GSH resynthesis models used for the oral gavage models previously developed for EA (Frederick *et al.* 1992), and the model for AA distribution within nasal tissues (Frederick *et al.* 1998). Additional or new features of the Frederick *et al.* (2002) EA model included renal metabolism, urinary excretion of AA, and a model for oronasal breathing (in humans). In a manner similar to the earlier work (Frederick *et al.* 1992), the model was validated against GSH depletion data in liver and lung tissues after inhalation exposures. Also, the model was validated against nasal deposition data at four EA concentrations with and without metabolic inhibitors

A similar model was later used to determine RfCs for EA (Sweeney *et al.* 2004). This model updated the Frederick *et al.* (2002) model for EA with additional *in vitro* data and modified the GSH resynthesis model to improve the fit to GSH data.

5.4.3 Epichlorohydrin

Epichlorohydrin (1-chloro-2,3-epoxypropane) causes nasal tumors and tumors of the rat forestomach. Unlike the other reactive vapors described here, it is an epoxide compound and requires no bioactivation to a more toxic form. It was suspected to be directly genotoxic, and nasal tumors were reported in the respiratory epithelium. This observation may be due to its ability to serve as a direct alkylator. With EPI, cytotoxicity may not be as important a component of the carcinogenicity mode of action as it appears to be with VA or formaldehyde. These concerns about direct genotoxicity contributed to the somewhat different type of PBPK model used for dosimetric analysis of EPI (Ginsberg *et al.* 1996). This model was developed to predict AUC and peak dose metrics for EPI in tumor sites in rat forestomach and nasal tissue.

In many respects, the model by Ginsberg *et al.* (1996) drew from earlier work described above. The oral model structure was based on the Frederick *et al.* (1992) model for EA for oral gavage and was extended to allow consumption of drinking water. A 14-compartment structure was used, as in the previous EA model. Metabolism via GSH conjugation and hydrolysis by epoxide hydrolase was included along with binding to tissue macromolecules. A GSH resynthesis submodel was included with induction and GSH concentrations were calculated. The drinking-water model simply required altered dosing regimens to be incorporated. The data for oral gavage in a mouse study was compared to model output after adjusting physiological parameters in the model. The inhalation portion of the model was also based on the earlier work. The Kimbell *et al.* (1993) nasal structure and airflow data were used. Parameter values for EPI were taken from various sources, with the exception of the fitted parameters for reaction rates in nasal tissue and diffusivity in mucus.

The model compared tumor incidence data with several dose metrics: peak EPI concentration, AUC for EPI, and an AUC averaged over lifetime. Bioassays data were available from studies using three routes of exposure (gavage, drinking water, and inhalation), all of which had no observable adverse effect levels, lowest observable adverse effect levels, and good dose responses. The tumor response data did not correlate well with either AUC metric or external concentration; it did appear to correlate better with the peak concentration. Another factor worth considering was GSH depletion. While not reported in nasal tissues, depletion of GSH was very significant in forestomach, and it might be useful as an alternate dose metric with EPI.

5.5 METHYL METHACRYLATE

MMA is another organic ester that causes nasal toxicity in olfactory epithelium in rodents. Due to generally lower rates of overall metabolism in nasal tissues, MMA extraction is much lower than that of VA. The steady-state uptake of MMA in rats has been analyzed with physiologically based clearance extraction (PBCE) models. This model formulation differed from other PBPK models in that it included only the tissues of interest (i.e., the nasal tissue compartments) and was solved at steady state. The authors justified this simplified approach since MMA reaches steady state quickly and systemic concentrations of MMA remain low due to rapid clearance. This PBCE model for MMA model used the air-phase mass transfer coefficient as the input into the uppermost tissue/mucus compartment.

The MMA model included (a) three tissue regions with the same two airflow pathways described by Morris *et al.* (1993) and (b) tissue stacks that included the airway lumen, mucus, epithelial tissue, and submucosa (Fig. 5.7). Because the analysis was restricted to steady-state conditions, the equations could be solved algebraically. Air stream "clearance" was defined as the volumetric flow (liters per hour)

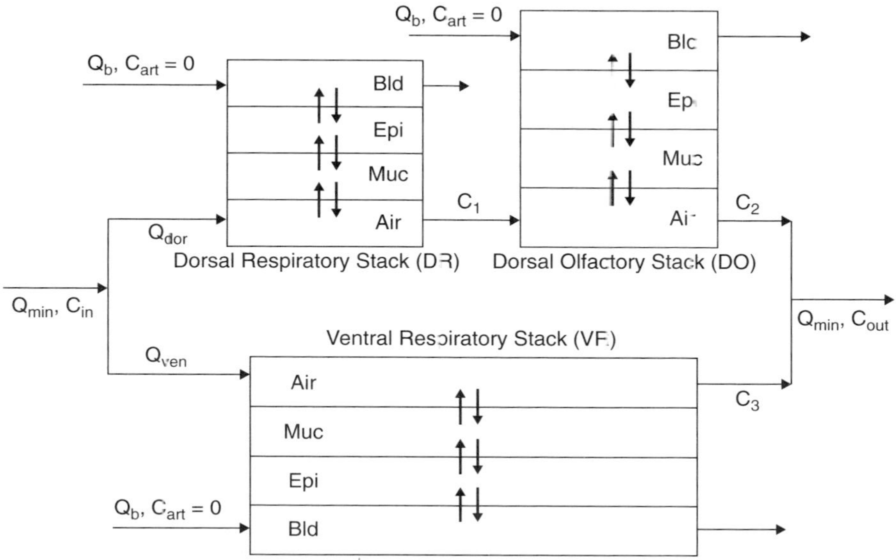

Figure 5.7 Conceptual model of the PBCE model for MMA (Andersen *et al.* 1999). Each mucosal region contained a stack of mucus and epithelial tissue adjacent to the airway lumen (Muc, mucus; Epi, epithelium; Bld, blood). Q_{min} (total airflow) was apportioned between the dorsal medial pathway (Q_{dor}) passing over the dorsal respiratory mucosa (DR) at the exposure concentration (C_{in}), and then (Q_{ven}) over the dorsal olfactory mucosa (DO) at concentration C_1, and the ventral respiratory pathway (VR) passing over respiratory mucosa at C_{in}. Q_b represents the blood flow rate, also apportioned between the two mucosal compartments. Reproduced with permission from *Inhalation Toxicology*.

that was cleared of the chemical under different operational conditions (Andersen and Sarangapani 2001). Clearance and extraction terminology are frequently applied in studies of the physiological behavior of the liver and kidney as they remove materials from the bloodstream by metabolism and filtration, respectively.

The first PBCE model for MMA presented algebraic expressions that represented the steady-state clearance in one to four compartment models (Andersen and Sarangapani 1999). For example, clearance in a two-compartment model that included just air and tissue was

$$\mathrm{CL}_{\mathrm{sys}} = \frac{\mathrm{CL}_t H_{t:a} \mathrm{PA}_{\mathrm{gas}} Q}{\mathrm{CL}_t H_{t:a}(Q + \mathrm{PA}_{\mathrm{gas}}) + \mathrm{PA}_{\mathrm{gas}} Q} \tag{5.2}$$

where

$\mathrm{CL}_{\mathrm{sys}}$ = clearance from airstream

CL_t = clearance in tissue

$\mathrm{H}_{t:a}$ = nasal tissue:air partition coefficient

$\mathrm{PA}_{\mathrm{gas}}$ = air-phase diffusional clearance

Q = air-stream flow

Steady-state clearance, for this simple model, is related to a small number of model parameters. Depending on the compound, metabolic characteristics and exposure concentration, individual parameters may be very important or insignificant. For most metabolized vapors, CL_t is the rate limiting factor at higher concentrations where metabolism becomes saturated. At low concentrations, CL_t may be much larger; PA_{gas} then becomes more important in determining clearance and extraction.

This model was used to reassess the dosimetric adjustment factor (DAF) for MMA (Andersen *et al.* 1999). The default approach (USEPA) to calculating a DAF for MMA resulted in a DAF of about 0.15, based on the ratios of ventilation rates and nasal surfaces areas in rats and humans. When physiological processes were taken into account by this PBCE model, the estimated DAF based on MMA metabolism per unit volume of target tissues increased to about 3. In other words, based on amount of chemical metabolized per tissue volume per unit time, humans are expected to have only 5% of the dose that was suggested by a direct comparison with default methods. This interspecies difference in delivered dose is due in large part to lower metabolism in nasal epithelial tissues from humans compared to those from rats. In a later version of the model (Andersen and Sarangapani 2001), DAFs were calculated for three enzyme distributions in humans: a worst case with all enzyme activity in the target tissue (i.e., the sustentacular cell population); an even distribution between sustentacular cells and submucosa; and a distribution assuming that humans have a distribution similar to the rat. These different distributions led to estimates of DAF ranging from 1.6 to 8.0 at the benchmark dose of 28.4 ppm.

VA extraction at steady state was also evaluated using this steady-state PBCE model (Andersen and Sarangapani 1999). At low concentrations, VA extraction was largely dependent on airflow; diffusivity and metabolism became more important at higher concentrations. At low inhaled concentrations, VA extraction is between 0.9 and 1.00. This high extraction appears to occur with VA due to metabolism in the uppermost tissue compartments throughput the nose.

5.6 FORMALDEHYDE

HCHO causes nasal cancer in rodents in the respiratory epithelium. HCHO is highly cytotoxic at concentrations above 5 ppm and also forms DNA-protein cross-links (DPX) in respiratory tissues. The mode of action as a carcinogen may contain dose-dependent contributions from both cytotoxicity and the DNA-reactivity of HCHO. *In vivo* studies showed that DPX varied in different regions of the nose, and the DPX has been considered to serve as a dose metric that extends the range of observed toxicity to an exposure region below concentrations with increased tumor incidence.

Much of the experimental data for HCHO and DPX have been collected in rats, although other data have been collected in monkeys in order to improve the scale-up to humans. An early pharmacokinetic model (Casanova *et al.* 1991) was developed for DPX formation in the monkey. This model was developed with data from nose-only studies of DPX formation that used a somewhat different structure than the more recent models. Three nasal compartments were included: a middle turbinate compartment, an anterior lateral wall/septum compartment, and a nasopharyngeal compartment. Each compartment had one tissue layer that was well mixed. Chemical transferred into each compartment in accordance with the local concentration of chemical in the airway. In each tissue, HCHO was metabolized by one saturable pathway, one first-order pathway, and could also form DPX as a first-order rate process. Curve fitting was used to estimate model parameters using nonlinear regression with three pharmacokinetic equations that described disposition in tissues. According to this model, the regional differences in DNA binding were caused more by uneven distribution of uptake into the different regions than by any differences in metabolic parameters in the different regions.

Improvements in the model predictions of HCHO uptake in differing nasal regions were then made by first using a CFD submodel to estimate regional fluxes and then using these regional fluxes as input for models that accounted for regional DPX formation. In this work, the CFD model (Kimbell *et al.* 1993) was first used to compare regional fluxes with the distribution of HCHO-induced nasal tumors. Then a hybrid CFD-PBPK model was assessed to describe the DPX tissue concentrations (Hubal *et al.* 1997). This tissue-phase model used a similar structure for HCHO metabolism and DNA reactivity as in the initial tissue description (Casanova *et al.* 1991). Fitted parameters again included metabolic constants for saturable and nonsaturable metabolism in the nasal tissue. These parameters were determined by fitting against total uptake for the whole nose and DPX concentrations. The model was then used to predict DPX in high-tumor regions of the respiratory mucosa at four exposure levels between 0.7 and 15 ppm. A close correspondence was obtained between the predicted and measured DPX (Fig. 5.8).

This model was later refined and scaled-up to monkeys and humans (Conolly *et al.* 2000). This work also extended the rat nasal model to include low-tumor regions in the nose as well as high tumor regions, while the description of the monkey nose was still limited to three compartments as in the earlier analysis (Casanova *et al.* 1991). CFD simulations again provided the fluxes of HCHO to each region of the rat, monkey, and human nasal passage. An excellent correspondence was obtained between model predictions and the measured DPX data for both low and high-tumor regions in rats and for all three regions of the monkey nose. Similar nonlinear dose–response curves, for DPX concentration versus inhaled HCHO concentrations, were found in all three species. These models were further refined in subsequent publications (Kimbell *et al.* 2001a,b).

A simpler model (Overton *et al.* 2001) was used to estimate uptake of HCHO in tissues in the tracheobronchial region of the human airway. In this model, the lung was described as a symmetric, bifurcating respiratory tract. This bifurcating model was constrained so that the same airflow rates and uptake rates were used in each area. Four levels of workload were simulated: sleeping, sitting, light exercise and heavy exercise. This analysis found that wall flux was expected in the first few generations of the lower respiratory tract and then declined rapidly for subsequent generations.

A noncancer risk assessment for HCHO was performed with the hybrid CFD-PBPK model based on tissue cytotoxicity. The cytotoxicity endpoint was observed

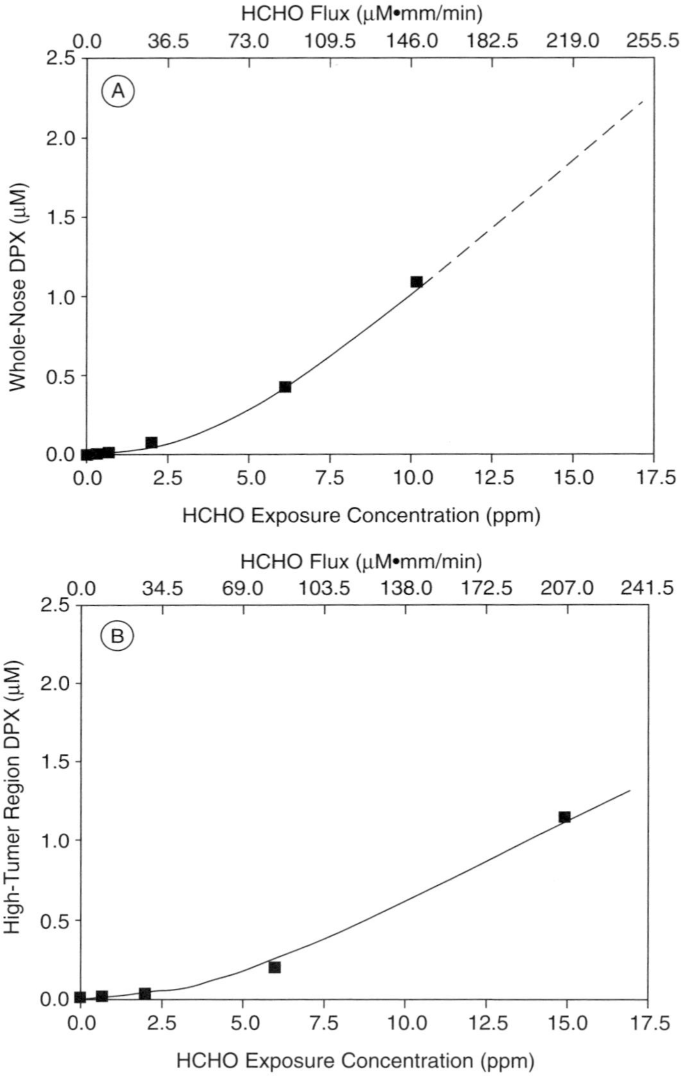

Figure 5.8 Experimental DPX data (symbols) and model predictions (curves) of DPX in the (A) whole-nose and (B) high-tumor regions of the rat (Hubal *et al.* 1997). The model used CFD simulations of airflow, coupled to a PBPK model of the rat nose describing uptake, metabolism, and cross-linking of protein and DNA. Reproduced with permission from *Toxicology and Applied Pharmacology.*

in rodent studies and had been quantified by regional rates of cell proliferation. In a risk assessment for these effects (Conolly *et al.* 2002), a CFD model of the human nose (Subramaniam *et al.* 1998) was coupled to the typical path model previously described (Overton *et al.* 2001) and regions with different predicted fluxes were binned (Kimbell *et al.* 2001a). The dose–response curve was then determined from average flux and regenerative cell proliferation. This dose–response curve predicted from the hybrid CD-PBPK model was then used for human risk assessment.

These hybrid CFD-PBPK models have also provided useful input for risk calculations based on benchmark dose approaches (Schlosser *et al.* 2003) and a biologically based dose–response model for carcinogenicity based on a two-stage clonal growth model for carcinogenesis (Conolly *et al.* 2003).

5.7 HYDROGEN SULFIDE

H_2S also causes nasal toxicity. As with many other reactive vapors, the olfactory epithelium is more sensitive than respiratory epithelium. Moulin *et al.* (2002) developed a PBPK model for H_2S that incorporates many of the same features of the models developed for HCHO. Specifically, the H_2S model included CFD calculations of airflow and chemical delivery within the nasal passages which were based on the rat nasal casts. The model was designed to permit calculation of the flux of H_2S into nasal epithelial tissue. Metabolism and systemic uptake were not described, and apparently the model did not need to include tissue layers and mucus in order to describe flux as a dose metric. Some uptake parameters could not be accurately specified, so the model was implemented with low, medium, and high estimates of uptake at the air:tissue boundary.

The modeling results indicated that high flux into both epithelial and olfactory tissues would be expected, corroborating the idea that olfactory tissues are much more sensitive to the effects of H_2S. Within the olfactory tissues, marginal to significant correlations between predicted fluxes and specific locations of pathology were found. However, some discrete smaller regions displayed a lack of correlation between calculated fluxes and observed responses.

5.8 SUMMARY

The development of models for reactivity of vapors and their metabolites within epithelial tissues in the nose exemplifies the iterative nature of PBPK modeling efforts for a specific target organ and on the factors that regulate toxicity for various vapors in this tissue. The sequential efforts of multiple investigators over the past 15 years have refined our understanding of steps necessary to provide a more seamless integration of airflow models with PBPK models for tissue reactivity of these chemicals and their metabolites. The hybrid CFD-PBPK models are still not completely integrated, but the hand-off between these models has become better defined by the applications with AA and HCHO. Although not as heavily emphasized in this chapter, the tissue dose metrics with these compounds varied considerably, depending on reactivity and modes of action. They included amount of metabolite formed per volume tissue (MMA), tissue acidification (VA and AA), and tissue GSH depletion (EA). The hybrid CFD-PBPK model with HCHO has produced the most comprehensive example to date of a biologically based dose–response model for chemical carcinogenesis and its application with human health risk assessment.

These hybrid CFD-PBPK models of the upper nasal regions should find many uses in assessing RfCs for inhaled compounds, especially for the compounds defined as hazardous air pollutants under the Clean Air Act.

NOTATION

AAld	acetaldehyde
AA	acrylic acid
AUC	area under the curve
CFD	computational fluid dynamic
CYP P450	cytochrome P450
DAF	dosimetric adjustment factor
DPX	DNA-protein cross-links
EA	ethyl acrylate
EPI	epichlorohydrin
GSH	glutathione
HCHO	formaldehyde
H_2S	hydrogen sulfide
IMTC	interfacial mass transfer coefficient
MMA	methyl methacrylate
PBCE	physiologically based clearance extraction
RfC	reference concentration
VA	vinyl acetate

REFERENCES

Aharonson, E. F., Menkes, H., Gurtner, G., Swift, D. L., and Proctor, D. F. (1974). Effect of respiratory airflow rate on removal of soluble vapors by the nose. *J. Appl. Physiol.* **37**, 654–657.

Andersen, M. E., and Sarangapani, R. (1999). Clearance concepts applied to the metabolism of inhaled vapors in tissues lining the nasal cavity. *Inhal. Toxicol.* **11**, 873–897.

Andersen, M. E., and Sarangapani, R. (2001). Physiologically based clearance/extraction models for compounds metabolized in the nose: An example with methyl methacrylate. *Inhal. Toxicol.* **13**, 397–414.

Andersen, M. E., Sarangapani, R., Frederick, C. B., and Kimbell, J. S. (1999). Dosimetric adjustment factors for methyl methacrylate derived from a steady-state analysis of a physiologically based clearance-extraction model. *Inhal. Toxicol.* **11**, 899–926.

Andersen, M., Sarangapani, R., Gentry, R., Clewell, H., Covington, T., and Frederick, C. B. (2000). Application of a hybrid CFD-PBPK nasal dosimetry model in an inhalation risk assessment: An example with acrylic acid. *Toxicol. Sci.* **57**, 312–325.

Bogdanffy, M. S., and Sarangapani, R. (2003). Physiologically-based kinetic modeling of vapours toxic to the respiratory tract. *Toxicol. Lett.* **138**, 103–117.

Bogdanffy, M. S., Sarangapani, R., Kimbell, J. S., Frame, S. R., and Plowchalk, D. R. (1998). Analysis of vinyl acetate metabolism in rat and human nasal tissues by an *in vitro* gas uptake technique. *Toxicol. Sci.* **46**, 235–246.

Bogdanffy, M. S., Sarangapani, R., Plowchalk, D. R., Jarabek, A., and Andersen, M. E. (1999a). A biologically based risk assessment for vinyl acetate-induced cancer and noncancer inhalation toxicity. *Toxicol. Sci.* **51**, 19–35.

Bogdanffy, M. S., Manning, L. A., and Sarangapani, R. (1999b). High-affinity nasal extraction of vinyl acetate vapor is carboxylesterase dependent. *Inhal. Toxicol.* **11**, 927–941.

Bush, M. L., Frederick, C. B., Kimbell, J. S., and Ultman, J. S. (1998). CFD-PBPK hybrid model for simulating gas and vapor uptake in the rat nose. *Toxicol. Appl. Pharmacol.* **150**, 133–145.

Casanova, M., Morgan, K. T., Steinhagen, W. H., Everitt, J. I., Popp, J. A., and Heck, H. D. (1991). Covalent binding of inhaled formaldehyde to DNA in the respiratory tract of rhesus monkeys: Pharmacokinetics, rat-to-monkey interspecies scaling, and extrapolation to man. *Fundam. Appl. Toxicol.* **17**, 409–428.

Conolly, R. B., Lilly, P. D., and Kimbell, J. S. (2000). Simulation modeling of the tissue disposition of formaldehyde to predict nasal DNA-protein cross-links in Fischer 344 rats, rhesus monkeys, and humans. *Environ. Health Perspect.* **108**(Suppl 5), 919–924.

Conolly, R. B., Kimbell, J. S., Janszen, D. B., and Miller, F. J. (2002). Dose response for formaldehyde-induced cytotoxicity in the human respiratory tract. *Regul. Toxicol. Pharmacol.* **35**, 32–43.

Conolly, R. B., Kimbell, J. S., Janszen, D., Schlosser, P. M., Kalisak, D., Preston, J., and Miller, F. J. (2003). Biologically motivated computational modeling of formaldehyde carcinogenicity in the F344 rat. *Toxicol. Sci.* **75**, 432–447.

Frederick, C. B., Bush, M. L., Lomax, L. G., Black, K. A., Finch, L., Kimbell, J. S., Morgan, K. T., Subramaniam, R. P., Morris, J. B., and Ultman, J. S. (1998). Application of a hybrid computational fluid dynamics and physiologically based inhalation model for interspecies dosimetry extrapolation of acidic vapors in the upper airways. *Toxicol. Appl. Pharmacol.* **152**, 211–231.

Frederick, C. B., Potter, D. W., Chang-Mateu, M. I., and Andersen, M. E. (1992). A physiologically based pharmacokinetic and pharmacodynamic model to describe the oral dosing of rats with ethyl acrylate and its implications for risk assessment. *Toxicol. Appl. Pharmacol.* **114**, 246–260.

Frederick, C. B., Gentry, P. R., Bush, M. L., Lomax, L. G., Black, K. A., Finch, L., Kimbell, J. S., Morgan, K. T., Subramaniam, R. P., Morris, J. B., and Ultman, J. S. (2001). A hybrid computational fluid dynamics and physiologically based pharmacokinetic model for comparison of predicted tissue concentrations of acrylic acid and other vapors in the rat and human nasal cavities following inhalation exposure. *Inhal. Toxicol.* **13**, 359–376.

Frederick, C. B., Lomax, L. G., Black, K. A., Finch, L., Scribner, H. E., Kimbell, J. S., Morgan, K. T., Subramaniam, R. P., and Morris, J. B. (2002). Use of a hybrid computational fluid dynamics and physiologically based inhalation model for interspecies dosimetry comparisons of ester vapors. *Toxicol. Appl. Pharmacol.* **183**, 23–40.

Gerde, P., and Dahl, A. R. (1991). A model for the uptake of inhaled vapors in the nose of the dog during cyclic breathing. *Toxicol. Appl. Pharmacol.* **109**, 276–288.

Ginsberg, G. L., Pepelko, W. E., Goble, R. L., and Hattis, D. B. (1996). Comparison of contact site cancer potency across dose routes: case study with epichlorohydrin. *Risk Anal.* **16**, 667–681.

Hubal, E. A., Schlosser, P. M., Conolly, R. B., and Kimbell, J. S. (1997). Comparison of inhaled formaldehyde dosimetry predictions with DNA-protein cross-link measurements in the rat nasal passages. *Toxicol. Appl. Pharmacol.* **143**, 47–55.

Kelly, J. T., Kimbell, J. S., and Asgharian, B. (2001). Deposition of fine and coarse aerosols in a rat nasal mold. *Inhal. Toxicol.* **13**, 577–588.

Kimbell, J. S., and Subramaniam, R. P. (2001). Use of computational fluid dynamics models for dosimetry of inhaled gases in the nasal passages. *Inhal. Toxicol.* **13**, 325–334.

Kimbell, J. S., Gross, E. A., Joyner, D. R., Godo, M. N., and Morgan, K. T. (1993). Application of computational fluid dynamics to regional dosimetry of inhaled chemicals in the upper respiratory tract of the rat. *Toxicol. Appl. Pharmacol.* **121**, 253–263.

Kimbell, J. S., Overton, J. H., Subramaniam, R. P., Schlosser, P. M., Morgan, K. T., Conolly, R. B., and Miller, F. J. (2001a). Dosimetry modeling of inhaled formaldehyde: Binning nasal flux predictions for quantitative risk assessment. *Toxicol. Sci.* **64**, 111–121.

Kimbell, J. S., Subramaniam, R. P., Gross, E. A., Schlosser, P. M. and Morgan, K. T. (2001b). Dosimetry modeling of inhaled formaldehyde: Comparisons of local flux predictions in the rat, monkey, and human nasal passages. *Toxicol. Sci.* **64**, 100–110.

Miller, F. J., Overton, J. H., Jr., Jaskot, R. H., and Menzel, D. B. (1985). A model of the regional uptake of gaseous pollutants in the lung. I. The sensitivity of the uptake of ozone in the human lung to lower respiratory tract secretions and exercise. *Toxicol. Appl. Pharmacol.* **79**, 11–27.

Morris, J. B., Hassett, D. N., and Blanchard, K. T. (1993). A physiologically based pharmacokinetic model for nasal uptake and metabolism of nonreactive vapors. *Toxicol. Appl. Pharmacol.* **123**, 120–129.

Moulin, F. J., Brenneman, K. A., Kimbell, J. S., and Dorman, D. C. (2002). Predicted regional flux of hydrogen sulfide correlates with distribution of nasal olfactory lesions in rats. *Toxicol. Sci.* **66**, 7–15.

Overton, J. H. (1990). A respiratory tract dosimetry model for air toxics. *Toxicol. Ind. Health* **6**, 171–180.

Overton, J. H. (2001). Dosimetry modeling of highly soluble reactive gases in the respiratory tract. *Inhal. Toxicol.* **13**, 347–357.

Overton, J. H., Graham, R. C., and Miller, F. J. (1987). A model of the regional uptake of gaseous pollutants in the lung. II. The sensitivity of ozone uptake in laboratory animal lungs to anatomical and ventilatory parameters. *Toxicol. Appl. Pharmacol.* **88**, 418–432.

Overton, J. H., Kimbell, J. S., and Miller, F. J. (2001). Dosimetry modeling of inhaled formaldehyde: The human respiratory tract. *Toxicol. Sci.* **64**, 122–134.

Pang, K. S., and Rowland, M. (1977). Hepatic clearance of drugs. I. Theoretical considerations of a "well-stirred" model and a "parallel-tube" model. Influence of hepatic blood flow, plasma and blood cell binding, and hepatocellular enzymatic activity on hepatic drug clearance. *J. Pharmacokinet. Biopharm.* **5**, 625–653.

Plowchalk, D. R., Andersen, M. E., and Bogdanffy, M. S. (1997). Physiologically based modeling of vinyl acetate uptake, metabolism, and intracellular pH changes in the rat nasal cavity. *Toxicol. Appl. Pharmacol.* **142**, 386–400.

Schlosser, P. M., Lilly, P. D., Conolly, R. B., Janszen, D. B., and Kimbell, J. S. (2003). Benchmark dose risk assessment for formaldehyde using airflow modeling and a single-compartment, DNA-protein cross-link dosimetry model to estimate human equivalent doses. *Risk Anal.* **23**, 473–487.

Subramaniam, R. P., Richardson, R. B., Morgan, K. T., and Kimbell, J. S. (1998). Computational fluid dynamics simulations of inspiratory airflow in the human nose and nasopharynx. *Inhal. Toxicol.* **10**, 91–120.

Sweeney, L. M., Andersen, M. E., and Gargas, M. L. (2004). Ethyl acrylate risk assessment with a hybrid computational fluid dynamics and physiologically based nasal dosimetry model. *Toxicol. Sci.* **79**, 394–403.

US EPA (1994). *Methods of Derivation of Inhalation Reference Concentrations and Application of Inhalation Dosimetry*, U.S. Environmental Protection Agency, Office of Research and Development, Office of Health and Environmental Assessment, Washington, D.C., EPA/600/8–90/066F.

CHAPTER 6

ALKANES, OXYHYDROCARBONS, AND RELATED COMPOUNDS

Robert T. Davidson and Micaela B. Reddy

6.1 INTRODUCTION

In organizing a contemporary contribution on PBPK modeling for broad classes of compounds, we faced challenges in placing compounds into convenient groupings. One possible organization reflected the metabolic steps involved in major clearance pathways for parent compounds and in downstream pathways for metabolites. The emphasis in Part I on PBPK model development for volatile organic compounds has focused on compounds metabolized by reductive dehalogenation (some anesthetics), by epoxidation (i.e., alkenes and aromatics), by oxidation of C–H bonds on carbon substituted with halogens (haloalkanes), and by conjugation with glutathione (trichloroethylene and PERC). This concluding chapter of Part I brings together a broader cross section of compounds characterized by some common pathways of biotransformation. In general, these compounds undergo C–H bond oxidation to produce oxygenated metabolites. As with chemicals in other chapters, these PBPK models are sometimes more focused on parent chemical and at other times on metabolites.

This chapter includes PBPK models for fours classes of chemicals. Models are presented for two alkanes (Table 6.1), many oxyhydrocarbons (Table 6.2), five alkylbenzenes (Table 6.3), and two siloxanes (Table 6.4). Some of these chemicals (e.g., toluene and hexane) are included in models for chemical mixtures, which are

Physiologically Based Pharmacokinetic Modeling, edited by Micaela B. Reddy, Raymond S. H. Yang, Harvey J. Clewell III, and Melvin E. Andersen.

included in Chapter 13. Although this is a diverse group of chemicals, a common theme throughout this work is the application of PBPK modeling techniques for studying the toxicity and understanding the risk from exposure to these chemicals. The development of PBPK models for chemicals in the alkane, alkylbenzene, oxyhydrocarbon, and siloxanes classes is of great importance because of the prevalence of these compounds as used in the world's societies with subsequent human exposure. Exposure to these chemicals ranges from deliberate human exposures to ethanol to inadvertent exposure to petroleum product solvents and fuel components such as hexane and toluene.

6.2 PURPOSES FOR PBPK MODEL DEVELOPMENT

The development of a PBPK model requires much information and many parameters. Development of PBPK models can be hampered by incomplete knowledge of the pharmacokinetics and pharmacodynamics of a process. However, the physiological basis of these models provides many advantages. To paraphrase the well-known fundamental principle of toxicology, the dose makes the poison. The toxic effect caused by a compound is related to some measure of dose (e.g., blood concentration) at the site of toxicity (e.g., the target organ). PBPK models provide a tool for studying the relationship between the observed toxic effect, the internal dose that caused the effect, and the external dose (e.g., the amount ingested) that caused the internal dose.

As an example of the utility of PBPK models in understanding the relationship between external and internal dose, DeJongh *et al.* (1998) used PBPK modeling to interpret LC50 values for 15 nonreactive volatile organic solvents including five from this chapter (i.e., hexane, methanol, acetone, toluene, and *p*-xylene). The PBPK model for each chemical also included, along with the lung, rapidly and slowly perfused tissue, and fat compartments typically found in PBPK models, a brain compartment so that concentrations of solvents in the lipoid brain fraction could be determined. For model parameterization, partition coefficients measured *in vitro* and parameters for metabolism in the liver from either *in vivo* or *in vitro* experiments reported by other investigators were used. Three measures of internal exposure (i.e., the simulated concentration of chemical in the venous blood, whole brain, and brain lipids) were examined as possible dose surrogates. The external exposure parameter (i.e., the concentration of solvent inhaled during the exposure times the duration of the exposure) resulting in the death of 50% of the exposed animals varied by a factor of 60. After applying the PBPK model to determine an estimate of the internal dose (i.e., based on the blood concentration of the solvent), there was more than a tenfold decrease in the toxic range of the 16 volatile organic compounds included in this study.

PBPK modeling is an iterative process (Fig. 6.1). The steps in PBPK model development include: (1) identifying the problem, (2) evaluating the literature, (3) formulating the model, (4) parameterizing the model, (5) performing the simulation, (6) comparing the model output to experimental data, and (7) conducting pertinent experiments to identify the cause of any discrepancies between the model simulation and the experimental data. In this way, a PBPK model can be used as a tool

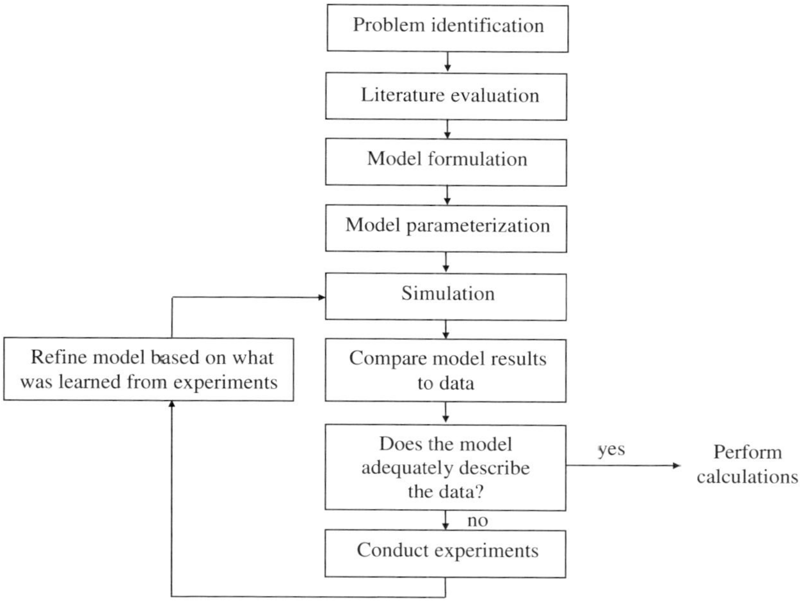

Figure 6.1 Schematic diagram depicting the iterative process of PBPK model development.

incorporating all known information of the system to determine if knowledge of the system is consistent with experimental data. Once a reliable PBPK model has been developed, it can be used for many applications in toxicology and risk assessment. The wide range of compounds included in this chapter provides the opportunity to present many examples of the utility of PBPK models.

Although the process of PBPK model development is involved (Fig. 6.1), PBPK modeling is frequently used in toxicology and risk assessment. PBPK models are useful in biological monitoring of human exposure to solvents [e.g., with cyclohexane (Perico *et al.* 1999) or toluene (Jang and Droz 1996; Thrall *et al.* 2001; Pierce *et al.* 1998)]. Additionally, PBPK modeling can be used to understand the relationship between dose and response [e.g., for ethylene oxide (Krishnan *et al.* 1992) or for volatile organic solvents (DeJongh *et al.* 1998)]. PBPK models can be used to understand risks to humans based on animal toxicity data [e.g., for 1,4-dioxane (Reitz *et al.* 1990)]. The many and varied uses of PBPK modeling in toxicology and risk assessment will be highlighted in this chapter.

6.3 PBPK MODELS FOR FOUR CLASSES OF COMPOUNDS

6.3.1 Alkanes

PBPK models have been developed for only two alkanes, the volatile organic solvents hexane and cyclohexane (Table 6.1). Cyclohexane can be used as a less toxic

TABLE 6.1 PBPK Models Developed for Hydrocarbons

Chemical	Structure	Species	Reference
Cyclohexane	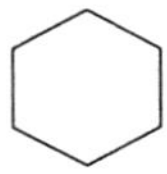	Human	Perico *et al.* (1999)
Hexane		Human	Perbellini *et al.* (1986)
		Human	Perbellini *et al.* (1990)
		Rat	McDougal *et al.* (1990)
		Rat	DeJongh *et al.* (1998)

alternative to the solvent hexane. The chemical 2,5-hexanedione is covered in this section instead of with the oxyhydrocarbons because 2,5-hexanedione is a toxic metabolite of hexane.

Hexane and 2,5-Hexanedione Hexane pharmacokinetics and metabolic interactions with other chemicals have been widely studied. There are numerous PBPK models investigating metabolic interactions of hexane with other chemicals and with its own metabolites (see Chapter 13). It has proven difficult to develop accurate hexane PBPK models despite the efforts of many researchers and the publication of several reports on PBPK modeling for hexane. To date, hexane PBPK models tend to underestimate blood hexane levels during simulations when compared to pharmacokinetics experimental datasets. This inability of hexane models to accurately describe blood time-course concentration data may be due to a metabolic product of hexane inhibiting metabolism. Another possibility is that hexane may be sequestered by proteins in the blood to a further extent than previously determined by headspace partition methods. After investigating the transportation and distribution of organic solvents in blood, Lam *et al.* (1990) reported that blood proteins are the major carrier for organic solvents in the blood in general and suggested that smaller-sized organic solvent molecules such as hexane may be able to interact more closely with some proteins.

An early PBPK model was developed for the hexane metabolite 2,5-hexanedione in rats by Angelo and Bischoff (1982) for the purpose of reconstructing tissue concentration histories following long-term, high-dose exposures. This model was described as a flow-limited model having plasma, liver, kidney, sciatic nerve, spinal cord, brain, and lean tissue compartments with mass balance equations describing the flow between compartments. Fat tissue was included with lean tissue because there was no difference in partitioning for 2,5-hexanedione observed in these tissues. Tissue : plasma partition coefficients were described as distribution ratios as the *in vivo* ratio of the respective concentrations during steady-state conditions following constant intravenous (*i.v.*) exposure. Metabolic and clearance rates were described by first-order terms calculated from *i.v.* infusion data. For intraperitoneal (*i.p.*) injection exposure data, the model simulation results were in very good agreement with experimental values of 2,5-hexanedione tissue and plasma concentrations for both a single exposure and for multiple exposures.

The first reported PBPK model for hexane was used to quantitatively describe and interpret experimentally derived data from human subjects occupationally exposed to hexane (Perbellini *et al.* 1986). Simulated venous hexane concentration values were about 10% lower than experimental values. This model also included the urinary-excreted metabolite 2,5-hexanedione. The metabolism of hexane was modeled using a metabolism and elimination scheme similar to one reported by Wagner (1967) and described in the liver by a disappearance of hexane with the formation of 2,5-hexanedione. The 2,5-hexanedione leaves the liver to a water compartment and is eliminated from the urine with a disappearance half-life of 13–14 hours. This same model was later used to generate a proposed biological exposure index for hexane in humans (Perbellini *et al.* 1990). PBPK modeling techniques have also been used to study hexane dermal absorption in the rat (McDougal *et al.* 1990) (see Chapter 14).

Cyclohexane Perico *et al.* (1999) used a PBPK model of cyclohexane to determine the suitability of using the urinary metabolites 1,2- and 1,4-cyclohexanediol for biological monitoring of industrial exposure. Urine samples were obtained from shoe and handbag factory workers exposed to this solvent because of its presence in glues. Environmental monitoring for individual cyclohexane exposures was performed using personal active samplers. Metabolism of cyclohexane to form 1,2- and 1,4-cyclohexanediol was incorporated into the cyclohexane inhalation PBPK model in the same manner that Perbellini *et al.* (1986, 1990) incorporated urinary hexane metabolites. Simulated metabolite concentrations in urine were similar to values reported in those occupationally exposed to cyclohexane. The authors concluded that both 1,2- and 1,4-cyclohexanediol are suitable for biological monitoring of cyclohexane exposures.

6.3.2 Oxyhydrocarbons

In this section, models for many types of organic compounds including alcohols, ketones, and furans are presented (Table 6.2).

Ethanol Derr (1993) developed his own form of PBPK model for orally administered ethanol in humans to compare ethanol pharmacokinetics and metabolic capabilities between four different populations: white American, white European, black American, and Japanese. This model utilized blood, lung, kidney, liver, stomach, intestinal lumen, and intestinal tissue compartments. Physiological parameters were for standard men from each nation. Metabolic parameters were found in the literature for each of the nationalities represented and included hepatic P450 metabolism of ethanol, hepatic alcohol dehydrogenase, and gastric alcohol dehydrogenase. No partition coefficients are reported, though they were used according to the model description. Model simulations generally underpredicted blood alcohol concentrations.

Pastino *et al.* (1996) developed a PBPK model for orally administered ethanol in mice to investigate the relative influence of gastric metabolism on ethanol bioavailability. The model was similar to that of Andersen *et al.* (1987) with the addition of a stomach compartment with gastric metabolism. Partition coefficients

TABLE 6.2 PBPK Models Developed for Oxyhydrocarbons

Chemical	Structure	Species	Reference
Acetone	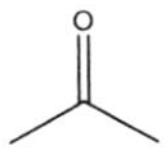	Rat	DeJongh *et al.* (1998)
		Human	Kumagai and Matsunaga (1995)
		Rat, human	Clewell *et al.* (2001)
2-Butoxyacetic acid	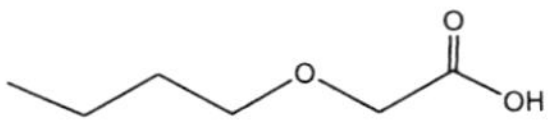	Rat, human	Corley *et al.* (1994)
		Rat, mouse	Lee *et al.* (1998)
2-Butoxyethanol (BE)	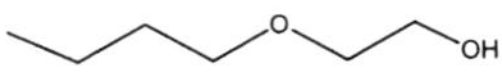	Human	Johanson (1986)
		Rat	Shyr *et al.* (1993)
		Rat, human	Corley *et al.* (1994)
		Rat, mouse	Lee *et al.* (1998)
1,4-Dioxane		Rat, mouse, human	Reitz *et al.* (1990)
Ethanol	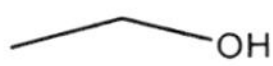	Dog	Rheingold *et al.* (1981)
		Human	Derr (1993)
		Mouse	Pastino *et al.* (1996)
		Mouse, rat, human	Pastino *et al.* (1997)
		Human	Ramchandani *et al.* (1999)
		Rat	Pastino and Conolly (2000)
		Rat	Morzarati *et al.* (2002)
		Human	MacDonald *et al.* (2002)
Ethylene oxide	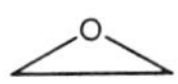	Rat	Krishnan *et al.* (1992)
Furan		Mouse, rat, human	Kedderis and Held (1996)
		Rat	Kedderis *et al.* (1993)
		Human	Price *et al.* (2003)
2,5-Hexanedione	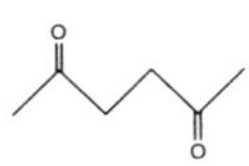	Human	Angelo and Bischoff (1982)
			Perbellini *et al.* (1986)
Isopropanol	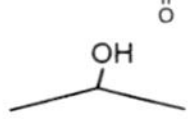	Rat, human	Clewell *et al.* (2001)
Methanol	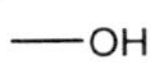	Rat	DeJongh *et al.* (1998)
		Monkey	Fisher *et al.* (2000)
		Rat, monkey, human	Horton *et al.* (1992)
MEK	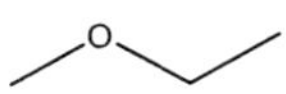	Human	Liira *et al.* (1990)
		Rat	Thrall *et al.* (2002a)
MTBE	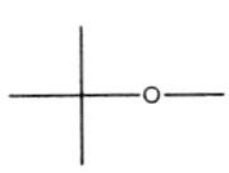	Rats	Borghoff *et al.* (1996)
		Human	Licata *et al.* (2001)
TAME	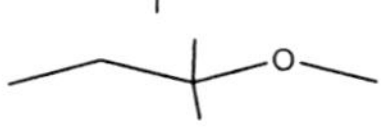	Rat	Collins *et al.* (1999)
tert-Amyl alcohol	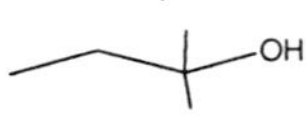	Rat	Collins *et al.* (1999)
tert-Butyl alcohol	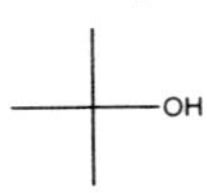	Rats	Borghoff *et al.* (1996)
Tetrahydrofuran		Human	Droz *et al.* (1999)

were determined using a headspace vial equilibration technique. The authors concluded that ethanol bioavailability was not significantly affected by gastric metabolism, based largely on an assumption that only 4% of the ethanol in the stomach is available for gastric metabolism.

Studies of the pharmacokinetics and toxicity of ingested ethanol are common, but studies of the inhalation pharmacokinetics and toxicity of ethanol are less common even though people are exposed to ethanol added to gasoline by the inhalation exposure route. For this reason, Pastino *et al.* (1997) adapted their earlier PBPK model for ethanol (Pastino *et al.* 1996) to examine inhalation exposures. The previous PBPK model in mice was scaled to rats and humans and modified to include the inhalation exposure route. Because ethanol is a water soluble compound, during inhalation exposures to ethanol some ethanol dissolves in the mucus lining of the upper respiratory tract, resulting in what is called the "washin–washout" effect. Ethanol in the mucus can diffuse from a region of higher concentration (lumen) to a region of lower concentration (epithelial cells and blood capillaries). The opposite occurs during exhalation because the ethanol concentration in air exhaled from the lower respiratory tract is low; ethanol in the mucus can pass back into the exhaled air. This "washin–washout" effect causes a significant reduction in the ethanol concentration in air reaching the alveoli of the lungs compared to the inhaled concentration. The concentration of ethanol in expired air is likewise higher than the concentration in the alveoli. This effect was accounted for in the PBPK model by adjusting the exposure concentration until simulation results matched experimental values. Model simulations agreed with blood ethanol concentrations following inhalation exposures in mice, rats, and humans. Based on model calculations, the blood ethanol concentrations resulting from ethanol vapors in gasoline are low enough that no behavioral or toxic effects are expected to result.

Pastino and Conolly (2000) investigated the relative contributions of gastric and hepatic metabolic clearance to oral bioavailability of ethanol in rats using a previous PBPK model (Pastino *et al.* 1997) extended to include oral administration of ethanol with gastric metabolic clearance. In calculating hepatic metabolic clearance, a distinction was made between ethanol entering the liver from the GI tract (first-pass metabolism) and recirculating ethanol entering the liver from the bloodstream. This distinction made it possible to calculate the bioavailability of ethanol from the orally administered route. The model described blood ethanol concentrations over a wide range of doses. The ability of the model to describe ethanol pharmacokinetics for a wide range of doses allowed the examination of the effect of dose on oral bioavailability. The model successfully simulated blood ethanol concentrations in rats pretreated with cimetidine (i.e., an inhibitor of gastric alcohol dehydrogenase activity) when the parameters for gastric metabolism were adjusted appropriately, which increases confidence in the ability of the model to describe gastric and hepatic metabolism. The PBPK model was able to quantitatively demonstrate the contributions of both gastric and hepatic first-pass metabolism of ethanol to ethanol bioavailability following oral administration.

In another application of PBPK modeling, Ramchandani *et al.* (1999) developed a PBPK model for ethanol adapted from the model of Rheingold *et al.* (1981) to compute the rate at which ethanol will be intravenously infused to maintain a particular blood or breath concentration. This method is useful because it supports the maintenance of steady blood levels for responses (e.g., early-phase neuroadaptive) to alcohol. This modeling application was later adapted by Morzorati *et al.* (2002), who developed a method for maintaining steady-state arterial ethanol concentrations in rats.

Methanol Horton *et al.* (1992) developed a PBPK model for methanol based on the method by Ramsey and Andersen (1984). This model was used to predict the methanol inhalation concentration range over which rats and monkeys exhibit quantitative similarities with humans. There were five compartments including blood (with lung gas exchange), liver (with metabolism of methanol to formaldehyde and formate), kidney (with excretion of methanol and a metabolite, formate), richly perfused, and slowly perfused tissues. Physiologic parameters were scaled on the basis of bodyweight according to Adolf (1949). Tissue : air and blood : air partition coefficients were determined experimentally by vial headspace equilibration (Fiserova-Bergerova 1975). Tissue : blood partition coefficients were then calculated by dividing tissue : air values by the blood : air value. Metabolic parameters for methanol, formaldehyde, and formate (methanol metabolites) in rats, monkeys, and humans were from the literature. Formate concentrations in the liver and blood were included in the model while formaldehyde was modeled in the liver. Renal extraction efficiencies for methanol and formate were determined by fitting simulation results to experimental values. To adequately simulate the experimental data in rats, the metabolism of methanol to formaldehyde by the enzyme catalase had to be augmented by a secondary enzyme pathway having a smaller V_{max} and higher affinity (K_m) for methanol. In both monkeys and humans, methanol metabolism was described using both catalase and alcohol dehydrogenase enzyme pathways with published kinetic values. For rats, kinetic values that resulted in the best simulation fit to experimental data were used. Simulation results reasonably predicted blood and urine concentrations of methanol and metabolites following inhalation or *i.v.* exposure to methanol.

Fisher *et al.* (2000) developed a PBPK model to quantitate respiratory uptake of methanol in monkeys. This study included data from a pharmacokinetic experiment in which Cynomolgus monkeys were exposed to 10–900 ppm ^{14}C-methanol for two hours. The model was based on the PBPK model by Ramsey and Andersen (1984). The fractional uptake of inhaled methanol vapors from the washin–washout effect was modeled using the method adopted by Pastino *et al.* (1997) to describe inhaled ethanol vapors in rodents and humans. Initial estimates of metabolic parameters for methanol were from the literature, but optimized values were determined by fitting data from blood time-course experiments to simulation results. This model generally overpredicted methanol blood concentration, though values were within experimental error ranges for high-exposure concentrations. Simulated concentrations of methanol in exhaled air were very close to experimentally determined values. The authors suggested that washin–washout effects of polar compounds might be accounted for by simply adjusting the inhaled concentration in PBPK models.

Acetone Kumagai and Matsunaga (1995) developed a PBPK model to predict acetone inhalation pharmacokinetics in humans for occupation health applications.

As acetone is a water-soluble compound and its interaction with mucus in the lungs might affect pharmacokinetics, Kumagai and Matsunaga proposed a method different from that of other authors to describe the washin–washout effect. In their model, separate compartments for the mucus layer of the inhaled air tract and for the mucus layer of the exhaled air tract were included. Because acetone could transfer between these two compartments, the interaction of acetone with mucus in the lungs could be modeled, despite treating the airflow in lungs as unidirectional. Modeling results were shown to be sensitive to the values selected for the volume of the mucus layer and the respiration rate. Model simulation results were successfully compared to previously data from occupational field studies. After verifying that the model results were consistent with the available data, the authors used the model to simulate occupational exposures to up to 2000 ppm acetone, and they used these simulations to examine and evaluate possible biological exposure indices. A PBPK model for acetone as a metabolite of isopropanol has also been developed in the rat and human, and this model will be discussed in the following section.

Isopropanol A PBPK model for isopropanol and its metabolite acetone has been developed for the rat and human (Clewell *et al.* 2001). This model describes *i.v.*, *i.p.*, oral, inhalation, and dermal exposures to isopropanol, and it describes oral and inhalation exposures to acetone. For isopropanol and acetone, the washin–washout effect (i.e., to account for the reservoir effect of the upper respiratory tract mucus layer) was incorporated using a model structure similar to that of Kumagai and Matsunaga (1995) for acetone. The goal of this modeling effort was to support a human assessment for isopropanol by allowing the integration of existing toxicity data.

2-Butoxyethanol, BE Johanson (1986) developed a PBPK model for inhaled BE in humans in order to simulate and invstigate the effects of physical exercise on the blood concentration profile of BE following the modeling scheme of Fiserova-Bergerova (1983). Muscle and skin were added as a separate compartment (i.e., they were split from the slowly perfused tissues compartment, where they are typically included) because of their wide variation in blood flow during physical activity (work). Physiological parameters for a human at both rest and work were obtained from Astrand (1983) and Fiserova-Bergerova (1983). The blood : air partition coefficient for BE was assumed to be greater than 1000. Biotransformation kinetic constants from perfused rat liver experiments were used to estimate metabolic parameters in humans. Another factor examined in this work was the effect of co-exposure to ethanol (without modeling ethanol pharmacokinetic effects on BE pharmacokinetics). There was good agreement between simulated and experimental blood values at different levels of physical work. However, the authors state that the model was created to demonstrate how a PBPK model may be used to explain the effects of physical activity on solvent exposure kinetics rather than to be a predictive model of BE pharmacokinetics in humans.

A PBPK model to describing the disposition and metabolism of BE in rats following dermal, oral, and inhalation exposures was used by Shyr *et al.* (1993) to provide a physiological basis for examining whether toxicities observed in rats might be seen in humans. The PBPK model, adapted from Ramsey and Andersen (1984), included skin as a separate compartment separated from the slowly perfused tissues compartment. Partition coefficients for BE were by Johanson and Dynesius (1988). Metabolism was determined to follow first-order Michaelis–Menton kinetics by fitting the simulation results to experimental data with the constraint that the sum of individual metabolic rates (for each metabolite) was similar to the experimentally derived value determined from perfused rat liver. The oral exposures simulated by Shyr *et al.* (1993) was rats ingesting BE in drinking water. Although rats drink water in a series of discrete events, BE intake from drinking water was modeled as a continuous input function consistent with patterns of nocturnal activity. For the inhalation exposures, it was determined that alveolar concentrations had to be reduced by one-third to accurately predict BE retention; the authors stated that this reduction was most likely necessary because of BE being detained in the nasal airways (i.e., the washin–washout effect). *In vitro* studies of BE uptake by the dermal route were used in model parameterizations, which indicated that ~20–25% of the applied dose (i.e., about half of the absorbed dose) was absorbed in the first 10 minutes of the exposure. The change of profiles of urinary metabolites exhibited with different exposure routes was consistent with model simulations after a minor pathway of metabolism in viable skin layers was incorporated in the model.

Corley *et al.* (1994) developed complete PBPK models for both BE and its major metabolite, 2-butoxyacetic acid, in rats and humans to compare differences in dose–response relationships and associated risk assessments between the two species. Their human PBPK model was based on the previously described PBPK model for BE human inhalation exposures by Johanson (1986), and this model was extended to include the physiology of rats. Additional routes of exposure and competing pathways for BE metabolism in the liver were added to the model. The metabolic parameters were from published sources or estimated by fitting simulation data to experimentally derived data. In order to adequately simulate the experimental data, a blood-protein binding parameter and a saturable elimination of 2-butoxyacetic acid by the kidneys was required. The model simulations were in good agreement with a wide variety of rat and human experimental data from different laboratories using different exposure routes. However, in rats for high, hemolytic-causing dose levels, the model tended to overpredict the amounts of 2-butoxyacetic acid excreted by the kidneys into the urine. The authors suggested that this overprediction may be due to the model including no diminished renal excretion or liver metabolism arising from the hemolytic effects of 2-butoxyacetic acid.

Lee *et al.* (1998) used a PBPK model to describe the toxicokinetics of BE and 2-butoxyacetic acid in rats and mice during a chronic 18-month inhalation exposure. The PBPK model was based on the models developed by Shyr *et al.* (1993) and Corley *et al.* (1994). Allometric scaling was used wherever possible to accommodate scaling to different species and to accommodate changes in body weight during long-term exposures. Pharmacokinetics differences observed between males and females for both mice and rats were included. The models were developed using data from short-term exposures (1 day to 3 weeks) and compared to data from long-term exposure (18 months). Several model parameters had to be adjusted for the chronic exposure model. Cardiac output decreased with aging and/or exposure, the fat compartment volume increased over time, the protein binding of 2-butoxyacetic acid increased over time, the rate of metabolism of 2-butoxymethanol to 2-

butoxyacetic acid increased over time, and the renal elimination rate of 2-butoxyacetic acid decreased over time. The final models incorporate these differences as time and/or exposure functions to successfully simulate blood and urine concentrations in mice, rats, and humans during short-term and long-term exposures.

Furan Kedderis *et al.* (1993) used two types of gas uptake studies to examine the inhalation pharmacokinetics of furan in rats and to develop a furan PBPK model for the inhalation exposure route. Tissue : air partition coefficients for furan were determined using vial equilibration techniques (Gargas *et al.* 1989). Kinetic parameters for furan metabolism were obtained by optimization of the model to experimental gas uptake data from rats. Simulations of the furan PBPK model were reasonably close to measured blood and liver concentrations following exposure and adequately predicted the concentrations of furan in the chamber air. Model predictions of furan metabolism following an oral dose were in agreement with previously published data for rats. This article also compared the biotransformation of furan in freshly isolated rat hepatocytes *in vitro* to biotransformation *in vivo* using a two-compartment model describing the *in vitro* vial incubation system. Because the PBPK model incorporating *in vitro* hepatocyte kinetic parameters accurately simulated *in vivo* pharmacokinetics, the authors suggest that the *in vitro* system used in their study (i.e., freshly isolated hepatocytes) may be useful for estimating metabolism parameters for PBPK models.

Kedderis and Held (1996) published another PBPK model for furan pharmacokinetics in mice, rats, and humans. This model was based on the model described in the preceding paragraph. Biological transformation parameters for the mouse and human were extrapolated from *in vitro* hepatocyte incubations. The *in vitro* studies of furan biotransformation indicated that furan metabolism is fast enough that the rate of metabolism is limited by the rate of blood flow to the liver. The human and mouse PBPK models were not validated against experimental data. The PBPK models were used to predict dosimetry values in mice and humans. Based on model simulations of 4-hour exposures to 10 ppm furan, the absorbed dose, and consequently the liver dose, of furan's toxic metabolite *cis*-2-butene-1,4-dial was approximately 3-fold less in humans than in rats and 10-fold less in humans than in mice. This comparison of the target-organ dose in three species led the authors to suggest that target organ concentration, rather than exposure concentration, is most appropriate for interspecies comparison of dose.

In an investigation of the differences between furan pharmacokinetics in children and adults, Price *et al.* (2003) developed a PBPK model for inhalation exposure incorporating age-specific differences in physiological parameters. Age-specific differences in body weight, tissue volumes, tissue water and lipid contents, alveolar ventilation rate, cardiac output, and tissue blood flow rates from a variety of different studies were compiled and incorporated into the furan model. An inhalation exposure to 1 μg/L furan in 6-, 10-, and 14-year-old children as well as adults was simulated. The simulated maximum arterial blood concentration of furan after a 30-hour exposure was highest in 6-year-olds, and it was about 40% higher in 6-year-olds than in adults. This modeling approach could be useful in assessing the risk to children from exposure to environmental contaminants and other chemicals.

Tetrahydrofuran Droz *et al.* (1999) used a PBPK model to predict tetrahydrofuran levels in the tissues and blood of humans following inhalation exposure. The PBPK model, patterned after a previous model by Droz *et al.* (1989), included seven compartments (i.e., lung, muscle and skin, fat, brain, kidneys, liver, and "others"). Partition coefficients were determined by headspace analysis for blood, water, and olive oil. A fat tissue : air partition coefficient was estimated as the sum of 70% of the oil : air partition coefficient and 30% of the blood : air partition coefficient. The fat tissue : blood partition coefficient, estimated by dividing the fat : air partition coefficient by the blood air partition coefficient, was 1.4. It was assumed all other tissues would have tissue : blood partition coefficients of one. Metabolic clearance was assumed to be flow-limited. Renal excretion was assumed to be in equilibrium with blood concentrations; the urine : air partition coefficient was estimated as the water : air partition coefficient, and urine production was estimated to be 1 mL/min. Model simulation results were quite similar to experimental values for sedentary subjects but not for subjects who performed physical activity (work) during exposure periods. The model was adequate for describing exposures in people at rest, but the authors stated that the model should be refined to include a better description of metabolism.

Methyl-*tert*-Butyl Ether (MTBE) and* tert*-Butyl Alcohol Borghoff *et al.* (1996) used a PBPK model to predict blood levels of methyl-*tert*-butyl ether, the gasoline additive referred to as MTBE, in rats following inhalation and oral exposures. The PBPK model for MTBE, based on that of Ramsey and Andersen (1984), included a kidney compartment. The major metabolite of MTBE, tert-butyl alcohol, was described by one of two secondary models. One of these secondary models used a PBPK model similar to that used to describe MTBE for describing *tert*-butyl alcohol pharmacokinetics, while the other secondary model was a simple two-compartment model involving total-body-water and liver compartments. In describing the metabolism of MTBE to *tert*-butyl alcohol, it was deemed necessary to include two saturable metabolic pathways in the liver, each with its own parameter values determined by fitting simulated blood concentrations to experimental data values. Simulated values for MTBE blood concentration and inhalation chamber concentrations were similar to experimentally measured values for all exposure routes. Simulated values for *tert*-butyl alcohol were similar to experimentally determined values when using the two-compartment model for *tert*-butyl alcohol pharmacokinetics at low exposure concentrations, but were not similar at higher exposure levels or when using the conventional PBPK model for *tert*-butyl alcohol. It was determined that further kinetic studies needed to be done in order to develop a PBPK model for *tert*-butyl alcohol.

The MTBE model of Borghoff *et al.* (1996) in rats was later adapted for use by Licata *et al.* (2001) in a study with the goals of verifying that a human MTBE PBPK model based on the rat model can adequately predict blood concentrations in humans following inhalation exposures and of examining the effect of variability in MTBE metabolism on blood levels. The rat PBPK model was scaled to humans by appropriately adjusting physiological parameters and by incorporating partition coefficients experimentally determined *in vitro* by vial equilibration. Two estimates of MTBE metabolic rate constants were employed by Licata *et al.* (2001). Metabolic rate constants for the two pathways of MTBE biotransformation had been previously determined experimentally using human liver microsomes *in vitro*, and they were extrapolated to describe metabolism *in vivo* in a human based on microsomal protein content and other factors. Metabolic parameters were also determined by fitting the model to previously reported *in vivo* human inhalation exposure data (i.e., MTBE concentrations in venous blood) in which volunteers were exposed to 1.7 ppm MTBE for 1 hour or to 4 or 40 ppm MTBE for 4 hours. To examine the effect of variability in MTBE metabolism on blood levels, the range of values in *in vitro* metabolism studies was used; it was determined that there was more variability in actual measured MTBE blood concentrations than predicted by PBPK model calculations performed with the range of *in vitro* metabolism parameters.

tert-*Amyl Methyl Ether (TAME) and* tert-*Amyl Alcohol (TAA)* Collins *et al.* (1999) developed a PBPK model for inhalation exposures to tert-amyl methyl ether (TAME) in the rat. The model included a description for the metabolite *tert*-amyl alcohol (TAA), that included compartments for the lung, liver, and total body water. The PBPK model for TAME and TAA, which included pathways for oxidation and glucuronide conjugation of TAA, underpredicted the experimental data (i.e., the concentration of TAA) at later times. Three hypotheses of the biological processes that could be causing the discrepancy between the model and the data were incorporated into the model to determine if any of these hypotheses were consistent with the data. The best model description of the data was provided by a model structure that incorporated nonspecific binding of TAA.

Ethylene Oxide As part of a biologically based approach to cancer risk assessment for ethylene oxide, Krishnan *et al.* (1992) suggest that three necessary steps include (1) developing a PBPK model to describe ethylene oxide pharmacokinetics, (2) developing a biologically based model of ethylene oxide carcinogenesis, and (3) developing a quantitative description of the mechanism of action of ethylene oxide. As the first step in this process, Krishnan *et al.* (1992) used a PBPK model to describe ethylene oxide pharmacokinetics as well as concentrations of urinary metabolite *N*-acetyl-*S*-(2-hydroxyethyl)-cysteine, hemoglobin and DNA adducts, and hepatic and extrahepatic glutathione. The conventional PBPK model of Ramsey and Andersen (1984) was modified to include brain and testes compartments because these tissues have been identified as target organs for ethylene oxide toxicities. Glutathione conjugation as well as hemoglobin and DNA binding were either estimated from published studies or experimentally determined by *in vivo* inhalation studies. Simulated values for concentrations of DNA and hemoglobin adducts, hepatic and extrahepatic glutathione, and urinary *N*-acetyl-*S*-(2-hydroxyethyl)-cysteine were similar to values obtained following inhalation exposures of 1.2 to 1200 ppm and *i.v.* administration of 1 to 100 mg/kg of ethylene oxide in rats. This ethylene oxide PBPK model characterized the individual metabolic pathways affecting ethylene oxide tissue dosimetry, which will be important for interspecies extrapolation and for understanding potential risks to humans based on animal toxicity data.

Methylethylketone (MEK) Liira *et al.* (1990) used a PBPK model to describe methylethylketone (MEK) pharmacokinetics in humans. This model was based on a conventional PBPK model by Johanson and Naslund (1988) that also included a gastrointestinal tract compartment. Simulated arterial blood concentrations were similar to actual blood concentrations at inhalation exposures of 25, 200, and 400 ppm MEK in two adult men.

In another modeling effort aimed at studying the pharmacokinetics of MEK to aid in the eventual development for a human PBPK model, Thrall *et al.* (2002a) developed a PBPK model for MEK in the rat following inhalation, oral, *i.v.*, and *i.p.* exposures. Partition coefficients used in the model were determined from tissue : air partition coefficients experimentally measured using vial equilibration methods. Values for V_{max} and K_m were determined by fitting the model to data from closed-chamber gas uptake studies, in which rats were exposed to initial concentrations of 100–2000 ppm MEK. For the highest inhalation exposure level, chamber concentrations could not be described without adding an additional first-order metabolic pathway to the PBPK model. To examine this hypothesis generated using a PBPK model, an additional study was performed in which rats were pretreated with pyrazole, which inhibits microsomal metabolism, before the inhalation exposure. Although the rate of metabolism was much lower in pretreated rats, some metabolism still occurred, which is consistent with MEK biotransformation occurring by two pathways.

To study oral, *i.v.*, and *i.p.* pharmacokinetics, Thrall *et al.* (2002a) administered aqueous MEK solutions to rats then placed them individually in a flow-though chamber to determine the amount of MEK exhaled. The PBPK model developed for inhalation exposures was first modified to include the *i.v.* exposure route. Initial model simulations of the *i.v.* exposure overestimated chamber air concentrations. To adequately describe chamber air concentrations, the alveolar ventilation rate was decreased by 30%. To compare the *i.p.* and oral pharmacokinetic data to model simulations, the oral and *i.p.* exposure routes were included in the model as first-order processes. For the oral and *i.p.* exposure routes, no reduction of the alveolar ventilation rate was necessary. This study of MEK pharmacokinetics demonstrated a stepwise process for building a multiexposure route PBPK model.

1,4-Dioxane Reitz *et al.* (1990) prepared a PBPK model for 1,4-dioxane rats, mice and humans for the inhalation, *i.v.* injection, and oral (i.e., by bolus gavage and drinking water) exposure routes. This model, which described the pharmacokinetics of the parent compound an the formation and urinary excretion of the principle metabolite β-hydroxyethoxyacetic acid, was developed in order to quantitatively estimate carcinogenic risk for human populations exposed to 1,4-dioxane based on rat and mouse toxicity data. This study included comparisons of simulated results and data for oral and inhalation exposures in rats for inhalation exposures in humans. Although no pharmacokinetics data was available for 1,4-dioxane exposures in the mouse, allometric scaling was used to develop a PBPK model in the mouse. Simulated blood concentrations were very similar to actual measured values in the rat and human. By integrating the results of the PBPK model analysis with appropriate animal bioassay data, the authors determined that people exposed to 740–3700 ppb 1,4-dioxane in air or 20,000–120,000 ppb 1,4-dioxane in water are unlikely to exhibit increased tumor frequencies as a result of the exposure.

6.3.3 Alkylbenzenes

PBPK models have been developed for five alkylbenzenes (Table 6.3): toluene, *m*-, *o*-, and *p*-xylene, and 1,2,4-trimethylbenzene (TMB). The model for *p*-xylene was discussed in Section 6.2 and will not be discussed here. These compounds are used as solvents among other applications. Models for related compounds, such as benzene and haloaromatics, are reviewed in Chapter 4.

Toluene Toluene has been the subject of many PBPK modeling efforts. The first published models involving toluene were investigations of metabolic and toxicokinetic interactions between toluene and other volatile organic chemicals (Purcell *et al.* 1990, Tardif *et al.* 1993) (see Chapter 13). Subsequent PBPK models involving toluene have been used for a variety of purposes. Jang and Droz (1996) used simulated blood concentrations to estimate a biological exposure index for toluene after comparing simulation results with data obtained from occupationally exposed workers.

TABLE 6.3 PBPK Models Developed for Alkylbenzenes

Chemical	Structure	Species	Reference
Toluene	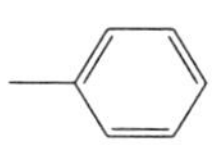	Rat	van Asperen *et al.* (2003)
		Rat	DeJongh and Blaauboer (1996)
		Human	Jang and Droz (1996)
		Human	Pierce *et al.* (1996)
		Human	DeJongh and Blaauboer (1997)
		Rat, human	Pierce *et al.* (1997)
		Rat, human	Benignus *et al.* (1998)
		Human	DeJongh *et al.* (1998)
		Human	Pierce *et al.* (1998)
		Human	Pierce *et al.* (1999)
		Human	Vicini *et al.* (1999)
		Human	Thrall *et al.* (2001)
		Human	Jonsson and Johanson (2001)
		Rat	Thrall and Woodstock (2002)
		Human	Thrall *et al.* (2002b)
1,2,4-Trimethylbenzene, TMB	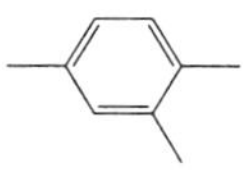	Human	Jarnberg and Johanson (1999)
m-Xylene	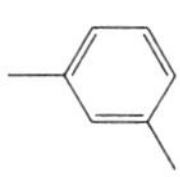	Human	Jang and Droz (1997)
		Human	MacDonald *et al.* (2002)
		Rat	Kaneko *et al.* (2000)
o-Xylene	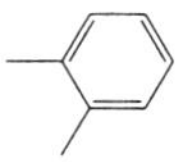	Rat, human	Thrall and Woodstock (2003)
p-Xylene		Rat	DeJongh *et al.* (1998)

Benignus *et al.* (1998) used a PBPK model to estimate toluene blood concentrations as a measure of internal dose in a meta-analysis to examine the relationship between internal dose and effect for toluene in rats and humans. As part of this study, the literature on behavioral effects of toluene in nonhuman (e.g., avoidance behavior) and human (e.g., choice reaction time) species was reviewed to uncover appropriate datasets for the meta-analysis. Rat model PBPK model parameters were from Purcell *et al.* (1990) and human model parameters were from (Tardif *et al.* 1995). The authors stated that, although the study has limitations such as uncertainties in performance of the PBPK models, expressing the independent variable in dose–response studies as internal tissue concentration rather than as external exposure parameters improved the concordance across studies.

Not many published PBPK models utilize metabolic parameters estimated from *in vitro* data. DeJongh and Blaauboer (1996) performed a study to determine if metabolic parameters reported in the literature from *in vitro* studies could be used in a PBPK model to describe *in vivo* experimental data from studies of rats exposed to toluene by inhalation. This study employed a PBPK model that included compartments for the lungs, fat, rapidly perfused tissue, slowly perfused tissue, liver, and brain. Eight sets of biotransformation parameters (V_{max} and K_m) were obtained from six *in vitro* and two *in vivo* experiments published in the literature. Simulations were performed using all eight sets of biotransformation parameters, and simulations results (i.e., calculated blood and brain toluene concentrations) were compared to experimentally determined values. Differences in simulation output were small for biotransformation parameters from the *in vivo* and *in vitro* studies. The authors concluded that the results from most *in vitro* toluene biotransformation studies can be used in a PBPK model to successfully predict toluene kinetics in the whole animal. DeJongh and Blaauboer (1997) essentially repeated this work using a conventional four-compartment PBPK model with the same conclusions.

Pierce *et al.* (1996) developed a PBPK model for human inhalation exposures to toluene to investigate interindividual differences in toxicokinetics. This study utilized data from 26 subjects exposed to a 100 ppm equimolar mixture of toluene and $^2H_8$-toluene for 2 hours. Systemic clearance of $^2H_8$-toluene was much greater than hepatic blood flow, and so metabolism of $^2H_8$-toluene in the lung was included in this model. To examine interindividual differences in $^2H_8$-toluene pharmacokinetics, simulations were performed with and without adjusting appropriate parameters (age, body weight, lean body weight, adipose tissue fraction and weight, blood/air partition coefficient, and alveolar ventilation rate) to the proper value for the individual subject. The use of subject-specific parameter values resulted in a 10% increase in the model's ability to explain data variation and a 38% increase in goodness-of-fit. A sensitivity analysis was used to show that the subject-specific parameters that had the greatest affect on simulated blood concentrations were the properties of the fat compartment, the alveolar ventilation rate, cardiac output, and the rate of lung metabolism. In a separate study, this toluene model of Pierce *et al.* (1996) was also used to determine background exposure to toluene and subsequently back-predict an individual's environmental dose hours after the initial exposure (Pierce *et al.* 1997). The toluene model was used in another study to investigate possible biological markers to toluene exposure (Pierce *et al.* 1998). The concentrations of toluene in blood and exhaled air and the urinary excretion rates of four metabolites (i.e., hippuric acid and *o*-, *m*-, and *p*-cresols) were examined to determine the best biological indicator of exposure. Of these, the concentration of toluene in exhaled air proved to be the most accurate and noninvasive indicator of exposure.

In using a PBPK modeling approach to investigate the differences in $^{1}H_{8}$-toluene and $^{2}H_{8}$-toluene toxicokinetics in men, Pierce *et al.* (1999) examined a dataset. This investigation used the previously published toluene model of Pierce *et al.* (1996) with some parameters such as fractional blood flow to adipose tissue and extrahepatic metabolism fitted to the $^{1}H_{8}$-toluene and $^{2}H_{8}$-toluene datasets. The PBPK model was then used to identify significant differences in kinetic parameters between the two datasets. This use of the PBPK model led to the discovery that $^{2}H_{8}$-toluene has a higher clearance rate than $^{1}H_{8}$-toluene, which may be due to an increased rate of ring oxidation in $^{2}H_{8}$-toluene than in $^{1}H_{8}$-toluene. In this instance, the use of a PBPK model allowed the researchers to ascribe observed isotope differences in solvent toxicokinetics to an underlying physiological mechanism. Vicini *et al.* (1999) used this same model incorporating empirical Bayes estimation techniques to demonstrate that the precision of a model's simulation results can be overly optimistic when parameters in the model are treated as known values having no error. By including an independent estimate of the measurement error in directly measured parameters, the model provided a better estimate of intra-individual variability compared to a least-squares approach.

In an evaluation of chemical dosimeters for workplace exposures, Thrall *et al.* (2001) used a previously published PBPK model for toluene (Tardif *et al.* 1993) to predict tissue concentrations corresponding to exhaled breath concentrations. In comparing the PBPK simulations to on-site breath analysis for toluene exposures, it was determined that the dosimeters underpredicted exposure during long exposures periods. A potential shortcoming of this study was that subjects in the study were exposed to both toluene and benzene, but the data were compared to simulation results from PBPK models designed for single chemicals instead of mixtures. The only simulated data comparison to actual occupational exposure in this article was to a single data point with approximately a 10% difference in between simulated and actual values.

Jonsson and Johanson (2001) used a toluene PBPK model to examine inter- and intra-individual variability in fat tissue blood flow with exercise. This PBPK model split the traditional fat tissue compartment into a subcutaneous compartment and a perirenal compartment. In this model, subcutaneous blood flow remained constant with exercise while perirenal blood flow increased with exercise. The model underpredicts the concentration of toluene in exhaled breath at rest and slightly overpredicts breath concentrations during exercise.

Thrall and Woodstock (2002) used a PBPK modeling approach in a study of the dermal absorption of toluene in rats. An aqueous solution of toluene was applied to an exposure patch (i.e., a hand-blown, glass cell with a needle hole that can be sealed with silicone) attached to the clipper-shaved back of rats, which were then placed in a small off-gassing chamber. The concentrations of toluene in chamber air were monitored. Additionally, the concentration of the original dosing solution and the concentration of the remaining solution in the exposure patch at the end of the study were determined so that the total absorbed dose could be measured. To simulate these dermal exposures, the toluene-*m*-xylene PBPK model of Tardif *et al.* (1993) was modified to include only toluene and to include the dermal exposure route. By analyzing the time-course data of chamber air concentrations with a PBPK model for toluene in rats, the rate and extent of toluene dermal absorption could be examined.

In a related study, Thrall *et al.* (2002b) evaluated the dermal absorption of toluene in human volunteers using similar methods. Volunteers were submerged to their neck in a hydrotherapy tub containing an initial concentration of 500 μg/L toluene, and the concentration of *o*-xylene in the tub water was determined every 5 minutes throughout the exposure, which lasted for 20–30 minutes. After the dermal exposure, the expired air was monitored for an additional 15–30 minutes. The rat PBPK model (Thrall and Woodstock 2002) was extended to humans by adjusting the physiological parameters appropriately and adjusting the Michaelis–Menten metabolic parameters to the values reported by Tardif *et al.* (1995) in a human toluene/xylene PBPK model. The blood : air partition coefficient was estimated using the data from human exposures. In this study, each subject underwent two experiments; in one experiment, subjects breathed purified air, while in the other study subjects breathed room air (i.e., they were exposed by the dermal and inhalation exposure routes). For the latter experiment, the relative contributions of the oral and inhalation exposure routes to the total body burden received from the combined exposure were calculated using the PBPK model. Based on model calculations and comparison of data from the two experiments, although the combined inhalation exposure resulted in a transient 50% increase in toluene concentrations in exhaled air, the inhalation exposure only resulted in a small increase in body burden.

van Asperen *et al.* (2003) used the previously described model of DeJongh and Blaauboer (1996) for toluene to investigate the effects of exposure conditions on toxicokinetics and behavior in rats. Physiological parameters and partition coefficients were not changed from the previously published model, but the liver metabolism parameters were taken from the literature specifically for male Wistar rats. Model simulations slightly overpredicted both blood and brain toluene concentrations at maximal toluene concentrations for constant and intermittent inhalation exposure while slightly underpredicting for lower exposure concentrations. The authors felt the model provided reliable predictions of toluene at different exposure conditions.

m-*Xylene* A PBPK model of *m*-xylene has been used to explain differences in xylene metabolic rates between Caucasian and oriental people (Jang and Droz 1997). This model included the metabolism of *m*-xylene to methylhippuric acid (MHA), with simulated values of MHA compared to measured values of excreted MHA. Breath and blood concentrations (simulated versus measured) were also compared. It was determined that intrinsic metabolic clearance rates were indistinguishable between Caucasian and Oriental populations for *m*-xylene. This article also included styrene and perchloroethylene. Another PBPK model describing inhalation exposures to *m*-xylene was recently developed to examine the effects of co-exposure to ethanol and *m*-xylene on central-nervous-system toxicity (MacDonald *et al.* 2002).

Kaneko *et al.* (2000) also created a PBPK model of *m*-xylene in rats that included metabolism, metabolite excretion, and a tail compartment for blood sampling. Metabolic parameters were obtained by fitting the simulation data to experimentally obtained values. Simulations underpredicted blood concentrations and overpredicted metabolic excretion.

***o*-Xylene** In a study similar to the toluene dermal absorption studies (Thrall *et al.* 2002b; Thrall and Woodstock 2002), Thrall and Woodstock (2003) evaluated the dermal absorption of *o*-xylene in rats and humans using a PBPK modeling technique. Rats were placed in an off-gassing chamber after dermal application of an aqueous xylene solution, and the concentrations in chamber air were monitored. For humans, volunteers placed their lower legs in an aqueous *o*-xylene solution, and the concentrations of *o*-xylene in this solution and in expired air were monitored. To simulate these dermal exposures, the toluene-*m*-xylene PBPK model of Tardif *et al.* (1993) for the rat was adapted to include only *m*-xylene and the dermal exposure route was incorporated. Partition coefficients for *o*-xylene were incorporated. This rat PBPK model was extended to humans by adjusting the physiological parameters appropriately. This PBPK model analysis allowed the comparison of rate and extent of *o*-xylene uptake during the rat and human dermal absorption studies.

1,2,4-Trimethylbenzene (TMB) A PBPK model for TMB inhalation exposures in humans was used to investigate the influence of factors such as the exposure concentration and workload on potential biomarkers of exposure (Jarnberg and Johanson 1999). This model included six compartments (i.e., lungs, rapidly perfused tissue, adipose tissue, liver, resting muscle, and working muscle). The clearance of the urinary metabolite 3,4-dimethylhippuric acid into the urine was included in the model because the levels of this metabolite in urine, along with TMB concentrations in venous blood and exhaled air, were studied as potential biomarkers of exposure. From a study of TMB inhalation exposures in 10 human volunteers, concentrations of TMB in arterial blood and exhaled air and data on the urinary excretion of 3,4-dimethylhippuric acid were used for model development. The effect of work on urinary excretion of metabolic product and blood and exhaled breath concentrations of TMB was simulated and compared to actual measured values. Based on model calculations, increasing the work load from rest to moderate exercise increased all biomarker concentrations and resulted in a fivefold increase in the urinary excretion of 3,4-dimethylhippuric acid through an overnight recovery period. The increases in biomarker concentrations with workload were due to the increased ventilation rate and high absorption into adipose tissue that comes with increased workload followed by the return of TMB from the adipose tissue during the recovery period, during which there is no exposure or increased workload. In simulations studying biomarker levels in blood during a typical work week with 5 days of 8-hour exposures to TMB, it was determined that biomarker levels at the end of the shift remain relatively constant while levels at the beginning of the shift increase gradually during the week.

6.3.4 Siloxanes

The two compounds included in this section (Table 6.4)—octamethylcyclotetrasiloxane (D4), and hexamethyldisiloxane (HMDS)—are used in a variety of consumer applications and can be used as intermediates in the production of silicone polymers. Silicone compounds have chemical properties that make them useful in a variety of applications, and which also lead them to have interesting chemical properties. In the following sections, it will become clear that these compounds also have interesting pharmacokinetic properties.

Octamethylcyclotetrasiloxane (D4) The pharmacokinetics of octamethylcyclotetrasiloxane (D4) has been extensively studied. For example, Plotzke *et al.* (2000) studied the disposition of D4 in male and female rats following single 6-hour inhalation exposures to 7, 70, or 700 ppm ^{14}C-D4 and multiple exposures to 7 or 700 ppm ^{14}C-D4; concentrations in up to 7 tissues as well as the rate of clearance by excretion in exhaled breath and urine were reported for up to 168 hours following the exposure. Pharmacokinetic studies for *i.v.*, dermal, and oral exposures were also performed (Sarangapani *et al.* 2003). The availability of this rich dataset provided an opportunity for the development of a detailed model for this silicone-based compound with the unusual combination of low blood:air and high fat:blood partitioning [i.e., $P_b = 4.3$ and $P_f = 485$ in rats *in vitro* (Andersen *et al.* 2001)].

The first D4 PBPK model published was developed for *i.v.* exposures in the rat, extrapolated to inhalation exposures in the rat, and then extrapolated to humans (Luu and Hutter 2001). Only a small subset of data was used to develop the rat model. For both the D4 *i.v.* and inhalation exposures, model predictions were only shown compared to fat and blood concentrations; for the inhalation exposure route, blood concentrations were not accurately described. In addition, the model was developed on the presumption that all radioactivity in blood was parent D4. Next, the rat inhalation model was extrapolated to develop a human inhalation PBPK model. For a limited validation of this model, a study in which 12 humans were exposed to 10 ppm D4 by inhalation during periods of rest and exercise was used

TABLE 6.4 PBPK Models Developed for Siloxanes

Chemical	Structure	Species	Reference
Hexamethyldisiloxane (HMDS)	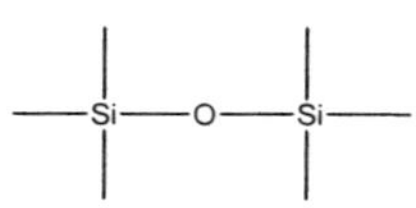	Rat	Dobrev *et al.* (2003)
Octamethylcyclotetrasiloxane (D4)	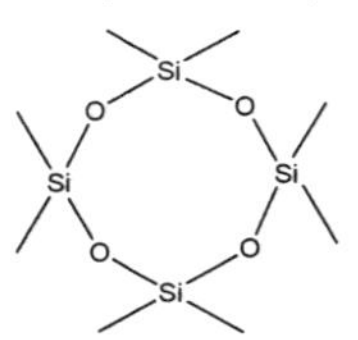	Rat Rat, human Rat Human	Andersen *et al.* (2001) Luu and Hutter (2001) Saragapani *et al.* (2003) Reddy *et al.* (2003)

(Utell 1998). Model predictions were shown to describe concentrations of D4 in blood plasma up to 24 hours after the exposure ended. Finally, the human inhalation model was extrapolated to include an additional exposure route (i.e., the seepage of D4 from polydimethylsiloxane breast implants, in which D4 can be found as a low-level impurity) (Luu and Hutter 2001).

A later PBPK model for D4 rat pharmacokinetics described D4 concentrations in the plasma, fat, lung, and liver as well as the rate of D4 exhaled and excreted following single and multiple inhalation exposures (Andersen *et al.* 2001). Using PBPK modeling techniques, it was found that some D4 appeared to be present in blood in a form that was not available for equilibration with end alveolar air. This discovery led to a model structure where an unavailable form of D4 was produced in the liver, transferred/stored in the blood, and cleared in the fat. The PBPK model also included deep-tissue compartments in the liver and lung and a second fat compartment. This rat inhalation PBPK model was later extended to describe D4 kinetics in the rat following oral, dermal, and *i.v.* exposures (Sarangapani *et al.* 2003). Interestingly, the multiexposure route PBPK model development revealed that D4 absorbed by the inhalation and dermal routes apparently enters the pool of available D4 in the blood (presumably corresponding to a free, molecular form of D4), while D4 absorbed during *i.v.* or oral exposures apparently enters the pool of unavailable D4 in the blood (presumably as a complexed form of D4 within chylomicrons or other lipoproteins). Thus, D4 pharmacokinetics following inhalation and dermal exposures was qualitatively different than D4 pharmacokinetics following oral and *i.v.* exposures.

The rat inhalation model structure was successfully adapted to describe D4 inhalation kinetics in humans (Reddy *et al.* 2003) using data from a later study of human inhalation pharmacokinetics of ^{14}C-D4 where six male volunteers were exposed to 10 ppm ^{14}C-D4 during periods of rest and exercise. During this study, concentrations of D4 were determined in blood and exhaled breath; additionally, urine samples were collected for up to 7 days and time-course amounts of individual metabolites in the urine were determined. As with efforts to describe D4 pharmacokinetics in the rat, the human inhalation PBPK model was not able to describe D4 concentrations in both blood and exhaled air unless the model included an unavailable form of D4 produced in the liver, transferred/stored in the blood, and cleared in the fat. Because the human inhalation PBPK model included the gross physiological changes that occur during exercise, the model successfully described the decreased retention of D4 that occurs with increased alveolar ventilation from exercise. The quantification of individual metabolites in the urine allowed the addition of a metabolism model to the PBPK model (i.e., a model structure for individual metabolites in blood was inferred from the urine data as well as from rates of metabolism of the linear siloxane metabolites).

Hexamethyldisiloxane (HMDS) A PBPK model describing hexamethyldisiloxane (HMDS) pharmacokinetics in the rat following closed-chamber and constant concentration inhalation exposures as well as oral and *i.p.* exposures has been developed (Dobrev *et al.* 2003). The model structure for HMDS, based on the model structure for D4, also indicated the presence of an unavailable form of HMDS in the blood. Significant lag times for HMDS appearance in the blood following oral exposures necessitated the use of a multicompartment description of the GI tract. Additionally, it was found that although for many volatile compounds closed-chamber inhalation exposure data are adequate for estimating the rate of metabolism, for HMDS they are not. Due to HMDS's combination of low blood : air and high fat : blood partitioning [i.e., $P_b = 1.4$ and $P_f = 186$ in rats *in vitro* (Dobrev *et al.* 2003)], additional constant concentration exposure data was necessary for determining the proper PBPK model structure and parameterization.

6.4 SUMMARY

This chapter has covered PBPK modeling for a diverse group of organic compounds. The primary interest with most of these compounds arises from their commercial uses or in relation to occupational exposures. For many there is the need to develop approaches to assess the best dose measures to correlate with toxicity or to use in risk assessments in developing exposure standards, such as reference concentrations (RfCs), reference doses (RfDs), or TLVs (threshold limit values). Another important application is to support biomonitoring, assessing the intensity of exposure and exposure patterns based on measurements of chemicals or metabolites in biological samples such as exhaled breath, urine, or blood. These PBPK models can be useful for assisting in developing standards based on more appropriate estimates of doses at target tissues rather than on the basis of exposure concentrations or oral doses. Some of the characteristics of particular groups of these compounds led to the need to alter model structures. For example, for the more water-soluble volatile compounds, washin–washout processes were required in the respiratory tract.

NOTATION

BE	2-butoxyethanol
D4	octamethylcyclotetrasiloxane
HMDS	hexamethyldisiloxane
i.v.	intravenous
i.p.	intravenous
$\mathbf{K_m}$	the Michaelis constant
MEK	methylethylketone
MHA	methylhippuric acid
MTBE	methyl-*tert*-butyl ether
$\mathbf{P_b}$	blood : air partition coefficient
$\mathbf{P_f}$	fat : blood partition coefficient
RfC	reference concentration
RfD	reference dose
TAA	*tert*-amyl alcohol
TAME	*tert*-amyl methyl ether

TLV threshold limit value
TMB 1,2,4-trimethylbenzene
V_{max} the enzymatic reaction velocity at maximal substrate concentration

REFERENCES

Adolf, E. F. (1949) Quantitative relations in the physiological constitutions of mammals. *Science* **109**, 579–585.

Andersen, M. E., Clewell, H. J., Gargas, M. L., Smith, F. A., and Reitz, R. H. (1987). Physiologically based pharmacokinetics and the risk assessment process for methylene-chloride. *Toxicol. Appl. Pharmacol.* **87**, 185–205.

Andersen, M. E., Sarangapani, R., Reitz, R. H., Gallavan, R. H., Dobrev, I. D., and Plotzke, K. P. (2001). Physiological modeling reveals novel pharmacokinetic behavior for inhaled octamethylcyclotetrasiloxane in rats. *Toxicol. Sci.* **60**, 214–231.

Angelo, M. J., and Bischoff, K. (1982). A physiologically-based pharmacokinetic model for 2,5-hexanedione. In: *Proceedings of the Twelfth Conference on Environmental Toxicology*, November 1981, AFAMRL-TR-81-149, pp. 250–286.

Astrand, I. (1983). Effect of physical exercise on uptake, distribution and elimination of vapors in man. In: *Modeling of Inhalation Exposure to Vapors: Uptake, Distribution and Elimination*, Vol. 1, V. Fiserova-Bergerova, ed., CRC Press, Boca Raton, FL, pp. 107–130.

Benignus, V. A., Boyes, W. K., and Bushnell, P. J. (1998). A dosimetric analysis of behavioral effects of acute toluene exposure in rats and humans. *Toxicol. Sci.* **43**, 186–195.

Borghoff, S. J., Murphy, J. E., and Medinsky, M. A. (1996). Development of a physiologically based pharmacokinetic model for methyl tertiary-butyl ether and tertiary-butanol in male Fischer-344 rats. *Fundam. Appl. Toxicol.* **30**, 264–275.

Clewell, H. J., Gentry, P. R., Gearhart, J. M., Covington, T. R., Banton, M. I., and Andersen, M. E. (2001). Development of a physiologically based pharmacokinetic model of isopropanol and its metabolite acetone. *Toxicol. Sci.* **63**, 160–172.

Collins, A. S., Sumner, S. C. J., Borghoff, S. J., and Medinsky, M. A (1999). A physiological model for *tert*-amyl methyl ether and *tert*-amyl alcohol: Hypothesis testing of model structures. *Toxicol. Sci.* **49**, 15–28.

Corley, R. A., Bormett, G. A., and Ghanayem, B. I. (1994). Physiologically-based pharmacokinetics of 2-butoxyethanol and its major metabolite, 2-butoxyacetic acid, in rats and humans. *Toxicol. Appl. Pharmacol.* **129**, 61–79.

DeJongh, J., and Blaauboer, B. J. (1996). Simulation of toluene kinetics in the rat by a physiologically based pharmacokinetic model with application of biotransformation parameters derived independently *in vitro* and *in vivo*. *Fundam. Appl. Toxicol.* **32**, 260–268.

DeJongh, J., and Blaauboer, B. J. (1997). Evaluation of *in vitro*-based simulations of toluene uptake and metabolism in rats. *Toxicol. Vitro* **11**, 485–489.

DeJongh, J., Verhaar, H. J. M., and Hermens, J. L. M. (1998). Role of kinetics in acute lethality of non-reactive volatile organic compounds (VOCs). *Toxicol. Sci.* **45**, 26–32.

Derr, R. F. (1993). Simulation studies on ethanol metabolism in different human populations with a physiological pharmacokinetic model. *J. Pharm. Sci.* **82**, 677–682.

Dobrev, I. D., Reddy, M. B., Plotzke, K. P., Varaprath, S., McNett, D. A., Durham, J., and Andersen, M. E. (2003). Closed-chamber inhalation pharmacokinetic studies with hexamethyldisiloxane in the rat. *Inhal. Toxicol.* **15**, 589–617.

Droz, P. O., Wu, M. M., Cumberland, W. G., and Berode, M. (1989). Variability in biological monitoring of solvent exposure. 1. Development of a population physiological model. *Br. J. Ind. Med.* **46**, 447–460.

Droz, P. O., Berode, M., and Jang, J. Y. (1999). Biological monitoring of tetrahydrofuran: Contribution of a physiologically based pharmacokinetic model. *Am. Ind. Hyg. Assoc. J.* **60**, 243–248.

Fiserova-Bergerova, V. (1975). *Biological-Mathematical Modeling of Chronic Toxicity*. AMRL-TR-75-5. Aerospace Medical Research Laboratory, Wright-Patterson Air Force Base, OH.

Fiserova-Bergerova, V. (1983). Physiological models for pulmonary administration and elimination of inert vapors and gases. In: *Modeling of Inhalation Exposure to Vapors: Uptake, Distribution and Elimination*, Vol. 1, V. Fiserova-Bergerova, ed., CRC Press, Boca Raton, FL, pp. 73–100.

Fisher, J. W., Dorman, D. C., Medinsky, M. A., Welsch, F., and Conolly, R. B. (2000). Analysis of respiratory exchange of methanol in the lung of the monkey using a physiological model. *Toxicol. Sci.* **53**, 185–193.

Gargas, M. L., Burgess, R. J., Voisard, D. E., Cason, G. H., and Andersen, M. E. (1989). Partition coefficients of low-molecular-weight volatile chemicals in various liquids and tissues. *Toxicol. Appl. Pharmacol.* **98**, 87–99.

Horton, V. L., Higuchi, M. A., and Rickert, D. E. (1992). Physiologically based pharmacokinetic model for methanol in rats, monkeys, and humans. *Toxicol. Appl. Pharmacol.* **117**, 26–36.

Jang, J. Y., and Droz, P. O. (1996). Simulation of toluene in venous blood with a physiologically based pharmacokinetic model: Its application to biological exposure index development. *Appl. Occup. Environ. Hyg.* **11**, 1092–1095.

Jang, J. Y., and Droz, P. O. (1997). Ethnic differences in biological monitoring of several organic solvents. 2. A simulation study with a physiologically based pharmacokinetic model. *Int. Arch. Occup. Environ. Health* **70**, 41–50.

Jarnberg, J., and Johanson, G. (1999). Physiologically based modeling of 1,2,4-trimethylbenzene inhalation toxicokinetics. *Toxicol. Appl. Pharmacol.* **155**, 203–214.

Johanson, G. (1986). Physiologically based pharmacokinetic modeling of inhaled 2-butoxyethanol in man. *Toxicol. Lett.* **34**, 23–31.

Johanson, G., and Dynesius, B. (1988). Liquid air partition coefficients of 6 commonly used glycol ethers. *Br. J. Ind. Med.* **45**, 561–564.

Johanson, G., and Naslund, P. H. (1988). Spreadsheet programming a new approach in physiologically based modeling of solvent toxicokinetics. *Toxicol. Lett.* **41**, 115–128.

Jonsson, F., and Johanson, G. (2001). Bayesian estimation of variability in adipose tissue blood flow in man by physiologically based pharmacokinetic modeling of inhalation exposure to toluene. *Toxicology* **157**, 177–193.

Kaneko, T., Horiuchi, J., and Sato, A. (2000). Development of a physiologically based pharmacokinetic model of organic solvent in rats. *Pharmacol. Res.* **42**, 465–470.

Kedderis, G. L., Carfagna, M. A., Held, S. D., Batra, R., Murphy, J. E., and Gargas, M. L. (1993). Kinetic analysis of furan biotransformation by F-344 rats *in vivo* and *in vitro*. *Toxicol. Appl. Pharmacol.* **123**, 274–282.

Kedderis, G. L., and Held, S. D. (1996). Prediction of furan pharmacokinetics from hepatocyte studies: Comparison of bioactivation and hepatic dosimetry in rats, mice, and humans. *Toxicol. Appl. Pharmacol.* **140**, 124–130.

Krishnan, K., Gargas, M. L., Fennell, T. R., and Andersen, M. E. (1992). A physiologically based description of ethylene-oxide dosimetry in the rat. *Toxicol. Ind. Health* **8**, 121–140.

Kumagai, S., and Matsunaga, I. (1995). Physiologically-based pharmacokinetic model for acetone. *Occup. Environ. Med.* **52**, 344–352.

Lam, C. W., Galen, T. J., Boyd, J. F., and Pierson, D. L. (1990). Mechanism of transport and distribution of organic solvents in blood. *Toxicol. Applied. Pharmacol.* **104**, 117–129.

Lee, K., Dill, J. A., Chou, B. J., and Roycroft, J. H. (1998). Physiologically based pharmacokinetic model for chronic inhalation of 2-butoxyethanolanol. *Toxicol. Appl. Pharmacol.* **153**, 211–226.

Licata, A. C., Dekant, W., Smith, C. E., and Borghoff, S. J. (2001). A physiologically based pharmacokinetic model for methyl *tert*-butyl ether in humans: Implementing sensitivity and variability analyses. *Toxicol. Sci.* **62**, 191–204.

Liira, J., Johanson, G., and Riihmaki, V. (1990). Dose-dependent kinetics of inhaled methylethylketone in man. *Toxicol. Lett.* **2**, 195–201.

Luu, H. M. D., and Hutter, J. C. (2001). Bioavailability of octamethylcyclotetrasiloxane (D4) after exposure to silicones by inhalation and implantation. *Environ. Health Perspect.* **109**, 1095–1101.

MacDonald, A. J., Rostami-Hodjegan, A., Tucker, G. T., and Linkens, D. A. (2002). Analysis of solvent central nervous system toxicity and ethanol interactions using a human population physiologically based kinetic and dynamic model. *Reg. Toxicol. Pharmacol.* **35**, 165–176.

McDougal, J. N., Jepson, G. W., Clewell, H. J. III, Gargas, M. L., and Andersen, M. E. (1990). Dermal absorption of organic chemical vapors in rats and humans. *Fund. Appl. Toxicol.* **14**, 299–308.

Morzorati, S. L., Ramchandani, V. A., Li, T. K., and O'Connor, S. (2002). Maintaining steady state arterial alcohol levels in rats by using a physiologically based pharmacokinetic model. *Alcohol* 28, 189–195.

Pastino, G. M., and Conolly, R. B. (2000). Application of a physiologically based pharmacokinetic model to estimate the bioavailability of ethanol in male rats: Distinction between gastric and hepatic pathways of metabolic clearance. *Toxicol. Sci.* **55**, 256–265.

Pastino, G. M., Sultatos, L. G., and Flynn, E. J. (1996). Development and application of a physiologically based pharmacokinetic model for ethanol in the mouse. *Alcohol Alcohol.* **31**, 365–374.

Pastino, G. M., Asgharian, B., Roberts, K., Medinsky, M. A., and Bond, J. A. (1997). A comparison of physiologically based pharmacokinetic model predictions and experimental data for inhaled ethanol in male and female B6C3F(1) mice, F344 rats, and humans. *Toxicol. Appl. Pharmacol.* **145**, 147–157.

Perbellini, L., Mozzo, P., Brugnone, F., and Zedde, A. (1986). Physiologicomathematical model for studying human exposure to organic-solvents—Kinetics of blood-tissue normal-hexane concentrations and of 2,5-hexanedione in urine. *Brit. J. Ind. Med.* **43**, 760–768.

Perbellini, L., Mozzo, P., Olivato, D., and Brugnone, F. (1990). Dynamic biological exposure indexes for *n*-hexane and 2,5-hexanedione, suggested by a physiologically based pharmacokinetic model. *Am. Ind. Hyg. Assoc. J.* **51**, 356–362.

Perico, A., Cassinelli, C., Brugnone, F., Bavazzano, P., and Perbellini, L. (1999). Biological monitoring of occupational exposure to cyclohexane by urinary 1,2- and 1,4-cyclohexanediol determination. *Int. Arch. Occup. Environ. Health* **72**, 115–120.

Pierce, C. H., Dills, R. L., Morgan, M. S., Nothstein, G. L., Shen, D. D., and Kalman, D. A. (1996). Interindividual differences in H-2(8)-toluene toxicokinetics assessed by a semiempirical physiologically based model. *Toxicol. Appl. Pharmacol.* **139**, 49–61.

Pierce, C. H., Dills, R. L., Lewandowski, T. A., Morgan, M. S., Wessels, M. A., Shen, D. D., and Kalman, D. A. (1997). Estimation of background exposure to toluene using a physiologically-based kinetic model. *J. Occup. Health* **39**, 130–137.

Pierce, C. H., Dills, R. L., Morgan, M. S., Vicini, P., and Kalman, D. A. (1998). Biological monitoring of controlled toluene exposure. *Int. Arch. Occup. Environ. Health* **71**, 433–444.

Pierce, C. H., Lewandowski, T. A., Dills, R. L., Morgan, M. S., Wessels, M. A., Shen, D. D., and Kalman, D. A. (1999). A comparison of H-1(8)- and H-2(8)-toluene toxicokinetics in men. *Xenobiotica* **29**, 93–108.

Plotzke, K. P., Crofoot, S. D., Ferdinandi, E. S., Beattie, J. G., Reitz, R. H., McNett, D. A., and Meeks, R. G. (2000). Disposition of radioactivity in Fischer 344 rats after single and multiple inhalation exposure to [^{14}C]octamethylcyclotetrasiloxane ([^{14}C]D4). *Drug Metab. Dispos.* **28**, 192–204.

Price, K., Haddad, S., and Krishnan, K. (2003). Physiological modeling of age-specific changes in the pharmacokinetics of organic chemicals in children. *J. Toxicol. Env. Health Pt. A* **66**, 417–433.

Purcell, K. J., Cason, G. H., Gargas, M. L., Andersen, M. E., and Travis, C. C. (1990). *In vivo* metabolic interactions of benzene and toluene. *Toxicol. Lett.* **52**, 141–152.

Ramchandani, V. A., Bolane, J., Li, T. K., and O'Connor, S. (1999). A physiologically-based pharmacokinetic (PBPK) model for alcohol facilitates rapid BrAC clamping. *Alcohol Clin. Exp. Res.* **23**, 617–623.

Ramsey, J. C., and Andersen, M. E. (1984). A physiologically based description of the inhalation pharmacokinetics of styrene in rats and humans. *Toxicol. Appl. Pharmacol.* **73**, 159–175.

Reddy, M. B., Dobrev, I. D., Jovanovic, M. L., Crofoot, S., McNett, D. A. Tobin, J. M., Utell, M. J., Morrow, P. E., Plotzke, K. P., and Andersen, M. E. (2003a). Physiological modeling of the inhalation kinetics of decamethylcyclopentasiloxane (D5) in rats and humans. *Toxicol. Sci.*

Reddy, M. B., Looney, R. J., Utell, M. J., Jovanovic, M. L., McMahon, J. M., McNett, D. A., Plotzke, K. P. and Andersen, M. E. (2003b) Physiological modeling of the dermal absorption of octamethylcyclotetrasiloxane (D4) and decamethylcyclopentasiloxane (D5). *Toxicol. Sci.*

Reitz, R. H., McCroskey, P. S., Park, C. N., Andersen, M. E., and Gargas, M. L. (1990). Development of a physiologically based pharmacokinetic model for risk assessment with 1,4-dioxane. *Toxicol. Appl. Pharmacol.* **105**, 37–54.

Rheingold, J. L., Lindstrom, R. E., and Wilkinson, P. K. (1981). A new blood-flow pharmacokinetic model for ethanol. *J. Pharmacokinet. Biopharmaceut.* **9**, 261–278.

Sarangapani, R., Teeguarden, J., Andersen, M. E., Reitz, R. H., and Plotzke, K. P. (2003). Route-specific differences in distribution characteristics of octamethylcyclotetrasiloxane in rats: Analysis using PBPK models. *Toxicol. Sci.* **71**, 41–52.

Shyr, L. J., Sabourin, P. J., Medinsky, M. A., Birnbaum, L. S., and Henderson, R. F. (1993). Physiologically based modeling of 2-butoxyethanol disposition in rats following different routes of exposure. *Environ. Res.* **63**, 202–218.

Tardif, R., Lapare, S., Krishnan, K., and Brodeur, J. (1993). Physiologically based modeling of the toxicokinetic interaction between toluene and *m*-xylene in the rat. *Toxicol. Appl. Pharmacol.* **120**, 266–273.

Tardif, R., Lapare, S., Charest-Tardif, G., Brodeur, J., and Krishnan, K. (1995). Physiologically-based pharmacokinetic modeling of a mixture of toluene and xylene in humans. *Risk Anal.* **15**, 335–342.

Thrall, K. D., and Woodstock, A. D. (2002). Evaluation of the dermal absorption of aqueous toluene in F344 rats using real-time breath analysis and physiologically based pharmacokinetic modeling. *J. Toxicol. Env. Health Pt. A* **65**, 2087–2100.

Thrall, K. D., and Woodstock, A. D. (2003). Evaluation of the dermal bioavailability of aqueous xylene in F344 rats and human volunteers. *J. Toxicol. Env. Health Pt. A* **66**, 1267–1281.

Thrall, K. D., Callahan, P. J., Weitz, K. K., Edwards, J. A., Brinkman, M. C., and Kenny, D. V. (2001). Design and evaluation of a breath-analysis system for biological monitoring of volatile compounds. *Am. Ind. Hyg. Assoc. J.* **62**, 28–35.

Thrall, K. D., Soelberg, J. J., Weitz, K. K., and Woodstock, A. D. (2002a). Development of a physiologically based pharmacokinetic model for methyl ethyl ketone in F344 rats. *J. Toxicol. Env. Health Pt. A* **65**, 881–896.

Thrall, K. D., Weitz, K. K., and Woodstock, A. D. (2002b). Use of real-time breath analysis and physiologically based pharmacokinetic modeling to evaluate dermal absorption of aqueous toluene in human volunteers. *Toxicol. Sci.* **68**, 280–287.

Utell, M. J., Gelein, R., Yu, C. P., Kenaga, C., Geigel, E., Torres, A., Chalupa, D., Gibb, F. R., Speers, D. M., Mast, R. W., and Morrow, P. E. (1998). Quantitative exposure of humans to an octamethylcyclotetrasiloxane (D_4) vapor. *Toxicol. Sci.* **44**, 206–213.

van Asperen, J., Rijcken, W. R. P., and Lammers, J. (2003). Application of physiologically based toxicokinetic modelling to study the impact of the exposure scenario on the toxicokinetics and the behavioural effects of toluene in rats. *Toxicol. Lett.* **138**, 51–62.

Vicini, P., Pierce, C. H., Dills, R. L., Morgan, M. S., and Kalman, D. A. (1999). Individual prior information in a physiological model of 2H8-toluene kinetics: An empirical Bayes estimation strategy. *Risk Anal.* **19**, 1127–1134.

Wagner, J. G. (1967). Method for estimating rate constants for absorption, metabolism and elimination from urinary excretion data. *J. Pharmacol. Sci.* **56**, 489–494.

PART II

PBPK MODEL DEVELOPMENT FOR ENVIRONMENTAL POLLUTANTS

CHAPTER 7

PESTICIDES AND PERSISTENT ORGANIC POLLUTANTS (POPs)

Carol J. Belfiore

7.1 INTRODUCTION

Persistent organic pollutants (POPs) are compounds that have long half-lives in soil, air, sediment, or biological organisms (Jones and de Voogt 1999). Typically lipophilic, they partition to solids in soil or aqueous systems, and they partition to lipids in organisms. The propensity of some POPs to volatilize allows redeposition over long distances. Bioaccumulation and magnification in the food chain can also be associated with persistence in the environment. POPs enter the environment through several sources: application to crops and soils (e.g., pesticides), release of manufactured compounds (e.g., polychlorinated biphenyls, PCBs), accidental formation or release (e.g., 2,3,7,8-tetrachlorodibenzo-*p*-dioxin, TCDD), or industrial activities (e.g., combustion products) (Jones and de Voogt 1999). Harmful effects of POPs are varied, and include impacts on reproduction such as oligospermia and decreased sperm motility noted following exposure to chlordecone (Faroon *et al.* 1995) as well as endocrine disruption by DDT, PCBs, and dioxins (Kavlock *et al.* 1996). Immunotoxicity (Chang *et al.* 1981), carcinogenicity (Safe 1994), and nervous system effects (Jacobson and Jacobson 1996) have been attributed as well to PCB exposure.

Physiologically Based Pharmacokinetic Modeling, edited by Micaela B. Reddy, Raymond S. H. Yang, Harvey J. Clewell III, and Melvin E. Andersen.

While many compounds have been identified as POPs, physiologically based pharmacokinetic (PBPK) models have primarily been designed to predict the disposition of pesticides and polychlorinated or polybrominated biphenyls in mammals and fish (Tables 7.1 to 7.3). Pesticides are defined by the United States Environmental Protection Agency (US EPA) as substances that are intended to prevent, destroy, repel, or mitigate any pest. Pests include undesirable insects, plants, mammals, fungi, or microorganisms. While the US EPA definition of pesticides is

TABLE 7.1 PBPK Models Developed for Insecticides and Insecticide Metabolites

Compound	Class	Species	Reference
Chlorofenvinfos	Organophosphate insecticide	Rat	Ikeda *et al.* (1992)
Chlorpyrifos	Organophosphate insecticide	Rat, human	Timchalk *et al.* (2002a)
		Human	Timchalk *et al.* (2002b)
Diisopropylfluorophosphate	Organophosphate insecticide	Rat, mouse	Gearhart *et al.* (1990)
		Rat, mouse, human	Gearhart *et al.* (1994)
Isofenphos	Organophosphate insecticide	Rat	Knaak *et al.* (1990)
		Rat	Knaak *et al.* (1994)
		Rat, dog, guinea pig	Knaak *et al.* (1996)
Malathion	Organophosphate insecticide	Pig (skin)	Carver *et al.* (1989)
		Human	Rabovsky and Brown (1993)
		Human	Dong *et al.* (1994)
		Human	Bouchard *et al.* (2003)
Parathion	Organophosphate insecticide	Pig (skin)	Carver *et al.* (1989)
		Mouse	Sultatos (1990)
		Rat	Gearhart *et al.* (1994)
Paraoxon	Metabolite of parathion	Rainbow trout	Abbas and Hayton (1997)
Soman	Organophosphate nerve agent	Rat	Maxwell *et al.* (1988)
Chlordecone (Kepone®)	Organochlorine insecticide (cyclodiene)	Rat	Bungay *et al.* (1981)
		Female rat	Heatherington *et al.* (1998)
		Rat	Belfiore *et al.* (2005)
Mirex	Organochlorine insecticide (cyclodiene)	Rat, Rhesus monkey	Belfiore *et al.* (2005)
p,p′-DDE	Metabolite of DDT (diphenyl aliphatic)	Rat dam, fetus	You *et al.* (1999)
Dieldrin, HEOD	Organochlorine insecticide (cyclodiene)	Mammals	Lindstrom *et al.* (1974)
		Mammals	Lindstrom *et al.* (1975)
		Mammals	Lindstrom *et al.* (1976)
Lindane, γ-BHC	Organochlorine insecticide (hexachlorocyclohexane)	Rat	DeJongh and Blaauboer (1997)

TABLE 7.2 PBPK Models Developed for Herbicides

Compound	Class	Species	Reference
2,4-Dichlorophenoxyacetic acid, 2,4-D	Organochlorine (phenoxyaliphatic acid)	Rat, rabbit	Kim *et al.* (1994)
		Rabbit	Kim *et al.* (1995)
		Rabbit dam, fetus	Kim *et al.* (1996)
		Rat	Kim *et al.* (2001)
Atrazine	Organochlorine (chlorotriazine)	Rat	McMullin *et al.* (2003)
Fluazifop-butyl	Organofluorine (aryloxyphenoxypropionate)	Human	Auton *et al.* (1994)
Trifluralin	Organofluorine (dinitrobenzenamine)	Fish species	Schultz and Hayton (1999)

TABLE 7.3 PBPK Models Developed for PCBs, PBBS, TCBTs, and DDS

Compound(s)	Species	Reference
4-Chlorobiphenyl,	Rat	Anderson *et al.* (1977)
4,4′-dichlorobiphenyl,	Rat	Lutz *et al.* (1977)
2,2′,4,5,5′-pentachlorobiphenyl, and	Mouse	Tuey and Matthews (1980b)
2,2′,4,4′,5,5′-hexachlorobiphenyl	Rat	Matthews and Dedrick (1984)
3,3′,5,5′-Tetrachlorobiphenyl	Rat	Tuey and Matthews (1977)
4,4′-Dichlorobiphenyl, 2,2′,4,4′,5,5′-hexachlorobiphenyl, 2,2′,3,3′,6,6′-hexachlorobiphenyl	Mouse, rat, dog, monkey	Lutz *et al.* (1984)
2,2′,5,5′-Tetrachlorobiphenyl	Fish species	Lien and McKim (1993)
209 congeners of chlorobiphenyls	Rat	Parham and Portier (1998)
2,2′,4,4′,5,5′-Hexachlorobiphenyl, 3,3′,4,4′,5-pentachlorobiphenyl	Mouse	Lee *et al.* (2002)
Tetrachlorobenzyltoluenes 87, 88, 94	Rat	Kramer *et al.* (2001)
2,2′,4,4′,5,5′-Hexabromobiphenyl	Rat	Tuey and Matthews (1980a)
p,p′-Dichlorodiphenylsulfone, DDS	Rat, mouse	Parham *et al.* (2002)

quite broad, exclusions include antibiotic drugs (covered in Chapter 10), plant health promoters such as fertilizers, biological control agents such as ladybugs, and products containing low-risk ingredients such as garlic. Because pesticides are designed to adversely affect living organisms, it is not surprising that they present potential risk of harm to humans and other nontarget animals and plants in the environment. The pesticides covered in this chapter include only those with the potential to accumulate in the environment or in an organism, often due to low metabolic clearance or high lipophilicity. In addition, a PBPK/PD model for the organophosphate soman, which was developed as a biological weapon, is reviewed among similar compounds. Tables 7.1 and 7.2, respectively, summarize the PBPK models for the insecticides and herbicides reviewed in Part II.

Insecticides comprise the largest class of pesticides currently studied using PBPK modeling. Of these, organophosphate compounds have received most attention, while the pharmacokinetics of several organochlorine insecticides has been modeled as well (Table 7.1). In addition, models for four chemically diverse organohalide herbicides have been developed (Table 7.2). Numerous models involving the POP TCDD, which occurs as a contaminant in herbicide mixtures of 2,4-di- and 2,4,5-trichlorophenoxyacetic acid, are reviewed in Chapter 8. Other pesticides such as attractants, repellants, and microbial pesticides are absent from the current modeling literature. However, PBPK models of several fungicides, along with disinfectants having characteristics common to other compounds, are reviewed in other chapters of this work. Hexachlorobenzene, utilized until the 1950s as a fungicide, is among alkenes included in Chapter 4. The fumigant ethylene dibromide is a volatile organic included in Chapter 2.

PCBs have been used commercially for decades. Mixtures of PCBs have been employed as plasticizers and adhesives, and they have also been used as dielectric fluid in capacitors and transformers. The pharmacokinetics of PCBs and PBBs have also been studied using PBPK modeling approaches, and as they share some distributional characteristics with pesticide POPs, models for these compounds are reviewed in this chapter (Table 7.3).

The PBPK models discussed in this chapter have served as tools to assess biochemical and physicochemical parameters, metabolic rates, exposure scenarios, polymorphic variability, and maternal–fetal exposure, as well as the role of the blood–brain barrier to toxin disposition. For some pesticides, these PBPK models have been extended to study the pharmacodynamic (PD) behavior of the compound. In the following sections, first general pharmacokinetic considerations for POPs are discussed. Next, PBPK models and their natural extension to PBPK/PD models for pesticides are reviewed. Finally, PBPK models for PCBs and PBBs are presented.

7.2 PESTICIDES

7.2.1 Chemical Classes of Pesticides

Organophosphates Generally, insecticide intoxication results in adverse neurological effects. The mechanism of toxicity, however, is characteristic of the chemical class of the compound. The toxicity of organophosphate insecticides arises from their ability to inhibit acetylcholinesterase (ACHE) activity in the target organism. Other esterases such as carboxylesterase (CAE) and butyrylcholinesterase (BCHE) are affected as well, but these interactions generally result in minimal toxicity. Inhibition of ACHE allows the neurotransmitter acetylcholine to accumulate at cholinergic synapses, producing continuous stimulation in these pathways within the central and peripheral nervous systems. Intoxication thus results in neurobehavioral, cognitive, and neuromuscular symptoms (Marrs 1993).

The PD aspect of these pesticides is emphasized by inclusion of anticholinesterase activity in PBPK models of several organophosphate insecticides. Gearhart *et al.* (1990) measured and modeled brain and plasma ACHE inhibition by diisopropylfluorophosphate (DFP). Inhibition was described by a model relating bimolecular rate of inhibition, regeneration of enzyme, aging, synthesis, and degradation of CAE and BCHE as well as ACHE. The model was extended to describe parathion pharmacokinetics and enzyme inhibition (Gearhart *et al.* 1994). Utilizing experimentally determined rates of ACHE and CAE inhibition, and hydrolysis of soman by DFPase, Maxwell *et al.* (1988) produced a PBPK/PD model describing tissue-specific activity of ACHE and CAE in soman-treated rats. Abbas and Hayton (1997), noting insignificant regeneration of ACHE bound by the parathion metabolite paraoxon, based the expression for free cholinesterase upon bimolecular inhibition of ACHE and CAE, enzyme synthesis, and degradation. Isofenphos (IF) inhibition of CAE in blood, liver, and brain, and of ACHE and BCHE in blood and brain, was represented by a bimolecular rate constant derived from *in vitro* assays (Knaak *et al.* 1994, 1996). Relating decreased blood concentration of toxicant to that in brain, Ikeda *et al.* (1992) attempted to clarify lower toxicity of chlorfenvinphos (CVP) following pretreatment with the same compound. While brain ACHE activity was measured, enzyme inhibition was not included in the model (Ikeda *et al.* 1992). Recently presented PBPK/PD describing the action of chlorpyrifos (CPF) in rats and humans (Timchalk *et al.* 2002a,b) address cholinesterase inhibition as well as distribution of parent compound and toxic metabolite.

Other models dealing with organophosphates parathion and malathion have focused on tissue levels rather than anticholinesterase activity. Such models necessarily include description of insecticide metabolism. Sultatos (1990) derived metabolic constants for both activation and detoxification of parathion from *in vitro* values reported earlier (Sultatos and Minor 1986). While malathion is metabolized both oxidatively and hydrolytically, biotransformation by the latter pathway predominates in humans (Dong *et al.* 1994). Thus, models of disposition in humans (Dong *et al.* 1994; Rabovsky and Brown 1993) utilized only hydrolytic metabolic constants. The toxicokinetic model of Bouchard *et al.* (2003) described the biotransformation of malathion to metabolites (i.e., mono- and dicarboxylic acid metabolites and phosphoric derivative metabolites) and also described the rate at which these metabolites are eliminated in the excreta. Finally, in a compartmental model of percutaneous absorption of malathion and parathion across an isolated porcine perfused skin flap, neither biotransformation nor anticholinesterase activity was addressed (Carver *et al.* 1989). Bioavailability was determined by rate constants fitted to observed absorption rates.

Organohalides Site and mechanism of toxicity of the organochlorines and organofluorines vary, but exposure to all the modeled compounds results in neurological symptoms. Distribution of lindane, a cyclohexane insecticide responsible for acute and chronic neurotoxic effects apparently through GABA(A) antagonism (Brannen *et al.* 1998; Ratra *et al.* 2001), has been modeled in rats (DeJongh and Blaauboer 1997). Simulation of brain and fat lindane concentrations suggested a lower biotransformation rate in females, corresponding to increased lethality relative to males after subchronic exposure (Eichler *et al.* 1983). In addition to neurological effects, several organochlorine insecticides affect the reproductive system of contaminated organisms. Dichlorodiphenyldichloroethylene (*p,p′*-DDE), a major metabolite of dichlorodiphenyltrichloroethane (DDT), exerts estrogenic effects (Galand *et al.* 1987) and possesses antagonistic activity at the androgen receptor (Kelce *et al.* 1995). You *et al.* (1999) developed a model for transplacental and lactational transfer of *p,p′*-DDE in rat, focusing on the finding that 50 times more of the compound administered to dams reaches pups via lactational exposure alone than *in utero*. Antiandrogenic effects of *in utero* exposure were confirmed, however, at a high dosing level. While dieldrin exposure can result in increased pup mortality in dogs and rats (Kitselman 1953; Treon and Cleveland 1955) and teratological effects in mice (Chernoff *et al.* 1975), the models of Lindstrom *et al.* (1974, 1975, 1976) have concentrated upon distribution of the lipophilic insecticide via lipid transport. Similarly, a model including a detailed description of enterohepatic circulation of chlordecone (Bungay *et al.* 1981) dealt with the chemical's unusual hepatic distribution without emphasis on its estrogenic or neurotoxic nature (Guzelian 1982). A subsequent PBPK model for chlordecone addressed dermal uptake of the insecticide relative to age in female rats (Heatherington *et al.* 1998), reflecting concern for the susceptibility of prepubescent agricultural workers to reproductive effects of exposure. The recent PBPK models for distribution of mirex in monkeys and Kepone (chlordecone) in rats (Belfiore *et al.* 2005) attribute dissimilarities in tissue distribution between the highly lipophilic compounds to reversible binding of Kepone in blood and liver.

A primary concern in assessment of herbicide pharmacokinetics has been bioaccumulation, as well as tissue distribution. A clearance-volume pharmacokinetic (CVPK) model of trifluralin (TF) accumulation has been developed to allow estimation of internal dose across several species of fish (Schultz and Hayton 1999). Dermal absorption of the herbicide fluazifop-butyl in humans was described in a model concentrating largely on skin compartments and their parameters (Auton *et al.* 1994). Modeling of 2,4-dichlorophenoxyacetic acid (2,4-D) pharmacokinetics was intended to clarify uptake of this and other organic anions across the intact blood–brain barrier of rat and rabbit (Kim *et al.* 1994). The model was refined to include six parallel brain compartments reflecting functional and anatomical differences (Kim *et al.* 1995) and dosimetry in fetal brain (Kim *et al.* 1996). The model for 2,4-D disposition (Kim *et al.* 1995) was recently assessed for application to low-dose, long-term exposures to organic acids (Kim *et al.* 2001). Finally, a preliminary model for the chlorotriazine atrazine was developed that utilized total radioactivity data and limited data on individual urinary metabolites to describe plasma time-course concentrations of chlorinated compounds separate from glutathione (GSH) conjugates, as well as binding to hemoglobin and plasma protein (McMullin *et al.* 2003).

7.2.2 Modeling Tissue Distribution

Partition Coefficients While elucidation and prediction of target tissue dosimetry is the ultimate goal of most pharmacokinetic models, application of a physiological basis implies realistic mechanisms of transport, delivery, and accumulation. Models have included detailed submodels of enterohepatic circulation, placental and mammary transfer, tissue or blood binding, lipid transport, and compartmentalization. In any modeling approach, accurate tissue : blood equilibrium partition coefficients are critical to describe distribution of chemicals in an organism.

Calculation of partition coefficients from *in vivo* tissue and blood concentration data is straightforward but assumes that the systems are at steady state. The brain : blood partition coefficient for lindane calculated by Oshiba (1972) from tissue concentrations was utilized for the PBPK model of lindane disposition in rats (DeJongh and Blaauboer 1997). Ratios of radiolabeled soman in tissues to that in plasma during the elimination phase (Reynolds *et al.* 1985) were used to construct the PBPD model of soman in rats (Maxwell *et al.* 1988). Chlordecone tissue and blood levels determined 7 and 45 days after intravenous or oral dosing were used to estimate distribution ratios for the enteric transport model of Bungay *et al.* (1981). Fat and liver ratios, representing organs of primary disposition, were confirmed by comparison to values determined from earlier data sets (Boylan *et al.* 1978; Egle *et al.* 1978). Abbas and Hayton (1997) similarly calculated partition coefficients for paraoxon in fish from blood and tissue concentrations determined by the steady-state method of Wagner (1974). In this method, a large volume of exposure water allowed paraoxon concentration to remain constant during approach to steady state (Abbas and Hayton 1997).

Partition coefficients determined by *in vitro* procedures allow estimation of a compound's affinity for tissue and blood, removed from potentially confounding physiological factors. A gas-phase vial equilibration technique generally applied to volatile compounds (Gargas *et al.* 1989) was used to generate partition coefficients for DFP (Gearhart *et al.* 1990). Tissue : blood partition coefficients were determined by dividing the tissue : air value by the blood : air value. Distribution coefficients of parathion were determined by equilibrium dialysis (Sultatos 1990), using the relationship developed by Lin *et al.* (1982) that considers binding (Sultatos *et al.* 1990). Distribution coefficients determined from perfusion studies were not significantly different from those determined by equilibrium dialysis (Sultatos *et al.* 1990). Partition coefficients for lindane determined by ultrafiltration (Jepson *et al.* 1994) were used by DeJongh and Blaauboer (1997). It was noted that the lumped compartment consisted primarily of muscle, clarifying use of the partition coefficient determined for muscle (Jepson *et al.* 1994). Ultrafiltration was utilized as well by Knaak *et al.* (1996), who used partition values previously reported for IF (Knaak *et al.* 1995). In addition, partition coefficients of five metabolites of IF were estimated based on their octanol : water partition coefficients ($K_{o/w}$'s) related to that of the parent compound (Knaak *et al.* 1996).

Calculation of tissue : blood partition from the $K_{o/w}$ provided parameters for the dermal absorption models for malathion (Dong *et al.* 1994; Rabovsky and Brown 1993). Here, the $K_{o/w}$ for malathion of 776.25 (Chiou *et al.* 1977) was applied to the regression equations of Fiserova-Bergerova and Diaz (1986) to obtain partition coefficients for all tissues modeled. Partition coefficients for mirex and Kepone were similarly estimated (Belfiore *et al.* 2005), assuming a fat : blood distribution ratio of 400, similar to hexachlorobiphenyl (Lutz *et al.* 1977). An algorithm based upon octanol : water partition coefficients and tissue lipid content (Poulin and Krishnan 1995) was used to estimate tissue partitioning of CPF and CPF-oxon (Timchalk *et al.* 2002b). Auton *et al.* (1994), modeling dermal absorption of fluazifop-butyl, estimated the stratum corneum : octanol partition coefficient to be 0.15, based on the estimated 15–20% lipid composition of this skin layer. Assuming transit via cellular water spaces of the viable tissue layer, the effective stratum corneum : viable tissues partition coefficient was set at $0.15 \times K_{o/w}$, or 4500. It was noted that this high partition coefficient represented the major barrier to absorption of fluazifop-butyl across skin. To represent the comparable stratum corneum : viable tissue partition coefficient for chlordecone, Heatherington *et al.* (1998) used the linear regression equation of Guy *et al.* (1985) and log $K_{o/w}$ for chlordecone of 5.41.

In situations in which measurement of tissue, blood, or fluid level is not physically possible, partition coefficients have been estimated by fitting the model to experimental time course data. Bungay *et al.* (1981) estimated gut lumen : tissue and gut tissue : blood partition by fitting to the model of enteric transport of chlordecone, explaining that the transient nature of their transport experiments precluded equilibrium conditions. An additional constraint was imposed, that there should be no large discontinuity in partitioning between adjacent segments of small intestine. While the model for fetal uptake of 2,4-D (Kim *et al.* 1996) contained some partition coefficients that were determined from data (Sandberg *et al.* 1996), several partition coefficients for fetal tissue were estimated by fitting to simulation curves. It is not noted whether these parameters were determined by fitting because appropriate tissue analyses were not available from the earlier study or as a result of the short duration of the experimental exposure (Sandberg *et al.* 1996). Ikeda *et al.* (1992) predicted liver : blood partitioning of CVP by fitting the coefficient to the model, noting a significant decrease in this estimated parameter following pretreatment with the same compound.

Submodels Frequently submodels of varying complexity have been incorporated into PBPK models. Models detailing fetal and brain physiology have been constructed to clarify target tissue distribution. Transport issues have been approached with regard to lactational, dermal, and enterohepatic uptake. Biotransformation of toxicant as well has been placed into submodels, particularly when multiple metabolic pathways interact.

The model describing uptake of 2,4-D by rabbit fetal brain (Kim *et al.* 1996) included a fetal unit comprised of amniotic fluid, venous and arterial blood, one body compartment, and three brain compartments: plasma, tissue, and cerebrospinal fluid (CSF). Transfer coefficients between brain compartments were estimated by fitting with the model. Toxicant in maternal blood entered the placenta via flow-limited uptake, where it was equilibrated prior to transport by the umbilical vein to fetal brain plasma and body compartments. The umbilical artery returned toxicant to the placenta. Exchange between fetal body and amniotic fluid and between amniotic fluid and placenta was represented by transfer constants (actually first-order rate constants) that were estimated by fitting to data (Kim *et al.* 1996). In this model, all fetuses were treated as one unit. In the fetal rat model for uptake of *p,p′*-DDE (You *et al.* 1999), each conceptus was represented by a yolk sac placenta, chorioallantoic placenta, and embryo/fetus. Transfer of *p,p′*-DDE between maternal blood and placenta was flow-limited; transfer between placenta and fetus was diffusion-limited. Diffusion coefficients from both placental units to fetus were equal, and they were different from those representing return of *p,p′*-DDE to the placenta; both coefficients were estimated by optimization using SIMUSOLV (Dow Chemical Co., Midland, MI). In addition to the gestational model, You *et al.* (1999) included a lactational model of *p,p′*-DDE uptake via milk. Here, each pup was represented with compartments similar to the dam. Milk volume and milk production rate values were obtained from the literature (Fisher *et al.* 1990; Knight *et al.* 1984), and diffusion rate of *p,p′*-DDE into milk was fitted to the data (You *et al.* 1999).

A PBPK model for 2,4-D dosimetry in rat and rabbit brain was developed by Kim *et al.* (1994) in which brain was represented by plasma, tissue, and CSF. Noting the complex nature of the brain both anatomically and functionally, the authors extended development of this model to produce a pharmacokinetic description of discrete areas of rabbit brain (Kim *et al.* 1995). Parallel brain compartments consisted of hypothalamus, caudate nucleus, hippocampus, forebrain, brainstem, and cerebellum. Each compartment was defined by a volume, blood flow, and partition coefficient. Transfer rate constants from brain plasma into tissue compartments were identical, while those governing exit of 2,4-D into CSF were characteristic for each compartment. Toxin entered brain plasma from systemic blood by a flow-limited process; transfer from CSF to venous circulation was via saturable clearance by the choroid plexus using previously determined kinetic constants (Kim *et al.* 1983).

Bungay *et al.* (1981) similarly divided the gut into blood, tissue, and lumenal compartments, with diffusion coefficients for movement of chlordecone between compartments determined by fitting to data. Gut was additionally divided sequentially into stomach, 15 small intestinal segments, cecum, and large intestine. According to the model, chlordecone diffused in both directions across the gut wall, allowing fecal excretion even in the absence of biliary exretion (Boylan *et al.* 1978). Additionally, chlordecone accumulation in liver tissue remained higher than that in fat due to enterohepatic circulation modeled in this way. The simpler description by Dong *et al.* (1994) of enterohepatic exchange appeared in the model for malathion distribution in dermally exposed humans. Transport from gastrointestinal tract to liver took place by a flow-limited process, while no biliary excretion was indicated. A simple enterohepatic model was included in the gestational model for *p,p′*-DDE to accommodate oral dosing of the dam (You *et al.* 1999). Absorption by liver, excretion by bile, and fecal excretion were represented by rate constants that were set by visual inspection of fit to the data.

The recent PBPK/PD models for CPF (Timchalk *et al.* 2002a,b) contained submodels describing the pharmacokinetics of the respective toxic metabolites, paraoxon and CPF-oxon. Michaelis–Menten kinetics determined entry of oxon into the submodel, and incorporation of specific tissue:blood partition coefficients allowed PD processes to reflect tissue dose of the actual toxin. Total radioactivity plasma data on atrazine equivalents and urinary metabolite data were used to develop an atrazine model in which two pools of compounds were tracked: chlorinated (atrazine and the chlorinated metabolites) and nonchlorinated (GSH conjugates.) Disposition was modeled using rates of conjugation and elimination, along with distinct partition coefficients describing each pool of compounds (McMullin *et al.* 2003).

As a route of exposure, dermal absorption is particularly pertinent for pesticides. The process of malathion (Dong *et al.* 1994; Rabovsky and Brown 1993) and IF (Knaak *et al.* 1994, 1996) absorption from skin was simulated simply by incorporation of loss due to evaporation. More highly detailed models of dermal uptake have been modeled for DFP, malathion, parathion, fluazifop-butyl, and chlordecone (Auton *et al.* 1994; Carver *et al.* 1989; Heatherington *et al.* 1998). Skin was divided into two to three layers including surface, stratum corneum, and viable tissue. While in all three models diffusion constants were estimated for chemical movement between layers, Auton *et al.* (1994) additionally estimated parameters of tissue depth and partition coefficients. Biotransformation of fluazifop-butyl in the viable tissue layer was included as well, with consideration of increased absorption accompanying decreased lipophilicity. One rate constant was used to simulate all losses from skin surface such as evaporation, adsorption, and washing; loss due to desquamation was also estimated in the fluazifop-butyl model. Additional analysis of these PBPK models appears in Chapter 14 of this work.

Lipophilicity Due to generally low solubility in plasma and intercellular water, lipophilic transport processes are potentially important determinants of pesticide disposition. A maternal lymph compartment was added to the gestational model for *p,p′*-DDE in rats (You *et al.* 1999), based on previous demonstration of 60% absorption of gavage dose in corn oil through the lymphatic system (Sieber 1976). The slow flow rate of lymph compared to blood necessitated assumption of a delay factor for appearance of chemical in the systemic circulation (You *et al.* 1999). The time delay and dual route of absorption resulted in simulated blood and lean tissue concentrations having two peaks, a phenomenon observed with other lipophilic compounds similarly administered (Gallo *et al.* 1993).

Lindstrom *et al.* (1974, 1975, 1976) described disposition of dieldrin in mammals based upon lipid-phase transport in a series of articles. The model was described in detail in the first article, while roles of auto-induction, growth, and compartmentalization were addressed in the second. The third article demonstrated, through simulations, features of transport and transfer. Dieldrin was assumed to be transported exclusively via lipids, and it transfers rapidly to tissue lipids with little excretion in feces or urine. Blood lipid flow and compartment lipid mass functions replaced organ perfusion rates and tissue mass; these physiological parameters were estimated by applying lipid fraction to tissue blood flow and tissue mass (Lindstrom *et al.* 1974).

Transport, uptake, or mobilization of lipophilic toxicants has been related to binding by lipid or protein components of tissue or blood. To describe disposition of CPF, the total mass absorbed was added to liver (Timchalk *et al.* 2002b). Holding protein binding constant, free CPF and metabolite in blood were then continuously recalculated; only unbound material was available for partitioning into tissue or for metabolism. Transport of dieldrin from liver to gut lumen via bile was expressed as the product of a bile lipid flow coefficient and an "effective" binding coefficient to the lipid portion of bile fluid (Lindstrom *et al.* 1974). The value of the binding coefficient was assumed to be compound-specific (Cole *et al.* 1970), while the bile lipid flow coefficient possessed species, age, and sex dependence related to feeding and anatomy (Lindstrom *et al.* 1974). Hepatic uptake of chlordecone (Heatherington *et al.* 1998) was described by Michaelis–Menten-type kinetics using values of V_{max} and K_m (i.e., the velocity at maximal substrate concentration and the Michaelis constant, respectively) estimated by fitting to data. The uptake kinetics were applied to the fraction bound to lipoproteins and albumin (Soine *et al* 1982), contrary to the conventional concept that only unbound ligand is available for PD interactions via receptors (Wright *et al.* 1996).

The contribution of *p,p′*-DDE binding to lipoproteins or fatty acids to its eventual disposition was included as well in the gestational/lactational model of You *et al.* (1999). Noting little placental but extensive lactational transfer of toxicant to offspring, the authors proposed that its association with fatty acids (Leighty *et al.* 1980) or very low density lipoprotein (Gallenberg and Vodichik 1987) during fat mobilization would aid transport of *p,p′*-DDE into milk. It was suggested, on the other hand, that association of *p,p′*-DDE with lipoprotein impeded its passage across the placenta. While bound fraction was not estimated, diffusion coefficients determining transfer of *p,p′*-DDE across the placenta and into milk reflected this proposed scenario (You *et al.* 1999).

Movement of *p,p′*-DDE into a deep fat compartment was included in both the gestational and lactational models (You *et al.* 1999) to account for unequal rates of uptake and release of chemical. Accordingly, a smaller diffusion coefficient was estimated for transfer out of the fat compartment. Discussion of similar compartmentalization due to specific binding was offered regarding the lipid-phase model of dieldrin distribution (Lindstrom *et al.* 1975). The model, however, did not include such partitioning of chemical.

In describing bioaccumulation of TF in fish, Schultz and Hayton (1999) proposed a clearance-volume pharmacokinetic (CVPK) model. An unsuccessful attempt was made to scale the model to other fish species by correlating the volume of peripheral storage of chemical with percent body lipid. It was suggested that the lipid composition of the organism should be considered in such a comparison (Ewald and Larsson 1994).

An additional observation was offered regarding disposition of lindane in rats (DeJongh and Blaauboer 1997). It was noted that less lindane was available for distribution in the body following intraperitoneal dosage compared to oral dosage. While this phenomenon was attributed to direct diffusion of part of the dose into intraperitoneal fat, it was unnecessary to either adjust absorption efficiency or incorporate a fraction of intraperitoneal (*i.p.*) dose directly into the fat compartment to obtain a suitable model.

As noted in the forgoing section, the effect of biotransformation of toxicant upon its lipophilicity was considered in the model of dermal uptake of fluazifop-butyl (Auton *et al.* 1994). Kinetic parameters were estimated by fitting the model to human *in vivo* data. It was noted that parameters thus derived were much lower than those determined in *in vitro* experiments (Clark *et al.* 1993); likely explanations for the difference were discussed. The relationship between kinetic parameters, concentration of parent compound in the viable tissue, and flux of penetrant was discussed briefly, with reference to a separate publication (Auton 1993) for complete solution of the pertinent differential equations (Auton *et al.* 1994).

7.2.3 Modeling Metabolism

Submodels While biotransformation of toxicant is commonly included in PBPK models by incorporation of metabolic constants into the liver mass balance equation, Knaak *et al.* (1994, 1996) designed a multipathway model for metabolism of IF in rat, dog, and guinea pig. As many as 17 pathways were represented in six passes through the body. Transformation via hepatic P450 enzymes and tissue CAEs and OP-hydrolases (Knaak *et al.* 1996), as well as A-esterases in rat (Knaak *et al.* 1994), was simulated using kinetic constants taken from previous studies by the authors or other investigators or estimated from urinary metabolite data (Knaak *et al.* 1996).

The submodel representing mammalian biotransformation of dieldrin (Lindstrom *et al.* 1974), based upon extensive metabolic studies performed by other investigators, described movement and biotransformation of liver dieldrin. The submodel consisted of two microsomal enzymes, one inducible, and one noninducible cytosolic enzyme. The destruction rate coefficient was determined by active microsomal and cytosolic enzyme mass. Induction was proposed to result from direct depression of genetic expression in the nucleus or from changes in microsomal structure associated with dieldrin's induction of other microsomal activities. While the compound-specific induction coefficients were included in liver mass balance equations, no auto-induction was simulated for presentation of the preliminary model (Lindstrom *et al.* 1974). However, the possibility that dieldrin might affect its own metabolism was pursued in an application of the preliminary model (Lindstrom *et al.* 1975). Comparison of adipose lipid-phase dieldrin concentrations in rats fed dieldrin at three levels (Walker *et al.* 1969) with values predicted by the model indicated close agreement only under conditions of weak auto-induction (Lindstrom *et al.* 1975).

Determination and Scaling of Kinetic Constants Generally, biotransformation of toxicant has been modeled using Michaelis–Menten kinetics for which V_{max} and K_m are either calculated from *in vitro* enzyme assays or estimated by fitting the parameters to data obtained from *in vivo* studies. To model the pharmacodynamics of soman, Maxwell *et al.* (1988) measured hydrolysis of the compound by plasma DFPase. In the PBPK/PD model for DFP by Gearhart *et al.* (1990), hydrolysis of DFP by A-esterase (DFPase) was measured in rat brain, blood, liver, and kidney tissue. A-esterase was noted to be present in nearly all tissues and was different from B-esterases such as ACHE, BCHE, and CAE. K_m was assumed to be similar for mice, while V_{max} values were scaled to (body weight)$^{0.7}$ (Ramsey and Andersen 1984). Kinetic constants representing both oxidative activation and detoxification of parathion in the mouse model of Sultatos (1990) were determined in mouse liver homogenates (Sultatos 1986); V_{max} values were corrected for microsome loss. A hepatic extraction ratio for parathion calculated from these parameters agreed with an extraction ratio determined by *in situ* liver perfusion studies (Sultatos and Minor 1986).

The model of bioaccumulation of TF in fish utilized kinetic parameters for TF *N*-depropylase determined in liver homogenates of individual fish species (Schultz and Hayton 1999). Intrinsic metabolic clearance, calculated from V_{max} and K_m values, was scaled to whole-animal equivalent based on relative liver mass and unbound fraction of TF as determined by equilibrium dialysis (Schultz and Hayton 1993). Kinetic constants for hydrolytic metabolism of malathion were similarly determined by *in vitro* analysis using rat liver microsomes (Mallipudi *et al.* 1980). The value for K_m was used in both the Rabovsky and Brown (1993) and Dong *et al.* (1994) human models. V_{max} was scaled from rat to human assuming a ratio of body weights to the 0.75 power (Travis 1988). While only hydrolytic liver metabolism was considered in either malathion model (Dong *et al.* 1994; Rabovsky and Brown 1993), kinetic parameters applying to oxidative liver metabolism were presented but not included in construction of the earlier model (Rabovsky and Brown 1993).

As noted earlier, the multipathway model for metabolism of IF (Knaak *et al.* 1994, 1996) included description of as many as seventeen pathways. Initially, metabolic parameters were limited to rat (Knaak *et al.* 1994). Values representing V_{max} for four different P450 enzymes were reduced from *in vitro* data (Knaak *et al.* 1993) to fit *in vivo* data obtained in an earlier percutaneous study (Knaak *et al.* 1990). Values of K_m determined *in vitro* (Knaak *et al.* 1993) were used unchanged. An unexplained discrepancy, however, should be noted in the K_m value given for conversion of IF to des-*N*-isopropyl IF [Tables II and V in Knaak *et al.* (1994)]. Enzyme parameters governing hydrolysis by A-esterases and CAE were estimated by fitting the model to elimination data obtained from the earlier percutaneous study (Knaak *et al.* 1990). The model was extended to dog and guinea pig by inclusion of results of *in vivo* and *in vitro* studies of IF metabolism. Metabolic rate constants applying to rat P450 metabolism were unchanged from the earlier model (Knaak *et al.* 1994), while those representing dog P450 metabolism were taken from the *in vitro* study (Knaak *et al.* 1993) and adjusted similarly to the rat values. Rat and dog K_m values were taken directly from the *in vitro* study (Knaak *et al.* 1993). V_{max} values for OP-hydrolases and CAEs were estimated by the model using urinary metabolite data obtained from current intravenous, intraperitoneal, and dermal studies (Knaak *et al.* 1996). K_m values for OP-hydrolases and CAEs were taken directly from *in vitro* studies of rat metabolism performed by earlier researchers. It was noted that the metabolic profile of guinea pig fell between that of the rat and dog; metabolic rate constants, however, were not listed (Knaak *et al.* 1996).

The apparent V_{max} obtained from *in vitro* studies of IF metabolism was consistently greater than the value used to fit *in vivo* data. Similarly, Auton *et al.* (1994) derived maximal velocity for dermal metabolism of fluazifop-butyl by fitting model predictions to measurement of *in vivo* absorption. The V_{max} value was much lower than that found in *in vitro* experiments performed with tissue homogenates (Clark *et al.* 1993). Noting issues of binding, partitioning, and proximity of enzyme to substrate, the authors suggested development of an *in vitro* system for measurement of metabolic parameters that would more closely model the structure of skin (Auton *et al.* 1994).

The model of dieldrin distribution (Lindstrom *et al.* 1974), as discussed earlier, utilized a destruction rate coefficient that changed over time. The differential equation from which the destruction rate coefficient was derived included hepatic microsomal and cytosolic enzyme mass and lipid mass, coefficients for liver cell protein catabolism and compound specific enzyme induction, and an unexplained compound specific proportionality coefficient. Induction was stated to be compound-, species, sex-, organ-, and enzyme-specific and to represent hormonal or physiological regulation.

Pertinent to this discussion is the proposal of Knaak *et al.* (2004), who noted a dearth of biological data on 31 organophosphate pesticides of interest. The authors stressed the need for parameters related to the transfer of pesticide residue across skin, partition coefficients, and both metabolic and inhibitory rate constants. It was noted that while biological parameters might be estimated using predictive quantitative structure–activity relationship (QSAR) models, there is at present insufficient data to develop such models. A program was recommended to require development of permeability and partition coefficients, metabolic parameters, and bimolecular rate constants to support both QSAR and PBPK/PD modeling technologies.

7.2.4 Summary of Individual Models

Chlorfenvinphos (CVP) Ikeda *et al.* (1990) reported protection against acute oral toxicity of CVP in rats by a single oral pretreatment dose. Ikeda *et al.* (1992), following the models of Colburn (1981) and Wedlund and Wilkinson (1984), designed a compartmental model to clarify the mechanism of this protection. The decreased plasma concentration of CVP observed following the second treatment was proposed to lead to reduction in brain cholinesterase inhibition. Three tissue compartments were included (rapidly and slowly equilibrating tissues, and liver), and gut was included as the oral absorption site. Elimination was via the liver. Using plasma and liver concentrations of CVP from rats with and without pretreatment, pharmacokinetic parameters were estimated by fitting simulations to the data points. Of 16 parameters estimated, pretreatment appeared to affect six by at least twofold. The affected parameters were primarily concerned with hepatic clearance, indicating an apparent increase in metabolic capacity and decrease in available fraction of CVP. It was noted additionally that the diminished absorption rate constant estimated following pretreatment could be largely responsible for the decrease in plasma CVP concentration.

The altered parameters produced good fit between simulations and data, and cogent physiological explanations were offered for such changes. In addition, this modeling approach supported the authors' previous conclusion regarding protection against toxicity by pretreatment with CVP (Ikeda *et al.* 1990). However, the parameter values should be confirmed by experimental measurement to validate the model.

Diisopropylfluorophosphate (DFP) Typical of organophosphate esters, DFP inhibits ACHE in the central and peripheral nervous systems (Koelle 1981). Noting the spectrum of acute neurotoxic symptoms resulting from subsequent accumulation of acetylcholine, Gearhart *et al.* (1990) sought to develop a PBPK model to relate DFP pharmacokinetics and ACHE inhibition, which would be useful for risk assessment. Disposition of DFP was determined by its hydrolysis by A-esterases, binding to esterases, tissue solubility, and volatility leading to exhalation. Binding of DFP to, and inhibition of, B-esterases including ACHE, butyrylcholinesterase (BCHE), and CAE represented the PD aspect of DFP exposure.

Of eight tissue, arterial, and venous blood compartments, all except for fat contained B-esterase activity. A-esterase (DFPase) activity was limited to brain, liver, kidney, rapidly perfused, and blood compartments. Michaelis–Menten kinetics governed hydrolysis of DFP, utilizing metabolic constants determined *in vitro* in rat tissue for this study. V_{max} was scaled by (body weight)$^{0.7}$ (Ramsey and Andersen 1984) for mouse tissue. Free (uninhibited) mass of each B-esterase was affected by rates of enzyme synthesis and degradation, inhibition by DFP, regeneration of free esterase from that bound, and aging of inhibited enzyme. Inhibition was determined *in vitro* in rat brain tissue by assaying decreases in ACHE, BCHE, and CAE activity in the presence of DFP (Jepson 1986). The resultant bimolecular rate constant (k_i), in units of $\mu M^{-1}\,hr^{-1}$, was calculated using the formula

$$-k_i = (\ln(V/V_0)/t)(K_m + [S])/K_m \times [\mathrm{DFP}] \tag{7.1}$$

where $[S]$ is the substrate concentration, V is the reaction velocity, V_0 is the esterase activity at time zero, t is time of incubation, K_m is the Michaelis–Menten constant, and [DFP] is the concentration of DFP. The same value of k_i was used for all tissue compartments and for both rat and mouse. Rates of enzyme regeneration, aging, and degradation in rat brain were obtained from previous mouse and rat studies by other investigators. Other rates of degradation and enzyme synthesis were determined from model optimization to data from subcutaneous dosing of rats and intravenous dosing of mice with DFP also obtained by other researchers.

The model successfully simulated plasma and brain concentrations of DFP following a single intravenous dose of ^{3}H-DFP in mice to 5 min. At later times the simulation underestimated the data, suggesting that ^{3}H activity might represent metabolites. Simulation of plasma ACHE activity determined in the same study reflected the data after optimization of enzyme synthesis rate. The model was also used to simulate brain and plasma ACHE activity determined during and following repeated subcutaneous doses in rats. Agreement between simulations and data was obtained after optimization of (a) rate of absorption of DFP into blood and (b) rates of synthesis and loss of ACHE. However, lack of agreement following cessation of dosing suggested downregulation of the ACHE synthesis rate.

The success of applying kinetic parameters determined *in vitro* to the modeled *in vivo* system was attributed to the soluble nature of the esterases and lack of direct dependence on cofactor concentration. Application of the model to investigate factors controlling DFP pharmacokinetics in rat and mouse was proposed. In addition, usefulness of the model in risk assessment was suggested, citing success in scaling kinetic parameters from rodent to human (Ramsey and Andersen 1984).

The PBPK model for disposition of DFP was further utilized (Gearhart *et al.* 1994) to describe cholinesterase inhibition in humans. Data of Grob *et al.* (1947) were compared to simulated plasma, RBC, and brain enzyme activities with good agreement. The authors noted that the model, thus validated, could be used to simulate the amount of ACHE inhibition in tissues for which no data will likely be available. A description of the dermal uptake model for DFP (Carver *et al.* 1989) is included with other malathion models.

Soman The PD model for inhibition of ACHE by soman (Maxwell *et al.* 1988) is reviewed in this chapter because it includes a detailed PBPK model describing soman biodistribution. At levels of toxin used in this study (≤1 LD_{50}), measurement of soman is difficult due to its rapid metabolism. However, to maintain usefulness of the model, parameters having minimal impact upon PD model response (tissue:blood partition coefficients, DFPase activity) were retained due to their potential importance in explaining pharmacokinetic phenomena at higher dose levels.

Similar to the model of Gearhart *et al.* (1990), the model for soman pharmacodynamics was based upon (a) bimolecular rate constants governing its irreversible binding to ACHE and CAE and (b) rate of hydrolysis by DFPase. Determination of kinetic parameters was complicated by the presence of four stereoisomers, two of which undergo rapid hydrolysis and are thus minimally toxic. Substrate concentrations and reaction times were chosen to eliminate interference by the nontoxic isomers; model code as well included correction for mass of toxic isomers of soman.

The flow-limited PBPK/PD model was designed to predict ACHE activity, as well as mass of soman in six tissues and plasma. Simulations of individual tissue ACHE activity over 30 minutes were representative of actual measured enzyme levels. In addition, inhibition of brain ACHE relative to the dose of soman showed close agreement between simulation and experimental measurements. Additional validity of the model was suggested by incremental sensitivity analysis that indicated overwhelming importance of plasma CAE concentration. Noting increased toxicity in older rats, the authors successfully predicted soman LD_{50} values using experimentally determined plasma CAE concentration. Utility of this PD model as a general model for systemic exposure to soman was suggested, citing similar patterns of maximal ACHE inhibition after equitoxic doses by different routes.

Chlorpyrifos (CPF) The PBPK/PD model describing distribution and action of CPF in rat and human (Timchalk *et al.* 2002b) incorporated elements of the models for DFP (Gearhart *et al.* 1990) and soman (Maxwell *et al.* 1988). Inhibition of esterases was represented by a bimolecular rate constant along with enzyme synthesis, degradation and regeneration, and aging of the CPF-oxon bound complex. As noted in other organophosphate models, the toxic moiety consisted of the CPF-oxon metabolite. In this model, generation of the oxon in the liver via Michaelis–Menten kinetics was followed by entry into a PBPK submodel that included sites of inhibition as well as degradation of CPF-oxon. Within the submodel, compartments containing esterase (B-EST) activity were characterized with kinetic parameters for all aspects of inhibition for each esterase, allowing prediction of all pertinent PD processes. Additionally, inhibition of red blood cell (RBC) ACHE was included in the PBPD model due to use of this measurement as a biomarker of exposure. The detoxification product 3,5,6-trichloro-2-pyridinol (TCP), produced from CPF or CPF-oxon, had no inhibitory effect and was described using a first-order rate of urinary elimination.

Exposure to CPF via dermal or oral routes was included in the model. Oral dosing was further described differently for gavage or dietary exposure using multiple compartments, rate constants, and fractional absorption to establish bioavailability. In addition, CPF and CPF-oxon were represented as bound or free after first pass in the liver; subsequent availability for inhibitory activity or metabolism was determined by mass of unbound material.

The model was extensively validated using data gathered from rats and humans in the course of this study, as well as from previously published investigations. While peak blood concentration of CPF in orally dosed rats was consistently underestimated by model output, overall representation as well as that of plasma, brain, and RBC esterase inhibition in the rat was within a factor of two. Simulation of cholinesterase inhibition following repeated dietary doses in the rat agreed well with previously reported data (Yano *et al.* 2000). Where necessary and possible, parameters were adjusted to model human data, reflecting differences in enzyme levels, elimination rate, and volume of distribution. The resultant model simulations closely reflected plasma CPF and TCP measured in human subjects administered 1 or 2 mg/kg body weight of CPF. TCP pharmacokinetics and plasma cholinesterase inhibition determined in an earlier study (Nolan *et al.* 1984) were also well represented by model simulations. Serum CPF concentrations from a human poisoning victim (Drevenkar *et al.* 1993) were simulated using the model, enabling an estimate of the acute toxic dose. Lack of detectable CPF-oxon, along with considerably higher levels of TCP than parent compound, suggested rapid metabolic clearance of CPF and slower elimination of the less toxic metabolite. The authors concluded that the model could predict the extent of esterase inhibition based on blood levels of parent or metabolite. Additional comments pertained to the possible effect of intestinal CYP450 metabolic capacity and P-glycoproteins upon bioavailability of environmentally relevant low-dose exposures. Current studies appear to confirm such intestinal influence on toxicity (Wu *et al.* 2002).

Further studies of the PBPK/PD model for CPF (Timchalk *et al.* 2002b) involved sensitivity and Monte Carlo variability analyses (Timchalk *et al.* 2002a). It was suggested that polymorphic variability in expression of PON1 (arylesterase) activity might influence susceptibility to organophosphate toxicity. Sensitivity analysis indicated that the model was more sensitive to PON1 activity at higher doses, suggesting that other esterase pathways might compensate for low PON1 activity. Analysis of variability involving Monte Carlo simulations of PON1 polymorphism confirmed greater sensitivity to changes in PON1 activity with increasing dose.

Isofenphos (IF) The PBPK model describing disposition of IF initially consisted of a straightforward six-compartment model that incorporated neither metabolic nor inhibitory processes (Knaak *et al.* 1990). This preliminary modeling effort was developed as a tool to predict IF dose in target tissue of humans when topical dose, skin exposure area, and permeability are varied. Data collected in cutaneous absorption studies with radiolabeled IF in male rats included ^{14}C-IF or ^{14}C-IF equivalent concentrations for air, skin, plasma, tissues, urine, and feces. These values, as well as tissue:blood partition coefficients derived from $K_{o/w}$ measurements were used to construct the model. The model fit was enhanced by varying elimination rate constants and partition coefficients, and a 32-hour delay was incorporated to accurately describe fecal elimination. While simulations closely approximated data points, the model was not validated by simulation of a different dataset. Scaling of parameters from rat to human was not attempted in this version of the model.

The IF model of Knaak *et al.* (1990) was expanded by incorporation of metabolic pathways for biotransformation of the pesticide, as well as phosphorylation and inhibition of cholinesterases (Knaak *et al.* 1994, 1996). Metabolic parameter estimation for both PBPK/PD models has been discussed. Tissue:blood partition coefficients were derived entirely from $K_{o/w}$ values for the earlier model (Knaak *et al.* 1994), while partition coefficients included in the later model (Knaak *et al.* 1996) characterizing parent compound were, as noted previously, determined by ultrafiltration (Knaak *et al.* 1995). Dermal exposure to the rat was represented initially (Knaak *et al.* 1994), and single-bolus intravenous dosing as well as dog and guinea pig parameters were included later (Knaak *et al.* 1996). *In vivo* enzyme inhibition data were not available for validation of the rat dermal model, but such data for all routes of administration and all species modeled provided acceptable agreement with later simulations (Knaak *et al.* 1996).

Malathion While excised whole skin, dermis, and epidermis have been used as *in vitro* media to investigate mechanisms of percutaneous absorption of xenobiotics (Guy *et al.* 1986), such studies have assumed that living processes have no effect on absorption processes. Noting present knowledge of cutaneous biotransformation and bioavailability via dermal tissues (Martin *et al.* 1987), Riviere *et al.* (1986) reported the development of the isolated perfused porcine skin flap as an alternative *in vitro* methodology for studying absorption processes in intact, viable skin. The PBPK model (Carver *et al.* 1989) was designed to determine whether *in vitro* data gathered using this skin flap model could accurately represent *in vivo* skin absorption. The authors also sought to validate the model and reported experimental techniques as a method to examine both diffusion and vascular uptake in skin.

To construct the model, one of several radioactive compounds, including malathion, parathion, and DFP, was applied to the surface of the skin flap. Samples from the arterial reservoir and venous drainage were collected hourly and assayed for radioactive xenobiotic, and absorption rate was calculated. The linear multicompartment model, adapted from a published model of Guy and Hadgraft (1982, 1985), consisted of three compartments: (1) skin surface, (2) stratum corneum, viable epidermis and dermis, and (3) capillary perfusate. Rate constants governing transfer of compound from the skin surface to the skin, between the skin and the capillary perfusate, and from the capillary perfusate to the blood, as well as permanent loss from skin surface, were estimated by fitting the model to the measured absorption rates. Bioavailability was calculated from the rate constants. The model was validated by comparison of this *in vitro* calculated bioavailability with fraction absorbed *in vivo*, determined from urinary excretion in pigs (Carver and Riviere 1989). It is noteworthy that the three organophosphate compounds tested were more accurately simulated by the model than were caffeine or benzoic acid.

The model structure included no assumptions about partitioning among layers or about the effect of molecular weight or skin structure on diffusion rates and mass transfer. Biotransformation within the skin was also not addressed. The model did, however, appear to accurately represent the currently available data and have potential to predict absorption via the dermal route.

The model of dermal uptake of malathion (Rabovsky and Brown 1993) is included in a comprehensive review of malathion metabolism and disposition in mammals by the same authors. The model described for butadiene disposition by Hattis and Wasson (1987) was modified, utilizing Stella 2.10 simulation software (isee systems, Inc., Lebanon, NH). Seven body compartments included skin surface and skin perfused, without the more recently introduced convention of dermal layers and their associated parameters. Malathion was biotransformed via hepatic hydrolytic metabolism; acid metabolites were excreted in urine.

Tissue:blood partition coefficients were calculated from the $K_{o/w}$ of malathion, and kinetic constants were adopted from animal data (Mallipudi *et al.* 1980) with scaling to human by (body weight)$^{0.75}$; both points have been discussed previously. Fecal and urinary excretion rate constants were estimated from radiolabel excretion data (Maibach *et al.* 1971; Wester *et al.* 1983). Simulation of these excretion data suggested the usefulness of three levels of permeability to predict disposition from different dermal sites, in other models, and in other-sized bodies (e.g., children). While inhibition of ACHE activity by the metabolite, malaoxon was mentioned, this aspect of malathion exposure was not addressed in the model. However, kinetic constants governing oxidative liver metabolism responsible for formation of malaoxon were included among model parameters, available for potential incorporation into the model.

The authors used a selection of acute dermal dose estimates from the California Department of Health Services health risk assessment (Kizer 1991) for comparison to model output. The selected dermal dose estimates were corrected for uptake efficiency by entering the deposited skin dose into the model and allowing the model to predict absorption, distribution, metabolism, and excretion using both high and low permeability parameters. The model typically predicted 35% absorption of applied dose, as well as 25% metabolism within 24 hours. It was noted that metabolism and excretion would be limited, at low environmental exposure levels, by skin uptake and flows and not by saturation of metabolic enzymes.

The physiological and biochemical parameters used to construct the model by Rabovsky and Brown (1993) were similarly employed in an attempt to estimate dermally absorbed malathion in potentially exposed human subjects (Dong *et al.* 1994). The model described by Dong *et al.* (1994) differed from that of Rabovsky and Brown (1993) in the expression of hepatic hydrolytic metabolism and by inclusion of enterohepatic circulation. Acid metabolites of malathion were measured in individual urine samples obtained within 48 hours of a potential exposure to malathion. The PBPK model was used to simulate predictions of cumulative urinary excretion of metabolites by varying the applied dose. When an acceptable approximation was reached between measured and predicted values, total absorbed dose of malathion was summed from simulated amounts for all internal tissues. The total absorbed dose of malathion was estimated by modeling for 11 subjects, including adult residents, agricultural workers, and children. The results were consistent with values available in the literature (Bradway *et al.* 1977; Feldman and Maibach 1974; Ross *et al.* 1991).

Informal sensitivity analysis revealed strong influence over model predictions by the skin:blood partition coefficient. It was noted that other parameters might be better represented by their ranges than their means; Monte Carlo simulation was suggested but not incorporated into the present PBPK model. Values assigned to "Other kinetic parameters" in Table II (Dong *et al.* 1994) were unclear due to typographical errors. The proper values are listed in the related poster presentation (Dong *et al.* 1993). Validation of the model appeared in a subsequent study (Dong *et al.* 1996) using data collected in two investigations into urinary excretion of malathion metabolites following dermal exposure in humans (Dary *et al.* 1994; Wester *et al.* 1983).

Bouchard *et al.* (2003) developed a toxicokinetic model to describe the urinary and fecal excretion kinetics of specific malathion metabolites. Although this model is not strictly a PBPK model, it is included here because of the detail with which it treats the metabolism of malathion. The model, developed to describe dermal and oral exposures, consisted of blood and tissue storage compartments for malathion, and it included descriptions of the total amount of metabolite and unconjugated metabolite in the body. Additionally, the model described the amount of five individual metabolites in the body (unconjugated and conjugated metabolites were described separately for each metabolite) and in the urine and feces. Transfer rate constants, established from published data, determined the kinetics of malathion metabolism and elimination. Simulated model output was compared to data collected in several studies in which urinary excretion was measured following dermal exposure to malathion (Maibach *et al.* 1971; Wester *et al.* 1983). Using the validated model, the authors proposed biological reference values for malathion metabolites in urine that would reflect a dermally absorbed dose. The risk from an exposure to malathion could thus be assessed by comparison of the amount of metabolites in a 24-hour urine sample with the proposed reference values.

Parathion and Paraoxon The validity of tissue:blood partition coefficients and metabolic constants estimated *in vitro* for use in PBPK models was assessed by Sultatos (1990), utilizing a model that described disposition of parathion in the mouse following an oral exposure. A PBPK model for the dermal uptake model for parathion has also been developed (Carver *et al.* 1989), but is reviewed with the malathion models. The model developed by Sultatos (1990), similar to that of Andersen *et al.* (1987), consisted of seven flow-limited compartments. Tissue:blood partition coefficients (K_p) were determined previously by equilibrium dialysis *in vitro* (Sultatos *et al.* 1990). It was concluded that such K_p values were similar to those determined in studies of single-pass perfused mouse liver *in situ*. Kinetic constants ($V_{\max}$, K_m) representing activation and detoxification were likewise estimated *in vitro* (Sultatos 1986). Following correction for microsome loss, these constants were used to calculate intrinsic clearance and hepatic extraction ratio. Agreement with the extraction ratio reported for the perfused mouse liver *in situ* suggested the validity of kinetic constants determined *in vitro* in this instance (Sultatos and Minor 1986). Incorporation of *in vitro*-determined values of $V_{\max}$, K_m, and K_p into the PBPK model allowed accurate prediction of tissue levels of parathion measured after intravenous dosing. The authors concluded that assessment of parameter validity by a method not dependent upon the model would contribute to confidence in the PBPK model.

Disposition and toxic activity of the active parathion metabolite, paraoxon, in trout was described in a PBPK/PD model relating tissue and water concentrations of toxicant, ACHE inhibition, and detoxification via binding to CAE (Abbas and Hayton 1997). In addition to production from parathion via P450-dependent monooxygenases, paraoxon can be produced in the environment by a nonenzymatic oxidative mechanism. The model was a modification of that of Nichols *et al.* (1990) describing disposition of waterborne organic chemicals in fish. Chemical was taken up and returned to the water by an uptake compartment (gills). The PD model of Gearhart *et al.* (1990), reviewed previously in this section, was modified so that the reactivation rate constant of paraoxon-bound ACHE was assigned the value of 0. Thus the model describing amount of free ACHE and CAE was based upon the bimolecular rate constants of inhibition and rates of synthesis and degradation of enzyme. Cholinesterase inhibition rates were adopted from published reports, while rates of enzyme synthesis and degradation were determined by fitting results of a dose–response study to a steady-state model (Abbas *et al.* 1994). Agreement was acceptable between model predictions and experimental data for tissue paraoxon concentration and ACHE activity inhibition in liver, heart, and brain.

Simulations lacking CAE inhibition overestimated ACHE inhibition, suggesting a protective effect of CAE by limiting availability of paraoxon to brain ACHE. Sensitivity analysis indicated a strong influence of the synthesis rate of brain ACHE upon simulated brain ACHE activity. Noting a new lower steady-state level of ACHE activity in the ongoing presence of paraoxon, the authors stressed the importance of this continued rapid synthesis of enzyme (Abbas and Hayton 1997).

The PBPK model that described the distribution of DFP (Gearhart *et al.* 1990) was adapted for the purpose of modeling parathion pharmacokinetics (Gearhart *et al.* 1994). According to the model, hydrolysis of parathion to paraoxon occurred in liver and kidney by Michaelis–Menten kinetics. Metabolic constants obtained from the literature were optimized to provide better model predictions. Using the model, time-course concentrations of parathion and paraoxon in the blood and in several tissues were successfully predicted following intravenous injection in the rat (Eigenberg *et al.* 1983). Simulation of parathion kinetics in fat required that uptake by the tissue be diffusion-limited. Inhibition of ACHE was not predicted by this preliminary model.

The parathion/paraoxon model (Gearhart *et al.* 1994) was further developed (Gentry *et al.* 2002) for the purpose of examining the effects of polymorphisms in the detoxification enzyme paraoxonase (PON1) on tissue dose. The same analyses were carried out using a PBPK model describing warfarin, an inhibitor of vitamin K-dependent clotting factors that has been used as an anticoagulant and rodenticide. The warfarin model of Luecke and Wosilait (1979) was further developed (Gentry *et al.* 2002) to examine the pharmacokinetic effect of polymorphisms in the gene CYP2C9, which is primarily responsible for metabolism of the more potent (*s*) enantiomer of this compound. Gentry *et al.* (2002), who noted that the possibility of such polymorphisms causing variability in human toxic responses has not been well characterized, demonstrated the use of Monte Carlo analysis with PBPK models for parathion and warfarin to examine the variability in the tissue dose that arises due to variability in metabolic enzyme systems. The goal was to use the analyses to estimate chemical-specific adjustment factors for human toxicokinetic variability of parathion and warfarin. It was concluded that for warfarin, significant *in vivo* variability in toxicokinetics is produced by polymorphism in the CYP2C9 gene, while for paraoxon, the critical metabolite of parathion, the PON1 polymorphism contributes minimally to variability.

Chlordecone (Kepone®) The enteric transport model of chlordecone in orally dosed rats, developed by Bungay *et al.* (1981), included a detailed gut transport model describing disposition and excretion of chlordecone, as well as that of a nonabsorbable tracer compound. Mass balance equations included in the rather complicated model were described clearly and explicitly. Calculations and rationale were included suggesting the efficacy of cholestyramine therapy for victims of chlordecone intoxication (Boylan *et al.* 1978). Additional comments concerning estimation of tissue:blood partition coefficients, along with description of the bidirectional transport submodel of chlordecone by gut tissue, appear in preceding sections.

The model for percutaneous absorption and disposition of chlordecone in female rats (Heatherington *et al.* 1998) is likewise reviewed later in this volume (Chapter 14). Estimation of partition coefficient values and diffusion constants, along with hepatic uptake of bound fraction, has been discussed previously. The PBPK model was designed to evaluate effects of age and dosage on absorption and disposition of radiolabeled chlordecone. The enterohepatic circulation used in the earlier model of Bungay *et al.* (1981) was not included in this model because gut was not a tissue of interest, and current blood data did not suggest recirculation (Heatherington *et al.* 1998). It is noteworthy as well, considering the lipophilicity of chlordecone, that fat did not appear separately among tissue compartments, but was apparently included in carcass measurement and simulation. Dermal penetration of the pesticide estimated using the PBPK model was found to be dosage-dependent, but was unaffected by age in this study.

The PBPK model developed by Belfiore *et al.* (2005) describing distribution of Kepone in rat was based on data of Egle *et al.* (1978). Liver, fat, and slowly perfused and rapidly perfused tissues were represented as flow-limited. Due to the extended time span over which pharmacokinetic data were collected, growth characteristics of each set of experimental animals were specified by table functions. Metabolism was not included in the model; elimination was represented by a clearance term, which was based on the concentration in liver blood. Binding to liver tissue was characterized by terms representing maximal binding and a dissociation constant, and binding to blood was indicated by a binding constant. As noted in an earlier section of this chapter, partition coefficients were calculated using the equations of Fiserova-Bergerova and Diaz (1986), assuming a fat:blood partition of 400 as reported for the lipophilic hexachlorobiphenyl (Lutz *et al.* 1977). The remarkable hepatic distribution of Kepone, compared to that of mirex, was successfully modeled only when both hepatic (Soine *et al.* 1982) and cytosolic (Soine and Blanke 1984) binding of Kepone were included in the model structure.

Mirex Mirex is structurally similar to Kepone, having only chlorine substitutions while Kepone has one ketone group. Both possess extremely low aqueous solubility. However, distribution of mirex is typical of a lipophilic compound, whereas, as noted earlier, hepatic concentrations of Kepone are higher than those of fat. The PBPK model describing distribution of mirex (Belfiore *et al.* 2005) was designed to explore the changes in model parameters required to allow description of Kepone distribution. Levels of mirex in tissues of Rhesus monkeys (Wiener *et al.* 1976) were adequately described using a flow-limited, four-compartment model along with tissue:blood partition coefficients representative of lipophilic compounds. To describe the distribution of the similarly lipophilic Kepone in rats (Egle *et al.* 1978), the same model required inclusion of liver and blood binding, as well as enhanced fecal elimination. The binding kinetics were estimated from published data describing preferential binding of Kepone to plasma proteins and lipoproteins (Soine *et al.* 1982) and to a porcine cytosolic protein (Soine and Blanke 1984).

Dichlorodiphenyldichloroethylene (p,p′-DDE) *p,p′*-DDE is a persistent metabolite of DDT. The PBPK model was developed by You *et al.* (1999) to elucidate tissue dose levels of *p,p′*-DDE received by dams and fetuses or nursing neonates. Maternal, fetal, and nursing pup models have been described in sections detailing submodels and lipophilicity. The model is also discussed in detail in Chapter 12 along with other PBPK models designed to describe the pharmacokinetics of toxicants that target developing organisms. The model was adapted from that of O'Flaherty *et al.* (1992), including changes in physiological parameters throughout pregnancy. Most pharmacokinetic parameters were optimized by visual inspection or use of SIMUSOLV simulation. Tissue:blood partition coefficients were based on values for hexachlorobiphenyl and DDT found in the literature. The model was accompanied by discussion of how association of *p,p′*-DDE with lipids and lipoproteins might affect transplacental and lactational transfer of the pesticide.

Dieldrin (HEOD) As discussed in preceding sections, the model describing disposition of dieldrin in mammals (Lindstrom *et al.* 1974, 1975, 1976) was based on (a) lipid-phase transport and (b) transfer to tissue lipids. The model consisted of eight tissues, characterized by blood lipid flows and tissue lipid masses. Uptake of dieldrin into the body was dependent upon (a) a source function relying on dietary concentration and (b) an absorption coefficient that was related to presence of lipid in the gut lumen. Metabolism was described by a destruction rate coefficient that was dependent upon enzyme mass, rather than by Michaelis–Menten kinetics. The predictive capability of the model was explored using simple hypothetical cases, such as a mature male rat and mature male human (Lindstrom *et al.* 1974). The model was later (Lindstrom *et al.* 1975) applied to cases of metabolic auto-induction, tissue growth, and comparison to data from pharmacokinetic studies. Results of these simulations suggested improvements involving membrane transport characteristics and variation of transfer rates (Lindstrom *et al.* 1976), illustrated by simulation of the blood–brain barrier. Discussions accompanying the developing model offer interesting insight into transport and disposition of lipophilic compounds in the mammalian body.

Lindane (γ-BHC) DeJongh and Blaauboer (1997) designed the PBPK model for simulation of lindane disposition in rat using five compartments. Brain, considered a primary target of neurotoxicity, was treated as a separate compartment. Elimination was represented exclusively by hepatic biotransformation, and uptake for either oral or *i.p.* route was directly into the liver by first-order rate constants. As noted previously, tissue : blood partition coefficients for all tissues except brain were derived *in vitro* by ultrafiltration; physiological parameters were taken from the literature as well. The first-order metabolic rate and oral absorption constants were adjusted to experimental *in vivo* blood concentration data. Simulation of blood and tissue lindane concentrations determined in alternate studies exhibited close agreement, with an interesting exception. Lindane disposition in brain and fat of female rats (Stein *et al.* 1980) was consistently underpredicted. Readjustment of the biotransformation rate resulted in better agreement between experimental and predicted data. Implications of this observation on sex-based differences for lindane metabolism were discussed.

2,4-Dichlorophenoxyacetic Acid (2,4-D) A partially water-soluble, strong acid, the herbicide 2,4-D, accumulates in brain at plasma levels at which the blood–brain barrier remains intact (Kim *et al.* 1988). In addition, 2,4-D has been implicated in central nervous system neurotoxicity (Casey and Collie 1984; Desi *et al.* 1962). Thus, PBPK models for disposition of 2,4-D were designed to describe the disposition of organic anions in the central nervous system (Kim *et al.* 1994, 1995, 1996, 2001). Initially, the model involved pharmacokinetics in rat and rabbit following oral or intravenous administration. Tissue compartments consisted of (a) a body with a deep compartment and (b) a brain, which included plasma, tissue, and CSF. Clearance from CSF to venous circulation, as well as renal clearance, was a saturable process represented by Michaelis–Menten kinetics. Neither metabolism nor plasma protein binding was incorporated into the model.

The model was modified to include discrete areas of rabbit brain (Kim *et al.* 1995), described earlier among other submodels. Further refinement of the model consisted of a single fetal unit with the original three-compartment brain (Kim *et al.* 1996). The fetal submodel has been described in a previous section as well. Finally, the model for disposition of 2,4-D in rabbit brain (Kelce *et al.* 1995) was modified to represent rat brain (Kim *et al.* 2001). Low-dose, long-term infusion was modeled by incorporation of a subcutaneous compartment. Close agreement of simulations to experimentally determined plasma and brain concentrations of 2,4-D were obtained after the subcutaneous absorption rate constant was adjusted. While model predictions have not yet been compared to independent data, the model was suggested as a noninvasive method to estimate actual brain dose for organic acids during low-dose, chronic exposure.

Atrazine The herbicide atrazine and its chlorinated metabolites (collectively termed Cl-TRIs) are found at low levels in water within agricultural areas. Producing neuroendocrine effects in rat, Cl-TRIs are considered by the US EPA to be of concern for human health risk (US EPA 2001). Only the chlorinated compounds appear to be biologically active (Eldridge *et al.* 1994). Thus, McMullin *et al.* developed a model aimed at providing estimates of net tissue exposure to Cl-TRIs, distinct from nonchlorinated GSH conjugates in rat (McMullin *et al.* 2003). Previous studies (Thede 1987; Timchalk *et al.* 1990) described total radioactivity in blood;

additionally, specific urinary metabolites were identified (Timchalk *et al.* 1990). Using the urinary data, the authors assigned relative rates of GSH-conjugation and excretion for Cl-TRI compounds. A GSH submodel was constructed that contained similar compartments (brain, liver, and remainder of body) as did the unconjugated chlorotriazine model. Partition coefficients for GSH conjugates were estimated by fitting to data.

Data obtained from a study in which rats received 10 daily oral doses of atrazine indicated persistence of bound Cl-TRI radioactivity in both plasma and RBCs (Thede 1987). Using these data, the authors estimated rate constants describing reactivity with hemoglobin and albumin. Rates of formation and degradation of albumin, but not of hemoglobin, were included. Because Cl-TRIs are extensively metabolized by GSH, the authors incorporated dose-dependent GSH depletion into the model. Simulations of tissue concentrations were not offered in this preliminary model due to a lack of existing data on tissue exposures. The authors noted that further description of the metabolic reaction rates, along with identification of bound forms of atrazine, will be necessary to create a true PBPK model integrating the radioactivity and metabolite speciation studies.

Fluazifop-butyl The model describing disposition of the herbicide fluazifop-butyl through human skin (Auton *et al.* 1994) was developed as a tool to investigate mechanisms of percutaneous absorption. Details of the model, including structure, physiological and physicochemical parameters, and biotransformation, have received consideration in earlier sections of this chapter, and additional analysis is included in Chapter 14. In addition to absorption processes within the skin, the model addressed the loss of most of the dose from skin, parameters for which were estimated using low-dose human volunteer data Data gathered from higher exposure levels were used to validate the model.

Trifluralin (TF) A clearance-volume pharmacokinetic model for the lipophilic herbicide TF was constructed by Schultz and Hayton (1999) for the purpose of predicting uptake clearance in a variety of fish species. Using a static water exposure system, uptake clearance, volume of distribution, and elimination clearance were estimated in trout, catfish, and sunfish. The uptake clearance rate was found to be predictable based on oxygen consumption rate, but not on lipid content. Metabolic clearance was predicted by *in vitro* biotransformation rate.

7.3 POLYCHLORINATED AND POLYBROMINATED BIPHENYLS (PCBs AND PBBs)

PCBs have been used commercially for decades. Mixtures of PCBs have been employed as plasticizers, as adhesives, and as dielectric fluid in capacitors and transformers. Recognized as widespread and persistent environmental pollutants, PCBs can be detected in normal components of human and animal diets. As a consequence, the compounds accumulate in fat tissue of organisms higher in the food chain, including nursing young contaminated through excretion of PCBs in milk. Accumulation of individual congeners (of which 209 exist) does not reflect the composition of the original mixture, as lipophilicity and chlorination affect their uptake, disposition, and persistence (Carpenter 1998). Evidence has linked PCB exposure to cancer (Cogliano 1998), immune suppression (Chang *et al.* 1981), nervous system effects (Jacobson and Jacobson 1996; Rogan *et al.* 1988), and endocrine disruption (Gray *et al.* 1993; McKinney and Waller 1994), in some cases mediated by Ah receptor activation (Carpenter 1998; Silkworth and Grabstein 1982). Several investigators have used PBPK modeling techniques to examine the pharmacokinetics of these POPs.

7.3.1 Modeling in Mammals

The initial PBPK model for various PCB congeners was presented in three articles (Anderson *et al.* 1977; Lutz *et al.* 1977; Tuey and Matthews 1977). The model described disposition in rat using six compartments, all flow-limited. A simple enterohepatic circulation included gut reabsorption and biliary excretion, both expressed as first-order rate constants. Physiological parameters consisted of values presented in the literature (Bischoff *et al.* 1971) or determined experimentally. Significantly, skin blood flow was decreased by a factor of 10 to obtain simulations that accurately represented the data. It was suggested that skin might not actually be flow-limited. A subsequent version of the model, used to predict metabolic parameters (Parham and Portier 1998), employed diffusion-limited uptake in all tissues. Later models as well have utilized diffusion-limited tissues to accurately reflect uptake in skin, muscle, and fat (Kramer *et al.* 2000; Lee *et al.* 2002).

The PBPK model was constructed with the primary objective of examining metabolism of selected chlorinated biphenyls. Pharmacokinetic studies included analysis of tissues, urine, feces, and blood for parent PCB and one metabolite per congener. Distribution of parent compound was predominantly to fat and skin, while the more hydrophilic metabolite was found primarily in liver. Studies of congeners containing 1, 2, 5, and 6 chlorines (denoted 1-CB, etc.) suggested that those containing fewer chlorines were more quickly biotransformed in the liver (Anderson *et al.* 1977; Lutz *et al.* 1977). Pharmacokinetics of 4-CB, however, indicated that position of chlorine substitution was important as well (Tuey and Matthews 1977). The presence of adjacent unsubstituted carbon atoms appeared to be an important determinant of metabolic potential. These points were further discussed in a review of the model (Matthews and Dedrick 1984). The authors noted that a depot of PCB in tissue (fat) would cause continued exposure of other tissues. They indicated as well the potential for extrapolation of animal data to predict disposition of PCBs in man. Accurate estimates of human metabolic parameters were viewed as the major limitation to construction of a PBPK model for disposition in humans.

The PBPK model just described was further modified (Tuey and Matthews 1980b) to describe disposition of 1-CB, 2-CB, 5-CB, and 6-CB in the mouse. The mouse-specific parameters were either (a) based upon experimental data obtained following exposure of mice to PCB congeners or (b) scaled from parameters previously applied to the rat model describing the same congeners (Lutz *et al.* 1977). Output of the model based upon experimentally derived parameters tended to agree more closely with data, as would be expected. Somewhat higher biliary clearance was seen in mice, but the major conclusions regarding persistence of the four congeners were found to apply to mice as well as to rats.

Lutz *et al.* (1984) scaled physiological parameters to represent dog and monkey to describe disposition of 2-CB and two congeners of 6-CB. Pharmacokinetic data regarding parent compound and metabolite had been published earlier (Sipes *et al.* 1980, 1982a,b). Metabolic and urinary clearance rate constants were calculated from the data. Biliary excretion consisted of a term that was proportional to concentration of metabolite in liver, along with a fractional preferential excretion of newly metabolized compound in liver (thus proportional to parent concentration in liver). Gut reabsorption was set at zero, unlike earlier versions of the model.

Tissue : blood distribution ratios for parent compound and metabolite were found to be similar in all four species. In addition, dog and monkey demonstrated similar preponderance of fecal to urinary excretion, and metabolism was necessary for excretion into either urine or feces. Support for the contention that two adjacent, unsubstituted carbons promote more rapid metabolism was lent by comparison of the two different 6-CB congeners.

Citing the goal of creating PBPK models describing disposition of all 209 PCB congeners, Parham and Portier (1998) used the model for rat to derive metabolic rates from previously reported *in vivo* data involving nine different PCBs. Metabolic rates were also calculated from *in vitro* data previously reported for metabolism of 21 different PCBs by rat liver microsomes. All the metabolic rates thus determined were used in a stepwise linear regression model. Seven descriptors were included in the model equation, including two indicating source of data and five related to chlorine substitution. The least-squares fit of the metabolic rates to the descriptors was then used to generate metabolic rates for each PCB congener. The estimated rates agreed closely with the original rates. Conversion of metabolic rates determined for rats to rates representative of humans was discussed, with the suggestion of allometric scaling as the only method presently available to make such a transformation.

The model as used for this exercise differed from the original (Lutz *et al.* 1977) by representation of tissues as diffusion-limited, rather than flow-limited. While physiological parameters were not listed in the model description, it is likely that skin perfusion was more accurately interpreted here than it was by the earlier model.

The original Lutz model (Lutz *et al.* 1977) was utilized again to model pharmacokinetics of PCBs 153 and 126 in female mice (Lee *et al* 2002). A brain compartment was included due to neurotoxicity toward adult and young animals. Fat tissue was represented as diffusion-limited, while all others were flow-limited. Additionally, coefficients of absorption and elimination consisted of different upper and lower GI tract components. Data collected following oral gavage indicated that disposition of PCB 153 was altered by co-administration with PCB 126 in nonpregnant mice: the concentration was higher in liver and lower in fat than in the case of mice exposed only to PCB 153. The fat diffusion permeation constant and liver partition coefficient were varied over time to accommodate the apparent pharmacokinetic interaction between the two congeners.

A mixture of compounds closely related to PCBs, Ugilec 141, was introduced in the early 1980s to replace PCBs. Use of Ugilec 141, which consists of tetrachlorobenzyltoluenes (TCBTs), was subsequently prohibited due to properties and accumulation similar to PCBs. A PBPK model for a three-component mixture of TCBTs in rats was designed by Kramer *et al.* (2001) based on the earliest PCB model (Lutz *et al.* 1977). Uptake by fat, muscle, and skin was modeled as a diffusion-limited process, while disposition to liver and the remainder of the body was flow-limited. Metabolic rates determined *in vitro* (Kramer *et al.* 2000) were found to be 2 to 3 times smaller than those obtained by optimizing the model to *in vivo* data. Sensitivity analysis indicated major roles of metabolic rates and liver blood flow in determining blood concentration and total amount of TCBT metabolized. Partition into liver tissue was much greater for TCBT 87 than for TCBT 88 or 94; metabolic rate was considerably lower for TCBT 87. The model demonstrated that the *in vivo* kinetics of TCBTs were comparable to those of PCBs, with highest levels of all three congeners in adipose tissue.

A compound that is structurally similar to PCBs, p,p′-dichlorodiphenylsulfone (DDS), has been used in preparation of plastics intended for high-temperature applications (Haglund *et al.* 1998). While feeding studies have produced no indication of carcinogenicity in rats or mice, evidence of hepatic toxicity has been apparent (NTP, 2000). Parham *et al.* (2002) further utilized the Lutz *et al.* (1977) model to develop a PBPK model describing DDS behavior in rat and mouse. Tissue compartments included a gut lumen divided into four parts to create a delay between ingestion and elimination. Distribution of the parent compound was diffusion-limited in all tissues, while that of the metabolite was flow-limited. Due to lack of data, the five known metabolites were treated as one. The short-term (rat) experimental data with which the authors constructed the model behaved in a nonlinear fashion. To accommodate this nonlinearity, absorption from the gut was modeled using Michaelis–Menten kinetics, and metabolism in the liver was modeled as a first-order, self-inducible process. Data obtained from chronic (feeding) studies in the rat and mouse were compared to model output; predictions were better for low doses than for high and were better for rats than for mice.

Hexabromobiphenyl (HBB) has found use as the major component of a commercial fire retardant (Firemaster BP-6, Michigan Chemical Company). Accidental introduction into livestock feed in the early 1970s led to its presence in the human population in Michigan (Kay 1977). HBB appears to bioaccumulate in adipose tissue and fat, and toxic effects of exposure appear to be similar to those of PCBs (Safe 1984).

Tuey and Matthews (1980a) modeled the fate of HBB by modifying the existing PBPK model for PCB (Tuey and Matthews 1980b). Model parameters were derived experimentally from intravenous and oral dose studies in Sprague–Dawley rats (Tuey and Matthews 1978). A further study in Fischer rats produced parameters necessary to scale the model down to represent the smaller animals. Growth of the adipose compartment was modeled, allowing appraisal of effect of changes in tissue volume upon tissue HBB concentration. Additionally, the gut compartment was represented as tissue and lumen, which permitted estimation of HBB transport in bile and through both gut lumen and tissue.

The model was further scaled to humans using parameters derived from reports in the literature and using tissue : blood partition coefficients determined for rat. Total exposure to HBB estimated in humans following ingestion of contaminated milk (Fries *et al.* 1978) was used as input to the scaled model, and simulations predicted adipose and blood levels for 5 years. A half-life in humans of approximately 6 years was estimated. Human tissue samples analyzed following the contamination incident were similar to predicted tissue concentrations and fat : serum ratios. Citing storage of HBB in fat rather than in liver, discussion was offered concerning the inappropriateness of cholestyramine therapy, which was previously used to lower the body burden of Kepone (Boylan *et al.* 1978).

7.3.2 Modeling in Nonmammalian Species

A PBPK model was designed to predict uptake of 2,2′,5,5′-tetrachlorobiphenyl (TCB) in two small fish species (Lien and McKim 1993). The intent was to predict bioconcentration of a waterborne, neutral, nonmetabolized xenobiotic compound. The model consisted of one body compartment and two exchange surfaces, branchial (gill) and cutaneous. Parameters representing water and blood flow exchange capacities and diffusion exchange capacities were based on physiological, morphological, and physicochemical properties. Without further adjustment of parameter values, the model was used to simulate bioconcentration of TCB in Japanese medaka and fathead minnows exposed to the compound in water. Comparison of simulations including uptake by gill only, as well as via both branchial and cutaneous surfaces, suggested that the observed accumulation results from absorption at both surfaces. The uptake route for larger fish, such as trout species, appears to be via gills alone. Predictive capabilities of the model for similar compounds and other fish species or sizes were proposed.

7.4 SUMMARY

The qualities of POPs that result in their persistence in the environment also lead them to share pharmacokinetic properties such as preferential storage in lipid compartments and slow biotransformation. For the two types of chemicals examined here, pesticides and PCBs/PBBs, differences in the mode of exposure, PD effects resulting from exposure, and distribution within organisms vary, necessitating distinctly different PBPK model structures. Neurotoxicity has been represented in several models by inclusion of ACHE inhibition or brain compartments of varying complexity. Percutaneous absorption has been incorporated into many of the PBPK models. The unlikely hepatic accumulation of chlordecone was modeled by an elaborate enteric transport compartment or hepatic tissue binding. A model representing the maternal, fetal, and nursing pup described pharmacokinetics of *p,p′*-DDE. Biotransformation rates of 209 PCB congeners have been described within the same PBPK model by including multiple descriptors in the model equation. Thus, a wide variety of approaches to pharmacokinetic modeling of POPs has been applied to apparently similar, but in effect disparate, compounds.

NOTATION

ACHE	acetylcholinesterase
A-esterase	nonspecific hydrolase present in nearly all tissues
BCHE	butyrylcholinesterase
Cl-TRIs	atrazine and its cholorinated metabolites
CAE	carboxylesterase
CPF	chlorpyrifos
CSF	cerebrospinal fluid
CVP	chlorfenvinphos
CVPK	clearance-volume pharmacokinetic
2,4-D	2,4-dichlorophenoxyacetic acid
DDT	dichlorodiphenyltrichloroethane
DDS	*p,p′*-dichlorodiphenylsulfone
DFP	diisopropylfluorophosphate
GSH	glutathione
HBB	hexabromobiphenyl
i.p.	intraperitoneal
IF	isofenphos
K_m	the Michaelis constant
K_p	partition coefficient
$K_{o/w}$	octanol : water partition coefficient
PBPK	physiologically based pharmacokinetic
PCBs	polychlorinated biphenyls
PD	pharmacodynamic
PON1	paraoxonase
***p,p′*-DDE**	dichlorodiphenyldichloroethylene
POPs	persistent organic pollutants
QSAR	quantitative structure–activity relationship
RBC	red blood cell
TCB	2,2′,5,5′-tetrachlorobiphenyl
TCBTs	tetrachlorobenzyltoluenes
TCDD	2,3,7,8-tetrachlorodibenzo-*p*-dioxin
TCP	3,5,6-trichloro-2-pyridinol
TF	trifluralin
V_{max}	the enzymatic reaction velocity at maximal substrate concentration

REFERENCES

Abbas, R., and Hayton, W. L. (1997). A physiologically based pharmacokinetic and pharmacodynamic model for paraoxon in rainbow trout. *Toxicol. Appl. Pharmacol.* **145**, 192–201.

Abbas, R., Hayton, W. L., and Schultz, I. R. (1994). Toxicodynamics of paraoxon (PO) in rainbow trout. *Toxicologist* **14**, 387.

Andersen, M. E., Clewell, H. J. III, Gargas, M. L., Smith, F. A., and Reitz, R. H. (1987). Physiologically based pharmacokinetics and the risk assessment process for methylene chloride. *Toxicol. Appl. Pharmacol.* **87**, 187–205.

Anderson, M. W., Eling, T. E., Lutz, R. J., Dedrick, R. L., and Matthews, H. B. (1977). The construction of a pharmacokinetic model for the disposition of polychlorinated biphenyls in the rat. *Clin. Pharmacol. Ther.* **22**, 765–773.

Auton, T. R. (1993). Concentration–flux relations for a multicellular biological membrane with metabolism. *Math. Biosci.* **115**, 103–117.

Auton, T. R., Westhead, D. R., Woollen, B. H., Scott, R. C., and Wilks, M. F. (1994). A physiologically based mathematical model of dermal absorption in man. *Hum. Exp. Toxicol.* **13**, 51–60.

Belfiore, C. K., Andersen, M. E., Chubb, L. S., Lohitnavy, M., Lohitnavy, O. S., and Yang, R. S. H. (2005). PBPK models of Kepone and mirex suggest binding to hepatic tissue. In Press.

Bischoff, K. B., Dedrick, R. L., Zaharko, D. S., and Longstreth, J. A. (1971). Methotrexate pharmacokinetics. *J. Pharm. Sci.* **60**, 1128–1133.

Bouchard, M., Gosselin, N. H., Brunet, R. C., Samuel, O., Dumoulin, M-J., and Carrier, G. (2003). A Toxicokinetic model of malathion and its metabolites as a tool to assess human exposure and risk through measurements of urinary biomarkers. *Toxicol. Sci.* **73**, 182–194.

Boylan, J. J., Egle, J. L., Jr., and Guzelian, P. S. (1978). Cholestyramine: Use as a new therapeutic approach for chlordecone (Kepone) poisonings. *Science* **199**, 893–895.

Bradway, D. E., Shafik, T. M., and Lores, E. M. (1977). Comparison of cholinesterase activity, residue levels, and urinary metabolite excretion of rats exposed to organophosphorus pesticides. *J. Agr. Food. Chem.* **25**, 1353–1358.

Brannen, K. C., Devaud, L. L., Liu, J., and Lauder, J. M. (1998). Prenatal exposure to neurotoxicants dieldrin or lindane alters tert-butylbicyclophosphorothionate binding to GABA(A) receptors in fetal rat brainstem. *Dev. Neurosci.* **20**, 34–41.

Bungay, P. M., Dedrick, R. L., and Matthews, H. B. (1981). Enteric transport of chlordecone (Kepone) in the rat. *J. Pharmacokinet. Biopharm.* **9**, 309–341.

Carpenter, D. O. (1998). Polychlorinated biphenyls and human health. *Int. J. Occup. Med. Environ. Health* **11**, 291–303.

Carver, M. P., and Riviere, J. E. (1989). Percutaneous absorption and excretion of xenobiotics after topical and intravenous administration to pigs. *Fundam. Appl. Toxicol.* **13**, 714–722.

Carver, M. P., Williams, P. L., and Riviere, J. E. (1989). The isolated perfused porcine skin flap. *Toxicol. Appl. Pharmacol.* **97**, 324–337.

Casey, P. H., and Collie, W. R. (1984). Severe mental retardation and multiple congenital anomalies of uncertain cause after extreme parental exposure to 2,4-D. *Clin. Lab Observations* **104**, 313–315.

Chang, K.-J., Hsieh, K.-H., Lee, T.-P., Tang, S.-Y., and Tung, T.-C. (1981). Immunologic evaluation of patients with polychlorinated biphenyl poisoning: Determination of lymphocyte subpopulations. *Toxicol. Appl. Pharmacol.* **61**, 58–63.

Chernoff, N., Kavlock, R. J., Kathrein, J. R., Dunn, J. M., and Haseman, J. K. (1975). Prenatal effects of dieldrin and photodieldrin in mice and rats. *Toxicol. Appl. Pharmacol.* **31**, 302–308.

Chiou, C. T., Freed, V. H., Schmedding, D. W., and Kohnert, R. L. (1977). Partition coefficient and bioaccumulation of selected organic chemicals. *Environ. Sci. Technol.* **11**, 475–478.

Clark, N. W. E., Scott, R. C., Blain, P. G., and Williams, F. M. (1993). Fate of fluazifop butyl in rat and human skin *in vitro*. *Arch. Toxicol.* **67**, 44–48.

Cogliano, V. J. (1998). Assessing the cancer risk from environmental PCBs. *Environ. Health Perspect.* **106**, 317–323.

Colburn, W. A. (1981). First-pass clearance of lidocaine in healthy volunteers and epileptic patients: Influence of effective liver volume. *J. Pharm. Sci.* **70**, 969–971.

Cole, J. F., Klevay, L. M., and Zavon, M. R. (1970). Endrin and dieldrin: A comparison of hepatic excretion in the rat. *Toxicol. Appl. Pharmacol.* **16**, 547–555.

Dary, C. C., Blancato, J. N., Castles, M., Reddy, V., Cannon, M., Saleh, M. A., and Cash, G. G. (1994). Dermal absorption and disposition of formulations of malathion in Sprague–Dawley rats and humans. In: *Biomarkers of Human Exposure to Pesticides*, Vol. 542, M. A. Saleh, J. N. Blancato, and C. H. Nauman, eds., ACS Symposium Series, Washington, D.C., pp. 231–263.

DeJongh, J., and Blaauboer, B. J. (1997). Simulation of lindane kinetics in rats. *Toxicology* **122**, 1–9.

Desi, I., Sos, J., Olasz, J., Sule, F., and Markus, V. (1962). Nervous system effects of a chemical herbicide. *Arch. Environ. Health* **4**, 101–108.

Dong, M. H., Draper, W. M., Papanek, P. J. J., Ross, J. H., Woloshin, K. A., and Stephens, R. D. (1994). Estimating malathion doses in California's Medfly eradication campaign using a physiologically based pharmacokinetic model. In: *Environmental Epidemiology—Effects of Environmental Chemicals on Human Health*, W. M. Draper, ed., *Advances in Chemistry*, Vol. 241, American Chemical Society, Washington, D.C., pp. 189–208.

Dong, M. H., Ross, J. H., Thongsinthusak, T., and Krieger, R. I. (1996). Use of spot urine sample results in physiologically based pharmacokinetic modeling of absorbed malathion doses in humans. In: *Biomarkers for Agrochemicals and Toxic Substances*, ACS Symposium Series, Vol. 643, American Chemical Society, Washington, D.C., pp. 229–241.

Dong, M. H., Ross, J. H., Thongsinthusak, T., Sanborn, J. R., and Wang, R. G. (1993). Physiologically-based pharmacokinetic (PB-PK) modeling for dermal absorption of pesticides (malathion) in man. *Toxicologist* **13**, 355.

Drevenkar, V., Vasilic, Z., Stengl, B., Frobe, Z., and Rumenjak, V. (1993). Chlorpyrifos metabolites in serum and urine of poisoned persons. *Chem.–Biol. Interactions* **87**, 315–322.

Egle, J. L., Jr., Fernandez, S. B., Guzelian, P. S., and Borzelleca, J. F. (1978). Distribution and excretion of chlordecone (Kepone) in the rat. *Drug Metab. Dispos.* **6**, 91–95.

Eichler, D., Heupt, W., and Paul, W. (1983). Comparative study on the distribution of α- and γ-hexachlorocyclohexane in the rat with particular reference to the problem of isomerization. *Xenobiotica* **13**, 639–647.

Eigenberg, D. A., Pazdernik, T. L., and Doull, J. (1983) Hemoperfusion and pharmacokinetic studies with parathion and paraoxon in the rat and dog. *Drug Metab. Dispos.* **11**, 366–370.

Eldridge, J., Fleenor-Heysor, D. G., Extrom, P. C., Wetzel, L. T., Breckenridge, C. B., Gillis, J. H., Luempert, L. G. I., and Stevens, J. T. (1994). Short-term effects of chlorotriazines on estrus in female Sprague–Dawley and Fischer 344 rats. *J. Toxicol. Environ. Health* **43**, 155–167.

Ewald, G., and Larsson, P. (1994). Partitioning of ^{14}C-labelled 2,2′,4,4′-tetrachlorobiphenyl between water and fish lipids. *Environ. Toxicol. Chem.* **13**, 1577–1580.

Faroon, O., Kueberuwa, S., Smith, L., and DeRosa, C. (1995). ATSDR evaluation of health effects of chemicals II. Mirex and chlordecone: Health effects, toxicokinetics, human exposure, and environmental fate. *Toxicol. Ind. Health* **11**, 1–195.

Feldman, R. J., and Maibach, H. I. (1974). Percutaneous penetration of some pesticides and herbicides in man. *Toxicol. Appl. Pharmacol.* **28**, 126–132.

Fiserova-Bergerova, V., and Diaz, M. L. (1986). Determination and prediction of tissue-gas partition coefficients. *Int. Arch. Occup. Environ. Health* **58**, 75–87.

Fisher, J. W., Whittaker, T. A., Taylor, D. H., Clewell, H. J. III, and Andersen, M. E. (1990). Physiologically based pharmacokinetic modeling of the lactating rat and nursing pup: A multiroute exposure model for trichloroethylene and its metabolite, trichloroacetic acid. *Toxicol. Appl. Pharmacol.* **102**, 497–513.

Fries, G. F., Marrow, G. S., and Cook, R. M. (1978). Distribution and kinetics of PBB residues in cattle. *Environ. Health Perspect.* **23**, 43–50.

Galand, P., Mairesse, N., Degraef, C., and Rooryck, J. (1987). *o,p′*-DDT (1,1,1-trichloro-2 (*p*-chlorophenyl) 2-(*o*-chlorophenyl)) ethane is a purely estrogenic agonist in the rat uterus *in vivo* and *in vitro*. *Biochem. Pharmacol.* **36**, 397–400.

Gallenberg, L. A., and Vodichik, M. J. (1987). Potential mechanisms for redistribution of polychlorinated biphenyls during pregnancy and lactation. *Xenobiotica* **17**, 299–310.

Gallo, J. M., Cheung, L. L., Kim, H. J., Bruckner, J. V., and Gillespie, W. R. (1993). A physiological and system analysis hybrid pharmacokinetic model to characterize carbon tetrachloride blood concentrations following administration in different oral vehicles. *J. Pharmacokinet. Biopharm.* **21**, 551–574.

Gargas, M. L., Burgess, R. J., Voisard, D. E., Cason, G. H., and Andersen, M. E. (1989). Partition coefficients of low-molecular-weight volatile chemicals in various liquids and tissues. *Toxicol. Appl. Pharmacol.* **98**, 87–99.

Gearhart, J. M., Jepson, G. W., Clewell, H. J. III, Andersen, M. E., and Conolly, R. B. (1990). Physiologically based pharmacokinetic and pharmacodynamic model for the inhibition of acetylcholinesterase by diisopropylfluorophosphate. *Toxicol. Appl. Pharmacol.* **106**, 295–310.

Gearhart, J. M., Jepson, G. W., Clewell, H. J., Andersen, M. E., and Conolly, R. B. (1994). Physiologically based pharmacokinetic model for the inhibition of acetylcholinesterase by organophosphate esters. *Environ. Health Perspect.* **102**(Suppl 11), 51–60.

Gentry, P. R., Hack, C. E., Haber, L., Maier, A., and Clewell, H. J., III. (2002). An approach for the quantitative consideration of genetic polymorphism data in chemical risk assessment: Examples with warfarin and parathion. *Toxicol. Sci.* **70**, 120–139.

Gray, L. E., Jr., Ostby, J., Marshall, R., and Andrews, J. (1993). Reproductive and thyroid effects of low-level polychlorinated biphenyl (Aroclor 1254) exposure. *Fundam. Appl. Toxicol.* **20**, 288–294.

Grob, D., Lilienthal, J. L., Jr., Harvey, A. M., and Jones, B. F. (1947). The administration of diisopropyl fluorophosphate (DFP) to man. *I. Bull.—Johns Hopkins Hosp.* **81**, 217–244.

Guy, A. H., and Hadgraft, J. (1982). Percutaneous metabolism with saturable enzyme kinetics. *Int. J. Pharm.* **11**, 187–197.

Guy, A. H., and Hadgraft, J. (1985). Pharmacokinetic interpretation of the plasma levels of clonidine following transdermal delivery. *J. Pharm. Sci.* **74**, 1016–1018.

Guy, R. H., Hadgraft, J., and Maibach, H. I. (1985). Percutaneous absorption in man: A kinetic approach. *Toxicol. Appl. Pharmacol.* **78**, 123–129.

Guy, R. H., Guy, A. H., Maibach, H. I., and Shah, V. P. (1986). The bioavailability of dermatological and other topically administered drugs. *Pharm. Res.* **3**, 253–262.

Guzelian, P. S. (1982). Comparative toxicology of chlordecone (Kepone) in humans and experimental animals. *Annu. Rev. Pharmacol. Toxicol.* **22**, 89–113.

Haglund, P., Leander, O., and Zook, D. (1998). Analysis of the recently detected environmental contaminant bis-(4-chlorophenyl)sulfone in temperature resistant polymers. *Organohalogen. Compd.* **35**, 427–429.

Hattis, D., and Wasson, J. (1987). *A pharmacokinetic/Mechanism-Based Analysis of the Carcinogenic Risk of Butadiene*, Center of Technology, Policy and Industrial Development, MIT. #NTIS/PB88-202817, Cambridge, MA.

Heatherington, A. C., Fisher, H. L., Sumler, M. R., Waller, C. L., Shah, P. V., and Hall, L. L. (1998). Percutaneous absorption and disposition of [^{14}C]chlordecone in young and adult female rats. *Environ. Res.* **79**, 138–155.

Ikeda, T., Kojima, T., Yoshida, M., Takahashi, H., Tsuda, S., and Shirasu, Y. (1990). Pretreatment of rats with an organophosphorus insecticide, chlorfenvinphos, protects against subsequent challenge with the same compound. *Toxicol. Appl. Pharmacol.* **14**, 560–567.

Ikeda, T., Tsuda, S., and Shirasu, Y. (1992). Pharmacokinetic analysis of protection by an organophosphorus insecticide, chlorfenvinphos, against the toxicity of its succeeding dosage in rats. *Fundam. Appl. Toxicol.* **18**, 299–306.

Isee Systems, Inc., 46 Centerra Parkway, Suite 200, Lebanon, NH 63766.

Jacobson, J. L., and Jacobson, S. W. (1996). Intellectual impairment in children exposed to polychlorinated biphenyls *in utero*. *N. Engl. J. Med.* **335**, 783–789.

Jepson, G. W. (1986). A kinetic model for acetylcholinesterase inhibition by diisopropylfluorophosphate in crude rat brain homogenate. M.S. Thesis, Wright State University.

Jepson, G. W., Hoover, D. K., Black, R. K., McCafferty, J. D., Mahle, D. A., and Gearhart, J. M. (1994). A partition coefficient determination method for nonvolatile chemicals in biological tissues. *Fundam. Appl. Toxicol.* **22**, 519–524.

Jones, K. C., and de Voogt, P. (1999). Persistent organic pollutants (POPs): State of the science. *Environ. Pollut.* **100**, 209–221.

Kavlock, R. J., Daston, G. P., DeRosa, C., Fenner-Crisp, P., Gray, L. E., Kaattari, S., Lucier, G., Luster, M., Mac, M. J., Maczka, C., Miller, R., Moore, J., Rolland, R., Scott, G., Sheehan, D. M., Sinks, T., and Tilson, H. A. (1996). Research needs for the risk assessment of health and environmental effects of endocrine disruptors: A report of the U.S. EPA-sponsored workshop. *Environ. Health Perspect.* **104**, 715–740.

Kay, K. (1977). Polybrominated biphenyls (PBB) environmental contamination in Michigan, 1973–1976. *Environ. Res.* **13**, 74–93.

Kelce, W. R., Stone, C. R., Laws, S. C., Gray, L. E., Kemppainen, J. A., and Wilson, E. M. (1995). Persistent DDT metabolite *p,p′*-DDE is a potent androgen receptor antagonist. *Nature* **375**, 581–585.

Kim, C. S., O'Tuama, L. A., Mann, J. D., and Roe, C. R. (1983). Saturable accumulation of the anionic herbicide, 2,4-dichlorophenoxyacetic acid (2,4-D), by rabbit choroid plexus: Early developmental origin and interaction with salicylates. *J. Pharmacol. Exp. Ther.* **225**, 699–704.

Kim, C. S., Keizer, R. F., and Pritchard, J. B. (1988). 2,4-Dichlorophenoxyacetic acid intoxication increases its accumulation within the brain. *Brain Res.* **440**, 216–226.

Kim, C. S., Gargas, M. L., and Andersen, M. E. (1994). Pharmacokinetic modeling of 2,4-dichlorophenoxyacetic acid (2,4-D) in rat and in rabbit brain following single dose administration [published erratum appears in *Toxicol. Lett.* (1995) **76**(2), 185]. *Toxicol. Lett.* **74**, 189–201.

Kim, C. S., Slikker, W., Jr., Binienda, Z., Gargas, M. L., and Andersen, M. E. (1995). Development of a physiologically based pharmacokinetic model for 2,4-dichlorophenoxyacetic acid dosimetry in discrete areas of the rabbit brain [published erratum appears in *Neurotoxicol. Teratol.* (1995), **17**(3), 359–360]. *Neurotoxicol. Teratol.* **17**, 111–120.

Kim, C. S., Binienda, Z., and Sandberg, J. A. (1996). Construction of a physiologically based pharmacokinetic model for 2,4-dichlorophenoxyacetic acid dosimetry in the developing rabbit brain. *Toxicol. Appl. Pharmacol.* **136**, 250–259.

Kim, C. S., Sandberg, J. A., Slikker, W., Jr., Binienda, Z., Schlosser, P. M., and Patterson, T. A. (2001). Quantitative exposure assessment: Application of physiologically-based pharmacokinetic (PBPK) modeling of low-dose, long-term exposures of organic acid toxicant in the brain. *Environ. Toxicol. Pharmacol.* **9**, 153–160.

Kitselman, C. H. (1953). Long term studies on dogs fed aldrin and dieldrin in sublethal dosages, with reference to the histopathological findings and reproduction. *J. Am. Vet. Med. Assoc.*, 28–30.

Kizer, K. W. (1991). *Health Risk Assessment of Aerial Application of Malathion-Bait*, California Department of Health Services.

Knaak, J. B., Al-Bayati, M. A., Raabe, O. G., and Blancato, J. N. (1990). *In vivo* percutaneous absorption studies in the rat: Pharmacokinetics and modelling of isofenphos absorption. In: *Prediction of Percutaneous Penetration: Methods, Measurements, Modelling*, R. C. Scott, R. H. Guy, and J. Hadgraft, eds., IBC Technical Service Ltd., London, pp. 1–18.

Knaak, J. B., Al-Bayati, M. A., Raabe, O. G., and Blancato, J. N. (1993). Development of *in vitro* $V_{\max}$ and K_{m} values for the metabolism of isofenphos by P-450 liver enzymes in animals and human. *Toxicol. Appl. Pharmacol.* **120**, 106–113.

Knaak, J. B., Al-Bayati, M. A., Raabe, O. G., and Blancato, J. N. (1994). Prediction of anticholinesterase activity and urinary metabolites of isofenphos. In: *Biomarkers of Human Exposure to Pesticides*, M. A. Saleh, J. N. Blancato, and C. H. Nauman, eds., ACS Symposium Series, Vol. 542, American Chemical Society, Washington, D.C., pp. 284–300.

Knaak, J. B., Al-Bayati, M. A., and Raabe, O. G. (1995). Development of partition coefficients, V_{max} and K_m values, and allometric relationships. *Toxicol. Lett.* **79**, 87–98.

Knaak, J. B., Al-Bayati, M. A., Raabe, O. G., and Blancato, J. N. (1996). Use of a multiple pathway and multiroute physiologically based pharmacokinetic model for predicting organophosphorus pesticide toxicity. In: *Biomarkers for Agrochemicals and Toxic Substances*, ACS Symposium Series, Vol. 643, American Chemical Society, Washington, D.C., pp. 206–228.

Knaak, J. B., Dary, C. C., Power, F., Thompson, C. B., and Blancato, J. N. (2004). Physicochemical and biological data for the development of predictive organophosphorus pesticide QSARs and PBPK/PD models for human risk assessment. *Crit. Rev. Toxicol.* **34**, 143–207.

Knight, C. H., Docherty, A. H., and Peaker, M. (1984). Milk yield in rats in relation to activity and size of the mammary secretory cell population. *J. Dairy Res.* **51**, 29–35.

Koelle, G. B. (1981). Organophosphate poisoning—An overview. *Fundam. Appl. Toxicol.* **1**, 129–134.

Kramer, H. J., vandenBerg, M., Delang, R. J., Brandsma, L., and DeJongh, J. (2000). Biotransformation rates of Ugilec 141 (tetrachlorobenzyltoluenes) in rat and trout microsomes. *Chemosphere* **40**, 1283–1288.

Kramer, H. J., Drenth, H., vandenBerg, M., Seinen, W., and DeJongh, J. (2001). Physiologically based pharmacokinetic model for tetrachlorobenzyltoluenes in rat: Comparison of *in vitro* and *in vivo* metabolic rates. *Toxicol. Sci.* **63**, 22–28.

Lee, S. K., Ou, Y. C., and Yang, R. S. H. (2002). Comparison of pharmacokinetic interactions and physiologically based pharmacokinetic modeling of PCB 153 and PCB 126 in nonpregnant mice, lactating mice, and suckling pups. *Toxicol. Sci.* **65**, 26–34.

Leighty, E. G., Fentiman, A. F., Jr., and Thompson, R. M. (1980). Conjugation of fatty acids to DDT in the rat: Possible mechanism for retention. *Toxicology* **15**, 77–82.

Lien, G. J., and McKim, J. M. (1993). Predicting branchial and cutaneous uptake of 2,2′,5,5′-tetrachlorobiphenyl in fathead minnows (*Pimephales promelas*) and Japanese medaka (*Oryzias latipes*): Rate limiting factors. *Aquatic Toxicol.* **27**, 15–32.

Lindstrom, F. T., Gillett, J. W., and Rodecap, S. E. (1974). Distribution of HEOD (Dieldrin) in mammals. I. Preliminary model. *Arch. Environ. Contam. Toxicol.* **2**, 9–42.

Lindstrom, F. T., Gillett, J. W., and Rodecap, S. E. (1975). Distribution of HEOD (Dieldrin) in mammals. II. Some applications of the preliminary model. *Arch. Environ. Contam. Toxicol.* **3**, 166–182.

Lindstrom, F. T., Gillett, J. W., and Rodecap, S. E. (1976). Distribution of HEOD (Dieldrin) in mammals. III. Transport-transfer. *Arch. Environ. Contam. Toxicol.* **4**, 257–288.

Luecke, R., and Wosilait, W. D. (1979). Drug elimination characteristics: Analysis using a mathematical model. *J. Pharmacokinet. Biopharm.* **7**, 629–641.

Lutz, R. J., Dedrick, R. L., Matthews, H. B., Eling, T. E., and Anderson, M. W. (1977). A preliminary pharmacokinetic model for several chlorinated biphenyls in the rat. *Drug Metab. Dispos.* **5**, 386–396.

Lutz, R. J., Dedrick, R. L., Tuey, D. B., Sipes, I. G., Anderson, M. W., and Matthews, H. B. (1984). Comparison of the pharmacokinetics of several polychlorinated biphenyls in mouse, rat, dog, and monkey by means of a physiological pharmacokinetic model. *Drug Metab. Dispos.* **12**, 527–535.

Maibach, H. I., Feldman, R. J., Milby, T. H., and Serat, W. F. (1971). Regional variation in percutaneous penetration in man. *Arch. Environ. Health* **23**, 208–211.

Mallipudi, N. M., Talcott, R. E., Ketterman, A., and Fukuto, T. R. (1980). Properties and inhibition of rat malathion carboxylesterases. *J. Toxicol. Environ. Health* **6**, 585–596.

Marrs, T. C. (1993). Organophosphate poisoning. *Pharmacol. Ther.* **58**, 51–66

Martin, R. J., Denyer, S. P., and Hadgraft, J. (1987). Skin metabolism of topically applied compounds. *Int. J. Pharmacol.* **39**, 23–32.

Matthews, H. B., and Dedrick, R. L. (1984). Pharmacokinetics of PCBs. *Annu. Rev. Pharmacol. Toxicol.* **24**, 85–103.

Maxwell, D. M., Vlahacos, C. P., and Lenz, D. E. (1988). A pharmacodynamic model for soman in the rat. *Toxicol. Lett.* **43**, 175–188.

McKinney, J. D., and Waller, C. L. (1994). Polychlorinated biphenyls as hormonally active structural analogues. *Environ. Health Perspect.* **102**, 290–297.

McMullin, T. S., Brzezski, J., Cranmer, B. K., Tessari, J. D., and Andersen, M. E. (2003). Pharmacokinetic modeling with disposition and time course studies with ^{14}C-atrazine. *J. Toxicol. Environ. Health Part A.* **66**, 941–964.

Nichols, J. W., McKim, J. M., Andersen, M. E., Gargas, M. L., Clewell, H. J. III, and Erickson, R. J. (1990). A physiologically based toxicokinetic model for the uptake and disposition of waterborne organic chemicals in fish. *Toxicol. Appl. Pharmacol.* **106**, 433–447.

Nolan, R. J., Rick, D. L., Freshour, N. L., and Saunders, J. H. (1984). Chlorpyrifos: pharmacokinetics in human volunteers. *Toxicol. Appl. Pharmacol.* **73**, 8–15.

NTP (2000). NTP technical report on the toxicology and carcinogenesis studies of *p,p′*-dichlorodiphenylsulfone (CAS No. 80-07-9) in F344/N rats and B6C3F1 mice (feed studies). NTP technical report 501. National Toxicology Program.

O'Flaherty, E. J., Scott, W., Schreiner, C., and Beliles, R. P. (1992). A physiologically based kinetic model of rat and mouse gestation: Disposition of a weak acid. *Toxicol. Appl. Pharmacol.* **112**, 245–256.

Oshiba, K. (1972). Experimental studies on the fate of β- and γ-BHC *in vivo* following daily administration. *Osaka Shiritsu daigaku Igaku Zasshi* **21**, 1–19.

Parham, F. M., Matthews, H. B., and Portier, C. J. (2002). A physiologically based pharmacokinetic model of *p,p′*-dichlorodiphenylsulfone. *Toxicol. Appl. Pharmacol.* **181**, 153–163.

Parham, F. M., and Portier, C. J. (1998). Using structural information to create physiologically based pharmacokinetic models for all polychlorinated biphenyls. II. Rates of metabolism. *Toxicol. Appl. Pharmacol.* **151**, 110–116.

Poulin, P., and Krishnan, K. (1995). An algorithm for predicting tissue: Blood partition coefficients of organic chemicals from *n*-octanol: Water partition coefficient data. *J. Toxicol. Environ. Health* **46**, 117–129.

Rabovsky, J., and Brown, J. P. (1993). Malathion metabolism and disposition in mammals. *J. Occup. Med. Toxicol.* **2**, 131–168.

Ramsey, J. C., and Andersen, M. E. (1984). A physiologically based description of the inhalation pharmacokinetics of styrene in rats and humans. *Toxicol. Appl. Pharmacol.* **73**, 159–175.

Ratra, G. S., Kamita, S. G., and Casida, J. E. (2001). Role of human $GABA_A$ receptor β3 subunit in insecticide toxicity. *Toxicol. Appl. Pharmacol.* **172**, 233–240.

Reynolds, M. L., Little, P. L., Thomas, B. F., Bagley, R. B., and Martin, B. R. (1985). Relationship between the biodisposition of [^{3}H]soman and its pharmacological effects in mice. *Toxicol. Appl. Pharmacol.* **80**, 409–420.

Riviere, J. E., Bowman, K. F., Monteiro-Riviere, N. A., Dix, L. P., and Carver, M. P. (1986). The isolated perfused porcine skin flap (IPPSF). *Fundam. Appl. Toxicol.* **7**, 444–453.

Rogan, W. J., Gladen, B. C., Hung, K.-L., Koong, S.-L., Shih, L.-Y., Taylor, J. S., Wu, Y.-C., Yang, D., Ragan, N. B., and Hsu, C.-C. (1988). Congenital poisoning by polychlorinated biphenyls and their contaminants in Taiwan. *Science* **241**, 334–336.

Ross, J. H., Thongsinthusak, T., Krieger, R. I., Frederickson, S., Fong, H. R., Taylor, S., Begum, S., and Dong, M. (1991). Human clearance of malathion. *Toxicologist* **11**, 160.

Safe, S. H. (1984). Polychlorinated biphenyls (PCBs) and polybrominated biphenyls (PBBs): Biochemistry, toxicology, and mechanism of action. *Crit. Rev. Toxicol.* **13**, 319–395.

Safe, S. H. (1994). Polychlorinated biphenyls (PCBs): Environmental impact, biochemical and toxic responses, and implications for risk assessment. *Crit. Rev. Toxicol.* **24**, 87–149.

Sandberg, J. A., Duhart, H. M., Lipe, G., Binienda, Z., Slikker, W., Jr., and Kim, C. S. (1996). Distribution of 2,4-dichlorophenoxyacetic acid (2,4-D) in maternal and fetal rabbits. *J. Toxicol. Environ. Health* **49**, 497–509.

Schultz, I. R., and Hayton, W. L. (1993). Toxicokinetics of trifluralin in rainbow trout. *Aquatic Toxicol.* **26**, 287–306.

Schultz, I. R., and Hayton, W. L. (1999). Interspecies scaling of the bioaccumulation of lipophilic xenobiotics in fish: An example using trifluralin. *Environ. Toxicol. Chem.* **18**, 1440–1449.

Sieber, S. M. (1976). The lymphatic absorption of *p,p′*-DDT and some structurally-related compounds in the rat. *Pharmacology* **14**, 443–454.

Silkworth, J. B., and Grabstein, E. M. (1982). Polychlorinated biphenyl immunotoxicity: Dependence on isomer planarity and the Ah gene complex. *Toxicol. Appl. Pharmacol.* **65**, 109–115.

Sipes, I. G., Slocum, M. L., Perry, D. F., and Carter, D. E. (1980). 4,4′-Dichlorobiphenyl: Distribution, metabolism, and excretion in the dog and the monkey. *Toxicol. Appl. Pharmacol.* **55**, 554–563.

Sipes, I. G., Slocum, M. L., Chen, H.-S. G., and Carter, D. E. (1982a). 2,3,6,2′,3′,6′-Hexachlorobiphenyl: Distribution, metabolism, and excretion in the dog and the monkey. *Toxicol. Appl. Pharmacol.* **62**, 317–324.

Sipes, I. G., Slocum, M. L., Perry, D. F., and Carter, D. E. (1982b). 2,2′,4,4′,5,5′-Hexachlorobiphenyl: Distribution, metabolism, and excretion in the dog and monkey. *Toxicol. Appl. Pharmacol.* **65**, 264–272.

Soine, P. J., and Blanke, R. V. (1984). Isolation of chlordecone binding proteins from pig liver cytosol. *J. Toxicol. Environ. Health* **14**, 305–317.

Soine, P. J., Blanke, R. V., Guzelian, P. S., and Schwartz, C. C. (1982). Preferential binding of chlordecone to the protein and high density lipoprotein fractions of plasma from humans and other species. *J. Toxicol. Environ. Health* **9**, 107–118.

Stein, K., Portig, J., Fuhrmann, H., Koransky, W., and Noack, G. (1980). Steric factors in the pharmacokinetics of lindane and α-hexachlorocyclohexane in rats. *Xenobiotica* **10**, 65–77.

Sultatos, L. G. (1986). The effects of phenobarbital pretreatment on the metabolism and acute toxicity of the pesticide parathion in the mouse. *Toxicol. Appl. Pharmacol.* **86**, 105–111.

Sultatos, L. G. (1990). A physiologically based pharmacokinetic model of parathion based on chemical-specific parameters determined in vitro. *J. Am. Coll. Toxicol.* **9**, 611–619.

Sultatos, L. G., and Minor, L. D. (1986). Factors affecting the biotransformation of the pesticide parathion by the isolated perfused mouse liver. *Drug Metab. Dispos.* **14**, 214–220.

Sultatos, L. G., Kim, B., and Woods, L. (1990). Evaluation of estimations *in vitro* of tissue/blood distribution coefficients for organothiophosphate insecticides. *Toxicol. Appl. Pharmacol.* **103**, 52–55.

Thede, B. (1987). *Study of 14-C-Atrazine Dose/Response Relationship in the Rat.* Ciba-Geigy Corporation, Greensboro, NC.

Timchalk, C., Dryzga, M. D., Langvardt, P. W., Kastl, P. E., and Osborne, D. W. (1990). Determination of the effect of tridiphane on the pharmacokinetics of [14C]-atrazine following oral administration to male Fischer 344 rats. *Toxicology* **61**, 27–40.

Timchalk, C., Kousba, A., and Poet, T. S. (2002a). Assessing the impact of human PON1 polymorphisms: Sensitivity and Monte Carlo analyses using a physiologically based pharmacokinetic/pharmacodynamic (PBPK/PD) model for chlorpyrifos. *Toxicol. Lett.* **135**, 51–59.

Timchalk, C., Nolan, R. J., Mendrala, A. L., Dittenber, D. A., Brsak, K. A., and Mattsson, J. L. (2002b). A physiologically based pharmacokinetic and pharmacodynamic (PBPK/PD) model for the organophosphate insecticide chlorpyrifos in rats and humans. *Toxicol. Sci.* **66**, 34–53.

Travis, C. C. (1988). Interspecies extrapolation. In: *Biological Based Methods for Cancer Risk Assessment*, C. C. Travis, ed., Plenum Press, New York, pp. 63–78.

Treon, J. F., and Cleveland, F. P. (1955). Toxicity of certain chlorinated hydrocarbon insecticides for laboratory animals, with special reference to aldrin and dieldrin. *J. Agricult. Food Chem.* **3**, 402–408.

Tuey, D. B., and Matthews, H. B. (1977). Pharmacokinetics of 3,3′,5,5′-tetrachlorobiphenyl in the male rat. *Drug Metab. Dispos.* **5**, 444–450.

Tuey, D. B., and Matthews, H. B. (1978). Pharmacokinetics of hexabromobiphenyl disposition in the rat. *Toxicol. Appl. Pharmacol.* **45**, 337.

Tuey, D. B., and Matthews, H. B. (1980a). Distribution and excretion of 2,2′,4,4′,5,5′-hexabromobiphenyl in rats and man: Pharmacokinetic model predictions. *Toxicol. Appl. Pharmacol.* **53**, 420–431.

Tuey, D. B., and Matthews, H. B. (1980b). Use of a physiological compartmental model for the rat to describe the pharmacokinetics of several chlorinated biphenyls in the mouse. *Drug Metab. Dispos.* **8**, 397–403.

US EPA (2001). *Revised preliminary human health risk assessment on atrazine*. Registration Branch 3, Health Effects Division, Office of Pesticide Programs.

Wagner, J. G. (1974). A safe method for rapidly achieving plasma concentration plateaus. *Clin. Pharmacol. Ther.* **16**, 691–700.

Walker, A. I. T., Stevenson, D. E., Robinson, J., Thorpe, E., and Roberts, M. (1969). The toxicology and pharmacodynamics of dieldrin (HEOD): Two-year oral exposures of rats and dogs. *Toxicol. Appl. Pharmacol.* **15**, 345–373.

Wedlund, P. J., and Wilkinson, G. R. (1984). Hepatic tissue binding and the oral first-pass effect. *J. Pharm. Sci.* **73**, 422–425.

Wester, R. C., Maibach, H. I., Bucks, D. A. W., and Guy, R. H. (1983). Malathion percutaneous absorption after repeated administration to man. *Toxicol. Appl. Pharmacol.* **68**, 116–119.

Wiener, M., Pittman, K. A., and Stein, V. (1976). Mirex kinetics in the Rhesus monkey I. Disposition and excretion. *Drug Metab. Dispos.* **4**, 281–287.

Wright, J. D., Boudinot, F. D., and Ujhelyi, M. R. (1996). Measurement and analysis of unbound drug concentrations. *Clin. Pharmacokinet.* **30**, 445–462.

Wu, H., Kousba, A., Timchalk, C., and Poet, T. S. (2002). Hepatic and intestinal CYP450 metabolism of chlorpyrifos. *Toxicologist* **66**, 35.

Yano, B. L., Young, J. T., and Mattsson, J. L. (2000). Lack of carcinogenicity of chlorpyrifos insecticide in a high-dose, 2-year dietary toxicity study in Fischer 344 rats. *Toxicol. Sci.* **53**, 135–144.

You, L., Gazi, E., Archibecque-Engle, S., Casanova, M., Conolly, R. B., and Heck, H. D. (1999). Transplacental and lactational transfer of *p,p′*-DDE in Sprague–Dawley rats. *Toxicol. Appl. Pharmacol.* **157**, 134–144.

CHAPTER 8

DIOXIN AND RELATED COMPOUNDS

Yasong Lu

8.1 INTRODUCTION

The term "dioxins" or "dioxin compounds" refers to polychlorinated dibenzo-*p*-dioxins (PCDDs) and polychlorinated dibenzofurans (PCDFs), two groups of planar, tricyclic ethers with up to eight chlorine atoms attached to positions 1 to 4 and 6 to 9 (Fig. 8.1). These compounds, especially those congeners with substitutions in the lateral (2, 3, 7, and 8) positions of the aromatic ring, possess the properties of thermal and chemical stability, resistance to biotransformation, low volatility, and high lipophilicity, which results in their persistence in the environment and bioaccumulation in organisms, including humans. Therefore they are included in the list of persistent organic pollutants (POPs) (Fiedler 2003). Dioxins have been produced as byproducts or impurities from industrial-chemical processes (e.g., production of chlorinated hydrocarbons) and thermal processes (e.g., municipal solid waste incineration, combustion in automobile engines) (Byard 1987; Fiedler 2003). Compared to these anthropogenic sources, the natural sources (e.g., biological formation) are not significant (Fiedler 2003). Human exposure to dioxins is predominantly through

Physiologically Based Pharmacokinetic Modeling, edited by Micaela B. Reddy, Raymond S. H. Yang, Harvey J. Clewell III, and Melvin E. Andersen.

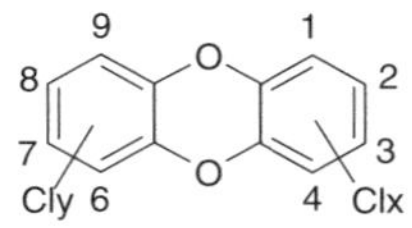

Polychlorinated dibenzo-*p*-dioxins (PCDDs)

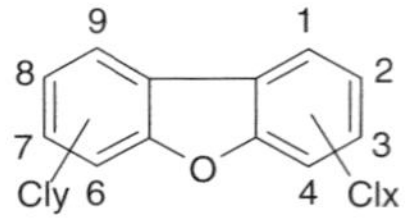

Polychlorinated dibenzofurans (PCDFs)

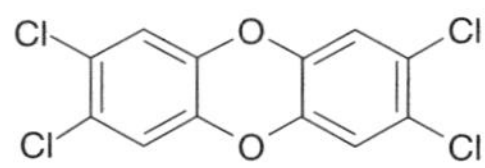

2,3,7,8-TCDD

Figure 8.1 Molecular structures of PCDDs, PCDFs, and TCDD. PCDDs and PCDFs have up to eight chlorine atoms attached to positions 1 to 4 and 6 to 9.

contaminated food (e.g., fish, diary products) (Liem *et al.* 2000), followed by inhalation and dermal absorption (Fiedler 2003). For breastfed infants, maternal milk can be an important exposure source (Lorber and Phillips 2002).

The pharmacokinetics, biochemical effects, and toxicity of dioxins have been widely studied in many species, and there are more than 25 PBPK models available for the most toxic and extensively studied dioxin congener, 2,3,7,8-tetrachlorodibenzo-*p*-dioxin (TCDD), in various species, such as wildlife (trout), livestock (cow), laboratory animals (mouse and rat), and humans (Table 8.1). TCDD (Fig. 8.1), a representative dioxin compound, is the primary focus of this chapter with brief discussions on other congeners also included. The development of PBPK models for various dioxin congeners and analogues has been informed by advances in TCDD model development (Table 8.2). This chapter is devoted to dioxins, and particularly to TCDD, because many of the dioxin PBPK models successfully incorporate information on mode of action and specific biochemical effects, which distinguishes them from PBPK models for other classes of chemicals, and because the evolution of TCDD models provides an excellent example of how modeling work and experimental studies were used together to improve our understanding of TCDD pharmacokinetics and pharmacodynamics.

8.2 TOXICITY

Dioxins can elicit a variety of biochemical and toxic effects in laboratory animals and humans. In laboratory animals (e.g., rats, mice, golden hamsters, guinea pigs, rabbits, or monkeys), the following effects have been observed: lethality (Christian *et al.* 1986), wasting syndrome (reduction of body weight and fat volume) (Christian *et al.* 1986), thymic atrophy (Gehrs and Smialowicz 1997), immunotoxicity (immunosuppresion) (Chastain and Pazdernik 1985; Nagarkatti *et al.* 1984), dermal lesions (hyperkeratinization or chloracne) (Brewster *et al.* 1988b), induction of the hepatic enzymes cytochrome P450 1A1, CYP1A1, and 1A2, CYP1A2 (Mills *et al.* 1993; Pegram *et al.* 1995; Sarkar *et al.* 2000), liver morphological alterations (Pohjanvirta *et al.* 1989), reproductive toxicity (Chahoud *et al.* 1992; Rier and Foster

TABLE 8.1 PBPK Models for TCDD

Species	Reference[a]
Mya arenaria	Moreno *et al.* (1992)
Cow	Derks *et al.* (1994)
Brook trout	Nichols *et al.* (1998)
Mouse	Leung *et al.* (1988)
Rat	Andersen *et al.* (1993, 1997a,c)
Rat	Emond *et al.* (2004)
Rat	Evans and Andersen (2000)
Rat	Kohn *et al.* (1993, 1994, 1996, 2001)
Rat	Leung *et al.* (1990a)
Rat	Portier *et al.* (1993)
Rat	Roth *et al.* (1993, 1994)
Rat	Santostefano *et al.* (1997, 1998)
Rat	Wang *et al.* (1996, 1997a,b)
Rat, mouse	Wang *et al.* (1997b, 2000)
Human	Andersen *et al.* (1997c)
Human	Aylward *et al.* (2004)
Human	Banks *et al.* (1993)
Human, rat, monkey	Carrier *et al.* (1995a,b)
Human	Clewell *et al.* (2004)
Human	Dankovic and Andersen (2002)
Human	Gentry *et al.* (2003)
Human	Kissel and Robarge (1988)
Human, including infants	Kreuzer *et al.* (1997)
Human, rat, and mice	Lawrence and Gobas (1997)
Human	Maruyama *et al.* (2002a,b, 2003)
Human	van der Molen *et al.* (1996)

[a] Related references are grouped together.

TABLE 8.2 PBPK Models for TCDD Congeners and Analogues

Chemical	Species	Reference
^{125}I-2-Iodo-3,7,8–trichlorodibenzo-*p*-dioxin	Mouse	Leung *et al.* (1990b)
2,3,7,8-Tetrabromodibenzo-*p*-dioxin	Rat	Kedderis *et al.* (1993)
2,3,7,8-Tetrachlorodibenzofuran	Rat, mouse, monkey	King *et al.* (1983)
Polychlorinated dibenzo-*p*-dioxins, polychlorinated dibenzofurans	Human	Maruyama *et al.* (2002a,b, 2003)
Polychlorinated dibenzo-*p*-dioxins, polychlorinated dibenzofurans	Rat, monkey, human	Carrier *et al.* (1995a,b)

2002; Silbergeld and Mattison 1987), developmental toxicity (cleft palate, hydronephrosis) (Birnbaum 1995; Couture *et al.* 1990), teratogenicity (Couture *et al.* 1990), and disturbance of lipid and glucose homeostasis (Brewster *et al.* 1988a; Enan and Matsumura 1994; Moran *et al.* 2004). Although TCDD does not directly cause mutation or DNA adducts, it can promote tumor formation in rats (Kociba *et al.* 1978; Luebeck *et al.* 2000), mice (Beebe *et al.* 1995), and hamsters (Rao *et al.* 1988). Several reviews on animal carcinogenesis of TCDD have been published (e.g., Huff *et al.* 1991; McGregor *et al.* 1998; Shu *et al.* 1987; US EPA 2000a).

In animal studies, induction of CYP1A1 and CYP1A2 in the liver is a frequently observed biochemical response to TCDD exposure. CYP1A1 induction may elevate the rate of metabolism of TCDD (Olson 1994; Olson *et al.* 1994), and CYP1A2 induction is responsible for the sequestration of TCDD in the liver in a dose-dependent manner (DeVito *et al.* 1998; Diliberto *et al.* 2001; Voorman and Aust 1989). Because of the ability of CYP1A1/2 induction to alter the pattern of TCDD tissue distribution, and probably the subsequent responses, many TCDD PBPK models incorporate the induction of either or both enzymes [i.e., have been extended to include pharmacodynamic (PD) response]. However, the relationship between this enzyme induction and toxic effects remains unclear.

In humans, the toxic effects of dioxins are mainly observed in populations subject to accidental and occupational exposure. Those effects usually include skin lesions (Dunagin 1984; Suskind 1985), elevated mortality (Michalek *et al.* 1998), disruption of lipid metabolism (Crow 1979), endocrine and reproductive toxicity (Egeland *et al.* 1994; Sever *et al.* 1997), and increased cancer risk (Crump *et al.* 2003; US EPA 2000a). Although its ability to cause cancer in humans is the subject of ongoing debate (Paustenbach 2002; Starr 2001), TCDD has been classified a "human carcinogen" by the United States Environmental Protection Agency, US EPA (US EPA 2000b), and a "carcinogenic to humans" by the International Agency for Research on Cancer, IARC (IARC 1997).

8.3 MODE OF ACTION

Most, if not all, of the effects of dioxins are mediated through their binding to the cytosolic aryl hydrocarbon receptor (AhR) and the subsequent interaction between the dioxin–AhR complex and specific response elements on the DNA. AhR, a transcription factor with a basic helix–loop–helix (bHLH) domain, is involved in signal transduction and transcriptional regulation. AhR is constitutively expressed in the liver and many extrahepatic tissues (e.g., lung, kidney, thymus, spleen, heart, brain, muscle, and placenta) (Gasiewicz and Rucci 1984; Li *et al.* 1994). In the inactive form, AhR complexes with several molecules of heat shock protein (hsp90) and other proteins. After entering a cell, dioxin reversibly binds to AhR, the associated proteins dissociate from the original complex, and the AhR–dioxin translocates into the nucleus where the complex is activated by binding to the aryl hydrocarbon receptor nuclear translocator (ARNT). Subsequently, the activated dioxin–AhR–ARNT complex can bind to a specific DNA response element—that is, a dioxin response element (DRE) or a xenobiotic response element (XRE). The interaction between

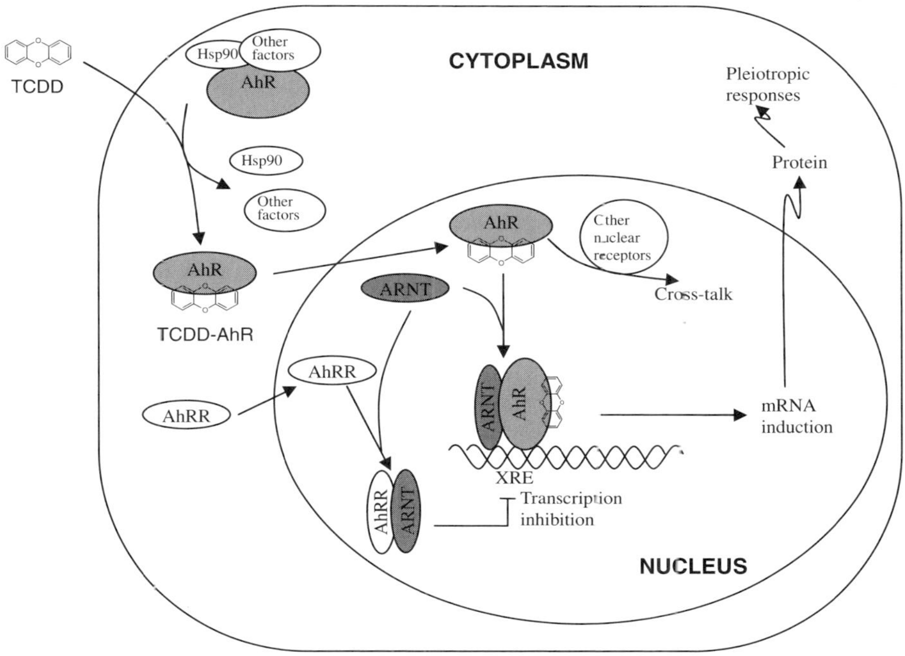

Figure 8.2 Schematic diagram of the AhR-mediated mode of action of TCDD-induced biochemical effects as well as the possibility of negative regulation by AhR repressor (AhRR) and cross-talk with other nuclear receptor signaling pathways.

dioxin–AhR–ARNT complex and a DRE or XRE alters the expression of various genes (e.g., the CYP1A1 gene) and subsequently causes related toxicity (Andersen and Dennison 2001; Birnbaum 1994; DeVito and Birnbaum 1994; Moran *et al.* 2004). This mode of action was solidified by the resistance of AhR knockout mice to TCDD-induced toxicity (Fernandez-Salguero *et al.* 1996; Vasquez *et al.* 2003).

The dioxin-activated AhR signaling pathway and gene transcription are negatively regulated by the AhR repressor (AhRR; Korkalainen *et al.* 2004). There is also cross-talk between the signaling pathway of AhR and other nuclear receptors, such as estrogen (Ohtake *et al.* 2003; Safe and Wormke 2003) or hypoxia inducible factors (Chan *et al.* 1999), which may result in alterations in the downstream of either pathway. Figure 8.2 is a schematic representation of the AhR-mediated mode of action of TCDD effects as well as negative regulation and cross-talk.

8.4 PHARMACOKINETICS

Extensive pharmacokinetic studies on TCDD provide essential information for PBPK modeling. For an oral exposure, TCDD is absorbed from the gastrointestinal tract and predominantly stored in the liver and adipose tissue. TCDD-induced

CYP1A2 in the liver plays a critical role in the distribution pattern. TCDD is slowly excreted via passive diffusion into the intestinal lumen as well as by metabolism. This section only briefly summarizes the available data; readers interested in more comprehensive information may refer to reviews by Hu and Bunce (1999), Olson (1994), van Birgelen and van den Berg (2000), and van den Berg *et al.* (1994).

8.4.1 Absorption, Metabolism, and Excretion

TCDD, well absorbed from the gastrointestinal tract, is usually administered to animals in their food or by gavage in oil vehicles. The bioavailability of TCDD, defined as the difference between administered dose (100%) and the unabsorbed percentage, is higher than 70% on average in Sprague–Dawley rats (Olson 1994). In guinea pigs and hamsters, bioavailabilities of 50% (Olson 1994) and 74% (Olson *et al.* 1980), respectively, were reported. High bioavailability was also observed in humans (80%) (Poiger and Schlatter 1986) and in infants (over 90%) (McLachlan 1993). Generally, a smaller percentage of an administered TCDD dose is absorbed through dermal absorption than through gastrointestinal absorption (Anderson *et al.* 1993; Banks and Birnbaum 1991; Banks *et al.* 1990).

In general, TCDD is not readily metabolized (Olson *et al.* 1980; Ramsey *et al.* 1982; Weber *et al.* 1997); especially in humans the metabolism rate and liver clearance are extremely low (Wendling *et al.* 1990). CYP1A1 is responsible for TCDD metabolism; as a CYP1A1 inducer, TCDD can induce its own metabolism (Olson *et al.* 1994). The slow metabolism may, in part, account for the long half-life for elimination of TCDD from the body.

Feces, rather than urine, is the major excretion route for TCDD (Gasiewicz *et al.* 1983). In studies with radiolabeled TCDD in the mouse and human, a considerable portion (over 50%) of the excreted radioactivity was attributable to metabolites (Gasiewicz *et al.* 1983; Wendling *et al.* 1990), suggesting the contribution of metabolism and biliary excretion. Unchanged TCDD in feces was proposed to be passively diffused into the intestinal lumen from blood and subsequently excreted (Olson 1986), a mechanism termed "exsorption" (Arimori and Nakano 1998), which also occurs with other lipophilic and metabolically resistant chemicals (Rozman *et al.* 1985). This mechanism was supported by the observation that ingestion of nondigestible dietary fat substitute significantly accelerated TCDD fecal excretion(Geusau *et al.* 1999), and it has been recognized to be essential for TCDD excretion (Moser and McLachlan 2002). TCDD has a long half-life in mice (20 days) (Koshakji *et al.* 1984), rats (31 days) (Rose *et al.* 1976), golden hamsters (15 days) (Olson *et al.* 1980), and humans (8.7 years) (Michalek *et al.* 1996). This characteristic, along with its toxicity, has attracted considerable attention to TCDD.

8.4.2 Distribution

In most animal species investigated so far, TCDD is predominantly distributed in adipose tissue and the liver, followed by adrenals, skin, lungs, and kidneys (Abraham *et al.* 1988; Diliberto *et al.* 2001; Viluksela *et al.* 1996). Based on distribution characteristics of other lipophilic chemicals, highly lipophilic TCDD is expected to be mainly distributed in adipose tissue and other fat-rich tissues. However, studies strikingly showed that the pattern of TCDD tissue distribution was dose- and time-dependent (Abraham *et al.* 1988; Diliberto *et al.* 2001).

Diliberto *et al.* (2001) dosed female B6C3F1 mice with ^{3}H-TCDD 5 days per week for 13 weeks, at 0.15, 0.45, 1.5, 4.5, 15, 45, 150, or 450 ng/kg body weight. The liver-to-fat concentration ratio ([Liver]/[Fat]) was 0.21 in the lowest dose group, and it increased with dose to 3.43 in the highest dose group (Fig. 8.3a). Similar results were reported by other groups in the same strain of mice (DeVito *et al.* 1998; van Birgelen *et al.* 1996) and in rats (Fig. 8.3b) (Abraham *et al.* 1988; DeVito *et al.* 1998; van Birgelen *et al.* 1996). Furthermore, pretreatment with TCDD increased [Liver]/[Fat] in mice (Curtis *et al.* 1990). These results led to a hypothesis that TCDD was sequestered by an inducible hepatic protein that was later demonstrated to be CYP1A2 (Diliberto *et al.* 1997; Santostefano *et al.* 1996). CYP1A2, a microsomal enzyme, is induced by TCDD and binds TCDD with affinity in the nanomolar range (Andersen *et al.* 1993). In the CYP1A2 gene knockout mouse, [Liver]/[Fat] of TCDD was approximately 0.2 even in a dose as high as 25 μg/kg (Diliberto *et al.* 1997), similar to the ratio at a very low body burden in normal mice (Diliberto *et al.* 2001), which strongly supported the role that CYP1A2 plays. With the increase in dose/body burden, CYP1A2 induction reaches saturation and more TCDD is stored in fat and hence [Liver]/[Fat] reduces (Andersen *et al.* 1993). Thus TCDD distribution varies with dose.

The pattern of TCDD distribution in tissues also varies with time. Abraham *et al.* (1988) administered a single dose of 300 ng TCDD/kg to female Wistar rats and determined the concentration in the liver and fat at various times. One day after dosing, [Liver]/[Fat] was about 10; 91 days after dosing, [Liver]/[Fat] had decreased to about 0.5 (Fig. 8.4). The phenomenon probably reflected the change in CYP1A2 protein level in the liver.

8.5 PBPK MODELS OF TCDD

Due to the characteristic dose-dependent distribution of TCDD in the body and the role of protein binding and induction in distribution, PBPK modeling has been important in predicting the tissue distribution and dose–response relationship at low dose levels. Much effort has been exerted on PBPK modeling of TCDD and its congeners and analogues, resulting in the availability of more than 25 models in mice, rats, humans, and other species. Since TCDD distribution in the body is inevitably related to specific biochemical responses (e.g., CYP1A2 induction), many of these models incorporate some PD information. The combination of a PBPK model and a PD model results in a PBPK/PD model that links external dose directly with biochemical responses to predict the dose–response relationship. Many of the models for TCDD presented here are actually PBPK/PD models.

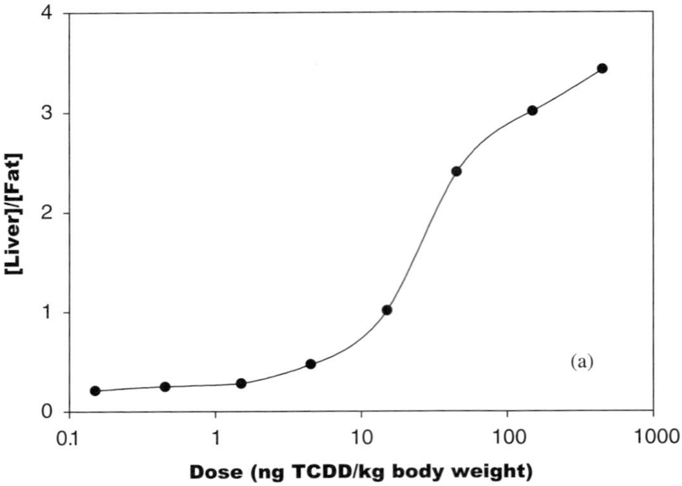

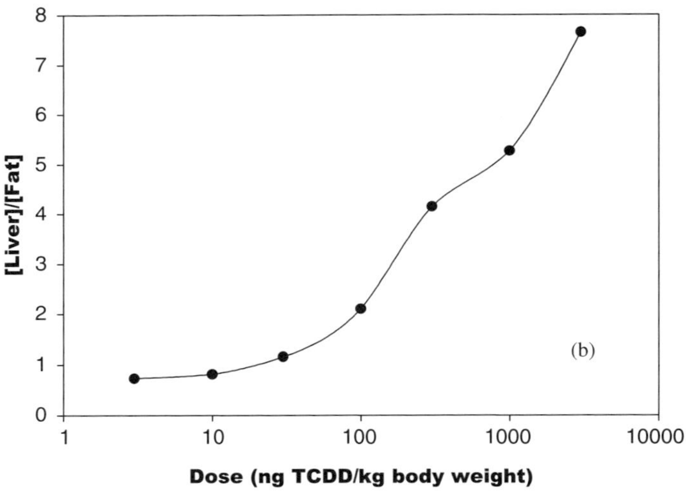

Figure 8.3 Liver-to-fat TCDD concentration ratio (i.e., [Liver]/[Fat]) as a function of dose in **(a)** female B6C3F1 mice administered with ^{3}H-TCDD by oral gavage 5 days/week for 13 weeks and **(b)** female Wistar rats administered with a single subcutaneous dose at various levels and sacrificed 7 days later. Parts a and b were reproduced from Diliberto *et al.* (2001) and Abraham *et al.* (1988), respectively.

8.5.1 PBPK Models of TCDD in Rodents

PBPK modeling of dioxins dates back to the early 1980s, when King *et al.* (1983) developed a model for a TCDD analogue, 2,3,7,8-tetrachlorodibenzofuran (TCDF), in the mouse, rat, and monkey. In this model, distribution of TCDF was based on tissue partitioning, and abnormally high values were designated to liver partition

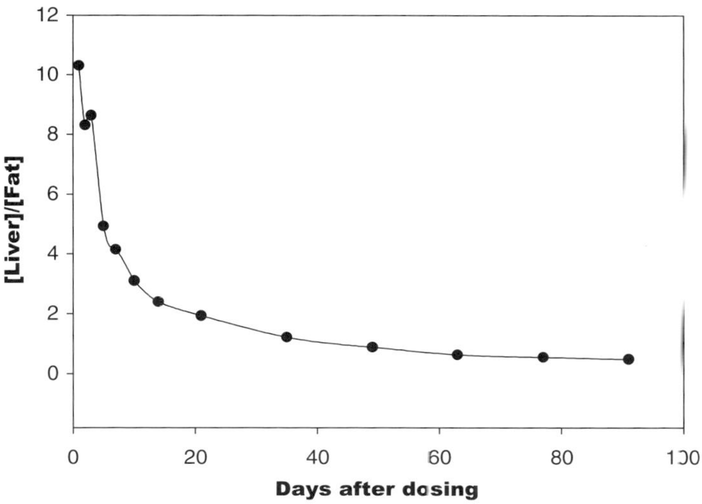

Figure 8.4 Liver-to-fat TCDD concentration ratio (i.e., [Liver]/[Fat]) in female Wistar rats as a function of time after a single TCDD dose of 300 ng/kg. Reproduced from Abraham *et al.* (1988).

coefficients to account for the chemical sequestration (not mechanistically understood then) in the liver. An increasing understanding of liver sequestration allowed Leung *et al.* (1988) to be the first to incorporate hepatic CYP1A2 binding in their model for TCDD. Thereafter, this concept was followed by most TCDD modelers, and TCDD PBPK models generally included adipose tissue and liver compartments with provision for tissue partitioning and CYP1A2 binding in liver. Figure 8.5a is a generic representation of these models.

To date, there are four levels of complexity in PBPK models that have been developed and used to study different aspects of TCDD pharmacokinetics and pharmacodynamics (US EPA 2000a) (Fig. 8.6). Level 1 models by Lawrence and Gobas (1997) and Leung *et al.* (1988) feature two specific binding proteins (i.e., cytosolic AhR and microsomal CYP1A2) in the liver, with or without consideration of CYP1A1/2 induction. The Leung *et al.* (1988) model was later subjected to modifications, such as incorporation of extrahepatic AhR-binding sites (muscle and skin) (Leung *et al.* 1990b) and CYP1A1/2 induction (Leung *et al.* 1990a). Next in complexity are the level 2 models incorporating Hill-type interactions between the TCDD–AhR complex and DREs and induction of CYP1A1/2 at the gene and protein levels (Andersen *et al.* 1993; Kohn *et al.* 1993; Santostefano *et al.* 1998; Wang *et al.* 1997a). In the Santostefano *et al.* (1998) model, TCDD–AhR–DRE interactions and induction of CYP1A1 gene expression and associated enzyme activity were simulated in multiple tissues (liver, lungs, kidneys, and skin). Level 3 models are represented by the Andersen *et al.* (1997a) model for zonal induction of CYP1A1/2 proteins and by the Kohn *et al.* (1993) model with extensive hepatic biochemistry incorporated. Level 4 models incorporate extrahepatic biochemical responses. They are represented by the model including TCDD-induced effects on

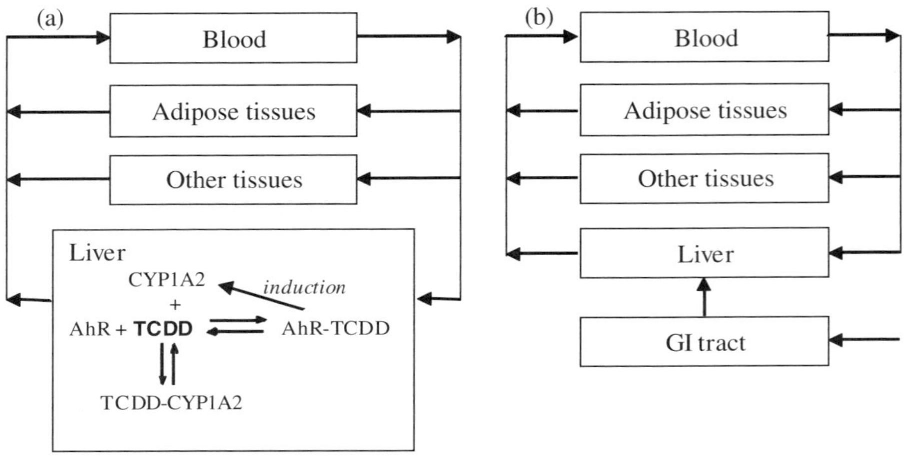

Figure 8.5 Generic representation of a TCDD PBPK model for **(a)** rodents and **(b)** humans.

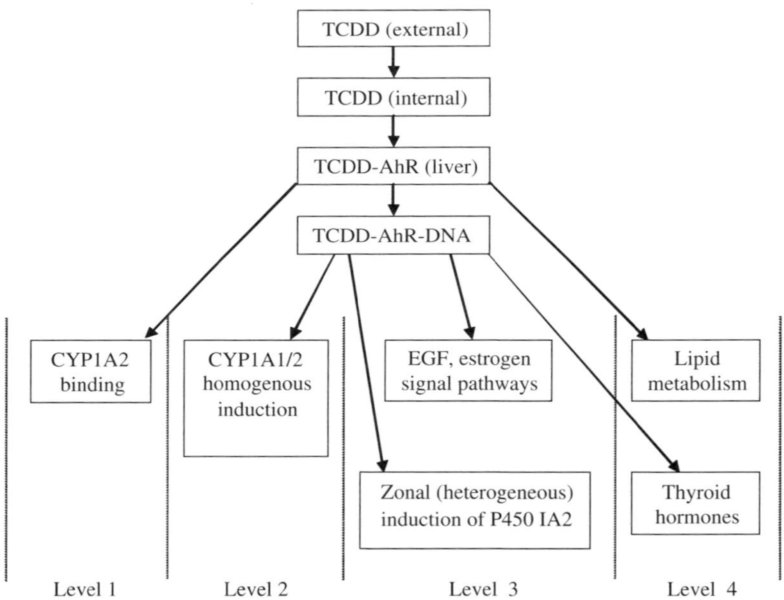

Figure 8.6 Diagram depicting four levels of TCDD PBPK models.

hormones (Kohn *et al.* 1996) and the model describing gastrointestinal tract uptake, lipoprotein transport, and mobilization of fat (Roth *et al.* 1994). Most of these models are applicable for acute or subchronic rather than chronic exposure scenarios. The following discussions predominantly cover representative TCDD PBPK models in rats and mice.

Level 1 Models: Binding of TCDD to AhR and CYP1A2 The development of PBPK models that included specific binding of TCDD in the liver began with the work by Leung *et al.* (1988). In order to explore the effects of various factors on TCDD distribution, Leung *et al.* (1988) developed PBPK models for C57BL/6J and DBA/2J mice. C57BL/6J mice are Ah-responsive with low body fat content, and DBA/2J mice are Ah-nonresponsive with high body fat content. The models consisted of blood, liver, fat, and slowly and rapidly perfused tissue compartments. In the blood, a linear binding process with equilibration between bound and free TCDD was incorporated. TCDD uptake into tissues was flow-limited. In the liver, besides uptake due to partitioning, two hepatic binding sites were included—that is, a high-affinity and low-capacity cytosolic receptor (AhR) site and inducible, low-affinity, and high-capacity microsomal protein sites. Therefore liver uptake depended on partitioning and specific binding to proteins. TCDD was eliminated only through metabolism in the liver. These models did not account for induction of the microsomal protein. A low rate constant for absorption of an intraperitoneal (*i.p.*) dose allowed the model to simulate increasing liver concentrations over the first days, but this may only have been necessary because of the absence of microsomal protein induction in the initial model (Leung *et al.* 1990a). Simulations revealed that the TCDD concentration in the liver was greatly influenced by the binding constant to the microsomal protein, but not by the body fat content.

To further explore the effect of microsomal protein induction on TCDD distribution, Leung *et al.* (1990b) conducted a pharmacokinetic experiment on ^{125}I-2-iodo-3,7,8–trichlorodibenzo-*p*-dioxin (ITCDD), which was developed as a substitute for TCDD, due to its ease of quantification even at very low doses. In their study, one group of female C57BL/6J mice was pretreated with TCDD, while a second group was pretreated with the vehicle only. The TCDD pretreatment dose was high enough (0.1 μmol/kg, *i.p.*) to maximally induce microsomal protein in the liver. Meanwhile, a specific dietary treatment was used to maintain the kinetic and dynamic behavior of ITCDD. Three days later the both groups were given *i.p.* ITCDD at a dose low enough to not cause microsomal protein induction. Tissue concentrations of ITCDD were determined by radiochemical methods measuring ^{125}I at various time points following the dosing. In the TCDD-pretreated mice, as expected, the concentration of ITCDD in the liver was much higher than in fat (Fig. 8.7a); in the vehicle-pretreated mice, however, the opposite was true (Fig. 8.7b), indicating that lipid content based partitioning predominated in the absence of microsomal protein induction.

Leung *et al.* (1990b) then modeled the distribution of that small dose of ITCDD affected by microsomal protein induction in the mice. Most parameters were chosen to be consistent with the previous models (Leung *et al.* 1988). One difference in the new model is that the slowly perfused tissue was divided into muscle and skin compartments, and both were described with cytosolic binding sites. To account for a delay in whole-body elimination from the intact mouse compared to the rate of tissue elimination of ITCDD, the model included a compartment for storing the metabolite before its complete elimination. In the TCDD pretreated mice, no competitive interaction occurred between TCDD and ITCDD. Furthermore, all microsomal protein, but no cytosolic protein, was available for ITCDD binding. The

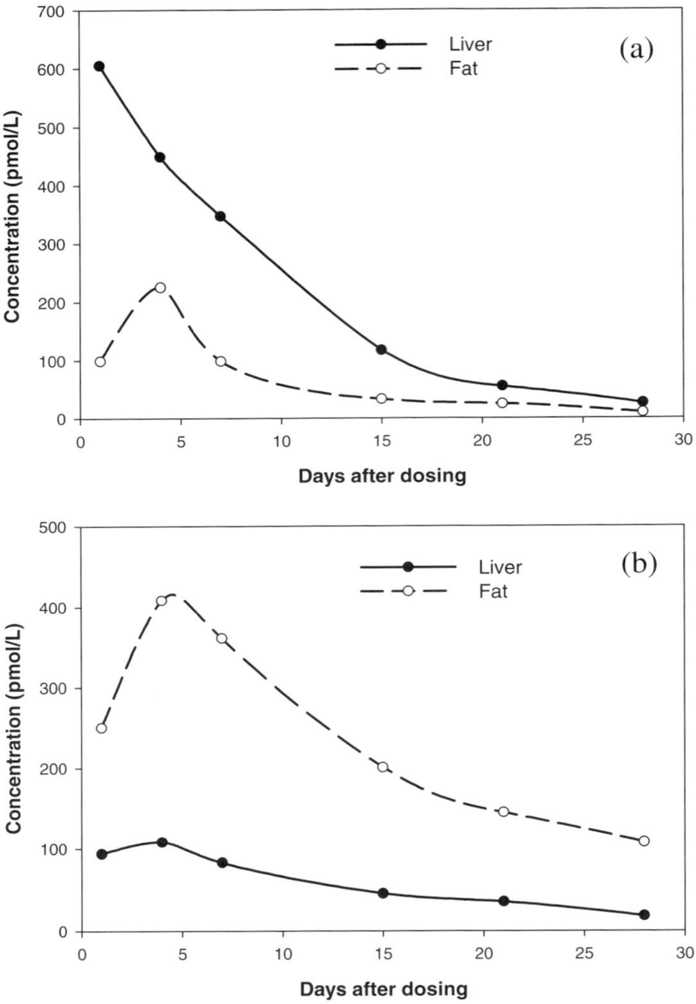

Figure 8.7 Time-course ITCDD concentrations in the liver and fat of **(a)** mice pretreated with TCDD sufficient for microsomal protein induction and **(b)** naive mice. These data are from Leung *et al.* (1990b).

induction of microsomal protein in the TCDD-pretreated mice was reflected by a high binding capacity for ITCDD. These experimental and modeling works strongly support the argument that the induction of the microsomal protein (later proved to be CYP1A2) affected TCDD distribution and should be included in subsequent models.

The mouse TCDD models were then extended to Sprague–Dawley rats. The resultant model, referred to as the Leung–Paustenbach–Murray–Andersen (LPMA) model, explicitly incorporated microsomal enzyme induction (Leung *et al.* 1990a). Both CYP1A1 and CYP1A2 induction were assumed to occur homogeneously throughout the liver in relation to fractional occupancy of the Ah receptor by TCDD.

The AhR binding process was described using Michaelis–Menten kinetics. Total microsomal binding capacity was apportioned between a basal level and an induced level. Metabolic clearance was a noninducible, first-order process. This model sufficiently described the McConnell *et al.* (1984) data for CYP1A1 enzyme induction, reflected by the activity of aryl hydrocarbon hydroxylase (AHH), in the female rats 6 days after a single oral dose (ranging from 0.04 to 5.5 μg/kg) of TCDD in corn oil. It also described the Rose *et al.* tissue distribution data from a single dose study (Rose *et al.* 1976) and Kociba *et al.* data from the 7- and 13-week multiple dose studies (Kociba *et al.* 1976). However, compared to the data from a 2-year feeding study (Kociba *et al.* 1978), the model underestimated TCDD concentrations in the liver and fat at the low dose and overestimated these concentrations at the high dose. The authors suggested that the overestimation might be due to toxicity-related alterations of the physiological parameters and a loss of microsomal binding sites in the chronically exposed rats. This work further confirmed the necessity of incorporation of microsomal induction and binding in dioxin modeling. In addition, it showed the capability of a TCDD PBPK model to predict biochemical effects, such as CYP1A1-dependent AHH activity.

Level 2 Models: Binding of the TCDD–AhR Complex to Regulatory Sites on DNA Binding of the TCDD–AhR complex to DNA is necessary for the modulation of gene expression. Kohn *et al.* (1993) first introduced DNA binding of the TCDD–AhR complex in a rat PBPK model using Hill equations (this model will be discussed in the section on Level 3 models). Andersen *et al.* (1993) also incorporated this characteristic concept into their model for TCDD distribution and enzyme induction in rats.

In the model of Andersen *et al.* (1993), each tissue compartment had separate blood and tissue subcompartments and tissue uptake of TCDD was diffusion-limited. In addition to induction of binding proteins, there was small induction of metabolic clearance in the liver. Induction of both proteins depended on the instantaneous occupancy of the DREs on DNA. This strategy greatly simplifies the true activation process of the AhR by TCDD. The concentration of DNA binding sites was much lower than that of the TCDD–AhR complex. The formation of the TCDD–AhR–DNA complex and the ensuing protein/enzyme induction were governed by Hill equations (Andersen *et al.* 1993):

$$\text{Induction of protein amount or enzyme activity} = \frac{\text{BM} \times (\text{TCDD} - \text{AhR})^n}{(\text{TCDD} - \text{AhR})^n + K_d^n} \qquad (8.1)$$

where BM is the maximal induction, K_d is the binding constant of TCDD–AhR to DNA, TCDD–AhR is the concentration of this complex, and n is the Hill exponent. The Hill equation defines cooperativity among the DNA binding sites and the shape of a dose–response relationship. If $n > 1$, there is positive cooperation and the dose–response curve is sigmoidal; if $n < 1$, there is negative cooperation and the curve is supralinear, climbing faster in the low-dose region than a straight line; if $n = 1$, there is no cooperation (or interaction) and the equation reduces to the familiar Michaelis–Menten equation (Portier *et al.* 1993; Toyoshiba *et al.* 2004).

Andersen *et al.* (1993) calibrated their model with the datasets from the literature (Abraham *et al.* 1988; Krowke *et al.* 1989). For the binding protein CYP1A2 induction, n was estimated to be 1.0, which indicated no interaction among those DREs involved in expression of the CYP1A2 gene. For induction of CYP1A1, however, a larger n (2.3) was required to fit the model to the data, indicating the presence of cooperativity among the DNA binding sites.

Evans and Andersen (2000) developed a steady-state version of this model and conducted sensitivity analyses on two composite variables (i.e., the [Liver]/[Fat] and the fraction of body burden contained in the liver), each as a function of total body burden. The curve shape of each variable was defined by an inflection and a maximum. The inflection was the body burden, where the derivative of the function reached a maximum value; and the maximum was the body burden, where the derivative was zero. These composite variables were described in a model for TCDD and related compounds by Carrier *et al.* (1995a,b). The sensitivity analyses of the steady-state model showed that the use of [Liver]/[Fat] was preferred for evaluating a wide variety of chlorinated dibenzodioxins, where binding affinities between the compounds and CYP1A2 varied over a wide range. In addition, the maximum and the inflection of the curves were sensitive to different parameters. The maximum was primarily sensitive to the fat partition coefficient, the CYP1A2 binding affinity, and the maximal induction of CYP1A2 by TCDD. The inflection point was primarily determined by the binding parameters of TCDD to AhR and the binding affinity of TCDD–AhR complex to DREs.

The level 1 and level 2 models enriched the knowledge of dose-dependent TCDD distribution and the involved factors such as AhR and CYP1A2. This knowledge is valuable for predicting the distribution pattern and the biochemical responses (e.g., CYP1A2 induction) in the low-dose range, where experimental investigation may be not practical. With more experimental findings, these models were extended to incorporate more details.

Level 3 Models: A. Heterogeneity of Protein Induction in the Liver Protein induction in the liver by a variety of enzyme inducers is nonuniform (Bars and Elcombe 1991; Mills *et al.* 1993). At low doses of TCDD, induction of CYP1A1/2 visualized with immunohistochemical staining occurs in the centrilobular regions of the liver. As the dose increases, the area of induction extends progressively further toward the periportal areas (Fig. 8.8) (Andersen *et al.* 1997a; Tritscher *et al.* 1992). A zonal induction model for TCDD was developed to describe this phenomenon (Andersen *et al.* 1997a).

The zonal induction model was added to the Andersen *et al.* models (Andersen *et al.* 1997b; Andersen *et al.* 1993). Compared to the earlier Andersen *et al.* (1993) model, this zonal induction model had several notable changes. First, a delay period of 12 hours occurred between the formation of TCDD–AhR complex and the appearance of new protein. Second, binding of the TCDD–AhR complex increased transcriptional rates of gene expression rather than leading to instantaneous changes in amounts of protein. Third, as described in a companion article (Andersen *et al.* 1997b), the liver was subdivided geometrically into five subcompartments. To create this geometric representation, the liver was regarded as a col-

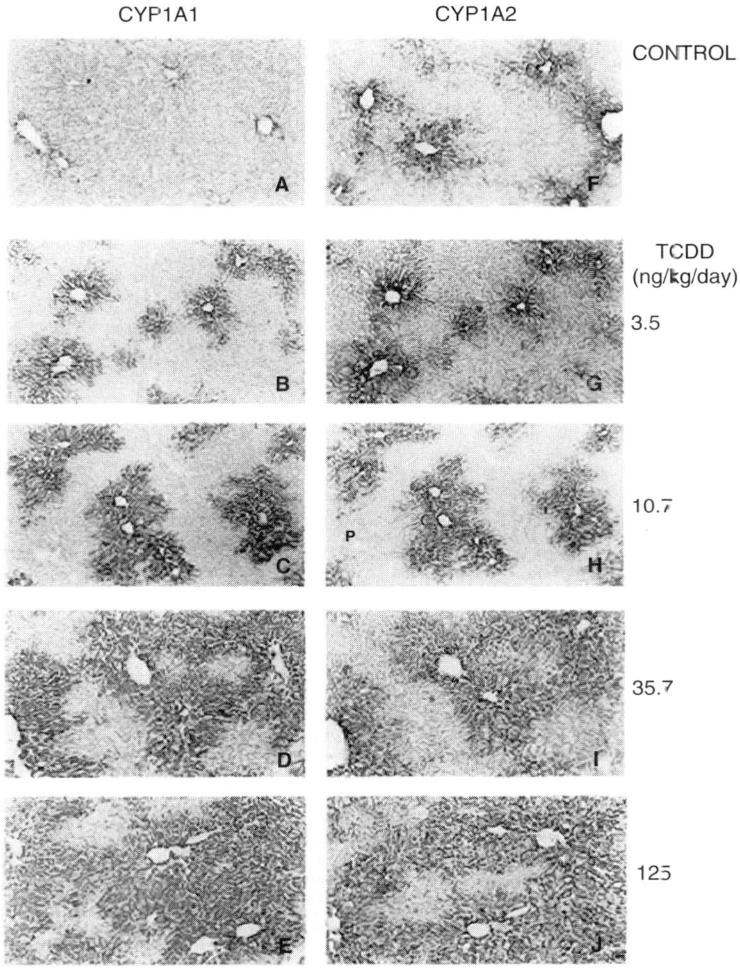

Figure 8.8 Immunohistochemical staining of CYP1A1 and CYP1A2 in the livers of rats treated for 30 weeks with 0, 3.5, 10.7, 35.7, or 125 ng/kg/day TCDD after initiation with diethylnitrosamine. CYP1A1: **A**, untreated control, **B–E**, dose–response. CYP1A2: **F**, untreated control, **G–J**, dose–response (Tritscher *et al.* 1992). Reproduced with permission from Cancer Research.

lection of identical hexagonal units. Each unit was composed of five distinct areas, moving from the portal triad (zone 1) to the central vein (zone 5) (Fig. 8.9). The volume fraction of each hexagonal stack was proportional to the corresponding area. The volume fractions of each subcompartment from the periportal to the centrilobular region were calculated to be 13.5%, 25.5%, 33.9% 20.3%, and 6.8% of the total liver (Andersen *et al.* 1997b). As a modeling strategy, the affinity of TCDD for AhR, K_b, was held constant in all the five subcompartments. Also, the TCDD–AhR–DRE dissociation constant, K_d, was varied by a constant factor between adjacent subcompartments, and the Hill exponent for induction in each sub-

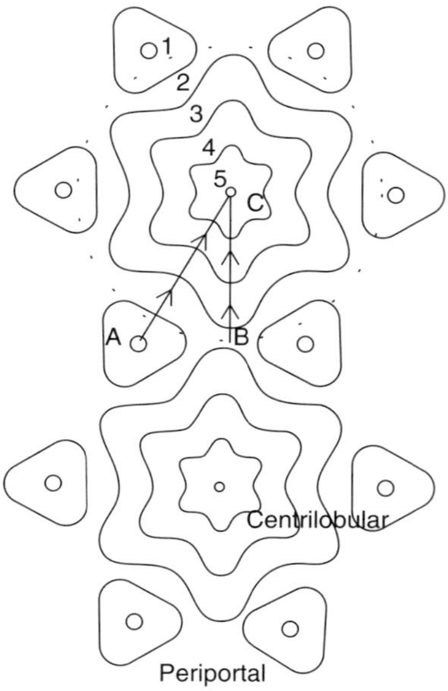

Figure 8.9 Schematic diagram of the geometrically subdivided liver subcompartments (Andersen *et al.* 1997b). Reprinted with permission from Elsevier, Copyright (1997).

compartment, n, had to be 4 or greater to match the observed regional enzyme results obtained by immunohistochemistry. The behavior of this model was compared to a homogeneous induction model in which K_b's and K_d's had the same value among each subcompartment and n's were 1.0. The prediction of TCDD distribution and total enzyme activity by the zonal induction model was generally as good as that by the simplified one. For predicting CYP1A1 mRNA at low doses (Vanden Heuvel *et al.* 1994) and enzyme induction pattern in each acinus, the five-subcompartment model had obvious advantages (Fig. 8.10). The behavior of the zonal induction model, in terms of prediction of demarcating regional induction, could be improved by increasing the number of subcompartments. A limitation to the approach of refining the resolution of the multicompartment liver model was the inherent difficulty in quantitatively comparing the results obtained with the idealized hepatic geometry to the more complex biological structures observed in the actual immunohistochemical staining (Andersen *et al.* 1997b).

The Hill relationship was assumed for CYP1A1/2 gene transcription in the zonal induction model, but there was no biological basis for this assumption. The authors also suggested an all-or-none response that occurred in the cells; that is, the cells could be either fully induced or completely noninduced. This behavior is termed "switch-like" or "binary" in which cell responses (e.g., gene transcription) have only two levels of activity: on and off, contrary to a "graded" behavior in which a cell responds in a continuous range of activity from fully on to fully off (Biggar and Crabtree 2001).

Two experiments have been performed to test whether or not the induction of CYP1A1 is switch-like using 3,3′,4,4′,5-pentachlorobiphenyl, a TCDD-like

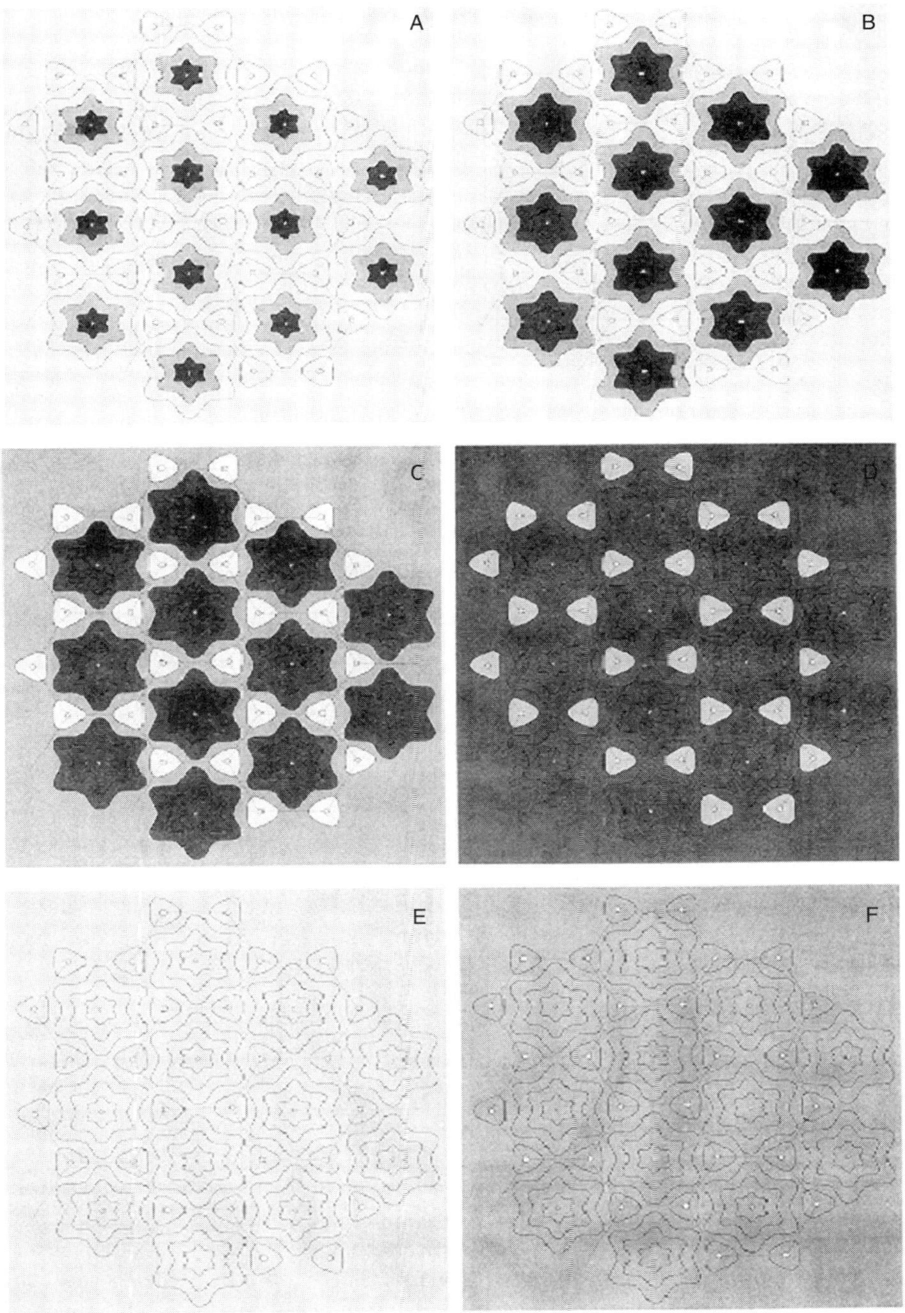

Figure 8.10 Predicted immunohistochemical localization of CYP1A1 in the livers of the rats treated for 30 weeks with various doses of TCDD after initiation with diethylnitrosamine. **A–D**: predictions of the five-compartment zonal induction model at each of the dose (3.5, 10.7, 35.7, and 125 ng/kg/day); **E**, **F**: Predicted uniform distribution derived from the one-compartment homogeneous induction model at the dose of 3.5 and 10.7 ng/kg/day, respectively (Andersen *et al.* 1997a). Reprinted with permission from Elsevier, Copyright (1997).

chemical. French *et al.* (2004) visualized the induction in rat primary hepatocytes of CYP1A1 gene mRNA with *in situ* hybridization and CYP1A1 with immunocytochemical staining at the level of the individual cell. Staining showed that the cells fell into two populations, induced and noninduced. Even in the highest concentration (2.5×10^{-7}M), some cells remained unresponsive. Hybridization analyses, however, did not show the two-population pattern: basal and fully induced. Instead, the induced state was variable and related to the treatment concentration. These results indicated that CYP1A1 induction was switch-like, and once the threshold was triggered, the induction was graded. Similar results were later reported in H4IIE rat hepatoma cells (Broccardo *et al.* 2004). This phenomenon was described as "a hybrid switching module where a switch works in concert with a rheostat, much like a dimmer on a light in a home" (French *et al.* 2004), or simply as a "dimmer switch" (Wilson 2004). Although its mechanism remains unclear, the switch behavior should profoundly affect the dose–response relationship in the low-dose range and, hence, TCDD risk assessment (Andersen *et al.* 2002).

Level 3 Models: B. Complex Biochemical Response-Based Models Estrogens and epidermal growth factor (EGF) signaling systems may be involved in TCDD-mediated hepatocarcinogenesis. Estrogens seem to be required for TCDD-mediated effects on altered expression of EGF-like peptides, such as transforming growth factor-α (TGF-α). EGF-like peptides, in turn, bind to the EGF receptor on cellular plasma membranes and internalize them, which subsequently leads to alterations in cellular regulation and mitotic activity, processes that may be related to hepatocarcinogenesis (Tritscher *et al.* 1994). In addition, cross-talk between the TCDD–AhR–DNA and estrogen signaling pathways has recently been uncovered (Ohtake *et al.* 2003). To obtain a more quantitative relationship between TCDD exposure and ensuing biochemical alterations in the liver, Kohn *et al.* (1993, 1994) developed a biochemical model, hereafter referred to as the Kohn–Portier–Tritscher (KPT) model, which included hepatic CYP1A1/2 induction and other responses such as AhR induction, alterations in estrogen metabolism, and stimulated TGF-α synthesis.

The KPT model was extended from the basic kinetic portion of the LPMA model, but it contained several notable modifications. TCDD was absorbed from the gut into the blood rather than directly into the liver. Protein binding in blood was considered, and the ratio between bound and free TCDD was reflected with an empirical function instead of a proportionality constant. TCDD uptake in all compartments was diffusion-limited. TCDD was metabolized in the liver, and metabolites were excreted either into urine from blood or to the gut and partitioning of the metabolites between liver and blood was the same as that of TCDD. In addition, some TCDD was eliminated from the liver due to TCDD-induced cell death and subsequent lysis. This clearance rate was assumed to increase as a hyperbolic function of the cumulative exposure to unbound liver TCDD.

In the KPT model, the expressions of the CYP1A1, CYP1A2, and AhR genes, due to binding of the TCDD–AhR complex, followed Hill kinetics. For AhR induction, a delay of 6 hours occurred between TCDD administration and increased gene expression. TGF-α was the only ligand of EGF receptor, and its gene expression was a result of synergistic interaction between TCDD–AhR and estrogen–estrogen receptor complex. All proteins were cleared by first-order catabolism. The metabolism of estradiol (catalyzed by CYP1A2) was competitively inhibited by the binding of TCDD to CYP1A2 (Voorman and Aust 1989). The synthesis of estrogen receptor was noncompetitively inhibited by the binding of the TCDD–AhR complex to DNA sites other than the estrogen receptor gene. The large number of parameters in this model was obtained or estimated from the literature. The values of the parameters not available in the literature were adjusted to make the model fit data from Tritscher *et al.* (1992) and Sewall *et al.* (1993). A generally good agreement was achieved between model predictions (i.e., tissue concentrations of TCDD, dose-dependent CYP1A1/2 levels, and maximum binding capacity and internalized fraction of EGF receptor) and experimental datasets (Abraham *et al.* 1988; Sewall *et al.* 1993; Tritscher *et al.* 1992). This pharmacodynamic model has been updated twice by incorporating more biological information reported in subsequent experiments (Kohn *et al.* 1996, 2001).

Level 4 Models: Thyroid Hormone Model Biochemical/toxic effects of TCDD take place not only in the liver, but also in extrahepatic tissues. TCDD is toxic to the thyroid and its signaling pathway. It enhances metabolism of thyroxine (3,5,3′,5′-tetraiodothyronine, T_4) by induction of UDP-glucuronosyltransferase (UGT) and hence reduces the serum concentration (Bastomsky 1977; Gorski *et al.* 1988). It also elevates the level of thyroid stimulating hormone, TSH (Sewall *et al.* 1995). TCDD itself is a potent thyroid hormone receptor agonist (McKinney *et al.* 1985). Furthermore, TCDD causes follicular-cell thyroid adenomas in male Osborne–Mendel rats and B6C3F1 mice (NTP 1982).

To quantitatively study the effects of TCDD on thyroid hormones as well as on the distribution of TCDD in the rat, Kohn *et al.* (1996) incorporated a thyroid compartment and thyroid hormone regulation loops in their model. For the distribution in tissues, hepatic metabolism of TCDD was described using the Hill equation rather than first-order or Michaelis–Menten equations the production and degradation of blood binding protein were incorporated. For modeling the effects on thyroid hormones, the following processes were included: the production of T_3 (3,5,3′-triiodothyronine), T_4, TSH, and other proteins and their interdependent regulation (i.e., hypothalamic–pituitary–thyroid axis); the uptake of T_3 and T_4 by the compartments; metabolism of T_3 and T_4; and the synthesis and degradation of UGT at the mRNA and protein levels. The fully assembled model consisted of 186 differential equations. The model was able to reproduce (1) observed TCDD blood, liver, and fat concentrations, (2) the dose response of hepatic proteins (CYP1A1/2), and (3) the blood hormone (i.e., T_3, T_4, and TSH) levels and hepatic UGT activity. Although there was success in fitting data from various experiments, several uncertainties persist, including uncertainty in the values of the fat-blood partition coefficient, the relative role or importance of extrahepatic metabolism of T_4, and challenges of interspecies extrapolation of model parameters and of mechanisms of thyroid hormone homeostasis.

8.5.2 PBPK Models of TCDD in Humans

The first PBPK modeling of TCDD in humans was performed by Kissel and Robarge (1988), and thereafter, more models in humans were developed (Kreuzer *et al.* 1997; van der Molen *et al.* 1996). Humans are generally exposed to TCDD at much lower levels than used in animal studies, and these low levels are not likely to be sufficient to induce CYP1A2. For this reason, most human models, with the exception of Carrier *et al.* (1995a,b), do not have inducible hepatic binding. Figure 8.5b represents a generic PBPK model for TCDD in humans. However, human exposures to high concentrations of TCDD and/or its analogues are not impossible. The use of Agent Orange, a defoliant mixture contaminated with TCDD, during the Vietnam War led to exposure of American soldiers and Vietnamese people. The explosion in a 2,4,5,-trichlorophenol plant near Seveso, Italy, in 1976 resulted in the highest TCDD exposure known in human residential populations. In the Yusho (oil disease) incident that occurred in Japan in 1968, people ingested PCDFs through contaminated rice oil. These accidents provided data on pharmacokinetics and toxic effects of TCDD and its analogues at a level higher than background (Masuda *et al.* 1998; Michalek *et al.* 2002). Some human PBPK models for TCDD and its analogues were built based on the data from these populations (Aylward *et al.* 2004; Carrier *et al.* 1995a). In the following sections, two human models for background-level TCDD exposure are presented in detail, and others are listed in Table 8.1 and 8.2.

Reference Man Lifetime Exposure Model To study TCDD internal distribution patterns in different species and their impact on risk assessment, Lawrence and Gobas (1997) developed PBPK models for rodents (rats and mice) and humans. The human model had blood, muscle, viscera, skin, fat, kidney, liver, and gut (divided into gut tissue and gut lumen) compartments. Intake of TCDD was exclusively through ingestion of TCDD-contaminated food. The dermal and respiratory exposure routes were not considered. Distribution of TCDD into tissues was governed only by blood flows and partition coefficients, and binding in the blood was not included in the model. In the liver, no specific binding to AhR or CYP1A2 was incorporated because these bindings may not significantly affect pharmacokinetics at the background level of human exposures. Parent TCDD was eliminated through urinary and biliary excretion, without provision for metabolism. Model calculations were performed for a 70-kg reference man exposed to a background TCDD level of 0.32 pg/kg/day for 70 years. Although the simulations covered a full lifetime, the model did not include physiological changes that occur during different life stages.

The model was calibrated by comparing model output with liver and adipose tissue specimens collected from autopsy patients. Predicted (6.7 pg/g) and observed (a weighted mean of 7.5 pg/g of 15 data sets) concentration in human adipose tissue were in good agreement. Similarly, predicted (0.56 pg/g) and measured (0.70 pg/g geometric mean of observation) liver concentration agreed well. On a lipid adjusted basis, liver concentrations were very similar to adipose tissue concentrations, indicating that the human liver accumulates TCDD mainly on the basis of solubility rather than by specific protein-based sequestration at background exposure levels. The model prediction showed that tissue concentrations did not increase significantly after 50 years old, which was to the contrary of the observation that tissue concentrations increased with age over a lifetime. This discrepancy may be due to the use of physiological parameters that did not change with time. Based on the comparison among simulation results, the authors stressed that the relationship between external dose and internal concentrations for TCDD differed between humans and rodents by orders of magnitude. They further suggested that body-weight-based interspecies scale-up may substantially under- or overestimate potential cancer risks of TCDD.

Fetus and Infant Exposure Model TCDD can be distributed from a mother to a fetus through the placenta and to an infant by maternal milk (Abbott *et al.* 1996; Nau *et al.* 1986) and cause toxicity (Couture *et al.* 1990; Gehrs and Smialowicz 1997). The pharmacokinetics of TCDD may vary across life stages—that is, fetus, infant, child, adult, and the elderly. Models were developed to describe internal dose metrics of TCDD, as well as other chemicals, in the first two stages of life (Gentry *et al.* 2003).

For the fetus, the model was a modification of an earlier pregnancy model for isopropanol (Gentry *et al.* 2002). It consisted of a maternal and a fetal component. The latter was divided into blood, liver, and body compartments. A diffusion-limited TCDD transfer process occurred between the placenta and the fetal blood compartment. For the infant stage, the fetal component was replaced with an infant component with a structure similar to that of the model for the mother. TCDD was transferred into the gastrointestinal tract of the infant through maternal milk in a zero-order process. No metabolism of TCDD was included in the fetal liver. TCDD uptake in each compartment was based on partitioning, and no protein binding was considered. The model simulation of maternal and fetal/infant blood levels actually required simulations to be performed with more than one PBPK model. First, a model by Clewell *et al.* (2001) was run for five years of exposure to 1 ng TCDD/kg/day in drinking water to get the input of maternal tissue levels for the fetus-stage model. Second, with the input as starting values and the same exposure regimen, the fetus-stage model was run for 274 days to get maternal and fetal tissue levels at the end of pregnancy. Third, the output of the second run was input of the infant-stage model which was run up to the first 6 months of life. The maternal and fetal/infant blood concentrations at the end of each trimester of pregnancy and at 1, 3, and 6 months of age were the dose metrics of concern.

The simulation results indicated that the fetal blood concentration peaked in the very early stages of pregnancy, decreased in the following 3 months to about 75% of the peak, and remained constant thereafter. The transition from exposure through the placenta to exposure via lactation did not significantly alter the blood concentration in either the mother or the offspring. During lactation, the infant blood concentration slightly declined; but if the infant was bottle-fed and exposed to TCDD via drinking water, the blood concentration declined faster. The blood concentration in the offspring was generally twofold higher than in the mother. This work suggested that exposure to TCDD during the perinatal and postnatal period is of great concern.

The Gentry *et al.* (2003) work was extended to examine the impact of age and sex on the dosimetry following a lifetime continuous exposure to 1 ng TCDD/kg/day (Clewell *et al.* 2004). The age-dependent changes in body weight, clearance, and volume and perfusion rate of the tissue compartments were incorporated. Sensitivity analyses demonstrated that the blood concentration of TCDD was sensitive to liver metabolic clearance and the size of fat compartment. While during adulthood the concentration was more sensitive to the clearance rate, the fat compartment size exerted more control over the TCDD blood concentration during childhood and adolescence. The TCDD blood concentration in males was higher than in females from the age of 7 through 50, paralleling with the larger fractional volume of fat in females.

8.6 SUMMARY

TCDD is the most toxic dioxin congener and has been extensively studied. Most biochemical and toxic effects of TCDD are AhR-mediated. It is predominantly distributed in the liver and adipose tissue in both animals and humans. Sequestration of TCDD in the liver due to CYP1A2 binding results in nonlinear disposition in the two tissues.

PBPK modeling of TCDD and its congeners provides a good example of the evolution of modeling strategies as new information becomes available on factors important in the disposition and biological activities of xenobiotics. An initial model for TCDF focused on tissue lipophilicity-based partitioning. Observations of dose-dependent sequestration in the liver led to new model structures that included blood binding and incrementally added binding to AhR, to CYP1A2, and to DREs in the genomic sequence. Detailed time-course datasets permitted the inclusion of diffusion-limited tissue uptake and estimates of tissue diffusional clearance parameters. Mechanistic studies supported extensions to PBPK/PD models of tumor-related hormones and thyroid hormone homeostasis and its perturbation by TCDD. While each of these models has uncertainties, ongoing work should continue to improve our understanding of the molecular mechanisms of induction and lead to new advances in modeling and insights for toxicology and risk assessment.

NOTATION

AHH	aryl hydrocarbon hydroxylase
AhR	aryl hydrocarbon receptor
AhRR	aryl hydrocarbon receptor repressor
ARNT	aryl hydrocarbon receptor nuclear translocator
BM	maximal induction of protein amount or enzyme activity due to the binding of TCDD–AhR complex to DNA

bHLH	basic helix–loop–helix
CYP1A1	cytochrome P450 1A1 protein
CYP1A2	cytochrome P450 1A2 protein
DRE	dioxin response element
EGF	epidermal growth factor
Hsp90	heat shock protein 90
IARC	International Agency for Research on Cancer
i.p.	intraperitoneal
ITCDD	$[^{125}I]$-2-iodo-3,7,8-trichlorodibenzo-*p*-dioxin
K_b	association constant of TCDD binding to AhR
K_d	dissociation constant of TCDD–AhR–DRE complex
KPT	Kohn–Portier–Tritscher
[Liver]/[Fat]	ratio of the TCDD concentration in the liver to the TCDD concentration in the fat
LPMA	Leung–Paustenbach–Murray–Andersen
n	Hill equation exponent
PCDDs	polychlorinated dibenzo-*p*-dioxins
PCDFs	polychlorinated dibenzofurans
PD	pharmacodynamic
POPs	persistent organic pollutants
T_3	3,5,3′-triiodothyronine
T_4	3,5,3′,5′-tetraiodothyronine
TCDD	2,3,7,8-tetrachlorodibenzo-*p*-dioxin
TCDF	2,3,7,8-tetrachlorodibenzofuran
TGF-α	transforming growth factor-α
TSH	thyroid stimulating hormone
UGT	UDP-glucuronosyltransferase
US EPA	United States Environmental Protection Agency
XRE	xenobiotic response element

REFERENCES

Abbott, B. D., Birnbaum, L. S., and Diliberto, J. J. (1996). Rapid distribution of 2,3,7,8-tetrachlorodibenzo-*p*-dioxin (TCDD) to embryonic tissues in C57BL/6N mice and correlation with palatal uptake *in vitro*. *Toxicol. Appl. Pharmacol.* **141**, 256–263.

Abraham, K., Krowke, R., and Neubert, D. (1988). Pharmacokinetics and biological activity of 2,3,7,8-tetrachlorodibenzo-*p*-dioxin. 1. Dose-dependent tissue distribution and induction of hepatic ethoxyresorufin *O*-deethylase in rats following a single injection. *Arch. Toxicol.* **62**, 359–368.

Andersen, M. E., and Dennison, J. E. (2001). Mode of action and tissue dosimetry in current and future risk assessments. *Sci. Total Environ.* **274**, 3–14.

Andersen, M. E., Mills, J. J., Gargas, M. L., Kedderis, L., Birnbaum, L. S., Neubert, D., and Greenlee, W. F. (1993). Modeling receptor-mediated processes with dioxin: implications for pharmacokinetics and risk assessment. *Risk Anal.* **13**, 25–36.

Andersen, M. E., Birnbaum, L. S., Barton, H. A., and Eklund, C. R. (1997a). Regional hepatic CYP1A1 and CYP1A2 induction with 2,3,7,8-tetrachlorodibenzo-*p*-dioxin evaluated with a multicompartment geometric model of hepatic zonation. *Toxicol. Appl. Pharmacol.* **144**, 145–155.

Andersen, M. E., Eklund, C. R., Mills, J. J., Barton, H. A., and Birnbaum, L. S. (1997b). A multicompartment geometric model of the liver in relation to regional induction of cytochrome P450s. *Toxicol. Appl. Pharmacol.* **144**, 135–144.

Andersen, M. E., Gargas, M. L., Hays, S. M., Clewell, H. J., III, and Paustenbach, D. J. (1997c). Estimating minimally effective inducing doses (ED01s) of TCDD in livers of rats and humans using physiologically based pharmacokinetic modeling. *Organohalogen Compd.* **34**, 311–316.

Andersen, M. E., Yang, R. S., French, C. T., Chubb, L. S., and Dennison, J. E. (2002). Molecular circuits, biological switches, and nonlinear dose–response relationships. *Environ. Health Perspect.* **110**(Suppl 6), 971–978.

Anderson, Y. B., Jackson, J. A., and Birnbaum, L. S. (1993). Maturational changes in dermal absorption of 2,3,7,8-tetrachlorodibenzo-*p*-dioxin (TCDD) in Fischer 344 rats. *Toxicol. Appl. Pharmacol.* **119**, 214–220.

Arimori, K., and Nakano, M. (1998). Drug exsorption from blood into the gastrointestinal tract. *Pharm. Res.* **15**, 371–376.

Aylward, L. L., Brunet, R. C., Carrier, G., Hays, S. M., Cushing, C. A., Needham, L. L., Patterson, D. G., Gerthoux, P. M., Brambilla, P., and Mocarelli, P. (2005). Concentration-dependent TCDD elimination kinetics in humans: toxicokinetic modeling for moderately to highly exposed adults from Seveso, Italy, and Vienna, Austria, and impact on dose estimates for the NIOSH cohort. *J. Expo. Anal. Environ. Epidemiol.* **15**, 51–65.

Banks, H. T., Musante, C. J., and Tran, H. T. (1998). A dispersion model for the hepatic uptake and elimination of 2,3,7,8-tetrachlorodibenzo-*p*-dioxin. *Math. Comp. Mod.* **28**, 9–29.

Banks, Y. B., and Birnbaum, L. S. (1991). Absorption of 2,3,7,8-tetrachlorodibenzo-*p*-dioxin (TCDD) after low dose dermal exposure. *Toxicol. Appl. Pharmacol.* **107**, 302–310.

Banks, Y. B., Brewster, D. W., and Birnbaum, L. S. (1990). Age-related changes in dermal absorption of 2,3,7, 8-tetrachlorodibenzo-*p*-dioxin and 2,3,4,7,8-pentachlorodibenzofuran. *Fundam. Appl. Toxicol.* **15**, 163–173.

Bars, R. G., and Elcombe, C. R. (1991). Dose-dependent acinar induction of cytochromes P450 in rat liver. Evidence for a differential mechanism of induction of P450IA1 by beta-naphthoflavone and dioxin. *Biochem. J.* **277**(Pt 2), 577–580.

Bastomsky, C. H. (1977). Enhanced thyroxine metabolism and high uptake goiters in rats after a single dose of 2,3,7,8-tetrachlorodibenzo-*p*-dioxin. *Endocrinology* **101**, 292–296.

Beebe, L. E., Anver, M. R., Riggs, C. W., Fornwald, L. W., and Anderson, L. M. (1995). Promotion of N-nitrosodimethylamine-initiated mouse lung tumors following single or multiple low dose exposure to 2,3,7,8-tetrachlorodibenzo-*p*-dioxin. *Carcinogenesis* **16**, 1345–1349.

Biggar, S. R., and Crabtree, G. R. (2001). Cell signaling can direct either binary or graded transcriptional responses. *Embo. J.* **20**, 3167–3176.

Birnbaum, L. S. (1994). Evidence for the role of the Ah receptor in response to dioxin. *Prog. Clin. Biol. Res.* **387**, 139–154.

Birnbaum, L. S. (1995). Developmental effects of dioxins. *Environ. Health Perspect.* **103**(Suppl 7), 89–94.

Brewster, D. W., Bombick, D. W., and Matsumura, F. (1988a). Rabbit serum hypertriglyceridemia after administration of 2,3,7,8-tetrachlorodibenzo-*p*-dioxin (TCDD). *J Toxicol. Environ. Health* **25**, 495–507.

Brewster, D. W., Elwell, M. R., and Birnbaum, L. S. (1988b). Toxicity and disposition of 2,3,4,7,8-pentachlorodibenzofuran (4PeCDF) in the rhesus monkey (*Macaca mulatta*). *Toxicol. Appl. Pharmacol.* **93**, 231–246.

Broccardo, C. J., Billings, R. E., Chubb, L. S., Andersen, M. E., and Hanneman, W. H. (2004). Single cell analysis of switch-like induction of CYP1A1 in liver cell lines. *Toxicol. Sci.* **78**, 287–294.

Byard, J. L. (1987). The toxicological significance of 2,3,7,8-tetrachlorodibenzo-*p*-dioxin and related compounds in human adipose tissue. *J. Toxicol. Environ. Health* **22**, 381–403.

Carrier, G., Brunet, R. C., and Brodeur, J. (1995a). Modeling of the toxicokinetics of polychlorinated dibenzo-*p*-dioxins and dibenzofurans in mammalians, including humans. I. Nonlinear distribution of PCDD/PCDF body burden between liver and adipose tissues. *Toxicol. Appl. Pharmacol.* **131**, 253–266.

Carrier, G., Brunet, R. C., and Brodeur, J. (1995b). Modeling of the toxicokinetics of polychlorinated dibenzo-*p*-dioxins and dibenzofurans in mammalians, including humans. II. Kinetics of absorption and disposition of PCDDs/PCDFs. *Toxicol. Appl. Pharmacol.* **131**, 267–276.

Chahoud, I., Hartmann, J., Rune, G. M., and Neubert, D. (1992). Reproductive toxicity and toxicokinetics of 2,3,7,8-tetrachlorodibenzo-*p*-dioxin. 3. Effects of single doses on the testis of male rats. *Arch. Toxicol.* **66**, 567–572.

Chan, W. K., Yao, G., Gu, Y. Z., and Bradfield, C. A. (1999). Cross-talk between the aryl hydrocarbon receptor and hypoxia inducible factor signaling pathways. Demonstration of competition and compensation. *J. Biol. Chem.* **274**, 12115–12123.

Chastain, J. E., Jr., and Pazdernik, T. L. (1985). 2,3,7,8-Tetrachlorodibenzo-*p*-dioxin (TCDD)-induced immunotoxicity. *Int. J. Immunopharmacol.* **7**, 849–856.

Christian, B. J., Inhorn, S. L., and Peterson, R. E. (1986). Relationship of the wasting syndrome to lethality in rats treated with 2,3,7,8-tetrachlorodibenzo-*p*-dioxin. *Toxicol. Appl. Pharmacol.* **82**, 239–255.

Clewell, H. J., 3rd, Gentry, P. R., Gearhart, J. M., Covington, T. R., Banton, M. I., and Andersen, M. E. (2001). Development of a physiologically based pharmacokinetic model of isopropanol and its metabolite acetone. *Toxicol. Sci.* **63**, 160–172.

Clewell, H. J., 3rd, Gentry, P. R., Covington, T. R., Sarangapani, R., and Teeguarden, J. G. (2004). Evaluation of the potential impact of age- and gender-specific pharmacokinetic differences on tissue dosimetry. *Toxicol. Sci.* **79**, 381–393.

Couture, L. A., Abbott, B. D., and Birnbaum, L. S. (1990). A critical review of the developmental toxicity and teratogenicity of 2,3,7,8-tetrachlorodibenzo-*p*-dioxin: Recent advances toward understanding the mechanism. *Teratology* **42**, 619–627.

Crow, K. D. (1979). Lipid profiles in dioxin-exposed workers. *Lancet* **1**, 982.

Crump, K. S., Canady, R., and Kogevinas, M. (2003). Meta-analysis of dioxin cancer dose response for three occupational cohorts. *Environ. Health Perspect.* **111**, 681–687.

Curtis, L. R., Kerkvliet, N. I., Baecher-Steppan, L., and Carpenter, H. M. (1990). 2,3,7,8-Tetrachlorodibenzo-*p*-dioxin pretreatment of female mice altered tissue distribution but not hepatic metabolism of a subsequent dose. *Fundam. Appl. Toxicol.* **14**, 523–531.

Dankovic, D. A., and Andersen, M. E. (2002). Estimating previous serum TCDD concentrations in an occupationally exposed cohort using a PBPK model (Abstract). *Toxicologist* **50**.

Derks, H. J. G. M., Berende, P. L. M., and Olling, M. (1994). Pharmacokinetic modeling of polychlorinated dibenzo-*p*-dioxins and furans in cows. *Chemosphere* **28**, 711–715.

DeVito, M. J., and Birnbaum, L. S. (1994). Toxicology of dioxins and related chemicals. In: *Dioxin and Health*, A. Schecter, ed., Plenum Press, New York, pp. 139–162.

DeVito, M. J., Ross, D. G., Dupuy, A. E., Jr., Ferrario, J., McDaniel, D., and Birnbaum, L. S. (1998). Dose–response relationships for disposition and hepatic sequestration of polyhalogenated dibenzo-*p*-dioxins, dibenzofurans, and biphenyls following subchronic treatment in mice. *Toxicol. Sci.* **46**, 223–234.

Diliberto, J. J., Burgin, D., and Birnbaum, L. S. (1997). Role of CYP1A2 in hepatic sequestration of dioxin: Studies using CYP1A2 knock-out mice. *Biochem. Biophys. Res. Commun.* **236**, 431–433.

Diliberto, J. J., DeVito, M. J., Ross, D. G., and Birnbaum, L. S. (2001). Subchronic Exposure of [^{3}H]-2,3,7,8-Tetrachlorodibenzo-*p*-dioxin (TCDD) in female B6C3F1 mice: Relationship of steady-state levels to disposition and metabolism. *Toxicol. Sci.* **61**, 241–255.

Dunagin, W. G. (1984). Cutaneous signs of systemic toxicity due to dioxins and related chemicals. *J. Am. Acad. Dermatol.* **10**, 688–700.

Egeland, G. M., Sweeney, M. H., Fingerhut, M. A., Wille, K. K., Schnorr, T. M., and Halperin, W. E. (1994). Total serum testosterone and gonadotropins in workers exposed to dioxin. *Am. J. Epidemiol.* **139**, 272–281.

Emond, C., Birnbaum, L. S., and DeVito, M. J. (2004). Physiologically based pharmacokinetic model for developmental exposures to TCDD in the rat. *Toxicol. Sci.* **80**, 115–133.

Enan, E., and Matsumura, F. (1994). 2,3,7,8-Tetrachlorodibenzo-*p*-dioxin (TCDD)-induced changes in glucose transporting activity in guinea pigs, mice, and rats *in vivo* and *in vitro*. *J. Biochem. Toxicol.* **9**, 97–106.

Evans, M. V., and Andersen, M. E. (2000). Sensitivity analysis of a physiological model for 2,3,7,8-tetrachlorodibenzo-*p*-dioxin (TCDD): Assessing the impact of specific model parameters on sequestration in liver and fat in the rat. *Toxicol. Sci.* **54**, 71–80.

Fernandez-Salguero, P. M., Hilbert, D. M., Rudikoff, S., Ward, J. M., and Gonzalez, F. J. (1996). Aryl-hydrocarbon receptor-deficient mice are resistant to 2,3,7,8-tetrachlorodibenzo-*p*-dioxin-induced toxicity. *Toxicol. Appl. Pharmacol.* **140**, 173–179.

Fiedler, H. (2003). Dioxins and furans (PCDD/PCDF). In: *The Handbook of Environmental Chemistry*, Vol. 3, Part O, O. Hutzinger, ed., Springer-Verlag, Berlin, pp. 123–201.

French, C. T., Hanneman, W. H., Chubb, L. S., Billings, R. E., and Andersen, M. E. (2004). Induction of CYP1A1 in primary rat hepatocytes by 3,3′,4,4′,5-pentachlorobiphenyl: Evidence for a switch circuit element. *Toxicol. Sci.* **78**, 276–286.

Gasiewicz, T. A., Geiger, L. E., Rucci, G., and Neal, R. A. (1983). Distribution, excretion, and metabolism of 2,3,7,8-tetrachlorodibenzo-*p*-dioxin in C57BL/6J, DBA/2J, and B6D2F1/J mice. *Drug. Metab. Dispos.* **11**, 397–403.

Gasiewicz, T. A., and Rucci, G. (1984). Cytosolic receptor for 2,3,7,8-tetrachlorodibenzo-*p*-dioxin. Evidence for a homologous nature among various mammalian species. *Mol. Pharmacol.* **26**, 90–98.

Gehrs, B. C., and Smialowicz, R. J. (1997). Alterations in the developing immune system of the F344 rat after perinatal exposure to 2,3,7,8-tetrachlorodibenzo-*p*-dioxin II. Effects on the fetus and the neonate. *Toxicology* **122**, 219–228.

Gentry, P. R., Covington, T. R., Andersen, M. E., and Clewell, H. J., 3rd (2002). Application of a physiologically based pharmacokinetic model for isopropanol in the derivation of a reference dose and reference concentration. *Regul. Toxicol. Pharmacol.* **36**, 51–68.

Gentry, P. R., Covington, T. R., and Clewell, H. J., 3rd (2003). Evaluation of the potential impact of pharmacokinetic differences on tissue dosimetry in offspring during pregnancy and lactation. *Regul. Toxicol. Pharmacol.* **38**, 1–16.

Geusau, A., Tschachler, E., Meixner, M., Sandermann, S., Papke, O., Wolf, C., Valic, E., Stingl, G., and McLachlan, M. (1999). Olestra increases faecal excretion of 2,3,7,8-tetrachlorodibenzo-*p*-dioxin. *Lancet* **354**, 1266–1267.

Gorski, J. R., Muzi, G., Weber, L. W., Pereira, D. W., Arceo, R. J., Iatropoulos, M. J., and Rozman, K. (1988). Some endocrine and morphological aspects of the acute toxicity of 2,3,7,8-tetrachlorodibenzo-*p*-dioxin (TCDD). *Toxicol. Pathol.* **16**, 313–320.

Hu, K., and Bunce, N. J. (1999). Metabolism of polychlorinated dibenzo-*p*-dioxins and related dioxin-like compounds. *J. Toxicol. Environ. Health Pt. B Crit. Rev.* **2**, 183–210.

Huff, J. E., Salmon, A. G., Hooper, N. K., and Zeise, L. (1991). Long-term carcinogenesis studies on 2,3,7,8-tetrachlorodibenzo-p-dioxin and hexachlorodibenzo-*p*-dioxins. *Cell Biol. Toxicol.* **7**, 67–94.

IARC (1997). *IARC Monographs on the Evaluation of the Carcinogenic Risks to Humans. Polychlorinated Dibenzo-p-dioxin and Polychlorinated Dibenzofurans*, International Agency for Research on Cancer, Lyon, France, p. 666.

Kedderis, L. B., Mills, J. J., Andersen, M. E., and Birnbaum, L. S. (1993). A physiologically based pharmacokinetic model for 2,3,7,8-tetrabromodibenzo-*p*-dioxin (TBDD) in the rat: Tissue distribution and CYP1A induction. *Toxicol. Appl. Pharmacol.* **121**, 87–98.

King, F. G., Dedrick, R. L., Collins, J. M., Matthews, H. B., and Birnbaum, L. S. (1983). Physiological model for the pharmacokinetics of 2,3,7,8-tetrachlorodibenzofuran in several species. *Toxicol. Appl. Pharmacol.* **67**, 390–400.

Kissel, J. C., and Robarge, G. M. (1988). Assessing the elimination of 2,3,7,8-TCDD from humans with a physiologically based pharmacokinetic model. *Chemosphere* **17**, 2017–2027.

Kociba, R. J., Keeler, P. A., Park, C. N., and Gehring, P. J. (1976). 2,3,7,8-tetrachlorodibenzo-*p*-dioxin (TCDD): Results of a 13-week oral toxicity study in rats. *Toxicol. Appl. Pharmacol.* **35**, 553–574.

Kociba, R. J., Keyes, D. G., Beyer, J. E., Carreon, R. M., Wade, C. E., Dittenber, D. A., Kalnins, R. P., Frauson, L. E., Park, C. N., Barnard, S. D., Hummel, R. A., and Humiston, C. G. (1978). Results of a two-year chronic toxicity and oncogenicity study of 2,3,7,8-tetrachlorodibenzo-*p*-dioxin in rats. *Toxicol. Appl. Pharmacol.* **46**, 279–303.

Kohn, M. C., Lucier, G. W., Clark, G. C., Sewall, C., Tritscher, A. M., and Portier, C. J. (1993). A mechanistic model of effects of dioxin on gene expression in the rat liver. *Toxicol. Appl. Pharmacol.* **120**, 138–154.

Kohn, M. C., Lucier, G. W., and Portier, C. J. (1994). The importance of biological realism in dioxin risk assessment models. *Risk Anal.* **14**, 993–1000.

Kohn, M. C., Sewall, C. H., Lucier, G. W., and Portier, C. J. (1996). A mechanistic model of effects of dioxin on thyroid hormones in the rat. *Toxicol. Appl. Pharmacol.* **136**, 29–48.

Kohn, M. C., Walker, N. J., Kim, A. H., and Portier, C. J. (2001). Physiological modeling of a proposed mechanism of enzyme induction by TCDD. *Toxicology* **162**, 193–208.

Korkalainen, M., Tuomisto, J., and Pohjanvirta, R. (2004). Primary structure and inducibility by 2,3,7,8-tetrachlorodibenzo-*p*-dioxin (TCDD) of aryl hydrocarbon receptor repressor in a TCDD-sensitive and a TCDD-resistant rat strain. *Biochem. Biophys. Res. Commun.* **315**, 123–131.

Koshakji, R. P., Harbison, R. D., and Bush, M. T. (1984). Studies on the metabolic fate of [^{14}C]2,3,7,8-tetrachlorodibenzo-*p*-dioxin (TCDD) in the mouse. *Toxicol. Appl. Pharmacol.* **73**, 69–77.

Kreuzer, P. E., Csanady, G. A., Baur, C., Kessler, W., Papke, O., Greim, H., and Filser, J. G. (1997). 2,3,7,8-Tetrachlorodibenzo-*p*-dioxin (TCDD) and congeners in infants. A toxicokinetic model of human lifetime body burden by TCDD with special emphasis on its uptake by nutrition. *Arch. Toxicol.* **71**, 383–400.

Krowke, R., Chahoud, I., Baumann-Wilschke, I., and Neubert, D. (1989). Pharmacokinetics and biological activity of 2,3,7,8-tetrachlorodibenzo-*p*-dioxin. 2. Pharmacokinetics in rats using a loading-dose/maintenance-dose regime with high doses. *Arch. Toxicol.* **63**, 356–360

Lawrence, G. S., and Gobas, F. A. (1997). A pharmacokinetic analysis of interspecies extrapolation in dioxin risk assessment. *Chemosphere* **35**, 427–452.

Leung, H. W., Ku, R. H., Paustenbach, D. J., and Andersen, M. E. (1988). A physiologically based pharmacokinetic model for 2,3,7,8-tetrachlorodibenzo-*p*-dioxin in C57BL/6J and DBA/2J mice. *Toxicol. Lett.* **42**, 15–28.

Leung, H. W., Paustenbach, D. J., Murray, F. J., and Andersen, M. E. (1990a). A physiological pharmacokinetic description of the tissue distribution and enzyme-inducing properties of 2,3,7,8-tetrachlorodibenzo-*p*-dioxin in the rat. *Toxicol. Appl. Pharmacol.* **103**, 399–410.

Leung, H. W., Poland, A., Paustenbach, D. J., Murray, F. J., and Andersen, M. E. (1990b). Pharmacokinetics of [^{125}I]-2-iodo-3,7,8-trichlorodibenzo-*p*-dioxin in mice: Analysis with a physiological modeling approach. *Toxicol. Appl. Pharmacol.* **103**, 411–419.

Li, W., Donat, S., Dohr, O., Unfried, K., and Abel, J. (1994). Ah receptor in different tissues of C57BL/6J and DBA/2J mice: Use of competitive polymerase chain reaction to measure Ah-receptor mRNA expression. *Arch. Biochem. Biophys.* **315**, 279–284.

Liem, A. K., Furst, P., and Rappe, C. (2000). Exposure of populations to dioxins and related compounds. *Food. Addit. Contam.* **17**, 241–259.

Lorber, M., and Phillips, L. (2002). Infant exposure to dioxin-like compounds in breast milk. *Environ. Health Perspect.* **110**, A325–A332.

Luebeck, E. G., Buchmann, A., Stinchcombe, S., Moolgavkar, S. H., and Schwarz, M. (2000). Effects of 2,3,7,8-tetrachlorodibenzo-*p*-dioxin on initiation and promotion of GST-P-positive foci in rat liver: A quantitative analysis of experimental data using a stochastic model. *Toxicol. Appl. Pharmacol.* **167**, 63–73.

Maruyama, W., Yoshida, K., Tanaka, T., and Nakanishi, J. (2002a). Possible range of dioxin concentration in human tissues: Simulation with a physiologically based model. *J. Toxicol. Environ. Health Pt. A* **65**, 2053–2073.

Maruyama, W., Yoshida, K., Tanaka, T., and Nakanishi, J. (2002b). Determination of tissue–blood partition coefficients for a physiological model for humans, and estimation of dioxin concentration in tissues. *Chemosphere* **46**, 975–985.

Maruyama, W., Yoshida, K., Tanaka, T., and Nakanishi, J. (2003). Simulation of dioxin accumulation in human tissues and analysis of reproductive risk. *Chemosphere* **53**, 301–313.

Masuda, Y., Schecter, A., and Papke, O. (1998). Concentrations of PCBs, PCDFs and PCDDs in the blood of Yusho patients and their toxic equivalent contribution. *Chemosphere* **37**, 1773–1780.

McConnell, E. E., Lucier, G. W., Rumbaugh, R. C., Albro, P. W., Harvan, D. J., Hass, J. R., and Harris, M. W. (1984). Dioxin in soil: Bioavailability after ingestion by rats and guinea pigs. *Science* **223**, 1077–1079.

McGregor, D. B., Partensky, C., Wilbourn, J., and Rice, J. M. (1998). An IARC evaluation of polychlorinated dibenzo-p-dioxins and polychlorinated dibenzofurans as risk factors in human carcinogenesis. *Environ. Health Perspect.* **106**(Suppl 2), 755–760.

McKinney, J. D., Chae, K., Oatley, S. J., and Blake, C. C. (1985). Molecular interactions of toxic chlorinated dibenzo-*p*-dioxins and dibenzofurans with thyroxine binding prealbumin. *J. Med. Chem.* **28**, 375–381.

McLachlan, M. S. (1993). Digestive tract absorption of polychlorinated dibenzo-*p*-dioxins, dibenzofurans, and biphenyls in a nursing infant. *Toxicol. Appl. Pharmacol.* **123**, 68–72.

Michalek, J. E., Pirkle, J. L., Caudill, S. P., Tripathi, R. C., Patterson, D. G., Jr., and Needham, L. L. (1996). Pharmacokinetics of TCDD in veterans of Operation Ranch Hand: 10-year follow-up. *J. Toxicol. Environ. Health* **47**, 209–220.

Michalek, J. E., Ketchum, N. S., and Akhtar, F. Z. (1998). Postservice mortality of US Air Force veterans occupationally exposed to herbicides in Vietnam: 15-year follow-up. *Am. J. Epidemiol.* **148**, 786–792.

Michalek, J. E., Pirkle, J. L., Needham, L. L., Patterson, D. G., Jr., Caudill, S. P., Tripathi, R. C., and Mocarelli, P. (2002). Pharmacokinetics of 2,3,7,8-tetrachlorodibenzo-*p*-dioxin in Seveso adults and veterans of operation Ranch Hand. *J. Exp. Anal. Environ. Epidemiol.* **12**, 44–53.

Mills, J. J., Murphy, J., and Lyght, O. (1993). Time course and regional distribution of cytochrome P4501A in rat liver following exposure to 2,3,7,8-tetrachlorodibenzo-*p*-dioxin. *Toxicologist* **13**, 192.

Moran, F. M., Hendrickx, A. G., Shideler, S., Overstreet, J. W., Watkins, S. M., and Lasley, B. L. (2004). Effects of 2,3,7,8-tetrachlorodibenzo-*p*-dioxin (TCDD) on fatty acid availability and neural tube formation in cynomolgus macaque, *Macaca fascicularis*. *Birth Defects Res. Part. B. Dev. Reprod. Toxicol.* **71**, 37–46.

Moreno, M. D., Cooper, K. R., Brown, R. P., and Georgopoulos, P. (1992). A physiologically based pharmacokinetic model for *Mya arenaria*. *Marine Environ. Res.* **34**, 321–325.

Moser, G. A., and McLachlan, M. S. (2002). Modeling digestive tract absorption and desorption of lipophilic organic contaminants in humans. *Environ. Sci. Technol.* **36**, 3318–3325.

Nagarkatti, P. S., Sweeney, G. D., Gauldie, J., and Clark, D. A. (1984). Sensitivity to suppression of cytotoxic T cell generation by 2,3,7,8-tetrachlorodibenzo-*p*-dioxin (TCDD) is dependent on the Ah genotype of the murine host. *Toxicol. Appl. Pharmacol.* **72**, 169–176.

Nau, H., Bass, R., and Neubert, D. (1986). Transfer of 2,3,7,8-tetrachlorodibenzo-*p*-dioxin (TCDD) via placenta and milk, and postnatal toxicity in the mouse. *Arch. Toxicol.* **59**, 36–40.

Nichols, J. W., Jensen, K. M., Tietge, J. E., and Johnson, R. D. (1998). Physiologically based toxicokinetic model for maternal transfer of 2,3,7,8-tetrachlorodibenzo-*p*-dioxin in brook trout. *Environ. Toxicol. Chemi.* **17**, 2422–2434.

NTP (1982). *Carcinogenesis Bioassay of 2,3,7,8-Tetrachlorodibenzo-p-dioxin (CAS No. 1746-01-6) in Osborne–Mendel Rats and B6C3F1 Mice (Gavage Study)*, National Toxicology Program Technical Report Series, Vol. 209, pp. 1–195.

Ohtake, F., Takeyama, K., Matsumoto, T., Kitagawa, H., Yamamoto, Y., Nohara, K., Tohyama, C., Krust, A., Mimura, J., Chambon, P., Yanagisawa, J., Fujii-Kuriyama, Y., and Kato, S. (2003). Modulation of oestrogen receptor signalling by association with the activated dioxin receptor. *Nature* **423**, 545–550.

Olson, J. R. (1986). Metabolism and disposition of 2,3,7,8-tetrachlorodibenzo-p-dioxin in guinea pigs. *Toxicol. Appl. Pharmacol.* **85**, 263–273.

Olson, J. R. (1994). Pharmacokinetics of dioxins and related chemicals. In: *Dioxin and Health*, A. Schecter, ed., Plenum Press, New York, pp. 163–197.

Olson, J. R., Gasiewicz, T. A., and Neal, R. A. (1980). Tissue distribution, excretion, and metabolism of 2,3,7,8-tetrachlorodibenzo-*p*-dioxin (TCDD) in the golden Syrian hamster. *Toxicol. Appl. Pharmacol.* **56**, 78–85.

Olson, J. R., McGarrigle, B. P., Gigliotti, P. J., Kumar, S., and McReynolds, J. H. (1994). Hepatic uptake and metabolism of 2,3,7,8-tetrachlorodibenzo-*p*-dioxin and 2,3,7,8-tetrachlorodibenzofuran. *Fundam. Appl. Toxicol.* **22**, 631–640.

Paustenbach, D. J. (2002). The U.S. EPA Science Advisory Board Evaluation (2001) of the EPA dioxin reassessment. *Regul. Toxicol. Pharmacol.* **36**, 211–219.

Pegram, R. A., Diliberto, J. J., Moore, T. C., Gao, P., and Birnbaum, L. S. (1995). 2,3,7,8-Tetrachlorodibenzo-p-dioxin (TCDD) distribution and cytochrome P4501A induction in young adult and senescent male mice. *Toxicol. Lett.* **76**, 119–126.

Pohjanvirta, R., Kulju, T., Morselt, A. F., Tuominen, R., Juvonen, R., Rozman, K., Mannisto, P., Collan, Y., Sainio, E. L., and Tuomisto, J. (1989). Target tissue morphology and serum biochemistry following 2,3,7,8-tetrachlorodibenzo-*p*-dioxin (TCDD) exposure in a TCDD-susceptible and a TCDD-resistant rat strain. *Fundam. Appl. Toxicol.* **12**, 698–712.

Poiger, H., and Schlatter, C. (1986). Pharmacokinetics of 2,3,7,8-TCDD in man. *Chemosphere* **15**, 1489–1494.

Portier, C., Tritscher, A., Kohn, M., Sewall, C., Clark, G., Edler, L., Hoel, D., and Lucier, G. (1993). Ligand/receptor binding for 2,3,7,8-TCDD: Implications for risk assessment. *Fundam. Appl. Toxicol.* **20**, 48–56.

Ramsey, J. C., Hefner, J. G., Karbowski, R. J., Braun, W. H., and Gehring, P. J. (1982). The *in vivo* biotransformation of 2,3,7,8-tetrachlorodibenzo-*p*-dioxin (TCDD) in the rat. *Toxicol. Appl. Pharmacol.* **65**, 180–184.

Rao, M. S., Subbarao, V., Prasad, J. D., and Scarpelli, D. G. (1988). Carcinogenicity of 2,3,7,8-tetrachlorodibenzo-*p*-dioxin in the Syrian golden hamster. *Carcinogenesis* **9**, 1677–1679.

Rier, S., and Foster, W. G. (2002). Environmental dioxins and endometriosis. *Toxicol. Sci.* **70**, 161–170.

Rose, J. Q., Ramsey, J. C., Wentzler, T. H., Hummel, R. A., and Gehring, P. J. (1976). The fate of 2,3,7,8-tetrachlorodibenzo-*p*-dioxin following single and repeated oral doses to the rat. *Toxicol. Appl. Pharmacol.* **36**, 209–226.

Roth, W. L., Weber, L. W., Stahl, B. U., and Rozman, K. (1993). A pharmacodynamic model of triglyceride transport and deposition during feed deprivation or following treatment with 2,3,7,8-tetrachlorodibenzo-*p*-dioxin (TCDD) in the rat. *Toxicol. Appl. Pharmacol.* **120**, 126–137.

Roth, W. L., Ernst, S., Weber, L. W., Kerecsen, L., and Rozman, K. K. (1994). A pharmacodynamically responsive model of 2,3,7,8-tetrachlorodibenzo-*p*-dioxin (TCDD) transfer between liver and fat at low and high doses. *Toxicol. Appl. Pharmacol.* **127**, 151–162.

Rozman, T., Scheufler, E., and Rozman, K. (1985). Effect of partial jejunectomy and colectomy on the disposition of hexachlorobenzene in rats treated or not treated with hexadecane. *Toxicol. Appl. Pharmacol.* **78**, 421–427.

Safe, S., and Wormke, M. (2003). Inhibitory aryl hydrocarbon receptor–estrogen receptor alpha cross-talk and mechanisms of action. *Chem. Res. Toxicol.* **16**, 807–816.

Santostefano, M. J., Johnson, K. L., Whisnant, N. A., Richardson, V. M., DeVito, M. J., Diliberto, J. J., and Birnbaum, L. S. (1996). Subcellular localization of TCDD differs between the liver, lungs, and kidneys after acute and subchronic exposure: Species/dose comparisons and possible mechanism. *Fundam. Appl. Toxicol.* **34**, 265–275.

Santostefano, M. J., Wang, X., DeVito, M. J., Richardson, V. M., Ross, D. G., and Birnbaum, L. S. (1997). Multiple tissue pharmacokinetic analysis of TCDD-induced biochemical responses. *Organohalogen Compd.* **34**, 19–24.

Santostefano, M. J., Wang, X., Richardson, V. M., Ross, D. G., DeVito, M. J., and Birnbaum, L. S. (1998). A pharmacodynamic analysis of TCDD-induced cytochrome P450 gene expression in multiple tissues: Dose- and time-dependent effects. *Toxicol. Appl. Pharmacol.* **151**, 294–310.

Sarkar, S., Jana, N. R., Yonemoto, J., Tohyama, C., and Sone, H. (2000). Estrogen enhances induction of cytochrome P-4501A1 by 2,3,7, 8-tetrachlorodibenzo-*p*-dioxin in liver of female Long–Evans rats. *Int J Oncol* **16**, 141–147.

Sever, L. E., Arbuckle, T. E., and Sweeney, A. (1997). Reproductive and developmental effects of occupational pesticide exposure: The epidemiologic evidence. *Occup. Med.* **12**, 305–325.

Sewall, C. H., Lucier, G. W., Tritscher, A. M., and Clark, G. C. (1993). TCDD-mediated changes in hepatic epidermal growth factor receptor may be a critical event in the hepatocarcinogenic action of TCDD. *Carcinogenesis* **14**, 1885–1893.

Sewall, C. H., Flagler, N., Vanden Heuvel, J. P., Clark, G. C., Tritscher, A. M., Maronpot, R. M., and Lucier, G. W. (1995). Alterations in thyroid function in female Sprague–Dawley rats following chronic treatment with 2,3,7,8-tetrachlorodibenzo-*p*-dioxin. *Toxicol. Appl. Pharmacol.* **132**, 237–244.

Shu, H. P., Paustenbach, D. J., and Murray, F. J. (1987). A critical evaluation of the use of mutagenesis, carcinogenesis, and tumor promotion data in a cancer risk assessment of 2,3,7,8-tetrachlorodibenzo-*p*-dioxin. *Regul. Toxicol. Pharmacol.* **7**, 57–88.

Silbergeld, E. K., and Mattison, D. R. (1987). Experimental and clinical studies on the reproductive toxicology of 2,3,7,8-tetrachlorodibenzo-*p*-dioxin. *Am. J. Ind. Med.* **11**, 131–144.

Starr, T. B. (2001). Significant shortcomings of the U.S. Environmental Protection Agency's latest draft risk characterization for dioxin-like compounds. *Toxicol. Sci.* **64**, 7–13.

Suskind, R. R. (1985). Chloracne, "the hallmark of dioxin intoxication." *Scand. J. Work. Environ. Health* **11**, 165–171.

Toyoshiba, H., Walker, N. J., Bailer, A. J., and Portier, C. J. (2004). Evaluation of toxic equivalency factors for induction of cytochromes P450 CYP1A1 and CYP1A2 enzyme activity by dioxin-like compounds. *Toxicol. Appl. Pharmacol.* **194**, 156–168.

Tritscher, A. M., Clark, G. C., and Lucier, G. W. (1994). Dose–response effects of dioxins. In: *Dioxin and Health*, A. Schecter, ed., Plenum Press, New York, pp. 233–235.

Tritscher, A. M., Goldstein, J. A., Portier, C. J., McCoy, Z., Clark, G. C., and Lucier, G. W. (1992). Dose–response relationships for chronic exposure to 2,3,7,8-tetrachlorodibenzo-*p*-dioxin in a rat tumor promotion model: Quantification and immunolocalization of CYP1A1 and CYP1A2 in the liver. *Cancer. Res.* **52**, 3436–3442.

US EPA (2000a). Exposure and human health reassessment of 2,3,7,8-tetrachlorodibenzo-*p*-dioxin (TCDD) and related compounds, Part II: Health assessment for 2,3,7,8-tetrachlorodibenzo-*p*-dioxin (TCDD) and related compounds. September 2000 SAB Review Draft. EPA/600/P-00/001Be, US Environmental Protection Agency, Washington, D.C.

US EPA (2000b). Exposure and human health reassessment of 2,3,7,8-tetrachlorodibenzo-*p*-dioxin (TCDD) and related compounds, Part III: Integrated summary and risk characterization for 2,3,7,8-tetrachlorodibenzo-*p*-dioxin (TCDD) and related compounds. September 2000 SAB Review Draft, EPA/600/P-00/001Bg.

van Birgelen, A. P., and van den Berg, M. (2000). Toxicokinetics. *Food. Addit. Contam.* **17**, 267–273.

van Birgelen, A. P., Ross, D. G., DeVito, M. J., and Birnbaum, L. S. (1996). Interactive effects between 2,3,7,8-tetrachlorodibenzo-*p*-dioxin and 2,2′,4,4′,5,5′-hexachlorobiphenyl in female B6C3F1 mice: Tissue distribution and tissue-specific enzyme induction. *Fundam. Appl. Toxicol.* **34**, 118–131.

van den Berg, M., De Jongh, J., Poiger, H., and Olson, J. R. (1994). The toxicokinetics and metabolism of polychlorinated dibenzo-*p*-dioxins (PCDDs) and dibenzofurans (PCDFs) and their relevance for toxicity. *Crit. Rev. Toxicol.* **24**, 1–74.

van der Molen, G. W., Kooijman, S. A., and Slob, W. (1996). A generic toxicokinetic model for persistent lipophilic compounds in humans: An application to TCDD. *Fundam. Appl. Toxicol.* **31**, 83–94.

Vanden Heuvel, J. P., Clark, G. C., Kohn, M. C., Tritscher, A. M., Greenlee, W. F., Lucier, G. W., and Bell, D. A. (1994). Dioxin-responsive genes: Examination of dose–response relationships using quantitative reverse transcriptase-polymerase chain reaction. *Cancer. Res.* **54**, 62–68.

Vasquez, A., Atallah-Yunes, N., Smith, F. C., You, X., Chase, S. E., Silverstone, A. E., and Vikstrom, K. L. (2003). A role for the aryl hydrocarbon receptor in cardiac physiology and function as demonstrated by AhR knockout mice. *Cardiovasc. Toxicol.* **3**, 153–163.

Viluksela, M., Duong, T. V., Stahl, B. U., Li, X., Tuomisto, J., and Rozman, K. K. (1996). Toxicokinetics of 2,3,7,8-tetrachlorodibenzo-*p*-dioxin (TCDD) in two substrains of male Long–Evans rats after intravenous injection. *Fundam. Appl. Toxicol.* **31**, 184–191.

Voorman, R., and Aust, S. D. (1989). TCDD (2,3,7,8-tetrachlorodibenzo-*p*-dioxin) is a tight binding inhibitor of cytochrome P-450d. *J. Biochem. Toxicol.* **4**, 105–109.

Wang, X., Santostefano, M. J., Evans, M. V., Richardson, V. M., Diliberto, J. J., and Birnbaum, L. S. (1996). Receptor incorporated physiologically-based pharmacokinetic model for TCDD distribution in rat. *Organohalogen Compd.* **29**, 389–393.

Wang, X., Santostefano, M. J., Evans, M. V., Richardson, V. M., Diliberto, J. J., and Birnbaum, L. S. (1997a). Determination of parameters responsible for pharmacokinetic behavior of TCDD in female Sprague–Dawley rats. *Toxicol. Appl. Pharmacol.* **147**, 151–168.

Wang, X., Santostefano, M. J., DeVito, M. J., and Birnbaum, L. S. (1997b). Extrapolation of a previous PBPK model for TCDD across routes of exposure, genders, and from rats to mice. *Organohalogen Compd.* **34**, 38–42.

Wang, X., Santostefano, M. J., DeVito, M. J., and Birnbaum, L. S. (2000). Extrapolation of a PBPK model for dioxins across dosage regimen, gender, strain, and species. *Toxicol. Sci.* **56**, 49–60.

Weber, R., Schmitz, H. J., Schrenk, D., and Hagenmaier, H. (1997). Metabolic degradation, inducing potency, and metabolites of fluorinated and chlorinated–fluorinated dibenzodioxins and dibenzofurans. *Chemosphere* **34**, 29–40.

Wendling, J. M., Orth, R. G., and Poiger, H. (1990). Determination of [^{3}H]-2,3,7,8-tetrachlorodibenzo-*p*-dioxin in human feces to ascertain its relative metabolism in man. *Anal. Chem.* **62**, 796–800.

Wilson, C. L. (2004). Molecular switch circuits in toxicology: A dimmer switch for dioxin. *Toxicol. Sci.* **78**, 178–190.

CHAPTER 9

METALS AND INORGANIC COMPOUNDS

Julie Campain

9.1 INTRODUCTION

Metals make up the large majority of the 105 elements in the periodic table; of these, many have been shown to be toxic to humans in one form or another. Metals of major toxicological concern include arsenic (As) and arsine, beryllium, cadmium (Cd), chromium (Cr), lead (Pb), mercury (Hg), and nickel (Ni) (Goyer and Clarkson 2001). In addition, approximately 10 essential metals have the potential for toxicity under appropriate exposure conditions. Historically, concern over metals has primarily been due to their acute toxicity. However, as environmental and occupational standards become more rigorous, cases of acute metal toxicity are increasingly uncommon. The more subtle health effects of chronic or long-term exposure to low levels of metals have taken center stage in the toxicological arena; attempts to understand these processes have brought many challenges to light. Physiologically based pharmacokinetic (PBPK) modeling can aid in understanding the toxicology of metals through acquisition of quantitative information and model-directed experimentation. Some of the first PBPK models for inorganic compounds were for

Physiologically Based Pharmacokinetic Modeling, edited by Micaela B. Reddy, Raymond S. H. Yang, Harvey J. Clewell III, and Melvin E. Andersen.

chloride in the cat (Gabelnick *et al.* 1970), bromide in the rat (Pierson *et al.* 1978) and zinc in the rat and human (Jain and Gerlowski 1981). To date, PBPK models have only been developed for five potentially toxic metals: zinc, arsenic, nickel, lead, and chromium. Additionally, even in the fairly complete models, such as that for lead, a lack of reliable data upon which to assign certain critical parameter values is one significant problem.

In this review, we have attempted to describe the state of the field as it exists today, with some discussion as to areas most requiring further consideration. The topics covered in this chapter include: (1) a detailed description of the most complete PBPK models for metals (i.e., those developed for arsenic, nickel, lead, and chromium), with emphasis on the unique aspects of each; (2) a comparison of models for lead and chromium with a similar PBPK model developed for fluoride, another bone-seeking element, albeit nonmetal; and (3) brief references to metals such as iron or other endogenous compounds for which compartmental, but not PBPK, models have been described. Mercury, due to its primary action as a developmental toxin, has been reviewed in Chapter 12. Additionally, PBPK modeling efforts for perchlorate have been reviewed in Chapter 12. Tables 9.1 and 9.2 summarize models that have been highlighted in this review. Additionally, the need for PBPK models that address essential metals and homeostatic regulatory mechanisms and other research needs in this area are discussed.

9.2 PHYSIOLOGICALLY BASED MODELING OF METALS

Both occupational and environmental exposures to hazardous metals are significant toxicological concerns. Not only do these metals lead to acute toxicity at higher concentrations, but they may also mediate additional pathologic conditions such as developmental defects and neoplasia in individuals exposed chronically to low levels. Many of these metals are relevant to human exposure. Arsenic, cadmium, chromium, lead, and mercury are the top five metals in site frequency count by the Agency for Toxic Substances and Disease Registry (ATSDR) Completed Exposure Pathway (CEP) Site Count Report (ATSDR 2003). Among the metals, arsenic, lead, mercury, and cadmium are among the Comprehensive Environmental Response, Compensation, and Liability Act (CERCLA) Top 10 Priority Hazardous Substances (ATSDR 2003)—that is, those considered to pose the greatest hazard to human health. In addition, as confirmed by ATSDR using the HazDat database, metals most often occur together; they are present in 8 of 10 and 5 of 10 of the Top 10 Binary Combinations of Contaminants in soil and water, respectively (Fay and Mumtaz 1996).

PBPK modeling of metals requires special considerations as compared to most other chemical species such as drugs and solvents. This issue has been reviewed by O'Flaherty (1998b). Metals are often present in multiple media in the ambient environment including ground water, soil, and air, and chronic exposure is an intrinsic part of day-to-day life. Many of the factors that influence the uptake and disposition of metals differ significantly from those that control the pharmacokinetics of organic

TABLE 9.1 PBPK Models for Metals and Nonmetals

Metal	Reference (in order of discussion)
Arsenic	Menzel *et al.* (1994)
	Mann *et al.* (1996a)
	Mann *et al.* (1996b)
	Yu (1998)
	Yu (1999a)
	Yu (1999b)
	Kitchin *et al.* (1999)
	Gentry *et al.* (2004)
Nickel	Yu and Diu (1982)
	Menzel *et al.* (1987)
	Menzel (1987)
	Menzel *et al.* (1988)
	Hsieh *et al.* (1999a)
	Hsieh *et al.* (1999b)
Lead	O'Flaherty (1991c)
	O'Flaherty (1991a)
	O'Flaherty (1991b)
	O'Flaherty (1992)
	O'Flaherty (1993b)
	O'Flaherty (1995b)
	Inskip *et al.* (1996)
	Polak *et al.* (1996)
	O'Flaherty (1996)
	O'Flaherty (1998)
	Fleming *et al.* (1999)
	O'Flaherty (2000)
	Beck *et al.* (2001)
	Timchalk *et al.* (2001)
Chromium	Thomann *et al.* (1994)
	O'Flaherty (1993a)
	O'Flaherty (1996)
	O'Flaherty (2001)
	Paustenbach and Finley (2003)
Zinc	Jain and Gerlowski (1981)
Fluoride	Rao *et al.* (1995)
Bromide	Pierson *et al.* (1978)
Chloride	Gabelnick *et al.* (1970)
Manganese	Andersen *et al.* (1999)
Radon	Yu and Kim (2004)

compounds. Issues that are relevant to metals include: (1) the necessity for modeling long-term exposure, perhaps to fluctuating concentrations or multiple sources; (2) the impacts of diet on metal kinetics and toxicity; (3) their metabolism, which does not involve the classical cytochrome P450 systems, but is mainly due to

TABLE 9.2 Compartmental Models for Metals and Nonmetals

Compound	Reference (in order of discussion)
Fluoride	Hall *et al.* (1977)
	Charkes *et al.* (1978)
	Charkes *et al.* (1979)
Iron	Nathanson *et al.* (1985)
	Nathanson and McLaren (1987)
	Berzuini *et al.* (1978)
Lead	Leggett (1993)
	US EPA (1994)
	Hogan *et al.* (1998)
	White *et al.* (1998)
	Hassanien and Horvath (1999)
	Mason (2000)
Nickel, zinc, uranium, strontium	ICRP (1979)
	ICRP (1991)
	ICRP (1993)
	ICRP (1994)
	ICRP (1995)
Uranium, thorium	Silk *et al.* (1995)
Uranium	Wrenn *et al.* (1989)
	Wrenn *et al.* (1994)
Strontium	Moreas *et al.* (1991)
	Sips *et al.* (1996)
Cadmium	Choudary *et al.* (2001)
	Kjellstrom and Nordberg (1978)
	Nordberg and Kjellstrom (1979)
Potassium, rubidium, cesium	Leggett and Williams (1988)
	Leggett and Williams (2003)
^{14}C, ^{3}H	Whillans (2003)

changes in oxidation status and other metal-specific alkylation/dealkylation reactions; (4) metal-specific patterns of tissue accumulation, which are more often capacity-limited with extended residence times; (5) limited bioavailability, which is dependent upon solubility and particle size and chemical characteristics; (6) protein and other ligand binding; (7) extensive differences among species in metal metabolism; and (8) metal–metal interactions and interactions between metals and organics, which may influence multiple pharmacokinetic parameters (Goyer and Clarkson 2001; O'Flaherty 1998b; Zalups and Koropatnick 2000). Frequently, speciation of individual metals leads to multiple chemical forms with differing targets and/or toxicities. Thus, with many metals, chemical mixture issues are relevant. In addition, many metals preferentially target to bone, which is a complex and specialized tissue, with pharmacokinetic considerations all its own. Due to this complexity, metals have been addressed individually in this review.

9.2.1 Arsenic

Humans have been exposed to arsenic in the environment and workplace, and its acute toxicity has been recognized, for centuries. Common sources of concern for arsenic include drinking water supplies in many countries and occupational exposure, primarily to arsenic trioxide, in such industries as the manufacture of pesticides and during smelting operations. Arsenic is a known human carcinogen, likely contributing to cancer of the skin, lungs, liver, bladder, and kidney through inhalation and/or oral exposure. However, development of a reliable animal model system for arsenic-induced carcinogenicity has been difficult. Only in the last decade has the metal been demonstrated to cause cancer in animals under specific exposure scenarios (Kitchin 2001; Wang *et al.* 2002). The difficulties in this area could be due to species-specific differences in detoxification, metabolism, or uptake or accumulation in target tissues. As shown in Fig. 9.1, humans are potentially exposed to

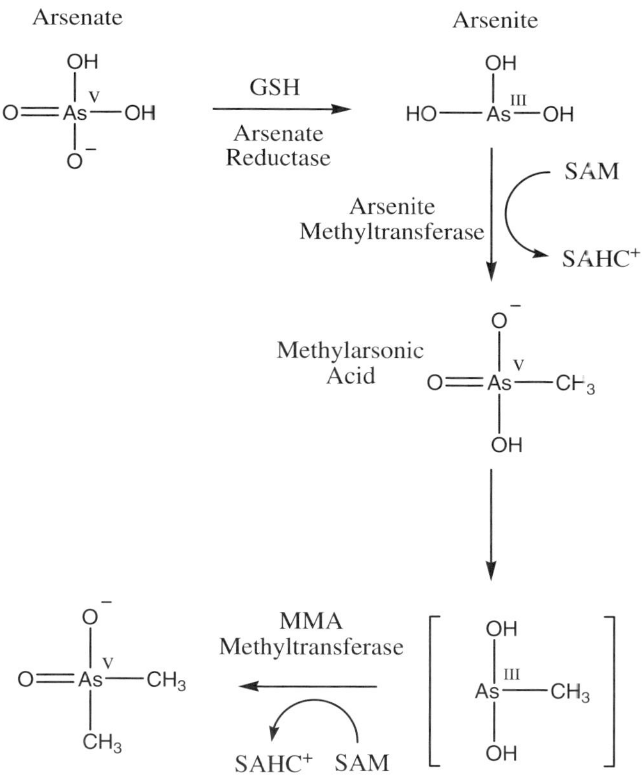

Figure 9.1 Pathway for the biotransformation of inorganic arsenic. Reprinted from Zakharyan, R. A., Ayala-Fierro, F., Cullen, W. R., Carter, D. M., and Vasken Aposhian, H. (1999). Enzymatic methylation of arsenic compounds. VII. Monomethylarsonous acid (MMAIII) is the substrate for MMA methyltransferase of rabbit liver and human hepatocytes. *Toxicol. Appl. Pharmacol.* **158**, 9–15, Copyright 1999, with permission from Elsevier.

multiple valence forms or metabolites of arsenic. As(V) and As(III) exhibit very different toxicities and biokinetics, as do the methylated metabolites, methyl arsenic acid (monomethyl arsonate or MMA) and dimethyl arsenic acid (DMA). The mode of action for arsenic-mediated effects may well depend on the animal species and chemical form under consideration. Physiologically based modeling allows one to look at the pharmacokinetic differences among species that are sensitive or resistant to the carcinogenic effects of arsenic and to derive mechanistic hypotheses that can be further explored *in vivo* and *in vitro*. Accurate estimation of the risks posed by arsenic exposure to human beings requires effective integration of epidemiological and other human studies with this type of controlled laboratory experimentation in animals. Through this mechanism we can potentially gain a clearer picture of the relationship between administered and target tissue dose and the resulting toxic effects in animals and humans.

The simplest PBPK model for arsenic came from Yu (1998). This investigator, using short-term oral exposures in the rat and mouse, modeled movement of inorganic arsenic, As(i), and did not differentiate between As(V) and As(III). In addition, the metabolism of arsenic through methylation was only briefly considered, with MMA and DMA modeled as excreted metabolites whose movement was not accounted for as active arsenic species in the blood or tissue groups. Monte Carlo analysis was used to determine the effects of parameter uncertainty (e.g., uncertainty in partition coefficients, metabolic constants, and rate constants for uptake and elimination of arsenic) on model output. Predicted values of fecal and urinary excretion of inorganic arsenic, as well as the amounts of DMA and MMA produced and excreted in the urine, depended strongly on the species of animal investigated and the chemical form of arsenic (i.e., As(i), MMA, DMA) being analyzed. In subsequent work, the model was expanded to more closely fit the human child, while including all arsenic species, As(V), As(III), MMA, and DMA, and considering both reductive metabolism and methylation (Yu 1999b). The author noted that reduction of As(V) to As(III) is, most likely, a second-order process, dependent on the concentrations of both As(V) and glutathione (GSH). The author discusses the potential use of a GSH synthesis/depletion submodel linked to the primary kinetic model through the process of arsenic reduction. This approach has been previously undertaken by other investigators (D'Souza *et al.* 1996; Johanson and Filser 1993). However, given the high initial tissue concentrations of GSH adopted by this investigator in his models, the reduction reaction actually followed first-order kinetics. Sensitivity analysis was carried out on a similar model that had been scaled to a human adult (Yu 1999a). From these studies, the input parameters that most significantly affected the output of the model were the V_{max} of the methylation reaction, the levels of GSH for determination of the reduction rate of As(V) to As(III), and the urinary excretion constants.

A similar arsenic pharmacokinetic model by Menzel *et al.* (1994) considered all four major forms of arsenic in submodels linked through metabolic processes. This model had several unique aspects, one of which was its steady-state approach to estimating the blood-to-organ ratio of arsenic and its metabolites. Additionally, the model considered arsenic metabolism through a detailed series of linked nonenzymatic and enzymatic reactions involving both GSH and *S*-adenosyl methionine

(SAM) and, potentially, glutathione-*S*-transferase (GST). Within the model, the As(V) and As(III) tissue concentrations themselves were considered to be the determinants of the velocity of reduction and methylation reactions. Lastly, these investigators approached the potential carcinogenicity of arsenic in the bladder via treatment of the tissue as a holding compartment, not influenced by blood flow, with a time-weighted average arsenic concentration in the urine and periodic emptying. These investigators specifically discuss important uncertainties inherent in their pharmacokinetic model, including the actual mediators, tissue-specific localization, and dependence on heritable or dietary factors of arsenic reduction and methylation.

Mann and co-workers also developed a PBPK model for arsenic in hamsters and rabbits, which was subsequently scaled to humans (Mann *et al.* 1996a,b). This model included detailed consideration of inhalation exposure and deposition of arsenic particles in three lung compartments (nasopharynx, tracheobronchial, pulmonary), the use of transit compartments for absorption into the gastrointestinal (GI) tract, and diffusion-limited distribution of arsenic to the tissues, which emphasizes properties of the capillaries and capillary surface area. The use of such detailed inhalation/deposition calculations allowed more accurate simulation of exposure to arsenic-containing aerosols of different sizes and under conditions of variable physical work load (i.e., breathing frequency). These investigators also added desquamation of skin as a means of arsenic elimination. The scaled model was tested by comparison of predicted cumulative excretion of arsenic and metabolites with experimental data from several human oral and inhalation studies (Buchet *et al.* 1981; Vahter *et al.* 1986; Valentine *et al.* 1979).

Recently, Gentry and colleagues extended the model developed by Mann *et al.* (1996a,b) to the mouse (Gentry *et al.* 2004). These investigators analyzed data from several published studies on experimental arsenic-mediated carcinogenesis in multiple strains of laboratory mice (Kanisawa and Schroeder 1967; Mass 1998; Moser *et al.* 2000; Ng *et al.* 1999; Waalkes *et al.* 2000). The ultimate goal of this exercise was to correlate differences in tissue dosimetry and metabolism of arsenic to strain-specific carcinogenic potency of the metal. In addition to the utilization of parameter values fit to the mouse, these investigators modified the previous model in several notable ways, including (a) a description of "colocality" for uptake and metabolism of MMA to DMA in the liver and (b) addition of uncompetitive inhibition by arsenite of this methylation step. The resulting model was tested for its ability to accurately predict arsenic kinetics in different mouse species and following both acute and chronic exposure. The model, built and parameterized using data from acute exposures in B6C3F1 mice, was also able to adequately describe the kinetics of arsenic in C57Bl/6N mice following both acute and long-term chronic exposure. The failure to identify striking pharmacokinetic differences for arsenic among mouse strains and exposure scenarios in these studies highlighted important data gaps in this area. As noted by the authors, more complete strain-specific pharmacokinetic data will be important, as will identification of the appropriate dose-metric from among the various forms of arsenic, methylated and unmethylated, and in the III+ and V+ valence states.

Although not actually describing development of a true PBPK model, recent work by Kitchin *et al.* (1999) links the pharmacokinetics of arsenic to its toxicological effects. These investigators carried out detailed time-course studies in rats on the relationship among administered dose of sodium arsenite, tissue dose of As(i) in the liver and kidney, and the biological endpoint of heme oxygenase induction. These carefully designed experiments simultaneously provided valuable data on the fate and effects of orally administered arsenic in animals. In addition, this is one of the first studies to combine kinetic data with a biological response that is "easy to measure and capable of considerable responsiveness," paving the way for development of an integrated pharmacokinetic–pharmacodynamic model for arsenic risk assessment. Furthermore, increased expression of heme oxygenase is viewed as a protective response to oxidative stress and is thought to aid in cellular restoration of a more normal redox balance. As induction of this enzyme is a common property to many other metals, among them cadmium, chromium, nickel, and lead, the studies described in this manuscript have bearing in analyses of these other contaminants as well (Sunderman 1987).

9.2.2 Nickel

Nickel is a known human carcinogen, leading primarily to lung and nasal tumors in humans exposed occupationally in smelting and refining operations. In addition, nickel exposure via inhalation, ingestion, and dermal contact to environmental sources is also of concern. Nickel has been shown to be carcinogenic in experimental animals. *In vitro*, nickel compounds are potent transforming agents; this activity is, however, dependent on the chemical and physical properties of the nickel compound. The most carcinogenic forms of nickel are insoluble and particulate compounds such as crystalline nickel subsulfide and nickel sulfide. Localization of the nickel compound within the cell likely plays an important role in the selective carcinogenic activity. Although the mechanism(s) of nickel carcinogenesis is uncertain, DNA appears to be one important target. Nickel can be highly genotoxic, inducing a variety of DNA lesions, including chromosomal aberrations, sister chromatid exchanges, DNA single-strand breaks, and DNA protein cross-links.

Humans are exposed to a variety of nickel compounds with different toxicological potencies through multiple routes, which complicates risk assessment for this metal. PBPK modeling is one approach to addressing this complication and has been carried out for nickel by Menzel (Menzel 1987, 1988; Menzel *et al.* 1987). This investigator and colleagues described mathematically the dose of nickel to specific organs from defined exposure scenarios; in these studies, the kinetic behavior of only the soluble salts of nickel (i.e., chloride, nitrate, and sulfate) were examined. The insoluble nickel compounds would likely have exhibited much different kinetic properties. PBPK modeling by these investigators was carried out using a two-pronged approach in rats as shown in Fig. 9.2. Inhalation exposure was examined in detail, taking into account both the structure of the respiratory tract in rats and the aerodynamic properties of the nickel-containing aerosol to estimate deposition efficiency in the nasopharynx and lung. Clearance to the blood and lymph from the lung was determined in experiments carried out in rats exposed to nickel via a head-only system developed by this investigator and his colleagues (Francovitch *et al.* 1987). Unexpectedly, clearance from the lung was characterized by saturable

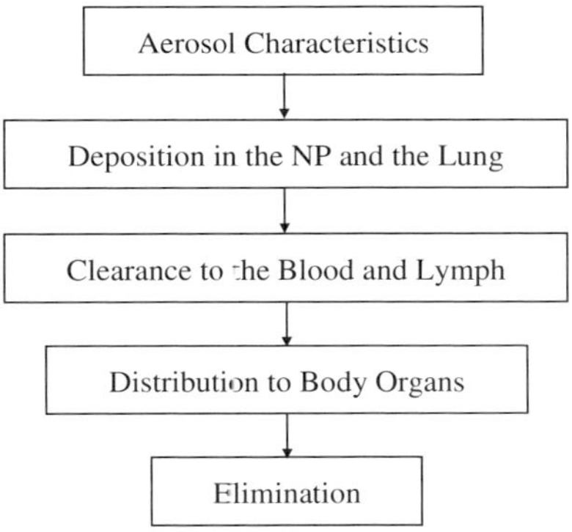

Figure 9.2 A schematic representation of the steps leading to an integration of inhalation and internal organ PBPK dosimetry models. The physical characteristics of an aerosol are used to predict the amount of the aerosol deposited in a region of toxicological interest. A PBPK lung dosimetry model is used to predict the blood burden of the toxicant leaving the lungs. The distribution of the toxicant to the internal organs and its metabolism and excretion are predicted by an internal PBPK dosimetry model. Integration of PBPK models describing the route of entry kinetics with PBPK models describing organ burdens provides a means of relating multiple sources of exposure to health effects. NP stands for nasopharynx. Reprinted from Menzel, D. (1988). Planning and using PB-PK models: An integrated inhalation and distribution model for nickel. *Toxicol. Lett.* **43**, 67–83. Copyright 1988, with permission from Elsevier.

Michaelis–Menten kinetics, which suggested that this process occurred via an ion channel or other carrier mechanism. Subsequent distribution of nickel to the internal organs of the body from the blood and lymph and further elimination were estimated using a flow-limited multicompartment PBPK model developed separately.

Menzel emphasized the iterative nature of the PBPK modeling process (Menzel 1988). Experiments described in the study were planned around a previously conceived mathematical model for nickel, and the model itself was modified as deemed appropriate to fit the experimental data as it was generated in the lab. In this way, the author suggests, model-guided experimentation allowed conservation of time and scarce scientific resources, especially research animals. Specifically, the authors discuss the use of physiologically based mathematical models for generating realistic dose ranges for subsequent studies in the areas of metabolism, tissue/organ dosimetry, and carcinogenesis, among others.

Since the lung is the primary physiological target for the toxic effects of nickel, Hsieh, Yu, and co-workers developed a detailed pharmacokinetic model on the deposition and clearance of inhaled nickel compounds in the lung from rats and humans (Hsieh *et al.* 1999a,b; Yu and Diu 1982). These investigators addressed several issues, including: (1) three mechanisms for regional nickel deposition in the lung, namely, impaction, sedimentation, and diffusion (important when examining nickel compounds with particles of variable size and solubility); (2) two different mechanisms for alveolar clearance of nickel, mechanical clearance by macrophages and dissolution; (3) mouth and nose breathing modes at several different levels of work;

and (4) structure and size of the airway, as well as breathing parameters, in both species. Although the human version has not yet been validated with experimental data, in the future these models will permit comparisons between dosimetry data from the rat and human and will also allow calculation of nickel burden in the human lung at various exposure and ventilation conditions for risk assessment purposes.

9.2.3 Lead

Lead has been, and will continue to be, the subject of a great deal of environmental concern due to its potential toxicity in both adults and children. Public awareness led to a substantial decrease in the use of lead in such materials as paint, gasoline, and solder. As a result, blood lead concentrations have dropped accordingly since the early 1970s. However, today it is recognized that exposure to even low levels of lead may lead to such diverse effects as renal functional deficits and hypertension in adults and serious cognitive defects in children. In addition, the potential for high intake of lead still exists in an occupational setting or in older housing. Outside of the occupational setting, dusts and soils contaminated with lead are the predominant source of exposure; thus, lead contamination is widespread and poses a particularly large threat to children. Current studies have shown that nearly a million children in the United States are living with blood lead levels exceeding 10 μg/dL, the level of concern for health effects in children identified by the U.S. Centers for Disease Control and Prevention, the United States Environmental Protection Agency, and the ATSDR (ATSDR 1988; CDC 1991; US EPA 1990). As the major risk factors for lead uptake are exposure to urban dust and living in housing with lead-based paint, specific populations of children, including African American and low-income children, are at greatest risk.

The need for accurate characterization of the biological behavior of lead in humans has been the impetus behind development of many biokinetic models for the toxic metal. Lead acts as a representative for a group of elements that preferentially accumulate in the bone, including the alkaline earth elements, strontium and uranium. The importance of bone as an internal source is illustrated by the fact that it can contribute up to 75% of the blood lead at any one time (Gulson *et al.* 1995). In addition, more than 90–95% of the body burden of lead in adult humans resides within the bone (Barry 1975). Models derived for bone-seeking elements are particularly cumbersome due to the necessity of accurately describing long residence times in the body and the complex dynamics of bone growth and metabolism. Most of the earlier models for lead were classical compartmental models. Major biokinetic models developed for lead that attempt to incorporate more biological characteristics are the International Commission for Radiation Protection (ICRP) model (Leggett 1993) and the Integrated Exposure Uptake Biokinetic (IEUBK) model (US EPA 1994; White *et al.* 1998).

The ICRP model was originally developed to address disposition of radioisotopes of lead and calculate doses of radiation from exposure; the model was subsequently modified to more accurately analyze lead as a chemical toxin in adults and children (Leggett 1993). This model described complex bone compartmentalization, including cortical and trabecular bone, each further divided into three separate compartments of surface, exchangeable, and nonexchangeable bone. Work by Mason emphasized the applicability of the ICRP model in estimating the accumulation of lead in the developing fetus as a result of maternal transfer during pregnancy and lactation (Mason 2000).

The IEUBK model employed four distinct functional components of exposure, uptake, biokinetics, and probability distributions for interindividual variability, to translate lead uptake as a function of environmental levels to predicted blood lead levels in children (US EPA 1994; White *et al.* 1998). The exposure component of this model was detailed, with media-specific consideration of lead absorption from air, water, diet, and soil/dust. Exposure was modeled as either constant or age-dependent and could be individualized by the user through application of alternative input models instead of defaults. There was minimal emphasis on bone as a dynamic and long-term storage compartment in this model. Validation of the IEUBK model has recently been carried out using epidemiological data on blood lead from children with a range of environmental lead exposures, including those residing in a polluted region of Hungary (Hassanien and Horvath 1999; Hogan *et al.* 1998).

Physiologically based modeling of lead has primarily been carried out by O'Flaherty and colleagues in a series of articles published between 1991 and 2001. This work, initiated with development of a model describing bone growth and metabolism as a function of age and weight in rats, was designed specifically for use in determining the pharmacokinetics of bone-seeking elements with extended residence times in the body (O'Flaherty 1991a,c). Important components of the skeletal growth model were the volumes, densities, and weights of bone and its various fractions, blood flow rate to and structure of bone, and equations describing the dependence of the growing rat skeleton on body weight and age.

O'Flaherty subsequently used the bone growth and metabolism model in her development of a PBPK model for lead disposition in rats (O'Flaherty 1991c). Processes outlined by this investigator as being involved in the uptake and removal of lead from bone included bone growth (accretion), resorption, and rapid and slow exchange; the importance of these ongoing processes in the whole-body kinetics of lead differ depending on the species, the type of bone, and the age of the individual as shown in Fig. 9.3. Bone modeling that occurs in the growing organism (juvenile) involves primarily accretion, while remodeling in young and adult individuals (specifically human as discussed below) requires both resorption and accretion and results in no net bone growth. Additional elements of the toxicokinetic model were the use of equations describing the age- and weight-dependence of tissue volumes, blood flow rates and clearances, age-dependence of lead absorption from the GI tract, and elimination of lead in bile and urine. The relationship between plasma lead and total blood lead was modeled as a capacity-limited function associated with binding to erythrocytes. Datasets used to either fit parameters or validate the model were from chronic and short-term lead studies in rats carried out primarily by the same investigator. The rat PBPK model for lead has more recently been used by Polak *et al.* (1996) to estimate bioavailability of lead from various sources, including mine wastes, soils contaminated with mine wastes, and soluble lead acetate, using intravenous dosing as a reference of 100% bioavailability.

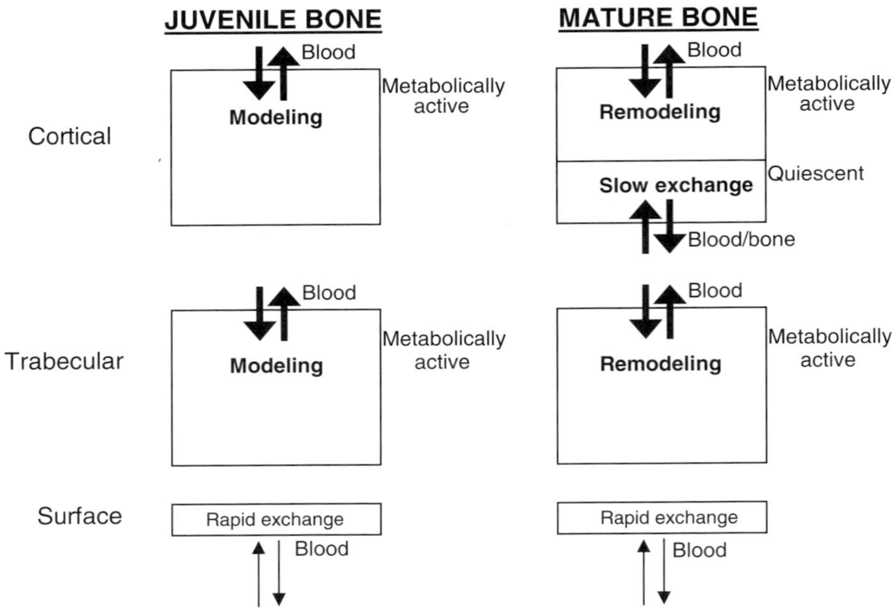

Figure 9.3 Processes involved in the kinetics of lead in bone compartments. Two compartments designated by O'Flaherty account for lead in bone, corresponding to the cortical and trabecular bone types. In an additional small surface compartment of bone (in equilibrium with blood), lead undergoes rapid exchange with the plasma. In juvenile-type bone, both cortical and trabecular bone have been designated as "metabolically active" (O'Flaherty 1991b, 1993b; O'Flaherty *et al.* 1998). That is, the primary process important in lead kinetics in growing individuals is bone accretion and resorption of lead during modeling. Slow (diffuse) exchange of lead within the bone volume and between bone and blood is relatively insignificant in the very young, increasing proportionally as an individual ages. However, in the mature individual, compared to the relatively small degree of bone remodeling that occurs, this slow exchange component assumes more importance in the whole-body kinetics of lead. By age 25, bone is assumed to be 100% of the mature type.

The next article in this series described the development of a bone growth and metabolism model for humans from birth to maturity (O'Flaherty 1991b). As discussed by the authors, the skeletons of large, long-lived and small animals are dissimilar in several anatomic and metabolic respects from one another; for example, structural remodeling of bone is rare in small animals and does not figure substantially in bone metabolism. As a result, the skeletal model previously developed for rats had to be modified to accurately describe human physiology. Features incorporated into this bone model for humans include a detailed description of the mature skeleton, including its composition and blood flow, and application of scaling equations to describe the body-weight-dependent changes in the human skeleton from birth to maturity. This model was tested by comparison of its predictions with several published analyses of different kinds of human bone at a variety of ages and was subsequently incorporated into a human PBPK model for lead disposition modified from the model for the rat (O'Flaherty 1991c, 1993b). Model validation was carried out using data from several major experimental and epidemiological studies of lead kinetics in humans in the literature (Cools *et al.* 1976; Gross 1979; Moore *et al.* 1977; O'Flaherty 1986; Rabinowitz *et al.* 1976; Sherlock *et al.* 1982; Williams *et al.* 1969).

In the human kinetic model, the default level for lead concentration in ambient air, food, and water is dependent on the date of the study. In addition, the magnitude of exposure from any particular source shifts with age and is also dependent on occupation and lifestyle. This phenomenon is examined in detail by O'Flaherty (1995b), where she used the previously developed kinetic model for lead in humans to examine absorption and disposition in children. Modifications to the generic human pharmacokinetic model for lead primarily took into account sources of exposure that are specific for infancy and childhood, including formula and food consumption equations as well as dust and soil ingestion equations. Additionally, updates were made to the expression for bone formation as a function of age, incorporating data from newer studies and accounting for increasing localization of bone turnover that takes place during childhood. Limitations to the study discussed by the author included poorly characterized functions such as the age-dependent fractional absorption of lead from the GI tract, ingestion rate of soils and dusts, and bioavailability of lead from different sources.

An important limitation to development of most PBPK models is the lack of large epidemiological databases upon which to test model predictions. However, due to the historical recognition of the susceptibility of infants and children to lead exposure, many epidemiological studies were carried out in this area between the late 1970s and the mid-1980s, thus providing a rich source of valuable data. In her work with lead, O'Flaherty (1995b) made extensive use of this substantial body of literature on the relationship between blood lead and environmental sources of exposure in infants and children. This investigator selected several large and fairly complete epidemiological studies for validation of her childhood lead model as described above (Bornschein *et al.* 1985; Chisholm *et al.* 1985; Lacey *et al.* 1985; Rabinowitz *et al.* 1985; Ryu *et al.* 1983). In one dataset used for validation, chelation therapy and environmental ablation were attempted for urban children with abnormally high blood lead concentrations. In her model, O'Flaherty simulated chelation as a reduction in the lead partition coefficients for soft tissues and bone surface for the duration of the child's hospital stay. In most cases, the predicted and observed blood lead levels in the infants and children in these studies matched well. However, the O'Flaherty model was unable to accurately predict blood lead in children at high levels of exposure; dose-dependent fractional absorption was suggested by the author as one parameter potentially important in this regard. Another issue raised was the fact that there are likely different bone remodeling rates in males and females as they reach and pass through puberty. Due to paucity of data, this issue has not yet been addressed in the PBPK model for lead in humans of any age.

In complementary work by Fleming *et al.* (1999), evaluation of the O'Flaherty model for lead kinetics in adult humans was carried out using data on workers exposed to lead as a result of employment in a smelting operation located in New Brunswick, Canada. In these studies, measurement of bone lead was quantified via noninvasive X-ray fluorescence and correlated with blood lead levels in individuals and populations with defined exposure histories. Initial studies utilized a subset of 20 workers in this comparison to refine parameters identified by sensitivity analysis as having substantial impact on output of the O'Flaherty model. Parameters describing transfer of lead from plasma to bone, red blood cell lead-binding coefficients, and the cortical and trabecular bone mineral formation rates were ultimately modified to increase the accuracy of model output when comparing cortical bone lead concentrations and blood lead levels in the selected workers. Upon adoption of the optimized model, simulations were performed for nearly 400 Brunswick workers for whom historical blood lead levels and bone lead measurements were available. Issues raised in this validation study include: (1) the difficulty in accurately predicting lead kinetics over time in individuals due to substantial and uncharacterized differences in physiology; (2) polymorphisms in the δ-aminolevulinate dehydratase gene which may, in human populations, influence lead metabolism through alteration of red blood cell binding; and (3) the heterogeneous nature of bone lead metabolism in different areas of the skeleton, even within the cortical and trabecular classes, and the inaccuracies arising from trying to model and predict blood lead as a function of localized bone lead or vice versa.

In related studies, O'Flaherty, Inskip, and colleagues, using data derived in the laboratory on lead disposition in the cynomolgus monkey, scaled the human model to the nonhuman primate (Inskip *et al.* 1996; O'Flaherty *et al.* 1996, 1998). Changes made in the primate model include alterations in the percentages of cortical and trabecular bone to fit the monkey physiology and species-specific bone turnover rates. Using isotopic tracer studies to allow differential labeling of endogenous (bone) and exogenous (dietary and environmental) lead in adult female monkeys, these investigators estimated the relative contributions to blood lead of the two sources as a function of exposure pattern. Both the length of exposure and past and current exposure rates were important for determining the fractional contribution of bone lead to blood lead.

Using data from recent studies, it has been estimated that 45–70% of blood lead in adult women exposed over a lifetime to low or moderate lead levels originates from bone (Gulson *et al.* 1995). The primary source of lead derived from bone is during ongoing remodeling and age-related bone loss. The kinetics of bone, including peak bone mass and rate of bone loss, is a complex function, dependent on such variables as gender, hormonal and nutritional status, exercise, and genetic makeup (Garn 1972; Hamdy *et al.* 1994; Johnston and Slemenda 1995; Stevenson *et al.* 1989). In women, basal rates of both cortical and trabecular bone loss are substantially accelerated during a 5- to 10-year span of the postmenopausal period (Krolner and Nielsen 1982; Mazess 1982; Meunier *et al.* 1973; Nilas and Christiansen 1988; Smith *et al.* 1976; Stepan *et al.* 1985). Not only may the decrease in absolute bone mass during this time lead to greater frequency of fractures in aging women, but a rapid rate of resorption may be detrimental through release of significant stores of lead and other bone-seeking, and potentially toxic, elements to the blood.

To address this issue, O'Flaherty further refined her model for human skeletal and bone growth to allow description of the bone loss that accompanies normal aging and the development of osteoporosis in men and women (O'Flaherty 1993b, 1995b, 2000). As part of her study, this investigator initially simulated the cumulative lifetime lead exposure of a woman born in 1918 (who would have been 60 years old in 1978) and a woman born 10 years later in 1928 (50 years old in 1978). Environmental factors contributing to the lead exposure in these "representative average" women—air, water, and dietary intake—were adjusted as appropriate for the date of the study. It was assumed that these women had no exposure to excessive lead from either their place of residence or their occupation. Model-predicted blood lead levels were then compared with data previously published for pre- and postmenopausal women between 1976 and 1980 (Silbergeld *et al.* 1988; National Center for Health Statistics 1984). This validation demonstrated that the model predicted values of blood lead for the two average reference women, ages 50 and 60 years, were somewhat higher than those actually measured in the study group. A similar overestimation of blood lead by the model was observed for men when compared. Explanations discussed by the author were that either (a) the model is structured in such a way as to overestimate lifetime blood lead concentrations resulting from various exposures or (b) the actual historical exposures themselves were lower than those assumed as model default values. Simulations aimed at exploring the latter through adjustment of historical exposure parameters suggested that this was not the explanation alone and that an underestimation of lead clearance in adults may be involved. However, there was excellent agreement between the model-predicted and measured magnitude of increase in bone loss and blood lead concentration during the immediate postmenopausal period in women.

Timchalk *et al.* (2001) have begun to address the need for a portable analytical system for monitoring lead levels in exposed humans in real-time and on-site. These investigators developed a microfluidics/electrochemical device for rapidly measuring lead levels in saliva via square-wave anodic stripping voltametry (SWASV); studies in rats acutely exposed to lead acetate were used to validate the microanalytical system against the standard analytical method for quantitation of lead, inductively coupled plasma mass spectrometry (ICP-MS). As a tool for better understanding the relationship between blood lead and saliva lead, the PBPK model developed by O'Flaherty was modified by these investigators to include a salivary gland compartment with the appropriate parameterization (O'Flaherty 1991a,c). Meticulous comparison of model predictions with lead measured in blood and saliva by ICP-MS and the microanalytical system demonstrated the feasibility of this approach in that: (1) the SWASV microanalytical system gave a linear response to lead in either body fluid over a wide concentration range (0–2000 ppb); (2) the microanalytical system was only slightly less responsive than the ICP-MS, averaging 75–85% of the results obtained from the latter; and (3) the modified PBPK model was capable of predicting blood lead and saliva lead based on data from a limited study.

Most recently, Beck *et al.* (2001) have extended the work by O'Flaherty's group through development of a prototype Monte Carlo module that uses distributions, specified as time-invariant or dependent on age, for many of the model parameters describing lead exposure, absorption, and pharmacokinetics in children. The overall approach taken by these investigators is schematized in Fig. 9.4. Exposure

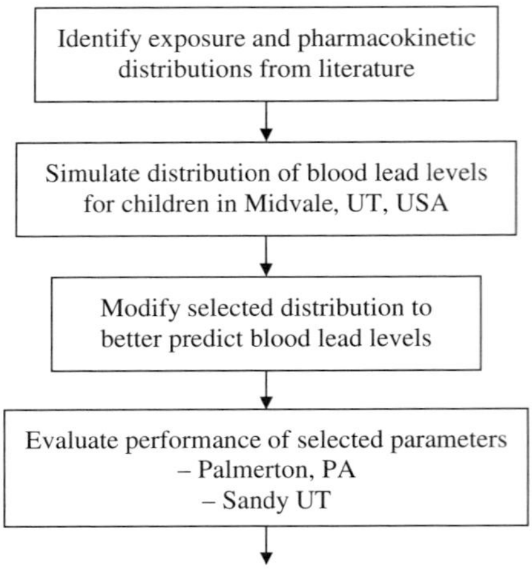

Figure 9.4 Overall approach to development and evaluation of a Monte Carlo PBPK model for lead. Reprinted from Beck, B., Mattuck, R., Bowers, T., Cohen, J., and O'Flaherty, E. (2001). The development of a stochastic physiologically based pharmacokinetic model for lead. *Sci. Total Environ.* **274**, 15–19. Copyright 2001, with permission from Elsevier.

parameters include soil and dust concentrations, daily soil ingestion rate, water lead concentration and ingestion rate, air lead concentration and inhalation rate, and dietary intake. Variable pharmacokinetic parameters are primarily concerned with red blood cell binding characteristics. The authors calibrated and then analyzed the performance of the completed model using blood lead levels measured in children from several geographical locations in Utah and Pennsylvania, where environmental contamination from smelting operations was a concern (Succop *et al.* 1998). These studies were useful in that they highlighted parameters that contribute substantially to the uncertainty and/or variability in model output—in this case, blood lead levels. However, as the author points out, lack of population data for model parameterization is one issue that will need to be addressed in future studies of this type.

Several reviews have been published in the area of PBPK modeling of lead. O'Flaherty (1992) reviewed important aspects of bone mineral metabolism, with special reference to uptake and incorporation of calcium and lead. In addition, she later addressed modeling issues pertaining to lead, uranium, and chromium, three bone-seeking elements of environmental concern (O'Flaherty 1995a). A third review by this author summarized the premises upon which her lead PBPK model was based, and it illustrated potential applications of her model to specific research questions and briefly compared it to the ICRP and IEUBK compartmental lead models (O'Flaherty 1998a). Pounds and Leggett (1998) also briefly reviewed the major aspects of the ICRP, IEUBK, and O'Flaherty models for lead in a comparative analysis; these investigators used the ICRP model to simulate cases of occupational exposure and a lead inhalation experiment in adults, as well as to examine the kinetics of lead during poisoning followed by chelation therapy in a child. The approach to chelation used by these investigators in their modeling was to increase the deposi-

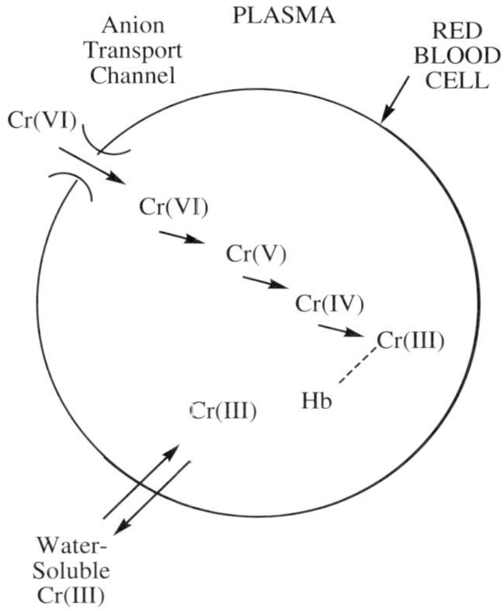

Figure 9.5 Cr(VI) and Cr(III) uptake in red blood cells. Cr(VI) readily enters the RBC as chromate, which can mimic phosphate and sulfate. In the cell it is reduced to short-lived reactive intermediates, Cr(V) and Cr(IV), and finally to Cr(III), which binds to hemoglobin (Hb) and remains part of the RBC for its entire life span. Conversely, Cr(III) compounds tend to enter the cell only via much slower diffusion. Adapted from Paustenbach and Finley (2003).

tion fraction of lead in urine and to uniformly decrease the total transfer rate from plasma to all other tissues. A similar review was published by Rabinowitz (1998).

9.2.4 Chromium

Exposure to high levels of chromium results in both acute and chronic health effects, including lung cancers and other pulmonary dysfunctions, dermatitis, GI problems, and damage to the liver and kidney. Exposure to chromium occurs both in occupational settings (welding, manufacturing of alloys and other chemical compounds, and in chrome-processing industries) and, as a result of environmental contamination, particularly at hazardous waste sites. Inhalation, oral, and dermal exposure routes are important in the toxicology of chromium. The cytotoxic action of chromium is mediated through intracellular reduction of Cr(VI), which is readily taken up into cells by general anion carrier transport, to Cr(III). Reactive free radical species that are the intermediates in this reaction are known to bind to intracellular macromolecules such as DNA, RNA, and proteins and are likely responsible for the carcinogenicity of Cr(VI) (Fig. 9.5). As a result, Cr(VI) compounds are mutagenic in appropriate assay systems (Cohen *et al.* 1993). Cr(VI) is also reduced with variable efficiency in extracellullular locations such as in the GI tract and blood; reduction acts as the predominant detoxification mechanism for Cr(VI) (Fig. 9.6). The resulting Cr(III), which is assumed to be relatively nontoxic because it does not readily enter cells, is primarily excreted by the kidneys. Incorporation of both the acute toxicity and carcinogenicity of chromium in risk assessments requires a quan-

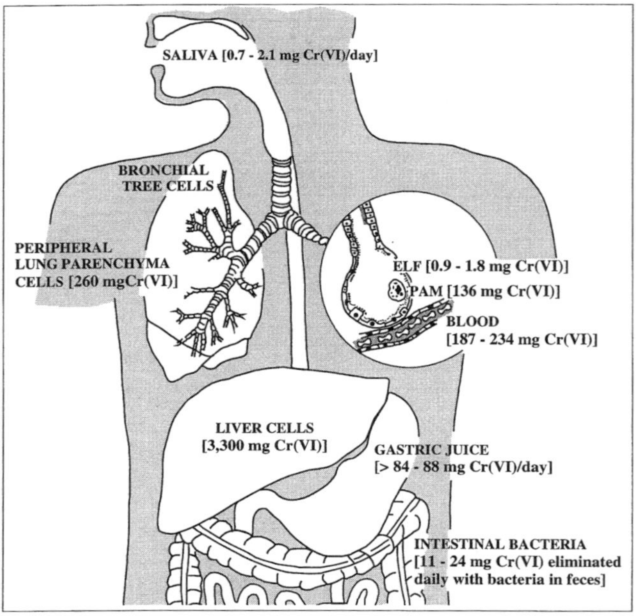

Figure 9.6 Estimates of the amount of Cr(VI) sequestration or reduction by human body organs, fluids, and cell populations. Reprinted from De Flora, *et al.* 1997.

titatively accurate understanding of the uptake, distribution, and excretion of both of these critical ionic species; physiologically based modeling is one potential approach.

Specific issues that will ultimately need to be addressed in chromium modeling are speciation-dependent uptake and passage into cells and solubility of Cr(III) versus Cr(VI) compounds, treatment of chromium as a bone-seeking element, and the essential nature of Cr(III) for normal physiological function, which requires some consideration of internal homeostatic mechanisms for the nutrient (see Section 9.5.1). The efficiency of the reductive processes that convert Cr(VI) to Cr(III) is also central to understanding the ultimate tissue dosimetry and biological effects of the metal.

Physiologically based modeling of chromium pharmacokinetics was first addressed by Thomann *et al.* (1994). These investigators developed a conceptually simple model using only three compartments—liver, kidney, and spleen—as the sites of chromium excretion linked by the blood (as the input compartment of oral administration) to a major long-term storage compartment composed of bone, skin, hair, and muscle. The primary emphasis of the model was on chronic oral exposure conditions and it did not deal with inhalation. Model simulations were compared with experimental data from controlled exposures in rats. These studies demonstrated differential accumulation of chromium in body tissues over time. For some compartments, such as bone, the investigators observed lengthy kinetics of chromium depuration, implying that the bone potentially acted as a storage site that slowly released chromium to such well-perfused tissues as the liver, kidney, and spleen. Author-noted limitations to this study included the facts that only a single dose of chromium was used, the metal was only administered orally, and the model did not take speciation into account, because only total chromium was measured. Additionally, the authors discussed the fact that the use of only one age of rat in the study would not elucidate age-dependent processes (which are important in the kinetics of bone-seeking elements). These studies were, however, instrumental in demonstrating the importance of such issues in the whole-body kinetics of chromium.

O'Flaherty and colleagues have also addressed PBPK modeling of chromium (O'Flaherty 1993a, 1996; O'Flaherty *et al.* 2001). The general structure of the chromium model initially developed by this investigator for rats was similar to that in her previous work on lead (O'Flaherty 1991a,c). Chromium was, however, treated as a minimally bone-seeking element that undergoes rapid surface exchange and incorporation into bone through new growth, but does not undergo slow diffusion through the entire volume of the bone. The model incorporated age-dependent rat parameters related to body growth, as well as bone growth and metabolism. The author accounted for both Cr(VI) and Cr(III) in her model through linkage by reduction processes governed by first-order rate constants.

In the earliest chromium model, O'Flaherty assigned reduction processes to all tissues except bone. The GI tract and alveolar space were also assumed to be sites of Cr(VI) reduction (O'Flaherty 1993a). Later versions of the model included reduction in the bone and, based on data from human studies, assumed no reduction in the plasma (Gray and Sterling 1950; Korallus *et al.* 1984; O'Flaherty 1996). As noted by the author, parameterization of the reduction process was a "gross" biological oversimplification, because reduction of Cr(VI) to Cr(III) is likely carried out by a number of enzymatic and nonenzymatic processes, some of which are influenced strongly by nutritional status. Exchange of Cr(VI) was modeled as both flow-limited (O'Flaherty 1993a) and diffusion-limited (O'Flaherty 1996) in these two versions of the model. However, in the latter case the author notes that movement of Cr(VI) into and out of tissues is so rapid that blood flow is actually rate-limiting as to its distribution. Cr(III), with its poor mobility within tissues, was assumed to be diffusion-limited in both cases. Multiple published datasets dealing with the intravenous, oral, and intratracheal exposure routes were subsequently utilized to fit/calibrate parameters and validate the chromium models (Bragt and van Dura 1983; Edel and Sabbioni 1985; Hopkins 1965; Langard *et al.* 1978; MacKenzie *et al.* 1958, 1959; Merts *et al.* 1965; Visek *et al.* 1953; Weber 1983).

In 2001, O'Flaherty and co-workers published a PBPK model for chromium in humans (O'Flaherty *et al.* 2001); this modification of the rat model incorporated the most updated physiologically based model describing human body and bone growth and took advantage of a series of studies of chromium kinetics after controlled oral exposure(s) in adult humans (Finley *et al.* 1997; Kerger *et al.* 1996; O'Flaherty 1993a, 1996, 2000; Paustenbach *et al.* 1996). Refinements in this model included parameters specific to the human (estimated from the experimental studies noted above), differential absorption of Cr(III) and Cr(VI), which is likely a function of variable reduction of the latter in the GI tract, and specific reduction constants for the GI tract, red cells, plasma, liver and kidney, and all other tissues.

Additionally, based on the available data, the author incorporated a concentration-dependent and capacity-limited renal process that conserves nearly all the filtered chromium at the ambient range of 0.05–0.15 μg/mL, but allows substantially greater excretion at higher plasma chromium concentrations. This was the first attempt in any PBPK model to address potential homeostatic mechanisms in the body that may operate on essential metals. As discussed by the authors in this series of publications, there are several additional complicating issues in the toxicokinetics of chromium that have not yet been adequately addressed. Among these are (a) the nature of the chromium salt and method of its administration and (b) the GI tract contents and the nutritional status of the individual on the efficiency of chromium reduction, absorption, and disposition. Second, a realistic description of the kinetics of chromium within the human lung compartment is lacking. Finally, the extent of deposition of either Cr(III) or Cr(VI) in human bone and the importance of this compartment as a long-term chromium reservoir and source of continual exposure are areas of high priority for future studies.

The utility of PBPK modeling, among other tools, in the risk assessment process has been illustrated recently for chromium by Paustenbach and Finley (2003). These investigators used data from nine studies and the human PBPK model for chromium developed by O'Flaherty *et al.* (2001) to estimate health risks posed by Cr(VI) in tap water (Corbett *et al.* 1998; Finley *et al.* 1996, 1997; Kerger *et al.* 1996, 1997; Kuykendall *et al.* 1996; O'Flaherty *et al.* 2001; Paustenbach *et al.* 1996; Proctor *et al.* 2002). Among the issues addressed through simulation approaches were: the differential absorption and excretion of Cr(VI) and Cr(III), the reduction of Cr(VI) in different body fluids and tissues, and tissue dosimetry for oral exposures. Dosimetry estimates for chromium from oral, dermal, and inhalation exposures were then compared to current regulatory guidelines to estimate the acute and chronic health risks to exposed humans.

9.3 PBPK MODELS FOR NONMETALS

9.3.1 A PBPK Model for Fluoride, a Bone-Seeking Nonmetal

In a manner similar to lead and chromium, fluoride is a bone-seeking element characterized by long residence times in the body. The vast majority of fluoride (up to 99% of the body burden) is typically found in the bone, where it has an extremely slow turnover. The element can act in a beneficial manner to strengthen teeth and bone, and yet can also cause detrimental responses. Cancer formation and altered bone strength have been issues of concern with long-term fluoride exposure in humans. Exposure to fluoride occurs chronically at low levels from its natural or supplemented presence in drinking water and many foods or to its use in dental products such as toothpaste, rinses, and other oral gels. Occupational exposure to fluoride in the workplace (e.g., aluminum and fertilizer industries, as well as arc welders) usually occurs by inhalation. Lastly, therapeutic uses of fluoride, such as for the treatment of osteoporosis, can add substantially to the total intake of fluoride.

There has been a great deal of interest in understanding the physiological distribution and effects of fluoride in the body since the 1940s, when water fluoridation was first considered. To this end, several data-based compartmental models were derived to describe short-term fluoride biokinetics (Charkes *et al.* 1978, 1979; Hall *et al.* 1977). Only one physiologically based pharmacokinetic model has been developed thus far for fluoride (Rao *et al.* 1995). This model, which considered both oral and inhalation exposure in rats and humans, emphasized the bone compartment as both a storage and a target tissue in fluoride biokinetics. This is in contrast to studies with lead and chromium, where bone was considered only from the standpoint of a long-term reservoir and source of endogenous exposure for the metals. The approach used by these investigators to model bone growth and metabolism was similar to previous work by O'Flaherty and co-workers (O'Flaherty 1991a–c). Forcing functions and input data files were used to describe physiological processes that change with time from birth to sexual maturity in both sexes. Bone was modeled as two subcompartments in series: (a) a small, rapidly exchangeable surface component that makes up approximately 1% of the total bone and (b) an inner nonexchangeable bulk bone component that contains nearly the entire body content of bound fluoride. The dependence of bone resorption (in part a function of solubility) on the fluoride content of bone as described initially by Turner *et al.* was also addressed in the model through use of reference data files that link age and bone fluoride content with the rate of fluoride return from the bone to the plasma (Okazaki *et al.* 1981; Turner *et al.* 1993). Model validation was carried out by comparing simulations with data from multiple long-term carcinogenicity studies in rats and one pharmacokinetic study in humans receiving oral fluoride therapy (Boivin *et al.* 1988; Maurer *et al.* 1990; National Toxicology Program 1990); the model fairly accurately predicted the average bone fluoride levels in rats and humans as a function of age and total exposure. Limitations to this model as discussed by the author included: (1) lack of kinetic data on age-specific metabolism of fluoride and, thus, large uncertainties in related parameterization; and (2) failure to account for structural differences between compact cortical and cancelous trabecular bone, which may lead to differential fluoride distribution within specific regions of the skeleton. This latter issue has also been raised during development and parameterization of PBPK models for lead (Fleming *et al.* 1999).

9.3.2 PBPK Models for Other Nonmetals

Yu and Kim (2004) have developed a PBPK inhalation model for radon to aid in exposure assessment of a human indoors. One interesting aspect of this work was the linkage of the PBPK model to a three-compartment model describing the transfer and distribution of radon released from groundwater in a house through showers, washing dishes, and flushing toilets. These investigators used the linked models to predict the concentration of radon in the lung target tissue resulting from several different indoor exposure scenarios. One possibility discussed by the authors was that, based on their modeling predictions, individuals spending the majority of their time in the home, including women, children, and the elderly, may be potentially susceptible to increased risk for radon-induced lung cancer.

9.4 COMPARTMENTAL MODELS FOR MISCELLANEOUS INORGANIC AND/OR ENDOGENOUS CHEMICALS

A variety of compartmental models have been developed for other metals and endogenous chemicals. The ICRP has described and subsequently modified models for several radioactive isotopes of metals including nickel, zinc, uranium, and strontium (International Commission for Radiation Protection 1979, 1991, 1993, 1994, 1995a,b). Silk *et al.* (1995) compared the original and revised ICRP models to determine the doses in ores and mineral sands of uranium and thorium; the latter version of these models entails more detailed descriptions of the respiratory tract and inhalation exposure and generally was found to give substantially lower dose estimates than the former. In addition, Wrenn *et al.* (1989, 1994) described a compartmental model for uranium that is based upon extensive data in animals and humans administered uranium through either the oral or inhalation routes. The pharmacokinetics of strontium is of interest not only because it is an important nuclear fallout product and, thus, likely represents a health risk to exposed individuals, but also because of its utility as a marker for the movement and metabolism of calcium in humans. Compartmental models for strontium have been described by Moreas *et al.* (1991) and Sips *et al.* (1996). Several investigators have addressed biokinetic modeling of cadmium (Choudhary *et al.* 2001; Kjellstrom and Nordberg 1978; Nordberg and Kjellstrom 1979). Leggett and co-workers have developed quite detailed biokinetic models for potassium, rubidium, and cesium (Leggett and Williams 1986, 1988; Leggett *et al.* 2003). Whillans has recently modified earlier ICRP biokinetic models for ^{14}C and organically bound ^{3}H to include aspects of human carbon metabolism and data from direct measurements of human excretion (International Commission for Radiation Protection 1989, 1998; Whillans 2003).

9.5 RESEARCH NEEDS

9.5.1 The Need for Physiologically Based Modeling for Essential Metals

Without doubt, the majority of emphasis in PBPK model development for the chemicals covered in this chapter is placed on strictly toxicological aspects; that is, how do we relate exposure levels to associated health risks in different populations? However, one additional consideration has not yet been adequately addressed in any PBPK model. Many metals, including cobalt, copper, iron, magnesium, manganese, molybdenum, selenium, and zinc, are essential micronutrients for humans; the trivalent form of chromium is also considered an essential nutrient for its role in carbohydrate metabolism. Toxicological analysis of these metals is complicated in that "for essential trace elements, risk assessment requires consideration of both toxicity from excess exposures and health consequences as a result of deficiencies" (Goyer and Clarkson 2001). Any standards that are set for exposure must be developed within this framework. Essential metals are subject to sophisticated homeostatic regulatory mechanisms within the body, which likely influences both their pharmacokinetic and pharmacodynamic interactions. These will undoubtedly be issues of future consideration as the data on metals becomes more complete and various models are more readily linked through a clearer mechanistic understanding of the processes regulating uptake, distribution, and excretion of these important elements.

Nathanson and colleagues began to specifically address the normal homeostatic mechanisms that rigorously maintain the total quantity and distribution of iron in the body over time (Nathanson and McLaren 1987; Nathanson *et al.* 1985). These investigators used ^{55}Fe and ^{53}Fe tracer measurements, together with a compartmental model of iron kinetics in beagle dogs (Berzuini *et al.* 1978), to examine the rate constants for several steps involved in iron absorption by intestinal epithelial cells. These studies revealed that under conditions of anemia (i.e., systemic iron deficiency), rate constants for uptake of iron by mucosal (and other affected) tissues increased with concomitant decreases in incorporation of iron into storage pools. These rate increases were attributed to increased levels of cell surface transferrin receptors and an increased affinity of the intestinal brush border for iron as has been demonstrated in other studies of this type (Louache *et al.* 1984; Muir *et al.* 1984; Ward *et al.* 1982). It will be interesting to compare these results with those obtained from a PBPK model for iron when one becomes available.

In a preview of work to come, Andersen *et al.* (1999) have begun to conceptualize potential PBPK models for the essential bone-seeking metal manganese. This element is interesting in that deficiency leads to a variety of symptoms including defects in lipid and carbohydrate metabolism and impaired reproduction. In excess amounts, manganese is toxic to the central nervous system (CNS). Within the body, manganese pharmacokinetics is complex, influenced by valence state, the route of exposure, and its complexation with carrier proteins, primarily transferrin. Although no model for manganese has yet been developed, these investigators analyzed the literature on manganese pharmacokinetic studies and discussed data gaps that will need to be addressed with targeted research prior to successfully modeling the uptake, distribution, and elimination of this complex element. These areas include: (1) the role played by the liver in regulating circulating concentrations of protein-bound manganese; (2) characterization of the dose-dependency and cellular mechanisms responsible for uptake of ingested manganese; (3) the rate at which systemically absorbed manganese is oxidized and complexed with transferrin; and (4) the contribution of CNS uptake of manganese via neurons in the nose in regard to inhalation of manganese-containing dusts.

9.5.2 Other Research Needs

Consideration of the current state of PBPK modeling of metals reveals many areas for which only minimal data are available in the literature; for other areas there are no data at all. In general, areas of high priority for future studies are similar for all the metals addressed in this review. In the following discussion, however, we have emphasized specific metals for which these issues are particularly central.

1. Lack of mechanistic knowledge of the toxicologically "active" metal species and the most relevant physiological endpoint. An illustration of this point is arsenic, where at least four species of the metal are formed by reduction and GSH metabolism. Each of these metabolites has biological activity and is likely to have different kinetics and toxicological targets. Developing useful PBPK models for target tissue dosimetry for human health risk assessments depends on knowing which specific compound gives rise to the various biological endpoint(s).

2. The need for more complete characterization of the metabolism/oxidative changes of metals. As metals are taken up into cells and/or are excreted in a manner that is dependent on their chemical or oxidative species, this is a crucial data gap in better predicting tissue dosimetry. For metabolic processes such as methylation (i.e., of metals such as arsenic), the biochemical mediators and reaction rates must be delineated. Additionally, determination of the cellular and extracellular locations where metabolism/reduction occurs will be important, as will accurate quantification of these processes under a variety of physiological conditions (e.g., nutritional status). An illustration of this important principle is with chromium. Extracellular reduction of Cr(VI), the more toxic species, to Cr(III), which is readily excreted, competes directly with cellular uptake of Cr(VI).

3. Characterization of bioavailability of metals in the gut and deposition in lung. Both the bioavailability of metals and the extent of their deposition in the lung is dependent on the chemical form, solubility, and particle size to which an individual is exposed. To further complicate the picture, bioavailability of metals is also influenced by age, general nutritional status, and gut contents at the time of exposure. Although these parameters are generally relevant to PBPK modeling of all metals, lead (oral bioavailability) and nickel (lung deposition) provide particularly striking examples of the variability of bioavailability under different exposure situations and its importance to the overall modeling process. With essential elements, such as chromium and manganese, important homeostatic processes controlling uptake of the metal from the diet may not influence control of uptake by routes of administration where the compounds enter the bloodstream without passing through intestinal and hepatic tissues before reaching the systemic circulation. Thus, differences in route of administration may significantly alter dose–response relationships for tissue accumulation or other kinetic endpoints.

4. Need for an accurate description of region-specific bone growth and metabolism and an understanding of how these processes are affected by age, hormonal status, and genetics. Improving our understanding in these areas is important for PBPK models for bone-seeking elements such as lead, chromium, and fluoride. Additionally, mechanistic descriptions of the impacts of fluoride on bone as a target tissue, particularly with regard to its rate of metabolism, will also be crucial.

5. Lack of large epidemiological datasets with statistical power. There is a need for a variety of population data to support development of models using distributions of parameters instead of defined default parameter values. This issue is important for accurate PBPK model development for all the elements discussed in this review. However, with current regulations, which substantially decrease the number of individuals exposed to metals and their exposure levels, these data will be even harder to come by than it has been in the past. As a result, a great deal of foresight and creativity will be required by scientists in optimizing the amount and utility of data gathered when individuals and/or populations are exposed.

5. Lastly, the necessity for mechanistic data allowing linkage of pharmacodynamic models to tissue dosimetry. This need is apparent for all metals (and this is equally true for all other groups of compounds). Only when we understand how a specific metal, at a variety of target tissue doses, perturbs crucial cellular processes in the body can we improve our risk assessments for these high priority contaminants. It bears some emphasis that PBPK modeling does not simply represent an interesting research discipline. The most important contribution of PBPK modeling has been the focus it provides on understanding and predicting tissue dose, permitting the correlation of tissue dose with toxic responses and the development of realistic pharmacodynamic models that appropriately consider tissue doses that initiate various responses.

9.6 SUMMARY

The studies described in this review are only the beginning of the utilization of biologically based modeling in risk assessment and of other studies requiring accurate dosimetry of potentially toxic metals. However, the work described in the highlighted publications is groundbreaking for the field of PBPK modeling in many ways. For example, comparisons can be made among modeling approaches described by different investigators for the same metal; testing of the respective models for their performance against a variety of datasets as they become available will indicate which approaches are more biologically realistic The ability to evaluate specific parameters for their impact on output in the different models should encourage more refined parameterization, and will facilitate the development of hybrid models, perhaps utilizing theoretical aspects from multiple investigators. In several cases, increasingly sophisticated models are available for a single metal from the same investigator(s). These models have been developed using the iterative process and have become increasingly grounded in biological knowledge with each iteration. In this process, parameterization becomes more accurate, and more is understood about the underlying physiological effects of the metals on biological systems. Lastly, these studies have been instrumental in identifying high-priority areas for future studies, both in the laboratory and in conceptual model development.

NOTATION

As	arsenic
As(i)	inorganic arsenic
ATSDR	Agency for Toxic Substances and Disease Registry
CEP	Completed Exposure Pathway
CERCLA	Comprehensive Environmental Response, Compensation, and Liability Act

Cd	cadmium
CNS	central nervous system
Cr	chromium
DMA	dimethyl arsenic acid
GI	gastrointestinal
GSH	glutathione
GST	glutathione-*S*-transferase
Hg	mercury
ICRP	International Commission for Radiation Protection
ICP-MS	inductively coupled plasma mass spectrometry
IEUBK	Integrated Exposure Uptake Biokinetic
Pb	lead
MMA	methyl arsenic acid (i.e., monomethyl arsonate)
Ni	nickel
NP	nasopharynx
PBPK	physiologically based pharmacokinetic
SAM	*S*-adenosyl methionine
SWASV	square-wave anodic stripping voltametry

REFERENCES

Andersen, M., Gearhart, J., and Clewell, H., III (1999). Pharmacokinetic data needs to support risk assessments for inhaled and ingested manganese. *Neurotoxicology* **20**, 161–172.

ATSDR (1988). *The Nature and Extent of Lead Poisoning in Children in the United States: A Report to Congress*, Agency for Toxic Substances and Disease Registry, Atlanta, GA.

ATSDR (2003). *2003 CERCLA Priority List of Hazardous Substances*, Agency for Toxic Substances and Disease Registry, Atlanta, GA.

Barry, P. (1975). A comparison of concentrations of lead in human tissues. *Br. J. Ind. Med.* **32**, 119–139.

Beck, B., Mattuck, R., Bowers, T., Cohen, J., and O'Flaherty, E. (2001). The development of a stochastic physiologically-based pharmacokinetic model for lead. *Sci. Total Environ.* **274**, 15–19.

Berzuini, C., Colli Franzone, P., Stefanelli, M., and Viganotti, C. (1978). Iron kinetics: Modelling and parameter estimation in normal and anemic states. *Comput. Biomed. Res.* **11**, 209–227.

Boivin, G., Chapuy, M., Baud, C., and Meunier, P. (1988). Fluoride content in human iliac bone: Results in controls, patients with fluorosis, and osteoporotic patients treated with fluoride. *J. Bone Miner. Res.* **3**, 497–502.

Bornschein, R., Hammond, P., Dietrich, K., Succop, P., Krafft, K., Clark, S., Berger, O., Pearson, D., and Que Hee, S. (1985). The Cincinnati prospective study of low-level lead exposure and its effects on child development. *Environ. Res.* **38**, 4–18.

Bragt, P., and van Dura, E. (1983). Toxicokinetics of hexavalent chromium in the rat after intratracheal administration of chromates of different solubilities. *Ann. Occup. Hyg.* **27**, 315–322.

Buchet, J., Lauwerys, R., and Roels, H. (1981). Urinary excretion of inorganic arsenic and its metabolites after repeated ingestion of sodium metaarsenite by volunteers. *Int. Arch. Occup. Environ. Health* **48**, 111–118.

CDC (1991). Preventing Lead Poisoning in Young Children: A Statement by the Centers for Disease Control, October 1991.

Charkes, N., Todd Makler, P., Jr., and Philips, C. (1978). Studies of skeletal tracer kinetics. I. Digital-computer solution of a five compartment model of [^{18}F] fluoride kinetics in humans. *J. Nucl. Med.* **19**, 1301–1309.

Charkes, N., Brookes, M., and Todd Makler, P., Jr. (1979). Studies of skeletal tracer kinetics. II. Evaluation of five-compartment model of [^{18}F]fluoride kinetics in rats. *J. Nucl. Med.* **20**, 1150–1157.

Chisholm, J., Mellits, E., and Quaskey, S. (1985). The relationship between the level of lead absorption in children and the age, type, and condition of housing. *Environ. Res.* **38**, 31–45.

Choudhary, H., Harvey, T., Thayer, W., Lockwood, T., Stiteler, W., Goodrum, P., Hassett, J., and Diamond, G. (2001). Urinary cadmium elimination as a biomarker of exposure for evaluating a cadmium dietary exposure–biokinetics model. *J. Toxicol. Environ. Health, Part A* **63**, 321–350.

Cohen, M., Kargacin, B., Klein, C., and Costa, M. (1993). Mechanisms of chromium carcinogenicity and toxicity. *Crit. Rev. Toxicol.* **23**, 255–281.

Cools, A., Salle, H., Verberk, M., and Zielhuis, R. (1976). Biochemical response of male volunteers ingesting inorganic lead for 49 days. *Int. Arch. Occup. Environ. Health.* **38**, 129–139.

Corbett, G., Dodge, D., O'Flaherty, E., Liang, J., Throop, L., Finley, B., and Kerger, B. (1998). *In vitro* reduction kinetics of hexavalent chromium in human blood. *Environ. Res.* **7**, 7–11.

D'Souza, R., Francis, W., and Andersen, M. (1996). Physiological model for tissue glutathione depletion and increased resynthesis after ethylene dichloride exposure. *J. Pharmacol. Exp. Ther.* **245**, 563–568.

De Flora, S., Camoirano, A., Bagnasco, M., Bennicelli, C., Corbett, G. E., and Kergr, B. D. (1997). Estimates of the chromium (VI) reducing capacity in human body compartments as a mechanism for attenuating its potential toxicity and carcinogenicity. *Carcinogenesis* **18**, 531–537.

Edel, J., and Sabbioni, E. (1985). Pathways of Cr(III) and Cr(VI) in the rat after intratracheal administration. *Hum. Toxicol.* **4**, 409–416.

Fay, M., and Mumtaz, M. (1996). Development of a priority list of chemical mixtures occurring at 1188 hazardous waste sites using the HazDat database. *Food Chem. Toxicol.* **34**, 1163–1165.

Finley, B., Kerger, B., Dodge, D., Meyers, S., Richter, R., and Paustenbach, D. (1996). Assessment of airborne hexavalent chromium in the home following use of contaminated tap water. *J. Expo. Anal. Environ. Epidemiol.* **6**, 229–245.

Finley, B., Kerger, B., Katona, M., Gargas, M., Corbett, G., and Paustenbach, D. (1997). Human ingestion of chromium(VI) in drinking water: Pharmacokinetics following repeated exposure. *Toxicol. Appl. Pharmacol.* **142**, 151–159.

Fleming, D., Chettle, D., Webber, C., and O'Flaherty, E. (1999). The O'Flaherty model of lead kinetics: An evaluation using data from a lead smelter population. *Toxicol. Appl. Pharmacol.* **161**, 100–109.

Francovitch, R., Deal, D., Tayyeb, M., Wolpert, R., Boger, I., JR, Valentini, J., and Menzel, D. (1987). A head only exposure system for controlled exposures of small rodents. *Toxicol. Lett.* **38**, 19–32.

Gabelnick, H. L., Dedrick, R. L., and Bourke, R. S. (1970). *In vivo* mass transfer of chloride during exchange hemodialysis. *J. Appl. Physiol.* **28**, 636–641.

Garn, S. (1972). The course of bone gain and the phases of bone loss. *Orthoped. Clin. North Am.* **3**, 520.

Gentry, P., Covington, T., Mann, S., Shipp, A., Yager, J., and Clewell, H., III (2004). Physiologically based pharmacokinetic modeling of arsenic in the mouse. *J. Toxicol. Environ Health, Part A* **67**, 43–71.

Goyer, R., and Clarkson, T. (2001). Toxic effects of metals. In: *Casarett & Doull's Toxicology*, C. Klaassen, ed., McGraw-Hill, New York, pp. 811–867.

Gray, S., and Sterling, K. (1950). The tagging of red cells and plasma proteins with radioactive chromium. *J. Clin. Invest.* **29**, 1604–1613.

Gross, S. (1979). *Oral and Inhalation Lead Exposures in Human Subjects (Kehoe Balance Experiments)*, Lead Industries Association, New York.

Gulson, B., Mahaffey, K., Mizon, K., Korsch, M., Cameron, M., and Vimpani, G. (1995). Contribution of tissue lead to blood lead in adult female subjects based on stable lead isotope methods. *J. Lab. Clin. Med.* **125**, 703–712.

Hall, L., Kilpper, R., Smith, F., Morken, D., and Hodge, H. (1977). Kinetic model of fluoride metabolism in the rabbit. *Environ. Res.* **13**, 285–302.

Hamdy, R., Anderson, J., Whalen, K., and Harvill, L. (1994). Regional differences in bone density of young men involved in different exercises. *Med. Sci. Sports Exerc.* **26**, 884–888.

Hassanien, M., and Horvath, A. (1999). Lead risk assessment for children in Hungary by predicting their blood lead levels using US EPA Integrated Exposure Uptake Biokinetic model. *Centr. Eur. J. Publ. Health.* **7**, 155–159.

Hogan, K., Marcus, A., Smith, R., and White, P. (1998). Integrated exposure uptake biokinetic model for lead in children: Empirical comparisons with epidemiologic data. *Environ. Health Perspect.* **106**, 1557–1567.

Hopkins, L., Jr. (1965). Distribution in the rat of physiological amounts of injected Cr51(III) with time. *Am. J. Physiol.* **209**, 731–735.

Hsieh, T., Yu, C., and Oberdorster, G. (1999a). A dosimetry model of Ni compounds in the rat lung. *Inhal. Toxicol.* **11**, 229–248.

Hsieh, T., Yu, C., and Oberdorster, G. (1999b). Modeling of deposition and clearance of inhaled Ni compounds in the human lung. *Regul. Toxicol. Pharmacol.* **30**, 18–28.

Inskip, M., Franklin, C., Baccanale, C., Manton, W., O'Flaherty, E., Edwards, C., Blenkinsop, J., and Edwards, E. (1996). Measurement of the flux of lead from bone to blood in a nonhuman primate (*Macaca fascicularis*) by sequential administration of stable lead isotopes. *Fundam. Appl. Toxicol.* **33**, 235–245.

International Commission for Radiation Protection (1979). *Limits for Intakes of Radionuclides by Workers*, Annals of the ICRP 2 (3/4), Pergamon Press, ICRP Publication 30.

International Commission for Radiation Protection (1989). *Age-Dependent Doses to Members of the Public from Intake of Radionuclides*, Part 1, Pergamon Press, ICRP Publication 56.

International Commission for Radiation Protection (1991). *Annual Limits on Intake of Radionuclides by Workers Based on the 1990 Recommendations*, Annals of the ICRP 21(4), Pergamon Press, ICRP Publication 61.

International Commission for Radiation Protection (1993). *Age-Dependent Doses to Members of the Public from Intake of Radionuclides*, Part 2: *Ingestion Dose Coefficients*, Annals of the ICRP 23(3/4), Pergamon Press, ICRP Publication 67.

International Commission for Radiation Protection (1994). *Human Respiratory Tract Model for Radiological Protection*, Annals of the ICRP 24(1/4), Pergamon Press, ICRP Publication 66.

International Commission for Radiation Protection. (1995a). *Age-Dependent Doses to Members of the Public from Intake of Radionuclides*, Part 3: *Ingestion Dose Coefficients*, Annals of the ICRP 25(1), Pergamon Press, ICRP Publication 69.

International Commission for Radiation Protection. (1995b). *Dose Coefficients for Intakes of Radionuclides by Workers: Replacement of ICRP Publication 61*, Annals of the ICRP 24(4), Pergamon Press, ICRP Publication 68.

International Commission for Radiation Protection (1998). *Radiation Dose to Patients from Pharmaceuticals*, Pergamon Press, ICRP Publication 80.

Jain, R. K., and Gerlowski, L. E. (1981). Kinetics of uptake, distribution and excretion of zinc in rats. *Ann. Biomed. Eng.* **9**, 347–361.

Johanson, G., and Filser, J. (1993). A physiologically based pharmacokinetic model for butadiene and its metabolite butadiene monoxide in rat and mouse and its significance for risk extrapolation. *Arch. Toxicol.* **67**, 151–163.

Johnston, C., Jr, and Slemenda, C. (1995). Pathogenesis of osteoporosis. *Bone* **17**, 19S–22S.

Kanisawa, M., and Schroeder, H. (1967). Life term studies on the effects of arsenic, germanium, tin, and vanadium on spontaneous tumors in mice. *Cancer Res.* **27**, 1192–1195.

Kerger, B., Paustenbach, D., Corbett, G., and Finley, B. (1996). Absorption and elimination of trivalent and hexavalent chromium in humans following ingestion of a bolus dose in drinking water. *Toxicol. Appl. Pharmacol.* **141**, 145–158.

Kerger, B., Finley, B., Corbett, G., Dodge, D., and Paustenbach, D. (1997). Ingestion of chromium(VI) in drinking water by human volunteers: Absorption, distribution, and excretion of single and repeated doses. *J. Toxicol. Environ. Health* **50**, 67–95.

Kitchin, K. (2001). Recent advances in arsenic carcinogenesis: Modes of action, animal model systems, and methylated metabolites. *Toxicol. Appl. Pharmacol.* **172**, 249–261.

Kitchin, K., Del Razo, L., Brown, J., Anderson, W., and Kenyon, E. (1999). An integrated pharmacokinetic and pharmacodynamic study of arsenite action. 1. Heme oxygenase induction in rats. *Teratogenesis, Carcinogenesis, Mutagenesis* **19**, 385–402.

Kjellstrom, T., and Nordberg, G. (1978). A kinetic model of cadmium metabolism in the human being. *Environ. Res.* **16**, 248–269.

Korallus, U., Harzdorf, C., and Lewalter, J. (1984). Experimental bases for ascorbic acid therapy of poisoning by hexavalent chromium compounds. *Int. Arch. Occup. Environ. Health* **53**, 247–256.

Krolner, B., and Nielsen, S. (1982). Bone mineral content of the lumbar spine in normal and osteoporotic women: Cross-sectional and longitudinal studies. *Clin. Sci.* **62**, 329–336.

Kuykendall, J., Kerger, B., Jarvi, E., Corbett, G., and Paustenbach, D. (1996). Measurement of DNA-protein cross-links in human leukocytes following acute ingestion of chromium in drinking water. *Carcinogenesis* **17**, 1971–1977.

Lacey, R., Moore, M., and Richards, W. (1985). Lead in water, infant diet and blood: The Glascow Duplicate Diet Study. *Sci. Total Environ.* **41**, 235–257.

Langard, S., Gundersen, N., Tsalev, D., and Gylseth, B. (1978). Whole blood chromium level and chromium excretion in the rat after zinc chromate inhalation. *Acta Pharmacol. Toxicol.* **42**, 142–149.

Leggett, R. (1993) An age-specific kinetic model of lead metabolism in humans. *Env. Health Perspect.* **101**, 598–616.

Leggett, R., and Williams, L. (1986). A model for the kinetics of potassium in healthy humans. *Phys. Med. Biol.* **31**, 23–42.

Leggett, R., and Williams, L. (1988). A biokinetic model for rubidium in humans. *Health Phys.* **55**, 685–702.

Leggett, R., Williams, L., Melo, D., and Lipsztein, J. (2003). A physiologically based biokinetic model for cesium in the human body. *Sci. Total Environ.* **317**, 235–255.

Louache, F., Testa, U., Pelicci, P., Thomopoulos, P., Titeux, M., and Rochant, H. (1984). Regulation of transferrin receptors in human hematopoietic cell lines. *J. Biol. Chem.* **259**, 11576–11582.

MacKenzie, R., Byerrum, R., Decker, C., Hoppert, C., and Langham, R. (1958). Chronic toxicity studies. II. Hexavalent and trivalent chromium administered in drinking water to rats. *AMA Arch. Ind. Health* **18**, 232–234.

MacKenzie, R., Anwar, R., Byerrum, R., and Hoppert, C. (1959). Absorption and distribution of Cr51 in the albino rat. *Arch. Biochem. Biophys.* **79**, 200–205.

Mann, S., Droz, P., and Vahter, M. (1996a). A physiologically based pharmacokinetic model for arsenic exposure. I. Development in hamsters and rabbits. *Toxicol. Appl. Pharmacol.* **137**, 8–22.

Mann, S., Droz, P., and Vahter, M. (1996b). A physiologically based pharmacokinetic model for arsenic exposure. II. Validation and application in humans. *Toxicol. Appl. Pharmacol.* **140**, 471–486.

Mason, H. (2000). A biokinetic model for lead metabolism with a view to its extension to pregnancy and lactation: Further validation of the original model for non-pregnant adults. *Sci. Total Environ.* **246**, 69–78.

Mass, M. (1998) *Wild Type and p53 Deficient Mice Exposed to Arsenical Compounds*, Environmental Protection Agency.

Maurer, J., Cheng, M., Boysen, B., and Anderson, R. (1990). Two-year carcinogenicity study of sodium fluoride in rats. *J. Natl. Cancer Inst.* **82**, 1118–1126.

Mazess, R. (1982). On aging bone loss. *Clin. Orthoped. Rel. Res.* **165**, 239–252.

Menzel, D. (1987). Physiological pharmacokinetic modeling. *Environ. Sci. Technol.* **27**, 324–350.

Menzel, D. (1988). Planning and using PB-PK models: An integrated inhalation and distribution model for nickel. *Toxicol. Lett.* **43**, 67–83.

Menzel, D., Deal, D., Tayyeb, M., Wolpert, R., Boger, I., JR, Shoaf, C., Sandy, J, Wilkinson, K., and Francovitch, R. (1987). Pharmacokinetic modeling of the lung burden from repeated inhalation of nickel aerosols. *Toxicol. Lett.* **38**, 33–43.

Menzel, D., Ross, M., Oddo, S., Bergstrom, P., Greene, H., and Roth, R. (1994). A physiologically based pharmacokinetic model for ingested arsenic. In: *Arsenic: Exposure and Health*, W. Chappell, C. Abernathy and C. Cothern, eds., pp. 209–218.

Merts, W., Roginski, E., and Reba, R. (1965). Biological activity and fate of trace quantities of intravenous chromium (III) in the rat. *Am. J. Physiol.* **209**, 489–494.

Meunier, P., Courpron, P., Edouard, C., Bernard, J., Bringuier, J., and Vignon, G. (1973). Physiological senile involution and pathological rarefaction of bone. Quantitative and comparative histological data. *Clin. Endocrinol. Metab.* **2**, 239–256.

Moore, M., Meredith, P., Campbell, B., Goldberg, A., and Pocock, S. (1977). Contribution of lead in drinking water to blood-lead. *Lancet*, 661–662.

Moreas, M., Aronson, J., and Grahame-Smith, D. (1991). Intravenous strontium gluconate as a kinetic marker for calcium in healthy volunteers. *Br. J. Clin. Pharmacol.* **31**, 423–427.

Moser, G., Goldsworthy, T., and Tice, R. (2000). *Sodium Arsenite Studies in p53+/– Mice*, AWWA Research Foundation and the American Water Works Association, Denver, CO.

Muir, W., Hopfer, U., and King, M. (1984). Iron transport across brush border membranes from normal and iron deficient mouse upper small intestine. *J. Biol. Chem.* **259**, 4896–4903.

Nathanson, M., and McLaren, G. (1987). Computer simulation of iron absorption: Regulation of mucosal and systemic iron kinetics in dogs. *J. Nutrition* **117**, 1067–1075.

Nathanson, M., Muir, A., and McLaren, G. (1985). Iron absorption in normal and iron-deficient beagle dogs: Mucosal iron kinetics. *Am. J. Physiol.* **249**, G439–G448.

National Center for Health Statistics (1984). *Blood Lead Levels for Persons Ages 6 Months–74 Years: United States 1976–1980*, Washington, D.C.

National Toxicology Program (1990). *Toxicology and Carcinogenicity Studies of Fluoride in F344/N Rats and B6C3F1 Mice (Drinking Water Studies)*, U.S. Department of Health and Human Services, Washington, D.C.

Ng, J., Seawright, S., Wi, L., Garnett, C., Cirswell, B., and Moore, M. (1999). Tumours in mice induced by exposure to sodium arsenite in drinking water. In: *Arsenic Exposure and Health Effects*, W. Chappell, C. Abernathy and R. Calderon, eds., Elsevier, New York, pp. 217–223.

Nilas, L., and Christiansen, C. (1988). Rates of bone loss in normal women: Evidence of accelerated trabecular bone loss after the menopause. *Eur. J. Clin. Invest.* **18**, 529–534.

Nordberg, G., and Kjellstrom, T. (1979). Metabolic model for cadmium in man. *Env. Health. Perspect.* **28**, 211–217.

O'Flaherty, E. (1986). The rate of decline of blood lead in lead industry workers during medical removal: The effect of job tenure. *Fundam. Appl. Toxicol.* **6**, 372–380.

O'Flaherty, E. (1991a). Physiologically based models for bone-seeking elements. II. Kinetics of lead disposition in rats. *Toxicol. Appl. Pharmacol.* **111**, 313–331.

O'Flaherty, E. (1991b). Physiologically based models for bone-seeking elements. III. Human skeletal and bone growth. *Toxicol. Appl. Pharmacol.* **111**, 332–341.

O'Flaherty, E. (1991c). Physiologically based models for bone-seeking elements. I. Rat skeletal and bone growth. *Toxicol. Appl. Pharmacol.* **111**, 299–312.

O'Flaherty, E. (1992). Modeling bone mineral metabolism, with special reference to calcium and lead. *NeuroToxicol.* **13**, 789–798.

O'Flaherty, E. (1993a). A pharmacokinetic model for chromium. *Toxicol. Lett.* **68**, 145–158.

O'Flaherty, E. (1993b). Physiologically based models for bone-seeking elements. IV. Kinetics of lead disposition in humans. *Toxicol. Appl. Pharmacol.* **118**, 16–29.

O'Flaherty, E. (1995a). PBK modeling for metals. Examples with lead, uranium, and chromium. *Toxicol. Lett.* **82/83**, 367–372.

O'Flaherty, E. (1995b). Physiologically based models for bone-seeking elements. V. Lead absorption and disposition in childhood. *Toxicol. Appl. Pharmacol.* **131**, 297–308.

O'Flaherty, E. (1996). A physiologically based model of chromium kinetics in the rat. *Toxicol. Appl. Pharmacol.* **138**, 54–64.

O'Flaherty, E. (1998a). A physiologically based kinetic model for lead in children and adults. *Env. Health Perspect.* **106**, 1495–1503.

O'Flaherty, E. (1998b). Physiologically based models of metal kinetics. *Crit. Rev. Toxicol.* **28**, 271–317.

O'Flaherty, E. (2000). Modeling normal aging bone loss, with consideration of bone loss in osteoporosis. *Toxicol. Sci.* **55**, 171–188.

O'Flaherty, E., Inskip, M., Yagminas, A., and Franklin, C. (1996). Plasma and blood lead concentrations, lead absorption, and lead excretion in nonhuman primates. *Toxicol. Appl. Pharmacol.* **138**, 121–130.

O'Flaherty, E., Inskip, M., Franklin, C., Durbin, P., Manton, W., and Baccanale, C. (1998). Evaluation and modification of a physiologically based model of lead kinetics using data from a sequential isotope study in cynomolgus monkeys. *Toxicol. Appl. Pharmacol.* **149**, 1–16.

O'Flaherty, E., Kerger, B., Hays, S., and Paustenbach, D. (2001). A physiologically based model for the ingestion of chromium(III) and chromium(VI) by humans. *Toxicol. Sci.* **60**, 196–213.

Okazaki, M., Aoba, T., Doi, Y., Takahashi, J., and Moriwaki, Y. (1981). Solubility and crystallinity in relation to fluoride content of fluoridated hydroxyapatites. *J. Dent. Res.* **60**, 845–849.

Paustenbach, D., and Finley, B. (2003). Human health risk and exposure assessment of chromium(VI) in tap water. *J. Toxicol. Environ. Health* **66**, 1295–1339.

Paustenbach, D., Hay, S., Brien, B., Dodge, D., and Kerger, B. (1996). Observation of steady state in blood and urine following human ingestion of hexavalent chromium in drinking water. *J. Toxicol. Environ. Health* **49**, 453–461.

Pierson, R. N., Jr., Price, D. C., Wang J., and Jain, R. K. (1978). Extracellular water measurements: Organ tracer kinetics of bromide and sucrose in rats and man. *Am. J. Physiol.* **235**(3), F254–F264.

Polak, J., O'Flaherty, E., Freeman, G., Johnson, J., Liao, S., and Bergstrom, P. (1996). Evaluating lead bioavailability data by means of a physiologically based lead kinetic model. *Fundam. Appl. Toxicol.* **29**, 63–70.

Pounds, J., and Leggett, R. (1998). The ICRP age-specific biokinetic model for lead: Validations, empirical comparisons, and explorations. *Env. Health Perspect.* **106**, 1505–1511.

Proctor, D., Otani, J., Finley, B., Paustenbach, D., Bland, J., Speizer, N., and Sargent, E. (2002). Is hexavalent chromium carcingenic via ingestion? A weight-of-evidence review. *J. Toxicol. Environ. Health* **65**, 701–746.

Rabinowitz, M. (1998). Historical perspective on lead biokinetic models. *Env. Health Perspect.* **106**, 1461–1465.

Rabinowitz, M., Leviton, A., Needleman, H., Bellinger, D., and Waternaux, C. (1985). Environmental correlates of infant blood lead levels in Boston. *Environ. Res.* **38**, 96–107.

Rabinowitz, M., Wetherill, G., and Kopple, J. (1976). Kinetic analysis of lead metabolism in healthy humans. *J. Clin. Invest.* **58**, 260–270.

Rao, H., Beliles, R., Whitford, G., and Turner, C. (1995). A physiologically based pharmacokinetic model for fluoride uptake by bone. *Regul. Toxicol. Pharmacol.* **22**, 30–42.

Ryu, J., Ziegler, E., Nelson, S., and Fomon, S. (1983). Dietary intake of lead and blood lead concentration in early infancy. *Am. J. Dis. Child* **137**, 886–891.

Sherlock, J., Smart, G., Forbes, G., Moore, M., Patterson, W., Richards, W., and Wilson, T. (1982). Assessment of lead intakes and dose-response for a population in Ayr exposed to a plumbosolvent water supply. *Hum. Toxicol.* **1**, 115–122.

Silbergeld, E., Schwartz, J., and Mahaffey, K. (1988). Lead and osteoporosis: Mobilization of lead from bone in postmenopausal women. *Environ. Res.* **47**, 79–94.

Silk, T., Kendall, G., and Phipps, A. (1995). Revised estimates of dose from ores and mineral sands. *J. Radiol. Prot.* **15**, 217–222.

Sips, A., van der Vijgh, W., Barto, R., and Netelenbos, J. (1996). Intestinal absorption of strontium chloride in healthy volunteers: Pharmacokinetics and reproducibility. *Br. J. Clin. Pharmacol.* **41**, 543–549.

Smith, D., Khairi, M., Norton, J., and Johnston, C., Jr. (1976). Age and activity effects on rate of bone mineral loss. *J. Clin. Invest.* **58**, 716–721.

Stepan, J., Tesarova, A., Havranek, T, Jodl, J., Formankova, J., and Paeovsky, V. (1985). Age and sex dependency of the biochemical indices of bone remodeling. *Clin. Chim. Acta* **151**, 273–283.

Stevenson, J., Lees, B., Devenport, M., Cust, M., and Ganger, K. (1989). Determinants of bone density in normal women: Risk factors for future osteoporosis? *BMJ* **298**, 924–928.

Succop, P., Bornschein, R., Brown, K., and Tseng, C. (1998). An empirical comparison of lead exposure pathway models. *Environ. Health Perspect.* **106**, 1577–1583.

Sunderman, W. (1987). Metal induction of heme oxygenase. *Ann. NY Acad. Sci.* **514**, 65–80.

Thomann, R., Synder, C., and Squibb, K. (1994). Development of a pharmacokinetic model for chromium in the rat following subchronic exposure. *Toxicol. Appl. Pharmacol.* **128**, 189–198.

Timchalk, C., Poet, T., Lin, Y., Weitz, K., Zhao, R., and Thrall, K. (2001). Development of an integrated microanalytical system for analysis of lead in saliva and linkage to a physiologically based pharmacokinetic model describing lead saliva secretion. *Am. Ind. Hyg. Assoc. J.* **62**, 295–302.

Turner, C., Boivin, G., and Meunier, P. (1993). A mathematical model for fluoride uptake by the skeleton. *Calcif. Tissue. Int.* **52**, 130–138.

US EPA (1990). *Air Quality Criteria for Lead: Supplement to the 1986 Addendum*, U.S. Environmental Protection Agency, Research Triangle Park, NC.

US EPA (1994). *Technical Support Document: Parameters and Equations Used in the Integrated Exposure Uptake Biokinetic (IEUBK) Model for Lead in Children*, Washington, D.C., pp. 1–56.

Vahter, M., Friberg, L., Rahnster, B., Nygern, A., and Nolinder, P. (1986). Airborne arsenic and urinary excretion of metabolites of inorganic arsenic among smelter workers. *Int. Arch. Occup. Environ. Health* **57**, 79–91.

Valentine, J., Kang, H., and Spivey, G. (1979). Arsenic levels in human blood, urine, and hair in response to exposure via drinking water. *Environ. Res.* **20**, 24–32.

Visek, W., Whitney, I., Kuhn, U., III, and Comar, C. (1953). Metabolism of Cr51 by animals as influenced by chemical state. *Proc. Soc. Exp. Biol. Med.* **84**, 610–615.

Waalkes, M., Keefer, L., and Diwan, B. (2000). Induction of proliferative lesions of the uterus, testes, and liver in Swiss mice given repeated injections of sodium arsenate: Possible estrogenic mode of action. *Toxicol. Appl. Pharmacol.* **166**, 24–35.

Wang, J., Qi, L., Moore, M., and Ng, J. (2002). A review of animal models for the study of arsenic carcinogenesis. *Toxicol. Lett.* **133**, 17–31.

Ward, J., Kushner, J., and Kaplan, J. (1982). Transferrin receptors of human fibroblasts. *Biochem. J.* **208**, 19–26.

Weber, H. (1983). Long-term study of the distribution of soluble chromate-51 in the rat after a single intratracheal administration. *J. Toxicol. Environ. Health* **11**, 749–764.

Whillans, D. (2003). Structure of a physiologically based biokinetic model for use in 14C and organically bound tritium dosimetry. *Radiat. Prot. Dosim.* **105**, 189–191.

White, P., Van Leeuwen, P., Davis, B., Maddaloni, M., Hogan, K., Marcus, A., and Elias, R. (1998). The conceptual structure of the integrated exposure uptake biokinetic model for lead in children. *Environ. Health Perspect.* **106**, 1513–1530.

Williams, M., King, E., and Walford, J. (1969). An investigation of lead absorption in an electric accumulator factory with use of personal samplers. *Br. J. Ind. Med.* **26**, 202–216.

Wrenn, M., Lipsztein, J., and Bertelli, L. (1989). Pharmacokinetic models relevant to toxicity and metabolism for uranium in humans and animals. *Radiat. Prot. Dosim.* **26**, 243–248.

Wrenn, M., Bertelli, L., Durbin, P., Singh, N., Lipsztein, J., and Eckerman, K. (1994). A comprehensive metabolic model for uranium metabolism and dosimetry based on human and animal data. *Radiat. Prot. Dosim.* **53**, 255–258.

Yu, C., and Diu, C. (1982). A comparative study of aerosol deposition in different lung models. *Am. Ind. Hyg. Assoc. J.* **43**, 54–65.

Yu, D. (1998). Uncertainties in a pharmacokinetic modeling for inorganic arsenic. *J. Environ. Sci. Health* **A33**, 1369–1390.

Yu, D. (1999a). A pharmacokinetic modeling of inorganic arsenic: A short-term oral exposure model for humans. *Chemosphere* **39**, 2737–2747.

Yu, D. (1999b). A physiologically based pharmacokinetic model of inorganic arsenic. *Regul. Toxicol. Pharmacol.* **29**, 128–141.

Yu, D., and Kim, J. (2004). A physiologically based assessment of human exposure to radon released from groundwater. *Chemosphere* **54**, 639–645.

Zakharyan, R. A., Ayala-Fierro, F., Cullen, W. R., Carter, D. M., and Vasken Aposhian, H. (1999). Enzymatic methylation of arsenic compounds. VII. Monomethylarsonous acid (MMAIII) is the substrate for MMA methyltransferase of rabbit liver and human hepatocytes. *Toxicol. Appl. Pharmacol.* **158**, 9–15.

Zalups, R., and Koropatnick, J., eds. (2000). *Molecular Biology and Toxicology of Metals*, Taylor & Francis, London.

PART III

PHARMACEUTICAL APPLICATIONS OF PBPK MODELS

CHAPTER 10

DRUGS

Tami S. McMullin

10.1 INTRODUCTION

Understanding the relationship between the pharmacokinetics of drugs and their pharmacodynamics for both therapeutic and toxic responses is essential for appropriate selection of potential drug candidates in early drug discovery and accurate optimization of a drug's therapeutic effect later in the drug development process. Classical pharmacokinetic (PK) models have been widely used to study kinetic behaviors of drugs. These models use curve-fitting techniques to estimate kinetic rate constants and other parameters from time-course curves for blood, tissue, or excreta. These compartmental PK models are limited as predictive tools, especially for interspecies extrapolation, because the curve-fitted parameters lack biological meaning.

As part of the high throughput process in early drug discovery, there is a pressing need to develop PK tools that relate the pharmacokinetics of drug candidates and underlying biological processes. Physiologically based pharmacokinetic (PBPK) models are becoming more widely used in the pharmaceutical field (Nestorov 2003) since they do offer promise in explaining macroscopic time-course behaviors on the basis of more fundamental parameters, such as binding affinities, metabolic rate constants, and tissue solubility. Anti-cancer agents were some of the first chemicals

Physiologically Based Pharmacokinetic Modeling, edited by Micaela B. Reddy, Raymond S. H. Yang, Harvey J. Clewell III, and Melvin E. Andersen.

studied with PBPK models (Chapter 11). The purported advantages of PBPK modeling in drug development are as follows: (1) They allow the prediction of concentration–time profiles of a drug in tissue and plasma prior to *in vivo* studies, (2) they elucidate the relationship between PK and pharmacodynamic (PD) effects of a drug and therefore aid in the appropriate selection of clinical drug candidates, and (3) they make possible the extrapolation of kinetic data across dose levels, route of administration, and species (Theil *et al.* 2003). Additionally, PBPK models guide more hypothesis-driven experimentation after potential drug candidates are selected. Nestorov (2003) recently summarized some interesting PBPK modeling approaches in drug research and discussed advantages and limitations of PBPK models for the drug industry.

The unbound (free) plasma concentration—or in the case of antibiotics, the unbound (free) interstitial fluid concentration—is a generally reliable estimate of target tissue time-course concentrations. Therefore, PBPK models that accurately estimate target tissue concentrations need to describe the physiological and PK factors that control the free plasma or interstitial fluid concentrations. This chapter focuses on the development of PBPK models for all drugs except chemotherapeutic agents. Because many PBPK models have been constructed for chemotherapeutic agents and because there are pharmacokinetic issues unique to this class of pharmaceutical agents, they were included in a separate chapter. Additionally, discussions on PBPK modeling approaches applied to volatile anesthetics (e.g., halothane and isoflurane) are found in Chapter 2, and those on the perinatal transfer of drugs (e.g., theophylline and tetracycline) are found in Chapter 12. This chapter examines those biological and physiochemical factors influencing tissue distribution and clearance of drugs that need to be considered when developing a PBPK model, and it will end with a discussion of some of the unique issues in PBPK model development for drugs.

10.2 DESCRIBING THE TISSUE DISTRIBUTION OF DRUGS

PBPK models for drugs to a large extent have focused on describing drug distribution in the body (Table 10.1). Two factors largely influence distribution of drug in the body: solubility and protein binding. To a great extent, these two factors are interrelated and determine the observed partitioning of drug in tissues and consequently the effectiveness of the drug at intended sites of action. Frequently, these factors are taken into account in a PBPK model by the tissue-to-plasma partition coefficient, $P_{t:p}$ (Tsuji *et al.* 1979). This parameter, a combination of drug solubility (apparent linear binding) and macromolecular binding (nonlinear binding) in tissue and plasma (Poulin and Theil 2000), has conventionally been determined experimentally in the animal by determining the ratio of *in vivo* tissue concentrations to plasma concentrations. However, because of the time-consuming nature of this approach, it is not useful for the pharmaceutical industry where high-throughput evaluation of the kinetic disposition of a drug is routine. Additionally, while this approach may be adequate for describing and predicting tissue time-course concentrations within a

TABLE 10.1 PBPK Models Emphasizing Factors that Influence Distribution of Drug

Drug	Reference	Species	Comments
β-Lactam antibiotics	Tsuji *et al.* (1979)	Rat	Describes differences in tissue distribution due to route of application. Uses a model predicted value for $P_{t:p}$ to describe accumulation of drug in tissues.
	Tsuji *et al.* (1983)	Rat	Comparison of binding estimates using descriptions of binding with $P_{t:p}$ (linear binding) or binding as occurring directly with albumin (nonlinear binding).
	Yoshiikawa *et al.* (1999)	Rat, dog, human	Separation of capillary bed, interstitial fluid and intracellular fluid compartments.
	Tsuji *et al.* (1985)	Rabbit, human	1983 model extended.
	Levitt (2003)	Human	General PBPK model to describe time-course concentrations of extracellular solutes.
Grepafloxacin, antimicrobial agent	Nakajima *et al.* (2000)	Rat, rabbit, monkey, dog, human	Extensive tissue distribution.
Tetraethylammonium ion	Mintum *et al.* (1980)	Rat	Active transport mechanisms to distribute drug.
CyA, antibiotic	Bernareggi and Rowland (1991)	Rat	Flow-limited distribution.
	Kawai *et al.* (1994)	Rat	Separation of model by Bernareggi and Rowland into three tissue subcompartments. Description of binding using a nonspecific, nonsaturable deep tissue pool that slowly and reversibly interacts with the drug.
	Kawai *et al.* (1998)	Rat, human	Extension of the 1994 model to include biological descriptions of tissue distribution. Nonlinear and linear binding in tissue and red blood cells.
	Tanaka *et al.* (1999)	Rat	Extension of model by Bernareggi and Rowland (1991) and Kawai *et al.* (1998) to describe nonlinear protein binding in tissue with specific receptor binding
Anesthetic gases	Fisherova-Bergerova (1992)	Human	PBPK model of a mixture of gases and their distribution into different fat compartments.
Thiopental	Bischoff and Dedrick (1968)	Dog	Description of processes leading to tissue redistribution of a drug.
	Chen and Andrade (1976)	Dog	Extension of the Bischoff and Dedrick model to include time-course concentrations in the brain.
	Ebling *et al.* (1994)	Rat	Whole-body PBPK model to describe drug distribution.
Lidocaine	Benowitz *et al.* (1974)	Monkey	Redistribution of drug from rapidly perfused tissue to skeletal muscle.

TABLE 10.1 ***continued***

Drug	Reference	Species	Comments
Barbiturates	Blakey *et al.* (1997)	Rat	Whole-body model that describes extensive distribution of drug in tissue and focuses on the relationship between drug redistribution and drug lipophilicity.
	Nestorov *et al.* (1997)		Sensitivity analysis of model by Blakey *et al.* (1997).
	Nestorov *et al.* (1998)		Relationship of drug lipophilicity in predicting drug distribution.
Phenobarbitol	Engasser *et al.* (1981)	Rat	Incorporation of *in vitro* parameters to describe protein binding.
Midazolam phenobarbitol thiopental hexobarbitol	Igari *et al.* (1982)	Rat	*In vitro* binding data to describe linear tissue and plasma protein binding. Compares the different kinetic properties of long-acting versus short-acting anesthetics.
Midazolam	Bjorkman *et al.* (2001)	Rat, human	Drug binding to albumin in different subcompartments within tissue to account for distribution of albumin throughout the body.
Diazepam	Igari *et al.* (1983)	Rat, human	Linear description of serum protein binding; description of interspecies differences in binding characteristics.
Quinidine	Harashima *et al.* (1986)	Rat, human	Interspecies differences in protein binding.
	Harashima *et al.* (1985)	Rat	Nonlinear tissue distribution due to protein albumin binding.
Cafalozin	Tsuji *et al.* (1985)	Rabbit, human	Scale-up of 1983 model using linear and nonlinear protein binding.
Sulfamethizole	Katayama *et al.* (1990)	Rabbit	Plasma protein binding.
	Bernareggi and Rowland (1991)	Rat	Saturable binding to red blood cells.
Digoxin	Harrison and Gibaldi (1977a)	Dog, Human	Interspecies differences in $P_{t:p}$ values.
Ceftibuten	Shimada *et al.* (1993)	Rat	Use of model by Tsuji *et al.* (1983) to determine the contribution of albumin to isomerization.
Nicotine	Plowchalk *et al.* (1992)	Rat	Nonlinear tissue distribution in brain, heart, and lung described by saturable binding.
Fentanyl, alfentanil	Bjorkman (2003)	Human	Examines methods for simplifying PBPK model descriptions.
Estradiol	Plowchalk and Teeguarden (2002)	Rat, human	Extensive tissue specific binding to estrogen receptor and plasma proteins.
Cefazolin, cephalexin, cephradine	Greene *et al.* (1978)	Human	The pharmacokinetics of three cephalosporin antibiotics are evaluated using a two-compartment PK model with and without plasma protein binding and a PBPK model.
Procaine	Smith *et al.* (1979)	Human	This PBPK model used an individual's characteristics (i.e., sex, age weight, height, hematocrit) to arrive at an individualized prediction of pharmacokinetics.

species, it is more limited when scaling a model up to humans because the extent to which solubility and protein binding contribute to the observed $P_{t:p}$ value can vary widely among different species (Harrison and Gibaldi 1977a). Thus, one of the main goals in early drug development is to develop predictive tools to determine the value of $P_{t:p}$ for new drug candidates in a variety of species (Poulin *et al.* 2001; Poulin and Theil 2002a,b). To evaluate the different approaches to determining the $P_{t:p}$ of a drug that exhibits interspecies variation in its tissue distribution, PBPK models using epiroprim as a model compound were developed. This rigorous assessment of several different methods indicated that the predictive approaches were superior to classical allometric scaling in describing the distribution of epiroprim in humans.

Because many drugs have very narrow concentration and time ranges in which the therapeutic benefits are achieved, detailed mechanistic and biological descriptions of tissue distribution may be required to accurately estimate tissue concentrations. For some drugs, investigators have chosen to develop PBPK models with a high degree of physiological detail. Blakey *et al.* (1997) developed a whole-body PBPK model for a series of structurally homologous barbiturates that included kinetic descriptions in 14 tissues. One unique PBPK modeling approach, termed piecewise PBPK modeling, was used to optimize parameters for a full-body physiological model for thiopental in which the extensive number of compartments made it difficult to simultaneously optimize for all parameters (Ebling *et al.* 1994). Separate submodels for each tissue compartment were created and the parameters for each submodel were independently optimized using constrained numerical deconvolution analysis. This method minimized the number of parameters that needed to be simultaneously optimized because the fit to each tissue was independent of the other tissue estimates. These independent submodels were reassembled together to make a whole-body PBPK model for the rat.

In whole-body PBPK models, distribution of drug into a tissue compartment is generally characterized as perfusion-limited with a high degree of mixing upon entrance into the compartment. However, some tissue profiles (as a function of time or position within a tissue) of drugs oscillate over time and therefore are more accurately described with an intermediate degree of mixing of drug in the tissue compartment. For this situation, dispersion models, where a unitless dispersion value characterizes the extent of dispersion of a drug within each organ, have proved more successful in describing both short- and long-term time-course concentrations of drugs in various tissues (Oliver *et al.* 2001).

Whole-body PBPK models may be necessary to sufficiently describe the kinetic disposition of some drugs. However, because the development of these models is time-consuming and requires extensive experimental work, it is advantageous to simplify PBPK models when possible. For example, the time-course concentrations of fentanyl and pethidine—two basic, lipophilic drugs—were successfully described in humans by simplifying an initially complex PBPK model. Successively simplified models were developed to investigate the effects of (1) using physicochemical data to predict $P_{t:p}$ rather than *in vivo* data, (2) lumping of tissue compartments, and (3) reducing the individual $P_{t:p}$ values for each tissue to two values that described partitioning in "fat" and "lean" tissues. All three of these methods for simplifying PBPK models proved successful. However, the final model required an additional separate $P_{t:p}$ for the lung compartment (Bjorkman 2003).

Because of the importance of solubility and protein binding in determining the therapeutic action of a drug, PBPK models have been developed to predict the relationship between the lipophilicity of a drug and its tissue distribution and thereby expedite the early drug discovery process (Nestorov *et al.* 1998). For example, a whole-body PBPK model on a homologous series of barbiturates describes how increasing lipophilicity increases drug deposition in muscle and, consequently, alters the volume of distribution for unbound drug (Blakey *et al.* 1997; Nestorov *et al.* 1997). Many drugs, especially central nervous system (CNS) agents and antibiotics, are designed to be highly lipid soluble in order to interact with membrane lipids and proteins, such as receptors, to produce the desired pharmacological effect. β-Lactam antibiotics (e.g., penicillin, cephalosporins) cause their pharmacological effect by binding to receptors on the cell wall and, thereby, selectively inhibiting bacterial cell wall synthesis. Estradiol is a steroid hormone that binds to estrogen receptors in various tissues to cause tissue-specific responses. PBPK modeling of the tissue distribution of this hormone in rats and humans included plasma protein binding (i.e., to albumin in the rat and to sex-hormone binding globulin in humans) and tissue-specific estrogen receptor binding. The PK profile of estradiol was highly dependent on the concentration of estrogen receptor in tissue (Plowchalk and Teeguarden 2002). The PK description of nicotine with a PBPK model is another example of incorporating saturable receptor binding in specific tissue to more accurately describe the nonlinear tissue distribution characteristics of nicotine (Plowchalk *et al.* 1992).

Because the chemical structure of many drugs favors protein binding, these compounds reversibly bind to various macromolecules, such as albumin. To investigate the effects of plasma protein binding on drug disposition, *in vitro* and *in vivo* techniques that determine the binding characteristics of drugs have been used and incorporated into PBPK models (Engasser *et al.* 1981; Igari *et al.* 1982; Harashima *et al.* 1985; Katayama *et al.* 1990). Bjorkman *et al.* (2001) developed a flow-limited model for midazolam that described different compartments within each tissue to account for albumin distribution in the rat. This model was scaled to humans and used to simulate various clinical conditions of different albumin disorders to characterize interindividual kinetic variations. Interspecies differences in protein binding also contribute to observed differences in unbound plasma concentrations and should be accounted for when performing interspecies scaling in PBPK models (Igari *et al.* 1983; Harashima *et al.* 1986).

Extracellular solutes, such as morphine-6-glucuronide, morphine-3-glucuronide, inulin, and β-lactam antibiotics, extensively distribute in interstitial spaces within tissues and blood vessels with the greatest partitioning in the kidney, gut wall (Tsuji *et al.* 1979), skin (Tsuji *et al.* 1983), and lung (Nakajima *et al.* 2000). Therefore, unlike the goal for most PBPK modeling (i.e., to simulate time-course blood concentrations), the objective for modeling the pharmacokinetics of these compounds is to predict the unbound time-course concentration in the interstitial fluid compartments.

To describe the distribution of these compounds into the interstitial spaces, Tsuji *et al.* (1983) developed a whole-body PBPK model for several β-lactam antibiotics, including penicillins. The majority of the model parameters were experimentally determined. In the equations describing plasma time courses, all protein binding in serum was to albumin. The theoretical predictions were compared with observed values of $P_{t:p}$ to determine the accuracy of the model description, especially for noneliminating organs. The values for $P_{t:p}$ were largely dependent on the volume of interstitial fluid and the capacity for drug binding to albumin in the interstitial fluid. Penicillins, except for Penicillin V, could be described by linear binding that correlated with drug lipophilicity. Utilizing species-specific information on tissue binding, elimination, and physiological parameters, this model was extended to describe the pharmacokinetics of cefazolin in the interstitial spaces of rabbits and humans. Each noneliminating organ was separated into three well-mixed compartments: capillary bed, interstitial fluid, and intracellular space. Antibiotics entered the capillary bed by flow-limited processes and diffused into interstitial fluid compartments (Tsuji *et al.* 1985). Splitting of each tissue into these three compartments has also been used in PBPK models for antibiotics such as cyclosporine A, CyA (Tsuji *et al.* 1983), and a derivative of CyA, SDZ IMM 125 (Kawai *et al.* 1994). The transfer of drug between capillary blood and interstitial water can be rapid, and so in many models the capillary blood and interstitial water are lumped into one extracellular fluid compartment—for example, as in the model for the ionic compound tetraethyl ammonium ion (Mintun *et al.* 1980).

To investigate the pharmacokinetics of extracellular solute distribution in the interstitial spaces, Levitt (2003) successfully developed a single PBPK model that described the pharmacokinetics of 11 different extracellular solutes, including the glucuronides of morphine and several different β-lactam antibiotics with various levels of albumin binding. The albumin concentration in the interstitial space and the unbound plasma fraction were the main determinants in describing the kinetics of the β-lactam antibiotics. All solutes could be described by a general PBPK parameter set adjusted for the renal clearance parameter.

Bernareggi and Rowland (1991) developed the first CyA PBPK model to describe tissue distribution using plasma data from a two-week subcutaneous infusion. Distribution of CyA into all organ compartments was described as perfusion-limited with each tissue modeled as a well-stirred compartment. Nonlinear tissue distribution of CyA occurred in red blood cells and adipose in addition to nonlinear elimination processes. To gain a greater understanding of the biological mechanisms underlying the tissue specific nonlinear behaviors, Kawai *et al.* (1994) extended this model for a CyA derivative, SDZ IMM 125, to evaluate tissue time-course concentrations in 14 organ compartments. Tissue time-course descriptions could not be modeled with a single, well-stirred flow-limited compartment for each tissue, especially in red blood cells where there was significant accumulation of the drug. Therefore, they incorporated membrane-limited uptake into tissue compartments by separating some organ compartments into three subcompartments, using an approach similar to that of Tsuji *et al.* (1983). The retention of CyA derivative in tissue compartments was described using a nonspecific, nonsaturable deep tissue pool that slowly and reversibly interacted with the drug. The rates of entrance and removal of the drug from the deep pool were described by first-order rate processes. By adjusting tissue distribution and clearance parameters, such as red blood cell binding rates and metabolism rates, this rat model was successfully scaled-up to simulate intravenous (*i.v.*) time-course data from human volunteers.

An alternative model to the one developed in 1994 described CyA disposition after a two-minute *i.v.* infusion (Kawai *et al.* 1998). This model assumed instantaneous mixing of drug with the interstitial fluid and subsequent specific saturable binding in the tissue to replace the nonspecific deep binding pool that accounted for the nonlinear time-course tissue profiles at therapeutic doses of CyA. Accumulation of drug in some of the tissues was described by a linear process, while other tissues followed more nonlinear binding behavior. Interestingly, the model-estimated dissociation constant was similar to the binding constant for CyA to its receptor, cyclophilin, determined *in vitro*.

To further examine the biological characteristics of CyA that contribute to its distribution, an extensive dataset on dose-dependent tissue distribution of CyA following a two-minute *i.v.* infusion in rat was developed to construct a comprehensive PBPK model that compared four different local organ models, including a model that accounted for the nonlinear binding with specific receptor binding (Tanaka *et al.* 1999). CyA distribution is highly tissue- and dose-dependent, and a thorough understanding of the biological determinants that contribute to CyA pharmacokinetics are required to appropriately extrapolate to humans. To evaluate the nonlinear blood and tissue distribution characteristics, drug accumulation in tissue was described in four ways: (1) linear distribution, (2) instantaneous, saturable distribution, (3) noninstantaneous, saturable distribution, and (4) saturable efflux of drug. The group of models developed for CyA and its derivative are good examples of the utility of PBPK models to aid in developing and testing biological hypotheses.

Several drugs tend to redistribute to tissues other than the primary site of action. Because tissue redistribution contributes to a loss of pharmacological effect of the drug, detailed descriptions of the processes that lead to redistribution into other tissues are necessary. In one of the first PBPK models developed, Bischoff and Dedrick (1968) investigated the physiological processes that lead to tissue redistribution of thiopental, a short-acting anesthetic agent, with a flow-limited PBPK model. Using the dog as an appropriate animal model for humans, this model described free plasma and tissue concentrations by assuming rapid transport of the drug across membranes and including binding to plasma and tissue proteins. Chen and Andrade (1976) extended this model to predict concentrations of thiopental in the target tissue (brain) while simultaneously describing time-course disposition in other tissues in the dog. Lidocaine disposition in the monkey is also affected by its redistribution from rapidly perfused tissues to skeletal muscle (Benowitz *et al.* 1974a).

CNS drugs are designed to distribute preferentially to the brain. However, mainly due to their solubility characteristics, redistribution of CNS drugs from the brain to other tissues is common. PBPK modeling efforts indicate that barbiturates redistribute from highly perfused tissue to adipose tissue according to their respective lipophilicities (Blakey *et al.* 1997). Generally, a more detailed description of disposition in fat tissue is required to accurately model the time-course concentrations of these types of drugs. Fiserova-Bergerova (1992) developed two human PBPK models, one with a single fat compartment and one that separated the fat compartment into subcutaneous and inner adipose tissue, to simulate experimental time-course data on uptake and elimination of a mixture of four fluorinated anesthetic agents (isoflurane, enflurane, halothane, and methoxyflurane). Although splitting the adipose tissue in the body into two compartments did not alter model simulations of the alveolar concentrations of anesthetic during the period of administration of these drugs, elimination by exhalation after drug treatment was more accurately described.

10.3 DESCRIBING METABOLISM AND OTHER CLEARANCE PROCESSES OF DRUGS

While metabolism of most drugs has been accurately described by Michaelis–Menten (M–M) saturable metabolism in the liver (Tanaka *et al.* 1999; Chen and Andrade 1976; Tsuji *et al.* 1979; El-Masri and Portier 1998), this description is not adequate for describing metabolic clearance of all drugs, especially some CNS acting agents. These drugs typically have concentration time-course profiles that show higher-than-expected elimination of drug within tissues if only metabolic clearance in the liver was occurring. For example, simulation of the metabolic clearance of drugs such as diazepam (Igari *et al.* 1983) and methoxyflurane (Fiserova-Bergerova 1992) required inclusion of extrahepatic metabolism to adequately describe their time-course plasma concentrations. SUN5555, a penem antibiotic, is also cleared by extrahepatic as well as hepatic metabolism, presumably due to metabolizing enzymes present within several tissues (Tsuji *et al.* 1990). An accurate PK description of plasma concentrations of this compound in rat, dog, and human included metabolic clearance processes within other tissues such as kidney, gut, and lung. Because metabolic clearance rates of estradiol tend to exceed hepatic blood flow, significant extrahepatic metabolism in the plasma compartment was added to describe estradiol concentrations in rats and humans using a PBPK model. The rate of extrahepatic metabolism was dependent on the clearance rate and total circulating estradiol concentrations (Plowchalk and Teeguarden 2002).

Williams *et al.* (1996) used a corollary approach to describe metabolism and predict the PK disposition of 2,2-dichloro-1,1,1-trifluoroethane, HCFC-123, and its metabolite, triflouracetic acid, in rats using halothane kinetic data. Initially, a PBPK model was developed to describe kinetics in rats. Because halothane is metabolized similarly in rats and humans, rat metabolic constants were used to develop a human PBPK model of halothane. Once a PBPK model for halothane was established, similarities were used to develop a PBPK model for HCFC-123 and its metabolite in rats with the ultimate goal of estimating dose–response behavior in humans.

The clearance of morphine and its metabolite, morphine-3-glucuronide, in the liver has been investigated using a PBPK model of an isolated liver compartment that describes diffusion-limited movement of the metabolite across hepatic sinusoidal and canalicular membranes (O'Brien *et al.* 1996). This model was extended to determine rate constants for the movement of this metabolite through the membranes and to determine the membrane permeability constants that lead to the distribution and accumulation of the metabolite in the liver compartments (Evans *et al.* 1999).

Capecitabine is a chemotherapeutic agent that was designed to be highly tissue-specific in its metabolism. A PBPK model was developed for this compound to aid in determining the key factors that contribute to the pharmacokinetics of the parent drug and its pharmacologically active metabolite, 5-fluorouracil. Upon characterization of the kinetic disposition of capecitabine, a pharmacodynamic model was developed to determine the clinical outcomes under various human conditions such as different levels of metabolizing enzyme (Blesch *et al.* 2003).

Metabolic clearance is dependent on the types and quantity of metabolizing enzymes present in the liver and other tissues and, therefore, can vary widely between and within species. Intra-individual variability in human metabolic capacity may lead to different observed sensitivities within a human population to the therapeutic effects of a drug or its toxic side effects. El-Masri and Portier (1998) developed a model that described plasma and brain time-course concentrations for primidone and its pharmacologically active metabolites, phenobarbital and phenylethylmalonamide (PEMA). The aim of this work was to investigate inter- and intraspecies metabolic differences between rats, mice, and humans that lead to different time-course concentration profiles. The ratios of the maximal velocity of the enzyme (V_{max}) to the binding affinity of drug to the enzyme (K_m), expressed as V_{max}/K_m, quantitatively indicated that metabolism varied greatly among the different species and between the human subjects. Therefore, risk of cancer from this compound could vary among different human populations.

Data on the polymorphisms in the genes that encode for different metabolizing enzymes is increasingly available for several environmental toxicants. PBPK models provide useful predictive tools to quantitatively assess tissue dose in different sensitive populations by linking polymorphism data to observed differences in plasma time-course profiles in human subjects. A PBPK modeling approach using Monte Carlo simulation analysis examined variability in tissue dose (estimated for area under the curve) in human populations with different polymorphisms for metabolizing the chemical warfarin, an anti-coagulant (Gentry *et al.* 2002). This quantitative method could prove useful in predicting variability in the tissue dose of extensively metabolized drugs in sensitive human populations.

In addition to metabolism, many drugs are cleared into the urine via the kidney. Although active tubular secretion may be significant with some drugs in some species, first-order (Tsuji *et al.* 1979) and saturable (Tsuji *et al.* 1985) clearance processes are usually sufficient to describe renal elimination. In modeling β-lactam antibiotics, an additional clearance process occurring in the bile described by active transport mechanisms was necessary (Tsuji *et al.* 1983; Kawai *et al.* 1994). Table 10.2 lists PBPK models for drugs that include descriptions of clearance processes.

TABLE 10.2 PBPK Models Emphasizing Metabolism/Clearance of Drug

Drug	Reference	Species	Comments
Thiopental	Chen and Andrade (1976)	Dog	Extension of PBPK model by Bischoff and Dedrick (1968) with saturable M-M metabolism in the liver.
Primidone	El-Masri and Portier (1998)	Rat, mouse, human	Metabolite model investigated intra- and interspecies differences in metabolic clearance.
β-Lactam antibiotics and derivatives	Tanaka *et al.* (1999)	Rat	M-M metabolism in the liver.
	Tsuji *et al.* (1979)	Rat	First-order renal clearance; M–M metabolism in liver.
	Tsuji *et al.* (1982)		Extension of (1979) model; clearance into bile by active transport mechanisms; first-order renal clearance.
	Kawai *et al.* (1994)	Rat, dog human	Clearance described with M–M metabolism, glomerular filtration and biliary secretion.
Midazolam	Igari *et al.* (1982)	Rat	Use of *in vitro* metabolic parameters to describe M–M metabolism.
Diazepam	Igari *et al.* (1983)	Rat, human	Extrahepatic metabolism; metabolism described using *in vitro* M–M metabolic parameters.
Methoxyflurane	Fisherova- Bergova (1992)	Human	Extrahepatic metabolism.
SUN5555 (antibiotic)	Tsuji *et al.* (1990)	Rat, dog human	Extrahepatic metabolism in several tissues including kidney, lung and gut.
Halothane (CNS acting anesthetic)	Williams *et al.* (1996)	Rat, human	Development of a human model for HCFC-123 and its metabolite, trifluoroacetic acid, using kinetic information on a structural analogue, halothane.
Cafalozin	Tsuji *et al.* (1985)	Rabbit, human	Saturable renal clearance.
Morphine	Obrien *et al.* (1996)	Rat	Model of an isolated perfused rat liver to examine biliary extraction ratios and effects of altered perfusate flow on the kinetics of morphine and its metabolites.
	Evan *et al.* (1999)		Extension of model by O'Brien *et al.* (1996) to describe cellular efflux kinetics into the sinusoidal and canalicular spaces of the liver.
AZT, antiretroviral agent	Chow (1997)	Mouse, human	Extensive biotransformation (glucuronidation) in humans and urinary excretion of glucuronidated species.
Capecitabine	Blesch *et al.* (2003)	Human	Hepatic and extrahepatic saturable metabolism.
Warfarin (anticoagulant)	Gentry *et al.* (2002)	Rat, human	Use of Monte Carlo simulation to quantitatively estimate tissue dose in human populations with different polymorphisms in genes encoding for the responsible metabolizing enzymes.
Nicotine, cotinine	Robinson *et al.* (1992)	Human	Nine-compartment PBPK models were developed for nicotine and cotinine separately, and then linked by the liver compartments so that the pharmacokinetics of the metabolite cotinine could be predicted based on administered nicotine.

10.4 OTHER ISSUES IN MODEL DEVELOPMENT FOR DRUGS

10.4.1 Altered Physiological States

PBPK models can be used to predict the disposition of a drug under altered physiological conditions, such as different disease states. Table 10.3 provides a summary of some of the articles that use PBPK modeling approaches for this purpose. For example, an episode of cardiac arrest dramatically alters hemodynamics and cardiac output. The kinetics of lidocaine, an anti-arrhythmic drug that is administered during cardiac arrest, were predicted under conditions such as hemorrhaging using a perfusion-limited PBPK model that altered physiological parameters such as the extent of mixing in the tissue compartments and the blood flow rate (Benowitz *et al.* 1974a,b). Another PBPK model for lidocaine was developed to specifically evaluate different dosing regimens and determine the optimal delivered dose to the patient. In this flow-limited model, blood flow parameters were altered to account for the physiological changes that occur during cardiac arrest (Benowitz *et al.* 1974a,b; Grillo *et al.* 2001). The influence of different basal cardiac output levels among different human populations on the kinetic disposition of a drug was also examined with thiopental using a PBPK model. The role of other factors such as age, gender, and obesity on the PK profile of thiopental were also considered. Of all these factors, it was determined that cardiac output had the most influence on the variability among human subjects (Wada *et al.* 1997). A similar evaluation has also been performed for the opiates, fentanyl and alfentanyl (Bjorkman *et al.* 1998). Renal failure is another common altered physiological state in patients receiving drugs. By adjusting the appropriate clearance parameters, a PBPK model successfully described the pharmacokinetic disposition of digoxin in ureter-ligated rats (Harrison and Gibaldi 1997a) and predicted the effects of renal failure on the kinetics in a dog (Harrison and Gibaldi 1977b). Tsuji *et al.* (1985) also used a PBPK model to estimate the disposition of cefalozin in rabbits with renal failure.

Weightlessness caused by extended time in orbit induces several physiological changes that can alter drug disposition. Since it is not feasible to study the pharmacokinetics of multiple drugs under these conditions using experimental methods, PBPK modeling is a perfect tool to examine the effectiveness of a drug under these altered conditions. To illustrate the utility of PBPK modeling for this purpose, a preliminary model that described the pharmacokinetics of acetaminophen in patients in low gravity focused on the effects of changes in physiological parameters such as blood volume and absorption (Srinivasan *et al.* 1994).

There are situations where the PD effects of a drug alter the drug pharmacokinetics. For example, many CNS drugs act to depress the CNS. This pharmacological effect alters other physiological processes, such as blood flow and cardiac output. PBPK models that examine the impact of these physiological changes on the pharmacokinetics of CNS-active drugs indicate the importance of considering these processes in PBPK model development (Nagata *et al.* 1990; Blakey *et al.* 1997). In a related example, inaperisone, a muscle relaxant, caused dose-dependent changes in blood flow in target tissues. The physiological changes that occur due to

TABLE 10.3 PBPK Models for Drugs Incorporating Altered Physiological States

Drug	Reference	Species	Comments
Lidocaine	Benowitz *et al*. (1974a,b)	Monkey, human	Pharmacokinetics under conditions of blood loss and in combination with sympathomimetic drugs.
	Grillo *et al*. (2001)	Human	Disposition during cardiac arrest using various dosing regimens to determine optimal therapy.
Acetominophen	Srinivasan *et al*. (1994)	Human	Assess drug disposition in low gravity (incorporates physiological changes associated with weightlessness).
Digoxin (cardiac agent)	Harrison and Gibaldi (1977b)	Rat	Comparison of kinetics in normal and bile-duct ligated rat; assess kinetic changes due to renal failure.
	Harrison and Gibaldi (1977a)	Dog	Modification of rat model (1977) to account for species differences.
Cafalozin	Tsuji *et al*. (1985)	Rabbit, human	Altered physiological parameters due to renal failure.
Inaperisone	Nagata *et al*. (1990)	Rat	Dose-dependent changes in blood flow rates due to drug concentrations at the site of action.
Thiopental	Wada *et al*. (1997)	Human	Changes in pharmacokinetics of thiopental due to altered physiological conditions.
Fentanyl, alfentanil	Bjorkman *et al*. (1998)	Human	Changes in drug disposition due to altered physiological conditions.

alteration of blood flow need to be accounted for in the PBPK model description of this drug (Nagata *et al.* 1990).

10.4.2 Drug Stereospecificity

The majority of synthetic drugs are chiral. Pharmacological action and kinetic disposition of similar compounds in the body can be different depending on the stereochemistry of the drug. To examine the role of chirality in drug disposition, Piotrovskij *et al.* (1994) developed a generic PBPK model to explore PK differences that might arise due to drug enantioselectivity. They used the model to predict changes in PK disposition due to drug–drug interactions, such as in the condition of administration of an enantiomeric mixture. Protein binding and drug metabolism parameters were the most important factors in determining the overall enantioselectivity of a drug.

Enantiomers can have differences in several important factors that determine drug disposition and time course, including protein binding, transport mechanisms, and metabolism. Racemic drugs bind to various proteins, such as specific receptors, plasma proteins, enzymes, and so on. PBPK models may also have to account for interconversion of enantiomers in the body. Albumin, for instance, has racemase activity *in vitro* (Shimada *et al.* 1993; Aso *et al.* 1990). PBPK models have been developed that mechanistically examine the role albumin has in isomerization and epimerization of chiral antibiotics and the consequent altered PK profiles of the different enantiomers. In general, the rate of isomerization of a drug is highly correlated with the percent of binding of the drug to albumin and contributes to the overall disposition of the drug in the body (Shimada *et al.* 1993). A PBPK model used to describe epimerization of the antibiotic moxalactam in dogs determined that drug binding to albumin in the serum accurately reflected the observed epimerization of the drug (Yoshikawa *et al.* 1999). These models demonstrated their predictive value in explaining the stereopharmacokinetics of a drug. Table 10.4 summarizes PBPK models that emphasize enantioselectivity.

10.4.3 Non-Steady-State Dynamics

Computer-controlled infusion pumps are widely used in anesthesiology to control the delivery of *i.v.* anesthetics prior to and during surgery. When these pumps are used, the approach to steady state of drug in plasma is generally assumed to be rapid. Yet, transient changes in blood concentrations are commonly encountered when administering inhaled and injected anesthesia because anesthesiologists are constantly adjusting dosing during surgery. As observed with the opiates fentanyl and alfentanil, flow-limited, steady-state PBPK models may not adequately describe initial uptake processes (Bjorkman *et al.* 1998). This type of model failure is especially evident during periods of rapid transient changes, such as the first few minutes of initial tissue uptake. Since recirculation of blood takes approximately one minute, transient side effects can occur during initial uptake and throughout the entire drug delivery time period even though plasma concentrations remain in the therapeutic range. To investigate the influence of these infusion pumps on plasma time course of drug, PBPK models have been used to examine transient changes in plasma concentrations (Wada and Ward 1994), and in target tissue and side effect site concentrations (Wada and Ward 1995) during use of a computer-controlled infusion pump. Equilibrium of a drug within a compartment cannot occur instantaneously. The PBPK models that provide a detailed description of the kinetic and biological processes that occur in tissues during these rapid transient changes have been useful in providing optimal therapy. Table 10.5 highlights some of the models developed to investigate non-steady-state dynamics. These models provide examples of specific applications of PBPK modeling to determine the recirculation dynamics that could lead to adverse side effects (Wada *et al.* 1997; Ebling *et al.* 1994).

10.4.4 Drug Interactions

Combination drug therapy is a common approach to treating many disease states. In addition, individuals are often treated for more than a single condition and take multiple drugs. Serious toxic side effects have occurred due to adverse drug interactions. Several PBPK models successfully predict PK behavior of drugs under conditions where interactions occur (Table 10.6). Two PBPK models were developed to evaluate the time courses of 2′,3′-dideoxyinosine (ddI) and pentamidine in AIDS patients where these drugs were co-administered. The model describing the kinetics of ddI and pentamidine included increased partitioning of ddI in the pancreas and muscle compartments compared to the model that described only ddI. These rat models were scaled-up to humans to determine which model more accurately estimated the disposition of these drugs in patients. Similar to rats, the time-course concentrations in patients receiving these drugs was better estimated with the model that included altered partitioning due to drug interactions (Kang *et al.* 1997).

Most drug interactions occur due to altered clearance of the drugs from the body by competitive inhibition of metabolic enzymes (e.g., as would occur when the drugs are metabolized by the same enzyme). When these interactions occur with one drug influencing the time course of another due to these interactions, the PBPK models need to include terms in the metabolism equations that account for metabolic inhibition. *In vitro* metabolic inhibition data obtained from liver microsomes have been used for extrapolation to *in vivo* PBPK models to describe competitive metabolic inhibition for some combinations of drugs, including triazolam by erythromycin (Kanamitsu *et al.* 2000) and simavastatin by itraconazole (Ishigam *et al.* 2001). Metabolites of a parent compound can also act as metabolic inhibitors if subsequent metabolism of the metabolites occurs. Drugs that have an altered kinetic disposition due to this phenomenon, such as ethoxybenzamide, can be described with a PBPK model that accounts for this inhibition in a similar way to describing inhibition due to a drug–drug interaction (Lin *et al.* 1984). Plasma protein binding can also be dramatically affected due to drug–drug interactions, such as in the case of altered plasma concentrations of tolbutamide in rats in the presence of sulfonamide (Sugita *et al.* 1982).

Co-administration of certain drugs can alter transport of drugs from compartments. For example, co-administration of cancer drugs with digoxin can cause inhibition of the protein, P-glycoprotein. In an interesting application of PBPK modeling, Kawahara *et al.* (1999) developed a PBPK model utilizing data from a

TABLE 10.4 PBPK Models Describing Drug Stereospecificity

Drug	Reference	Species	Comments
Moxalactam	Yoshikawa *et al.* (1999)	Rat, dog, human	Description of the effects of albumin on epimerization of moxalactam.
Ceftibuten antibiotic	Shimada *et al.* (1993)	Rat	Model uses Tsuji *et al.* (1983) model to determine the extent to which albumin contributes to isomerization of ceftibuten.
Various	Piotrovskij *et al.* (1994)		Several PBPK models used to predict pharmacokinetics of drug enantiomers.

TABLE 10.5 PBPK Models Describing Kinetics of Drugs under Non Steady-state Conditions

Drug	Reference	Species	Comments
Fentanyl, alfentanil	Bjorkman *et al.* (1998)	Human	Initial uptake processes when transient changes in blood concentrations are occurring; diffusion-limited model.
Alfentanil	Wada and Ward (1994)	Human	Description of a computer controlled infusion pump to predict transient changes in drug concentrations and recirculation dynamics.
	Wada and Ward (1995)		Extension of (1994) model to side effects due to transient changes in plasma.
Thiopental	Ebling *et al.* (1994)	Rat	Short-term, transient changes in blood concentrations of the drug.
	Wada *et al.* (1997)	Human	Diffusion-limited tissue uptake.

TABLE 10.6 PBPK Models Developed for Drug–Drug Interactions

Drug	Reference	Species	Comments
Tolbutamide, sulfonamide	Sugita *et al.* (1981)	Rat	Effects of sulfonamide on the elimination of tolbutamide by examining differences in total body clearance. Displacement of plasma protein binding versus metabolic inhibition is examined.
	Sugita *et al.* (1982)		Further development of the (1981) Model describing specific interactions of tolbutamide with sulfonamide. These interactions include plasma protein interactions, binding, metabolic inhibition.
2′,3′-Dideoxyinosine	Kang *et al.* (1997)	Rat, human	Interactions of dideoxyinosine with pentamidine and uses the model to scale-up to humans. Extrahepatic metabolism is incorporated into the model.
Triazolam	Kanamitsu *et al.* (2000)	Human	Use of *in vitro* metabolic data to predict *in vivo* interactions of triazolam with erythromycin by including enzyme inhibition.
Simavastatin	Ishigam *et al.* (2001)	Human	Using *in vitro* data to predict *in vivo* interactions on metabolic enzyme inhibition; interations of itraconazole on the PK disposition of simavastatin.
Lidocaine	Benowitz *et al.* (1974b)	Monkey	This lidocaine phamacokinetics in combination with sympathomimetic drugs.
Digoxin	Kawahara *et al.* (1999)	Mouse	Modification of model by Harrison and Gibaldi (1977b) to study the role of *p*-glycoprotein inhibition due to co-administration of anticancer agents with digoxin.
Ethoxybenzamide	Lin *et al.* (1984)	Rat, rabbit, human	Product inhibition to account for slower than expected clearance.
Bromosulfophthalein, warfarin	Luecke and Wosilait (1979)	Rat	The magnitude of drug elimination interactions between bromosulfophthalein and warfarin were predicted with a PBPK model to test an interaction model.

P-glycoprotein knockout mouse (mdr 1a –/–) to simulate inhibition of intercompartmental drug transport of P-glycoprotein. They initially used an existing PBPK model for digoxin (Harrison and Gibaldi 1977b) and adjusted model parameters to simulate loss of the transporter. Total body clearance and values of $P_{t:p}$ for heart and muscle were decreased in the mdr 1a (–/–) mice. In contrast the brain concentrations were 68-fold higher in mdr 1a(–/–) versus mdr 1a(+/+). To account for these differences, the brain compartment consisted of a well-mixed compartment and a deep tissue compartment referred to as an accumulative part. Active efflux from the accumulative part of the brain was zero in mdr 1a (–/–) mice.

10.4.5 Utilization of *In Vitro* Data

In vivo PK studies are costly and time-consuming. Early stages of drug discovery can be accelerated if it were possible to use *in vitro* data to predict the *in vivo* PK profiles of drug candidates prior to *in vivo* studies. Biochemical parameters for absorption, distribution, metabolism, and elimination (ADME) estimated *in vitro* can be integrated into PBPK models to determine kinetic disposition of a potential drug candidate during the early drug discovery process. These parameters include metabolic clearance (Kanamitsu *et al.* 2000; Lin *et al.* 1982, 1984; Ishigam *et al.* 2001), protein binding constants (Engasser *et al.*, 1981; Ichimura *et al.* 1985), and tissue partitioning (Mintun *et al.* 1980; Poulin and Theil 2000). These methodologies support more rational decision-making during selection of clinical drug candidates because they provide hypothesis driven approaches to developing relationships between the pharmacokinetics and pharmacodynamics of a drug candidate. This approach also aids in more accurate extrapolation to humans (Theil *et al.* 2003). Several of the models previously discussed in this chapter have used this approach (Kanamitsu *et al.* 2000; Ishigam *et al.* 2001; Mintum *et al.* 1980; Engasser *et al.* 1981; Igari *et al.* 1982; Igari *et al.* 1983).

To illustrate the use and reliability of integrating *in vitro* parameters into a predictive PBPK model, Poulin and Theil (2002b) integrated *in vitro* parameters into three generic PBPK models to estimate time-course behaviors of two lipophilic, basic drugs (diazepam, propranolol) and one neutral, hydrophilic drug (ethoxybenzamide) in plasma and in 10 different tissue compartments. *In vitro* estimated values of $P_{t:p}$ and *in vivo* intrinsic clearance rates that were estimated from studies in hepatocytes were used to predict distribution and metabolism of the drugs. Using this generic approach, plasma and most tissue compartments were reliably estimated. However, tissues that had more complex distribution and absorption behaviors, such as the brain, liver, and gut, could not be adequately described without modifications to the generic models. This approach highlights the value of PBPK models as tools to generate further hypotheses to understand the mechanisms regulating the disposition of drug candidates in those tissues where there are deviations from the expected time-course profiles.

While *in vitro* data are regularly used to develop PBPK models, it is important to consider the potential limitations of linking information from an isolated system to an integrated system (Mintun *et al.* 1980). Table 10.7 summarizes drugs

TABLE 10.7 PBPK Models Emphasizing *In Vitro* to *In Vivo* Extrapolation

Drug	Reference	Species	Comments
Ethoxybenzamide	Lin *et al.* (1982)	Rat, rabbit, human	Use of *in vitro* parameters to estimate *in vivo* time course of drug in plasma.
	Lin *et al.* (1984)		Extension of 1982 model to include product inhibition to account for slower than expected clearance.
Triazolam	Kanamitsu *et al.* (2000)	Human	*In vitro* metabolic data.
Simavastatin	Ishigam *et al.* (2001)	Human	Interaction with itraconazole, resulting in enzyme inhibition; *in vitro* metabolic data to predict *in vivo* interactions.
Phenobarbitol	Engasser *et al.* (1981)	Rat	Protein binding.
Thiopental	Bishoff and Dedrick (1968)	Human	*In vitro* metabolic parameters
Barbiturates	Poulin and Theil (2000a)	Rat	*In vitro* estimation of $P_{t:p}$.
Tetraethylammonium ion	Mintun *et al.* (1980)	Rat	*In vitro* estimation of $P_{t:p}$.
Diazepam, propanolol, ethoxybenzamide	Poulin and Theil (2002b)	Rat	A generic PBPK model using *in vitro* estimated clearance rates and $P_{t:p}$ values to simulate time-course concentrations of the drugs *a priori*.
Valproic acid	Ichimura (1985)	Rabbit	Protein binding.

where research has used *in vitro* parameters to develop PBPK models to describe *in vivo* ADME data.

10.5 FUTURE PERSPECTIVES

In 2002, a workshop, "Physiologically Based Pharmacokinetics in Drug Development and Regulatory Science," was held in Washington, D.C. Participants from pharmaceutical companies, academia and governmental organizations discussed the current contributions, future potential, and intrinsic limitations of PBPK modeling for drug development and discovery. An article on the outcome of the meeting (Rowland *et al.* 2004) provides a balanced view of the current state of PBPK modeling for pharmaceutical compounds. Some of the challenges in continuing growth of PBPK modeling in the pharmaceutical industry are nicely captured in a recent review, where Nestorov (2003) noted that "a substantial body of PBPK related research has been accumulated that can facilitate the PBPK modeling and implementation process. What is probably lagging behind is the expertise component, where the demand for appropriately qualified staff outreaches availability." Synergies may be possible to draw on the more rapid growth of expertise in PBPK modeling by environmental health scientists to synergize and augment development of internal expertise in the pharmaceutical industries.

10.6 SUMMARY

The field of PBPK modeling was initially developed through interest in the mechanistic determinants that govern the delivery of drugs to target sites and their persistence at these sites. Early work with anesthetic agents paved the way for the later applications with multiple types of drugs, especially chemotherapeutic compounds beginning in the 1960s. Nonetheless, PBPK modeling within the pharmaceutical industry has not become widespread. While PBPK model development is clearly useful in the drug discovery process, Nestorov (2003) explained that PBPK modeling approaches are not more widely used because they typically require "the investment of significant experience, effort, time and resources." Despite more limited use of PBPK modeling with drugs compared to environmental compounds, there is still a significant and rich body of work describing applications of PBPK models to investigate specific properties of drugs that affect their disposition within the body. The PBPK modeling applications highlighted in this chapter investigated (1) physicochemical factors that influence drug distribution, (2) processes that control drug clearance , (3) integration of *in vitro* data into PBPK models, (4) interspecies extrapolation from rodent to humans, (5) transient changes in tissue concentrations with intravenous infusion, (6) alterations in physiological or biochemical parameters within the model structure, and (7) drug–drug interactions on distribution, metabolism, and clearance.

NOTATION

ADME	absorption, distribution, metabolism, and elimination
CNS	central nervous system
HCFC-123	2,2-dicloro-1,1,1-trifluoroethane
i.v.	intravenous
K_m	binding affinity of a drug to an enyme
M–M	Michaelis–Menten
PBPK	physiologically based pharmacokinetic
PEMA	phenylethylmalonamide
PD	pharmacodynamic
PK	pharmacokinetic
$P_{t:p}$	tissue-to-plasma partition coefficient
$V_{\max}$	maximal velocity of an enzyme

REFERENCES

Aso, Y., Yoshioka. S., and Takeda, Y. (1990). Epimerization and racemization of some chiral drugs in the presence of human serum albumin. *Chem. Pharm. Bull. (Tokyo).* **38**, 180–184.

Benowitz, N., Forsyth, R. P., Melmon, K. L., and Rowland, M. (1974a). Lidocaine disposition kinetics in monkey and man. 1. Prediction by a perfusion model. *Clin. Pharmacol. Ther.* **16**, 87–98.

Benowitz, N., Forsyth, R. P., Melmon, K. L., and Rowland, M. (1974b). Lidocaine disposition kinetics in monkey and man. 2. Effects of hemorrhage and sympathomimetic drug administration. *Clin. Pharmacol. Ther.* **16**, 99–109.

Bernareggi, A., and Rowland, M. (1991) Physiologic modeling of cyclosporin kinetics. *J. Pharmacokinet. Biopharm.* **19**, 21–50.

Bischoff, K. B., and Dedrick, R. L. (1968). Thiopental pharmacokinetics. *J. Pharm. Sci.* **57**, 1346–1351.

Bjorkman, S. (2003). Reduction and lumping of physiologically based pharmacokinetic models: prediction of the disposition of fentanyl and pethidine in humans by successively simplified models. *J. Pharmacokinet. Pharmacodyn.* **30**, 285–307.

Bjorkman, S., Wada, D. R., and Stanski, D. R. (1998). Application of physiologic models to predict the influence of changes in body composition and blood flows on the pharmacokinetics of fentanyl and alfentanil in patients. *Anesthesiology* **88**, 657–667.

Bjorkman, S., Wada, D. R., Berling, B. M., and Benoni, G. (2001). Prediction of the disposition of midazolam in surgical patients by a physiologically based pharmacokinetic model. *J. Pharm. Sci.* **90**, 1226–1241.

Blakey, G. E., Nestorov, I. A., Arundel, P. A., Aarons, L. J., and Rowland, M. (1997). Quantitative structure–pharmacokinetics relationships: I. Development of a whole-body physiologically based model to characterize changes in pharmacokinetics across a homologous series of barbiturates in the rat. *J. Pharmacokinet. Biopharm.* **25**, 277–312.

Blesch, K. S., Gieschke, R., Tsukamoto, Y., Reigner, B. G., Burger, H. U., and Steimer, J. L. (2003). Clinical pharmacokinetic/pharmacodynamic and physiologically based pharmacokinetic modeling in new drug development: The capecitabine experience. *Invest. New Drugs* **21**, 195–223.

Chen, C. N., and Andrade, J. D. (1976). Pharmacokinetic model for simultaneous determination of drug levels in organs and tissues. *J. Pharm. Sci.* **65**, 717–724.

Ebling, W. F., Wada, D. R., and Stanski, D. R. (1994). From piecewise to full physiological pharmacokinetic modeling—applied to thiopental disposition in the rat. *J. Pharmacokinet. Biopharm.* **22**, 259–292.

El-Masri, H. A., and Portier, C. J. (1998). Physiologically based pharmacokinetic model of primidone and its metabolites phenobarbital and phenylethylmalonamide in humans, rats, and mice. *Drug Metab. Dispos.* **26**, 585–594.

Engasser, J. M., Sarhan, F., Falcoz, C., Minier, M., Letourneur, P., and Siest, G. (1981). Distribution, metabolism, and elimination of phenobarbital in rats—physiologically based pharmacokinetic model. *J. Pharm. Sci.* **70**, 1233–1238.

Evans, A. M., O'Brien, J., and Nation, R. L. (1999). Application of a loading wash-out method for investigating the hepatocellular efflux of a hepatically-generated metabolite morphine-3-glucuronide. *J. Pharm. Pharmacol.* **51**, 1289–1297.

Fiserova-Bergerova, V. (1992). Inhalation anesthesia using physiologically based pharmacokinetic models. *Drug Metab. Rev.* **24**, 531–557.

Gentry, P., Hack, C., Haber, L., Maier, A., and Clewell III, H. (2002). An approach for the quantitative consideration of genetic polymorphism data in chemical risk assessment: Examples with warfarin and parathion. *Toxicol. Sci.* **70**, 120–129.

Greene, D. S., Quintiliani, R., Nightingale, C. H. (1978). Physiological perfusion model for cephalosporin antibiotics I: Model selection based on blood drug concentrations. *J. Pharm. Sci.* **67**, 191–196.

Grillo, J. A., Venitz, J., and Ornato, J. P. (2001). Prediction of lidocaine tissue concentrations following different dose regimes during cardiac arrest using a physiologically based pharmacokinetic model. *Resuscitation* **50**, 331–340.

Harashima, H., Sawada, Y., Sugiyama, Y., Iga, T., and Hanano, M. (1985). Analysis of nonlinear tissue distribution of quinidine in rats by physiologically based pharmacokinetics. *J. Pharmacokinet. Biopharm.* **13**, 425–440.

Harashima, H., Sawada, Y., Sugiyama, Y., Iga, T., and Hanano, M. (1986). Prediction of serum concentration–time course of quinidine in human using a physiologically based pharmacokinetic model developed from the rat. *J. Pharmacobio-Dynam.* **9**, 132–138.

Harrison, L. I., and Gibaldi, M. (1977a). Physiologically based pharmacokinetic model for digoxin disposition in dogs and its preliminary application to humans. *J. Pharm. Sci.* **66**, 1679–1683.

Harrison, L. I., and Gibaldi, M. (1977b). Physiologically based pharmacokinetic model for digoxin distribution and elimination in rat. *J. Pharm. Sci.* **66**, 1138–1142.

Ichimura, F., Deguchi, Y., Yokogawa, K., and Yamana, T. (1985). Physiologically based pharmacokinetics of valproic acid in rabbits. *Int. J. Pharm.* **27**, 45–60.

Igari, Y., Sugiyama, Y., Awazu, S., and Hanano, M. (1982). Comparative physiologically based pharmacokinetics of hexobarbital, phenobarbital, and thiopental in the rat. *J. Pharmacokinet. Biopharm.* **10**, 53–75.

Igari, Y., Sugiyama, Y., Sawada, Y., Iga, T., and Hanano, M. (1983). Prediction of diazepam disposition in the rat and man by a physiologically based pharmacokinetic model. *J. Pharmacokinet. Biopharm.* **11**, 577–593.

Ishigam, M., Uchiyama, M., Kondo, T., Iwabuchi, H., Inoue, S., Takasaki, W., Ikeda, T., Komai, T., Ito, K., and Sugiyama, Y. (2001). Inhibition of *in vitro* metabolism of simvastatin by itraconazole in humans and prediction of *in vivo* drug–drug interactions. *Pharm. Res.* **18**, 622–631.

Kanamitsu, S., Ito, K., Green, C., Tyson, C., Shimada, N., and Sugiyama, Y. (2000). Prediction of *in vivo* interaction between triazolam and erythromycin based on *in vitro* studies using human liver microsomes and recombinant human Cyp3a4. *Pharm. Res.* **17**, 419–426.

Kang, H. J. K., Wientjes, M. G., and Au, J. L. S. (1997). Physiologically based pharmacokinetic models of 2′,3′-dideoxyinosine. *Pharm. Res.* **14**, 337–344.

Katayama, K., Ohtani, H., Murai, A., Kakemi, M., and Koizumi, T. (1990). Kinetic studies on drug disposition in rabbits. II. Dose dependent pharmacokinetics of sulfamethizole. *J. Pharmacobio-Dynam.* **13**, 108–119.

Kawahara, M., Sakata, A., Miyashita, T., Tami, I., and Tsuji, A. (1999). Physiologically based pharmacokinetics of digoxin in mdr1a knockout mice. *J. Pharm. Sci.* **88**, 1281–1287.

Kawai, R., Lemaire, M., Steimer, J. L., Bruelisauer, A., Niederberger, W., and Rowland, M. (1994). Physiologically-based pharmacokinetic study on a cyclosporine derivative, Sdz Imm-125. *J. Pharmacokinet. Biopharm.* **22**, 327–365.

Kawai, R., Mathew, D., Tanaka, C., and Rowland, M. (1998). Physiologically based pharmacokinetics of Cyclosporine A: Extension to tissue distribution kinetics in rats and scale-up to human. *J. Pharmacol. Exp. Ther.* **287**, 457–468.

Levitt, D. (2003).The pharmacokinetics of the interstitial space in humans. *BMC Cli. Pharmacol.* **3**, 3.

Lin, J. H., Sugiyama, Y., Awazu, S., and Hanano, M. (1982). Physiological pharmacokinetics of ethoxybenzamide based on biochemical data obtained *in vitro* as well as on physiological data. *J. Pharmacokinet. Biopharm.* **10**, 649–661.

Lin, J. H., Sugiyama, Y., Hanano, M., and Awazu, S. (1984). Effect of product inhibition on elimination kinetics of ethoxybenzamide in rabbits—Analysis by physiological pharmacokinetic model. *Drug Metab. Dispos.* **12**, 253–256.

Luecke, R. H., and Wosilait, W. D. (1979). Drug elimination interactions: Analysis using a mathematical model. *J. Pharmacokinet. Biopharm.* **7**, 629–641.

Mintun, M., Himmelstein, K. J., Schroder, R. L., Gibaldi, M., and Shen, D. D. (1980). Tissue distribution kinetics of tetraethylammonium ion in the rat. *J. Pharmacokinet. Biopharm.* **8**, 373–409.

Nagata, O., Murata, M., Kato, H., Terasaki, T., Sato, H., and Tsuji, A. (1990). Physiological pharmacokinetics of a new muscle-relaxant, inaperisone, combined with its pharmacological effect on blood flow rate. *Drug Metab. Dispos.* **18**, 902–910.

Nakajima, Y., Hattori, K., Shinsei, M., Matsunaga, N., Iizasa, H., Sasabe, H., Akiyama, H., Miyamoto, G., and Nakashima, E. (2000). Physiologically-based pharmacokinetic analysis of grepafloxacin. *Biol. Pharmacol. Bull.* **23**, 1077–1083.

Nestorov, I. (2003). Whole body pharmacokinetic models. *Clin. Pharmacokinet.* **42**, 883–908.

Nestorov, I. A., Aarons, L. J., and Rowland, M. (1997). Physiologically based pharmacokinetic modeling of a homologous series of barbiturates in the rat: A sensitivity analysis. *J. Pharmacokinet. Biopharm.* **25**, 413–447.

Nestorov, I., Aarons, L., and Rowland, M. (1998). Quantitative structure–pharmacokinetics relationships: II. A mechanistically based model to evaluate the relationship between tissue distribution parameters and compound lipophilicity. *J. Pharmacokinet. Biopharm.* **26**, 521–545.

O'Brien, J. A., Nation, R. L., and Evans, A. M. (1996). The disposition of morphine and morphine-3-glucuronide in the isolated perfused rat liver: Effects of altered perfusate flow. *J. Pharm. Pharmacol.* **48**, 498–504.

Oliver, R., Jones, A., and Rowland, M. (2001). A whole-body physiologically based pharmacokinetic model incorporating dispersion concepts: Short and long time characteristics. *J. Pharmacokinet. Biopharm.* **28**, 27–55.

Piotrovskij, V. K., Shvatchko, E. V., and Trnovec, T. (1994). The use of physiologically-based models to simulate enantioselective differences in pharmacokinetics. *Methods Find. Exp. Clin. Pharmacol.* **16**, 263–269.

Plowchalk, D., and Teeguarden, J. (2002). Development of a physiologically based pharmacokinetic model for estradiol in rats and humans: A biologically motivated quantitative framework for evaluating responses to estradiol and other endocrine-active compounds. *Toxicol. Sci.* **69**, 60–78.

Plowchalk, D., Andersen, M., and deBethizy, J. (1992). A physiologically based pharmacokinetic model for nicotine disposition in the Sprague–Dawley rat. *Toxicol. Appl. Pharmacol.* **116**, 177–188.

Poulin, P., and Theil, F. P. (2000). A priori prediction of tissue:plasma partition coefficients of drugs to facilitate the use of physiologically-based pharmacokinetic models in drug discovery. *J. Pharm. Sci.* **89**, 16–35.

Poulin, P., and Theil, F. (2002a). Prediction of pharmacokinetics prior to *in vivo* studies. 1. Mechanism-based prediction of volume of distribution. *J. Pharm. Sci.* **91**, 129–156.

Poulin, P., and Theil, F. P. (2002b). Prediction of pharmacokinetics prior to *in vivo* studies. II. Generic physiologically based pharmacokinetic models of drug disposition. *J. Pharm. Sci.* **91**, 1358–1370.

Poulin, P., Schoenlein, K., and Theil, F. P. (2001). Prediction of adipose tissue:plasma partition coefficients for structurally unrelated drugs. *J. Pharm. Sci.* **90**, 436–447.

Robinson, D. E., Balter, N. J., and Schwartz, S. L. (1992). A physiologically based pharmacokinetic model for nicotine and cotinine in man. *J. Pharmacokinet. Biopharm.* **20**, 591–609.

Rowland, M., Balant, L., and Peck, C. (2004). Physiologically based pharmacokinetics in drug development and regulatory science: A workshop report (Georgetown University, Washington, D.C., May 29–30, 2002). *AAPS PharmSci.* **6**, e6.

Shimada, J., Hori, S., Oguma, T., Yoshioka, T., Yamamoto, S., Nishikawa, T., and Yamada, H. (1993). Effects of protein binding on the isomerization of ceftibuten. *J. Pharm. Sci.* **82**, 461–465.

Smith, R. H., Hunt, D. H., Seifen, A. B., Ferrari, A., Thompson, D. S. (1979). Pharmacokinetic model for procaine in humans during and following intravenous infusion. *J. Pharm. Sci.* **68**, 1016–1022.

Srinivasan, R. S., Bourne, D. W. A., and Putcha, L. (1994). Application of physiologically-based pharmacokinetic models for assessing drug disposition in space. *J. Clin. Pharmacol.* **34**, 692–698.

Sugita, O., Sawada, Y., Sugiyama, Y., Iga, T., and Hanano, M. (1982). Physiologically based pharmacokinetics of drug–drug interaction—a study of tolbutamide–sulfonamide interaction in rats. *J. Pharmacokinet. Biopharm.* **10**, 297–316.

Tanaka, C., Kawai, R., and Rowland, M. (1999). Physiologically based pharmacokinetics of Cyclosporine A: Reevaluation of dose-nonlinear kinetics in rats. *J. Pharmacokinet Biopharm.* **27**, 597–623.

Theil, F. P., Guentert, T. W., Haddad, S., and Poulin, P. (2003). Utility of physiologically based pharmacokinetic models to drug development and rational drug discovery candidate selection. *Toxicol. Lett.* **138**, 29–49.

Tsuji, A., Miyamoto, E., Terasaki, T., and Yamana, T. (1979). Physiological pharmacokinetics of beta-lactam antibiotics—Penicillin-v distribution and elimination after intravenous administration in rats. *J. Pharm. Pharmacol.* **31**, 116–119.

Tsuji, A., Yoshikawa, T., Nishide, K., Minami, H., Kimura, M., Nakashima, E., Terasaki, T., Miyamoto, E., Nightingale, C. H., and Yamana, T. (1983). Physiologically based pharmacokinetic model for beta-lactam antibiotics. I. Tissue distribution and elimination in rats. *J. Pharm. Sci.* **72**, 1239–1252.

Tsuji, A., Nishide, K., Minami, H., Nakashima, E., Terasaki, T., and Yamana, T. (1985). Physiologically based pharmacokinetic model for cefazolin in rabbits and its preliminary extrapolation to man. *Drug Metab. Dispos.* **13**, 729–739.

Tsuji, A., Sato, H., Tamai, I., Adachi, H., Nishihara, T., Ishiguro, M., Ohnuma, N., and Noguchi, T. (1990). Physiologically based pharmacokinetics of a new Penem, Sun5555, for evaluation of *in vivo* efficacy. *Drug Metab. Dispos.* **18**, 245–252.

Wada, D. R., and Ward, D. S. (1994). The Hybrid Model—A new pharmacokinetic model for computer-controlled infusion pumps. *IEEE Trans. Biomed. Eng.* **41**, 134–142.

Wada, D. R., and Ward, D. S. (1995). Open-loop control of multiple-drug effects in anesthesia. *IEEE Trans. Biomed. Eng.* **42**, 666–677.

Wada, D. R., Bjorkman, S., Ebling, W., Harashima, H., Harapat, S., and Stanski, D. R. (1997). Computer simulation of the effects of alterations in blood flows and body composition on thiopental pharmacokinetics in humans. *Anesthesiology* **87**, 884–899.

Williams, R. J., Vinegar, A., McDougal, J. N., Jarabek, A. M., and Fisher, J. W. (1996). Rat to human extrapolation of HCFC-123 kinetics deduced from halothane kinetics: A corollary approach to physiologically based pharmacokinetic modeling. *Fundam. Appl. Toxicol.* **30**, 55–66.

Yoshikawa, T., Oguma, T., Ichihashi, T., Kinoshita, H., Hirano, K., and Yamada, H. (1999). Epimerization of moxalactam by albumin and simulation of *in vivo* epimerization by a physiologically based pharmacokinetic model. *Chirality* **11**, 309–315.

CHAPTER 11

ANTINEOPLASTIC AGENTS

Wendy Pott O'Brien

11.1 INTRODUCTION

The development of physiologically based pharmacokinetic (PBPK) models for antineoplastic agents arose from the need to balance therapeutic effect with toxicity—a need present for all pharmaceutical compounds, to some extent, but particularly important for antineoplastic agents because of their very narrow therapeutic margin. The fundamental issues governing drug toxicity and therapeutic effect include the physical aspects of drug distribution throughout the body and the effect of the drug on vulnerable targets (e.g., tumors or target organs of toxicity). With PBPK models, the dynamics of drug distribution can be predicted using basic information on physicochemical properties, transport, biotransformation, and excretion. PBPK models facilitate estimation of internal target tissue dose and lead to a better understanding of target tissue therapeutics and/or toxicity. PBPK models can account for interspecies and intraspecies differences with respect to drug distribution and metabolism, and they allow prediction of target tissue concentrations as a function of dose, dose schedule, and route of administration. In clinical applications, PBPK models can enhance the development of optimum treatment regimens; in the development of new antineoplastic agents, PBPK models offer reliable predictions that can be used to more effectively screen candidate drugs. This chapter, to a greater extent than others, emphasizes the compartmental structures of these PBPK models developed for antineoplastic drugs and the insights about drug disposition gained from use of these models.

Physiologically Based Pharmacokinetic Modeling, edited by Micaela B. Reddy, Raymond S. H. Yang, Harvey J. Clewell III, and Melvin E. Andersen.

11.2 PBPK MODELS FOR ANTINEOPLASTIC AGENTS

11.2.1 Methotrexate

Methotrexate (MTX) is one of the earliest examples of the use of PBPK models to investigate the distribution of an antineoplastic agent in mammals, and it represents perhaps the most frequently studied chemotherapeutic agent in this arena. Developed in 1970, the initial model consisted of seven compartments: liver, kidneys, gastrointestinal tract, muscle, plasma, bone marrow, and tumor (Bischoff *et al.* 1970). The bone marrow and tumor compartments were included as regions of interest, although neither significantly influenced the pharmacokinetics of MTX. The authors also considered alternative models with only one or two compartments in addition to a gut lumen. All versions of the model assumed flow-limited conditions. Only one intravenous dose level, 3 mg/kg, which falls in the range of clinical treatment levels, was evaluated in CDF-1 male mice.

Because only one dose level was studied, the conclusions were considered preliminary. In general, the model demonstrated good agreement between predicted values and experimental results. An initial drop in the plasma MTX concentration indicated rapid uptake, tissue localization, and excretion (MTX is not metabolized extensively in mice) and emphasized the significance of tissue uptake and biliary secretion in the pharmacokinetics of MTX. The importance of enterohepatic recycling was reflected by a rapid early increase in the concentration of MTX in the small intestine, with a peak gut lumen : plasma concentration ratio of about 100. The model incorporated a time-delay term to account for bile formation and holding time, and the same time-delay function was included in the gut absorption term, since hepatic biliary secretion must precede its absorption from the gastrointestinal tract. The model assumed zero-order absorption from the gut based on an apparent leveling of plasma concentration after 2–3 hours. Biliary clearance rates were estimated by bile duct cannulation experiments. To account for gastrointestinal travel time and fecal excretion time delays, time-delay functions were also applied in the gut lumen. Renal clearance was determined by comparison of the integrated plasma concentration data with cumulative urinary excretion data. Plasma protein binding was estimated at 25%.

Comparison of experimental data and model predictions showed that for very short time periods, the model predicted nearly instantaneous equilibration between all parts of the body except the liver. The model overpredicted gut lumen concentrations for time periods greater than 60 minutes, probably due to assumptions that implied abrupt infusion of concentrated bile into the gastrointestinal tract lumen at incremental time points. The authors noted that gastrointestinal tract absorption of MTX might be a zero-order process in some concentration ranges, presumably due to saturation of a transport mechanism.

In an extension of their 1970 model, Bischoff *et al.* (1971) studied a variety of dose levels of MTX in several species, including humans. Using a lumped compartmental design, the five-compartment PBPK model validated the previous model's features of (1) a rapid drop in MTX plasma concentrations and a corresponding increase in gut lumen concentrations, indicating the importance of tissue uptake and biliary secretion, (2) a peak gut lumen : plasma ratio of approximately 100, and (3) linear binding of MTX in tissues as plasma concentrations increased above 0.1 μg/mL, indicating nonlinear binding at low concentrations. The authors speculated that this phenomenon was probably due to strong binding to dihydrofolate reductase.

A key feature of the revised model was its use of "multicompartment" models for biliary secretion and gastrointestinal transport. The original model used a mathematical time-delay function (step function) to account for bile formation and secretion from the liver. Because of the abrupt introduction of bile into the gastrointestinal tract lumen, the predicted values overestimated actual data. In contrast, in the revised model, biliary secretion was represented by a three-compartment submodel in which a finite series of discrete compartments simulated a smoother "S-shaped" bile concentration efflux curve.

MTX transit through the gastrointestinal tract was modeled using an approach similar to that used for biliary secretion, with an additional provision for transport through the intestinal wall. Due to the tubular nature of the gut lumen, the assumption of uniform mixing was invalid, so the gastrointestinal tract was divided into four distinct regions. Transit of MTX through the gastrointestinal tract was complicated by significant intestinal metabolism at later time points and presumed binding to dihydrofolate reductase. The model therefore provided for both saturable and nonsaturable gastrointestinal tract absorption. In the absence of detailed data indicating otherwise, absorption characteristics were assumed to be the same for all regions, but specific location-dependent absorption characteristics could be included using this approach. An assumption of zero-order absorption from the gastrointestinal tract was adequate for the 3 mg/kg dose level but not for a wider range of dose levels.

Comparison of model predictions and experimental data indicated the model predicted MTX distribution reasonably well for all species at dose levels greater than or equal to 1 mg/kg. The authors concluded that gastrointestinal tract absorption was not readily saturable, and both saturable and nonsaturable absorption should be included in the model. In the case of very low doses, the authors concluded that strong, saturable nonlinear binding probably occurred in the liver and kidney. The greatest uncertainty in the model rested in the kinetics describing intestinal absorption.

Renal clearance of MTX was determined by comparing the time integral of plasma concentration with cumulative urine formation after intraperitoneal (*i.p.*) or intravenous (*i.v.*) administration. Tissue : plasma equilibrium ratios were derived from constant infusion experiments and/or the portion of the *i.v.* pulse injection curve after initial redistribution. Constant tissue : plasma equilibrium ratios at high concentrations indicated linear binding; at lower concentrations, equilibrium ratios were represented by the sum of linear nonspecific binding and strong binding (presumably associated with dihydrofolate reductase).

The MTX model represented a unique system for testing the robustness of the PBPK model among species. Using the flow-limited MTX model developed in mice, Zaharko *et al.* (1972) studied the distribution of MTX in sting rays (*Dasyatidae sabrine* and *D. sayi*), which have a significantly lower circulation velocity compared to mammals. MTX was administered intravenously and intraintestinally at doses of

0.3, 3.0, and 30 mg/kg. Liver flow rates were determined indirectly from plasma concentrations over time and MTX accumulation in the liver. Plasma flow rates to various organs were determined indirectly by proportional analysis with respect to liver flow rates. Plasma protein binding and MTX metabolism were determined via *in vivo* and *in vitro* experiments. Physiological parameters were estimated from available literature. The model predicted the experimental data moderately well.

In another application of the MTX PBPK model, MTX uptake by bone marrow, small intestine, and spleen of the rat was studied during time intervals when MTX plasma concentration was decreasing rapidly due to MTX redistribution and elimination (Dedrick *et al.* 1973b). To determine membrane resistance to transport of MTX in bone marrow, small intestine, and spleen, pulse *i.v.* injections were administered to rats at four doses ranging from 0.05 mg/kg to 25 mg/kg. The eight-compartment model incorporated blood flow, transport across cell membranes, and strong intracellular binding due to dihydrofolate reductase. Drug concentrations in various tissues were plotted as functions of plasma concentrations.

The model was based on previous MTX models with several modifications. Bile absorption kinetics were not required in the rat model because of the direct secretion of bile from the bile duct, eliminating enterohepatic circulation. Also, the skin compartment contributed more to MTX distribution than in previous models and served to maintain elevated plasma concentrations at long times due to its poor perfusion. The bone marrow, spleen, and small intestine compartments incorporated detailed descriptions of MTX transport across cell membranes. Because of the significant role of cell membranes in determining the rate of drug uptake by these three tissues, the assumption of flow-limited uptake, although appropriate for the remainder of the model, was not appropriate for these tissues.

In modeling the transport of MTX across cell membranes, the model assumed rapid equilibration across the capillary membrane and required estimates of (1) extracellular and intracellular volumes and plasma flow rate to the compartment, (2) transport parameters (both saturable and passive), and (3) binding constants. Extracellular volumes were based on tissue : plasma MTX ratios from short time periods following administration of the highest dose. Intracellular volumes were calculated by subtracting extracellular volume from total volume. Values for plasma flow rates were obtained from the literature. Transport parameters were determined based on the proportionality between dose and the amount of drug in tissue at the two lowest doses. Binding constants were estimated from tissue concentrations at longer time periods, which remained nearly constant.

The authors reported general agreement between model estimates and experimental data. At lower doses, the assumption of linearity was consistent with experimental data. At higher doses, however, model simulations and experimental data deviated significantly, demonstrating the saturable nature of MTX transport across membranes. Under conditions of high cell membrane permeability, the assumption of flow-limited uptake was appropriate. But, under conditions of limited permeability (i.e., permeability much less than tissue perfusion), membrane resistance to MTX transport was limited.

The MTX PBPK model developed by Bischoff *et al.* in 1971 was used to investigate the time course effect of MTX on DNA synthesis in femoral bone marrow, gastrointestinal tract, and tumor in mice bearing the Lewis lung carcinoma (Zaharko *et al.* 1974). Using the model, the relationship between plasma concentrations of MTX and rate of DNA synthesis was studied under conditions of a single *i.p.* injection (5 mg/kg,) and constant infusion (0.8 μg/hr, subcutaneously). The results supported the hypothesis that the rate of transport of MTX from the plasma into the cells of the bone marrow, small intestine, and solid tumor may be important in determining the relative toxicity of the drug to these tissues.

The MTX PBPK model was also used to study the rate of drug release from a controlled release device under both *in vitro* and *in vivo* conditions (Dedrick *et al.* 1974). The model did not include metabolism of MTX. Model simulations overpredicted the observed plateau concentration of MTX, potentially due to metabolism to a less reactive or inactive metabolite that may be important at very low infusion rates over long periods of time. Furthermore, the model predicted that 90% of the plateau concentration of MTX should be achieved within approximately 7.5 hours with a constant infusion rate of 1 μg/hr. Under *in vivo* experimental conditions, however, the actual time to achieve plateau concentration was actually 6 hours.

The MTX PBPK model for rats developed in 1973 was extended to study solid tumor MTX pharmacokinetics in dogs diagnosed with lymphosarcoma (Lutz *et al.* 1975). Dogs were administered MTX at *i.v.* doses of 0.01, 0.03, 0.1, 0.3, or 3.0 mg/kg. Regional blood flow to the involved lymph nodes, determined by a thermal dilution technique, was used as a surrogate for tumor blood flow rates. One limitation of the model was that it did not consider binding of MTX to plasma proteins.

Preliminary simulations of the model indicated that the tumor compartment was not flow-limited and suggested the existence of a diffusion-limited transport mechanism, similar to that reported for the bone marrow, spleen, and small intestine in the rat. MTX concentrations in tumor at early time points could not be simulated by the model unless a substantial fraction of the drug, rapidly and reversibly bound by a saturable process, was assumed to exist in the extracellular compartment. The authors proposed this represented MTX binding to cell membranes or interstitial proteins. Thus, the model was amended to incorporate a saturable transmembrane transport process and a saturable rapidly exchanging fraction in the extracellular space. In whole tissue experiments, if the rapidly exchangeable extracellular binding of MTX was omitted from the model, a sizable fraction of extracellular MTX could be interpreted as intracellular drug. Accurate representation of intracellular MTX concentrations in the model was important because maintenance of an adequate intracellular free concentration of MTX was necessary to inhibit DNA synthesis.

To evaluate MTX treatment schedules in human ovarian cancer, Dedrick *et al.* (1978) extended the 1973 rat PBPK model by adding an additional compartment—the peritoneal cavity—as well as by reducing biliary clearance, since it is less significant in humans compared to rodents. The model simulated a treatment schedule in which 4 liters of solution containing 1×10^{-5} M MTX was infused into the peritoneal cavity over a period of 4 hours. At 4 hours, and at subsequent 4-hour intervals for a total of 32 hours, the peritoneal cavity was drained and another 4 liters of MTX solution was infused. The model contained several approximating simplifications: (1) The peritoneal fluid volume was maintained at 4.5 liters throughout the calculations, and (2) the drain time was neglected.

More recently, Li and Gwilt (2002) developed a PBPK model to describe MTX transport into and out of effusion spaces. Clinically, malignant effusions are often associated with an increased plasma half-life of chemotherapeutic agents, resulting in myelosuppression and other toxicities. Increases in plasma half-life can result from alterations in effusion characteristics such as fluid volume, protein binding, and membrane permeability clearance. The PBPK model developed by Li and Gwilt examined the influence of effusate characteristics on the pharmacokinetics of MTX in plasma and evaluated the effect of disposition characteristics (e.g., volume of distribution) on the overall influence of an effusate on concentrations of MTX in plasma.

Li and Gwilt (2002) expanded previous PBPK models of MTX in humans (Bischoff *et al.* 1971; Dedrick *et al.* 1978) to include lung, heart, and pleural, peritoneal, and pericardial fluid spaces. Effusion spaces were represented using permeability-rate-limited drug transport processes. Values for permeability clearances, effusion site fluid volumes, and unbound MTX in plasma and effusion fluids were obtained from the literature. Membrane permeability was represented by a diffusive process characterized by permeability clearances, assuming instant and homogeneous distribution, instantaneous binding of drug, and crossing of the membrane by unbound drug only.

The model initially simulated concentrations of MTX in blood, tissue, and effusion fluid spaces in patients without effusions, and then it was extended to study the effects of cardiac, peritoneal, and pleural effusions on concentrations of MTX in plasma. Model predictions were compared with MTX concentrations in plasma and pleural fluid of patients without effusions as well as a single pediatric patient with a pleural effusion. In patients without effusions, model output and observed values were in general agreement. The authors also noted "remarkable" similarity between model predictions and observed values in a single pediatric patient with malignant pleural effusions administered 400 mg/kg MTX in a constant *i.v.* infusion over 6 hours. Additional simulations were performed to investigate the influence of site of effusion and alterations in effusion fluid characteristics on MTX pharmacokinetics. Results indicated that the rate of MTX sequestration in effusates was most affected by effusate volume, drug binding in effusate, and transport rate into and out of effusate space.

11.2.2 *cis*-Dichlorodiammine-platinum

A seven-compartment flow-limited model for *cis*-dichlorodiammine-platinum (II) (DDP) in the dog was developed using total platinum content data derived from *in vitro* ultrafiltration studies with canine plasma and *in vivo* studies performed in female beagle dogs (LeRoy *et al.* 1979). At the time, this PBPK model was relatively unique in its approach to dealing with a diverse range of possible metabolites: the authors did not determine the various parameters required by the model (e.g., metabolic rates, distribution coefficients, renal/biliary clearance values) for each metabolite. Instead, they described the rate of disappearance of parent compound DDP by reaction and eventual distribution to and excretion from the body.

DDP can undergo a diverse range of biotransformation reactions *in vivo* which involve a series of aquation–anation and protonation–deprotonation reactions. To simplify its complex biotransformation pathway, the model assumed that the parent compound was metabolized in the aqueous space of all compartments. The rate of DDP metabolic depletion was determined by *in vitro* ultrafiltration studies in which dog serum was incubated with the parent compound DDP. The ultrafiltration studies indicated first-order reaction kinetics. Additionally, the studies showed that parent DDP was not significantly bound to plasma protein, but subsequent to a nonenzymatic aquation step, 92–95% of metabolic products were protein-bound. By describing DDP biotransformation in terms of the rate of disappearance of parent compound rather than the rate of appearance of metabolites, the reaction scheme was simplified so that all metabolites could be represented kinetically by a single species. A hybrid rate constant, k_m, therefore represented the rate of conversion of DDP parent compound to all bound reaction species.

Elimination was assumed to occur via both urinary (parent compound and metabolites) and biliary (metabolites only) excretion routes. Renal clearance, estimated for both parent DDP and metabolites, was determined by plotting the cumulative amount of platinum in urine against the area under the curve (AUC) of platinum concentration in plasma over time and then taking the slope of the resulting line. In modeling the rate of excretion of the DDP parent compound and metabolites (primarily via the kidney), renal clearance at first occurred rapidly over a period of several hours and then decreased to a much slower rate over a period of nearly two weeks. The rapidly decreasing slope at earlier time points represented the initial clearance of platinum as parent compound (DDP). The more slowly decreasing slope at later times implied a decreasing clearance of platinum, which the authors speculated may have been due to the fact that either (1) metabolites of DDP dominated in the plasma and had lower clearance values than parent DDP or (2) metabolites were bound to plasma protein and thereby decreased the concentration of free platinum in plasma and increased platinum clearance values. A secondary clearance rate of 0.6 mL/min accurately simulated the long half-life of platinum to 12 days. Biliary excretion was included in the model but was considered insignificant compared to renal excretion.

The model adequately simulated tissue data with the exception of muscle, for which the model overpredicted concentrations at times less than 1 hour and underpredicted concentrations from 4 to 24 hours. The authors attributed these discrepancies to the assumption of flow-limited uptake and suggested the discrepancies may reflect a mode of transport other than flow-limited uptake, such as low cell membrane permeability. The model also allowed prediction of parent DDP and metabolite concentration, which is difficult to measure experimentally, in plasma, tissues, and excreta.

Another PBPK model of DDP studied its disposition in children and adolescents with carcinoma (Evans *et al.* 1982). The six-compartment model, based on the 1971 MTX model of Bischoff and Dedrick and the 1979 DDP model of LeRoy *et al.*, used volume terms, flow rates, and clearance terms based on human data and metabolic rate constants developed in canines. For comparison purposes, the authors also developed a simplified multicompartment operational model. While the multicompartment operational model accurately predicted parent DDP concentrations, it was less accurate in predicting total serum platinum when compared to the PBPK model. In contrast, the PBPK model accurately simulated serum concentrations of both DDP and total platinum through at least 48 hours post-infusion. Advantages of the PBPK model included improved accuracy, ability to simulate DDP disposition in individual extravascular tissues (tissues that cannot be assessed from clinical samples), and ability to predict alterations in DDP disposition induced by changes in selected physiological variables such as biliary obstruction and decreased muscle mass.

A subsequent PBPK model incorporated new data on the biotransformation of DDP (Farris *et al.* 1985). Female Sprague–Dawley rats bearing the Walker 256 carcinoma were used to study the pharmacokinetics of a single 4 mg/kg *i.v.* dose of DDP. The complex biotransformation of DDP was accounted for in terms of protein binding and mobile species. The model assumed irreversible DDP binding to low-molecular-weight nucleophiles and high-molecular-weight macromolecules (mostly proteins) in plasma and tissue compartments to form various platinum-containing metabolites. Binding was assumed to occur in all tissues at rates that were tissue-specific. All platinum-containing bound species were lumped into two groups: "fixed" metabolites and "mobile" metabolites. Fixed metabolites included all macromolecular platinum-containing species and remained in tissues until metabolized to form mobile species. Mobile metabolites included all low-molecular-weight platinum-containing species. Both categories represented "lumped" species rather than single chemical species. DDP and mobile metabolites followed flow-limited transport, freely traversed compartmental barriers, and partitioned equally in all compartments. The rate constant for fixed metabolite formation in plasma was determined from *in vitro* incubation data. Rate constants for fixed metabolite formation in tissue were fitted by matching simulations of tissue concentration versus time profiles to experimental data. The mobile metabolite formation rate was determined by scaling fixed metabolite formation rates to available protein and nonprotein binding sites.

The eight-compartment, flow-limited model, which, unlike previous models, did not include an ovary compartment, simulated concentrations of DDP and mobile and fixed metabolites in plasma, liver, gastrointestinal tract, skin, muscle, tumor, kidney, and carcass (bone, spleen, pancreas, fat, red blood cells), as well as excretion of DDP and mobile metabolites. Fixed platinum could be eliminated from a compartment only after biotransformation to a mobile metabolite. In most compartments, the rate of elimination corresponded closely with the average rate of protein turnover in that compartment. Both DDP and its mobile metabolites were excreted in urine and modeled as a linear process approximating the glomerular filtration rate. Biliary excretion, considered minor, was modeled as a linear process in the liver. In general, the model underpredicted total filterable platinum (DDP plus mobile metabolite) in plasma, possibly due to the assumption of flow-limited uptake.

In an extension of the 1985 rat PBPK model for DDP, King *et al.* (1986) modeled DDP pharmacokinetics in the rabbit, dog, and human. Intravenous dose levels were 2 mg/kg in the rabbit and 1 mg/kg and 3 mg/kg in the dog. DDP was also administered *i.p.* to dogs at a dose of 3 mg/kg. In humans, several dose levels, with varying durations of infusion, were studied. In rabbits, the model underpredicted total filterable platinum in plasma, possibly due to the assumption of flow-limited uptake. The dog model exhibited fairly good agreement between model predictions and experimental data, with the exception of *i.p.* dosing and tissues in the vicinity of the peritoneal cavity, perhaps because the model did not account for the distributed nature of DDP transport. The authors noted that the rate constant for DDP absorption from the peritoneal cavity would be necessary to accurately simulate the data. Human data were simulated using a parameter set estimated from a combination of published data and estimates based on animal data where limited human data existed. The modeled predictions for humans fit the limited experimental data reasonably well.

The DDP PBPK model was also extended to study mice dosed intravenously with 4 mg/kg DDP (King and Dedrick 1992). The model validated previous models in rats (Farris *et al.* 1985), rabbits, dogs, and humans (King *et al.* 1986) at a much lower range of body weights. Results showed good correlation between experimental data and predictions and supported the observation that body weight did not influence protein binding constants.

11.2.3 Actinomycin D

A PBPK model was developed to study actinomycin D pharmacokinetics in dogs (Lutz *et al.* 1977). The flow-limited model consisted of 14 compartments (plasma, lung, heart, spleen, stomach, intestines, liver, kidney, testes, salivary gland, thymus, bone marrow, pancreas, and muscle) and did not consider metabolism of the parent compound. Binding to DNA in all compartments was assumed to be saturable. Experimental data were derived from beagle dogs dosed with 0.5 mg/m^2 (0.03 mg/kg) and 2.7 mg/m^2 (0.135 mg/kg) actinomycin D.

The model predictions were consistent with the experimental data for all compartments except testes. Because the testes data suggested diffusion-limited uptake, an alternative model in which all compartments were diffusion-limited was developed for comparison with the flow-limited model. Membrane diffusion parameters were varied from low to high. The diffusion-limited model underpredicted the data at low K values. Increasing the value of K improved the simulations up to a point, although further increases in the value of K did not result in measurable actinomycin D uptake in the liver compartment. Consequently, the authors concluded that the simpler flow-limited model was most appropriate. The assumption of flow-limited uptake was verified by comparison of cell mass transfer coefficients with blood flow rate per unit volume of normal tissues. Both were of the same order of magnitude, and the authors concluded that overall, the flow-limited model was most appropriate.

However, the experimental data for testes did not parallel the declining plasma curve as normal tissues did but remained nearly constant over time. These data were best simulated by the diffusion-limited model with a diffusion parameter, K, of $0.2\,hr^{-1}$. The perfusion rate for testes was much larger than the mass transfer coefficient, suggesting that testes, like brain, have a low permeability membrane barrier. In general, transport of a drug is diffusion-limited only if the diffusion parameter is much smaller that the tissue perfusion rate per unit volume (Q/V). For actinomycin D in testis, Q/V was $1.0\,hr^{-1}$, which was much greater than the value for the diffusion parameter ($0.2\,hr^{-1}$), suggesting the presence of a significant barrier limiting the rate of actinomycin D uptake by testes. In a clinical setting, this could be an important consideration in the treatment of testicular cancer. The authors also suggested that alternatively, some resistant tumor cells may have lower permeability to actinomycin D than normal tissues.

11.2.4 2′-Deoxycoformycin (Pentostatin)

In 1981, King and Dedrick developed a PBPK model for 2′-deoxycoformycin that consisted of six compartments: liver, kidney, gut, blood, tumor, and carcass (King and Dedrick 1981). The assumption of flow-limited uptake was valid for all compartments except the tumor. Since 2′-deoxycoformycin was known to be nonuniformly distributed or bound in all tissues, the model included a linear term in the equilibrium binding equation to account for tissue distributions at high concentrations as well as a nonlinear saturable term to account for tight specific binding to adenosine deaminase at low concentrations. In this respect, the 2′-deoxycoformycin model was similar to the Bischoff *et al.* (1971) MTX model. Metabolism was not included in the 2′-deoxycoformycin model. Normal and leukemic mice were evaluated with 2′-deoxycoformycin *i.p.* doses of 0.25 and 2.5 mg/kg and an *i.v.* dose of 0.25 mg/kg. The tumor blood flow rate ($0.22\,min^{-1}$) was estimated by fitting the model to L1210 tumor data. In leukemic mice, experimental data indicated the level of strong binding to adenosine deaminase was higher than in normal mice. The model indicated that the primary means of drug excretion occurred through the urinary tract, but it was not able to predict the data well at high doses. Because the renal clearance of 2′-deoxycoformycin in the mouse was nearly twice the estimated glomerular filtration rate, a linear secretion term was included in the model to describe the urinary excretion process.

Pharmacokinetic parameters for normal tissues were determined by breaking the model down into several hybrid models that could be solved individually. Due to the large number of parameters that needed to be estimated for the global PBPK model and the lack of experimental data for the carcass compartment, a series of hybrid models were constructed and solved individually. Each hybrid model consisted of a mass balance equation for tissue and an empirically fitted forcing function for arterial blood concentrations versus time. The hybrid models had the advantageous capability of estimating unknown parameters one or two at a time. Binding and clearance parameters were determined from the hybrid models for use in the global PBPK model.

11.2.5 5-Fluorouracil

Collins *et al.* (1980) developed a two-compartment flow-limited hybrid model to study 5-fluorouracil (5FU) in humans. The model was "lumped" around the pulmonary system and incorporated saturable whole-body clearance. Dosage regimens included *i.v.* boluses over a range of doses and constant *i.v.* infusion. Apparent total-body clearances were based on the dosage regimen; route- and schedule-dependent drug clearances were also included. Hepatic and extrahepatic elimination were inferred from the model. The effect of arterial versus venous sampling was also evaluated. The model was extended to include *i.p.* and oral administration by addition of peritoneal fluid and liver compartments to make a four-compartment model. The model represented the disappearance of 5FU, in terms of both anabolic (anticancer effect) and catabolic elimination. The model adequately predicted *i.v.* bolus, *i.v.* infusion, and *i.p.* injection routes.

Several decades later, a PBPK model was developed by Tsukamoto *et al.* (2001a) to evaluate the influence of different factors on the therapeutic index of capecitabine (N^4-pentyloxycarbonyl-5′-deoxy-5-fluorocytidine), a triple prodrug of 5FU. Designed to exploit the tissue specificity of key metabolic enzymes, capecitabine is preferentially activated to 5FU in malignant tumors and as such limits exposure of the gastrointestinal tract and other healthy tissues to 5FU. The parent compound is metabolized to 5′-deoxy-5-fluorocytidine (5′DFCR) by carboxylesterase, primarily in the liver. 5′DFCR is metabolized in both liver and tumor to 5′-deoxy-5-fluorouridine (5′DFUR) by cytidine deaminase, then subsequently to the active moiety 5FU by thymidine phosphorylase (dThdPase) in the tumor. Through the enzymatic action of dihydropyrimidine dehydrogenase (DPD), 5FU is converted to dihydroxyfluorouracil (FUH_2). The pharmacokinetic behavior of capecitabine and its metabolites in humans is difficult to predict due to large interspecies, interpatient, and intrapatient variations in organ-specific metabolism (Blesch *et al.* 2003).

Tsukamoto *et al.* used a flow-limited model that incorporated sequential compartmental metabolism according to Michaelis–Menten kinetics. A submodel consisting of four compartments (liver, gastrointestinal tract, tumor, and noneliminating tissues such as skin and muscle) was constructed for the parent compound and each of three metabolites. Enzyme kinetic parameters and plasma/tissue binding data were obtained from *in vitro* experiments using human liver and intestinal tissue. The model accurately predicted blood concentrations of all four compounds in humans over time. Furthermore, the simulated blood AUCs of 5FU and 5′DFUR were consistent with clinical observations. No experimental data were available with respect to concentrations of parent compound or metabolites in the liver, gastrointestinal tract, or colorectal cancer tissue.

A sensitivity analysis identified factors that most influenced the pharmacokinetics of capecitabine and active metabolites of 5FU in tumor tissue, liver, and gastrointestinal tract. Enzyme activities (i.e., the maximal velocity of metabolism, V_{max}) were varied from 0.1- to 10-fold the mean values in the liver and gastrointestinal tract, and from 0.01- to 100-fold the mean values in tumor tissue. The systemic 5FU concentration was most affected by DPD hepatic activity. 5FU AUC in tumor tissue increased with increasing blood flow, due to increased supply of 5′DFCR and 5′DFUR (independent of tumor tissue enzyme activity). DPD saturation in tumor tissue was suggested by the nonlinearity of 5FU AUC in tumor tissue. The authors concluded that a greater pharmacologic effect of capecitabine should be observed in patients whose dThdPase activity in tumor tissue was high and whose DPD activity was low.

Comparisons between capecitabine and other fluoropyrimidines regarding therapeutic indices (ratio of 5FU AUC in tumor tissue to 5FU AUC in gastrointestinal tract) and gastrointestinal toxicity were also performed. Oral administration of doxifluridine or capecitabine was associated with significant improvement in 5FU accumulation in tumor tissue compared with administration of 5FU by itself. The tumor-to-blood ratio of 5FU AUC indicated that targeting efficiency was much higher after oral administration of capecitabine compared to doxifluridine. Because gastrointestinal toxicity was dose-limiting with 5FU, the ratio of 5FU exposure in tumor tissue to 5FU exposure in the gastrointestinal tract was evaluated. For oral administration of doxifluridine, the ratio of 5FU AUC in tumor tissue to that in the gastrointestinal tract was lower over the clinical dose range due to a nonlinear increase of 5FU AUC in the gastrointestinal tract. Nonlinearity, however, was not observed with capecitabine. Furthermore, model simulations revealed that the ratio of 5FU AUC in tumor compared to blood was comparable to that reported in humans in 11 colorectal patients treated with capecitabine at 125.5 mg/m^3. The therapeutic index of capecitabine was thus much higher than that of any other fluoropyrimidine, in accord with very low gastrointestinal toxicity of capecitabine at the same clinical dose of doxifluridine. Therefore, the dose of capecitabine could be increased to a level where saturation of 5FU disposition was attained. The high tumor-specific accumulation of 5FU after oral capecitabine administration was due to the nonlinear profile in 5FU AUC, whereas 5FU AUC ratios were nearly linear over the dose ranges of other fluoropyrimidines. These results agreed with the finding that dThdPase and DPD activities in tumor tissue were important factors governing 5FU AUC in tumor tissue, along with blood flow to tumor tissue.

Tsukamoto *et al.* (2001b) then extended their original PBPK model to (1) predict blood and tumor concentrations of capecitabine and metabolites, (2) predict the therapeutic index of capecitabine and compare it with that of 5FU and doxifluridine, (3) compare AUC in tumor tissue of human cancer xenograft models at the minimum effective dose with those estimated for humans at a clinical dose, and (4) investigate the preferential accumulation of 5FU in tumor tissue after oral administration of capecitabine in a mouse human tumor xenograft model. As with their original model, the revised model was flow-limited and considered capecitabine and three of its metabolites. Mouse biochemical parameters (enzyme kinetics, plasma protein binding, and tissue binding) were determined *in vitro*. An *in vitro–in vivo* scaling factor was taken into consideration to account for differences in V_{max} values. Blood flow to the liver consisted of three separate components: hepatic artery flow, mucosal flow, and gastrointestinal tract flow.

At early time points, the model underestimated concentrations of capecitabine and metabolites in blood and tumor tissue, probably due to underestimation of K_a

for capecitabine. There were large species differences in the tissue distribution of enzymatic activities. Higher accumulation of 5FU in tumor tissue was noted with capecitabine administration compared to other fluoropyrimidines. This was reflected in the predicted therapeutic index in human xenograft models, which was 5 and 3000 times greater than that of 5′DFUR and 5FU, respectively. These values were comparable to those reported in the literature. The model predicted a nonlinear increase in 5FU AUC in the gastrointestinal tract in the xenograft model, which was not observed in humans. The authors proposed that this may have been due to increased 5FU production in the mouse gastrointestinal tract due to higher activities of metabolizing enzymes compared to the human gastrointestinal tract, leading to saturation of 5FU elimination in the mouse gastrointestinal tract.

11.2.6 2-Amino-1,3,4-thiadiazole

King and Dedrick (1979) used a five-compartment flow-limited model to study 2-amino-1,3,4-thiadiazole (ATDA) in mice (*i.p.* dose of 100 mg/kg), dogs (*i.v.* dose of 3 mg/kg), and monkeys (*i.v.* doses of 0.3, 3.0, and 30 mg/kg). A lean compartment included muscle, skin, and bone marrow. For metabolism, arbitrarily assumed to occur in the liver, a saturable kinetic model was used for dogs and monkeys because zero-order elimination was reported for these species For mice, metabolism was represented as a first-order process. All three metabolites of the ATDA complex were represented as one species for purposes of analysis. Metabolic parameters were obtained by fitting the model to tissue concentration data. Plasma protein binding was considered negligible because the serum concentration was equivalent to the blood concentration. Linear tissue binding was assumed. Partition coefficients were determined from long-term experimental values of serum and tissue concentrations. Renal elimination represented the major (linear) route of excretion, and fecal elimination was considered negligible.

In general, model predictions were consistent with experimental tissue and serum data from the mouse, dog, and monkey. In mice and dogs, the model underpredicted the serum concentration of ATDA at early time points, suggesting transport of metabolites was not strictly or rigorously flow-limited in all tissues. The model tended to overpredict urinary excretion in the mouse. In monkeys, the model predicted experimental data well at the two lowest doses, but overpredicted serum concentrations for the highest dose, indicating the saturable metabolism/linear renal clearance model did not adequately represent the overall rate of clearance. The authors concluded that at very high concentrations of ATDA, relative reabsorption in the kidney may be decreased. The model also revealed that the major excretion pathway in the mouse occurred via renal clearance, whereas in dogs and humans, the saturable metabolic pathway dominated. The authors noted that because of interspecies differences, scale-up of the model to humans would be difficult for this particular drug.

11.2.7 1-β-D-Arabinofuranosylcytosine

One of the first PBPK models to utilize *in vitro* data in describing *in vivo* processes was developed by Dedrick *et al.* (1972) to predict plasma concentrations of 1-β-D-arabinofuranosylcytosine (Ara-C) in humans. Besides representing one of the first models incorporating *in vitro* data, their model is unique because it was first developed for humans, then extended to other mammals (usually, animal models precede the development of human models). To investigate why Ara-C, following *i.v.* administration, disappeared from blood more rapidly than accounted for by excretion, Dedrick *et al.* (1972) used *in vitro* metabolic data in a flow-limited model to predict plasma concentration of the drug in humans. No significant plasma protein binding was known to occur. The model consisted of seven compartments—blood, heart, liver, gastrointestinal tract, kidneys, bone marrow, and lean tissue—with bone marrow representing the critical site of toxicity. To facilitate the use of *in vitro* data regarding saturable deamination enzymatic reactions describing Ara-C distribution and metabolism to 1-β-D-arabinofuranosyluracil (Ara-U), the liver, kidneys, and heart were lumped into a single compartment. The model did not account for insignificant metabolism outside of the liver, heart, and kidney. Lean tissue acted as an important reservoir for parent drug and metabolite.

Experimental data were obtained from patients diagnosed with acute myelogenous leukemia and lymphoma but with normal hepatic and renal function. Patients were administered a single Ara-C *i.v.* dose of 1.2–86 mg/kg.

In general, model simulations accurately predicted experimental data. At early time points, the model overpredicted Ara-C plasma concentrations, but underpredicted them at extended time points. This suggested the existence of one or more additional "deep" compartments (e.g., in the bone cortex, brain, or skin). In another Ara-C PBPK model, Dedrick *et al.* (1973a) presented plasma and lean tissue concentrations of Ara-C and Ara-U in Swiss mice, monkeys, and dogs. Model predictions and experimental data agreed reasonably well for all species.

11.2.8 Adriamycin

Harris and Gross (1975) developed a flow-limited model to study adriamycin administration in rabbits at an *i.v.* dose of 3 mg/kg. The 10-compartment model did not consider metabolism of adriamycin. Target sites of toxicity were heart and bone marrow. Intestinal absorption, tissue/plasma equilibrium concentration ratios, plasma protein binding, and elimination clearance were determined by experimentation. Renal clearance was considered negligible. Intestinal reabsorption was not included in the model despite extensive excretion into bile. Experimental data were predicted by the model relatively well, although early time-course concentrations of adriamycin were overpredicted in the kidney, gastrointestinal tract, adipose tissue, lung, heart, and spleen, suggestive of membrane resistance to diffusion. The model also predicted adriamycin pharmacokinetic data in individual human patients reasonably well.

In another PBPK model, based on an application of the 1975 model of Harris and Gross, Chan *et al.* (1978) predicted the time-course plasma concentrations of adriamycin in 23 cancer patients and in biopsy tissues of nine surgery patients following a single *i.v.* bolus administration of 10–60 mg/m^2. The model revised the 1975 model's plasma protein binding fraction of 0.5 to 0.9. In general, model predictions of plasma time-course concentrations agreed well with experimental data from patients with no clinical evidence of impaired renal or hepatic function. Not surprisingly, the model, which assumed normal liver function and drug clearance, underpredicted plasma concentrations and plasma half-lives in patients with elevated serum bilirubin.

To optimize dosage regimens and schedules, particularly in patients exhibiting hepatic or cardiac compromise, another PBPK model of adriamycin was developed to enable a better understanding of its distribution and elimination (Sonneveld and Mulder 1981). In comparison with other PBPK models, the main objective of this 10-compartment, flow-limited model was not the simulation of concentration time curves in various tissue compartments, but rather the identification and interpretation of rate constants from experimental data. Rate constants were estimated based on experimental distribution data of adriamycin, and they were determined using the principle of model matching. Output from the mathematical model was fitted, as well as possible, to experimental data by selecting an optimal set of values for the unknown rate constants, resulting in maximum likelihood estimates of rate constants. Experimental data were derived from female Brown–Norway rats injected intravenously with 7.5 mg/kg adriamycin (comparable to 40 mg/m^2 in humans). The model did not consider enterohepatic circulation. Because the combined drug rate constants were derived from experimentally obtained drug concentrations, they were assumed to represent all the factors contributing to drug transport.

In general, model predictions agreed well with experimental data. The model overpredicted serum concentrations at early time points, suggesting that the assumption of flow-limited uptake may have been inappropriate. In spleen and bone marrow, the slopes of the model predictions differed compared to those for other organs. These tissues presented significant problems for modeling the distribution of adriamycin because of its tendency to bind strongly to nuclear structures of bone marrow cells in certain phases of the cell cycle, leading to a gradual increase in amount of drug present in these tissues. The model assumption of urinary and biliary excretion of adriamycin fit the experimental data well, although a slight underestimation of experimental data by the model may have been attributable to the exclusion of enterohepatic recirculation from the model. The authors noted that neither diffusion (membrane) nor flow-limited models adequately represented the disappearance or elimination of adriamycin from rapidly proliferating tissues. Such targets would represent an important clinical consideration with respect to prediction and prevention of toxicity.

In describing the pharmacokinetic behavior of adriamycin, the PBPK models of Harris and Gross (1975) and Chan *et al.* (1978) employed experimentally derived plasma : tissue concentration ratios to define tissue partitioning. In contrast, a model developed by Gustafson *et al.* (2002) incorporated experimentally measured biochemical parameters regarding tissue-specific metabolism, biliary/urinary elimination, and macromolecular binding to describe the tissue uptake and release of adriamycin. Because adriamycin is known to bind to DNA as well as cardiolipin, an anionic cardiac lipid, the model relied on specific terms to account for binding to these macromolecules.

The flow-limited model consisted of seven compartments: slowly perfused tissue, rapidly perfused tissue, bone marrow, heart, kidney, liver, and gut. Tissue volumes and blood flow values were obtained from the literature. Tissue DNA content was measured experimentally in mice, dogs, and humans, and cardiolipin levels were estimated from rat values published in literature. The concentrations of DNA and cardiolipin in tissues were modified by factors representing the number of molecules of DNA or cardiolipin binding one molecule of adriamycin. For cardiolipin the modifying factor was based on a molar ratio of 2 : 1 (adriamycin : cardiolipin), and for DNA a modifying factor of 500 : 1 (adriamycin : DNA) was assumed. Affinity constants describing the binding of adriamycin to DNA and cardiolipin were set at 200 and 400, respectively, consistent with *in vitro* binding data obtained from the literature. The model considered metabolism of adriamycin to doxorubinicinol and 7OH aglycone in the heart, liver, and kidney; values for K_m and $V_{\max}$ in mice, dogs, and humans were obtained from literature. Excretion of adriamycin was optimized in the model to best describe the tissue distribution data and fecal/urinary elimination kinetics.

Model output was compared with experimental data in mice, dogs, and humans. Mouse data were obtained from experiments in which female mice were administered adriamycin at *i.v.* dose levels of 6 and 10 mg/kg. Clinical data were obtained from dogs diagnosed with lymphoma receiving an *i.v.* infusion of adriamycin at a dose of 30 mg/m^2 (the dog model included an *i.v.* infusion term). Clinical human data were obtained from the literature. The model simulated the experimental data in mice reasonably well, and it generally agreed with canine plasma experimental data. In the human model, the excretory parameters and cardiolipin content were scaled from the mouse/dog model to account for differences in organ size. Compared with published human data, the model simulations agreed with the experimental data reasonably well.

11.2.9 Melphalan

Wu *et al.* (1997) developed a PBPK model to investigate the kinetics of melphalan administered during isolated limb perfusion (ILP) chemotherapy. ILP is a form of regional chemotherapy that involves exposure of a tumor-bearing limb to high concentrations of antineoplastic agents. The ILP technique spares the patient from systemic exposure and toxicity and includes a washing-out procedure that minimizes the undesirable escape of melphalan into the systemic circulation. The PBPK model was designed to determine the optimal composition of perfusate to be used in ILP by evaluating whether binding of melphalan in perfusate containing either albumin or dextran was a determinant of melphalan tissue concentrations and washout success.

The flow-limited model consisted of four compartments: skin, muscle, fat, and plasma. Experimentally derived quantitative data regarding the amount of drug uptake by limb tissues during perfusion were generated under different perfusate conditions. Tissue blood flow (skin, muscle, and fat) was determined using a radiolabeled microsphere method. To determine tissue vascular space, rat hindlimbs were perfused with ^{125}I-labeled albumin mixed with melphalan for 60 minutes followed by a washout period of 30 minutes. The concentration of albumin was then measured in perfusate inflow ($C_{perfusate}$, Dpm/mL) and in tissue (C_{tissue}, Dpm/mL per gram of tissue). The total weight of each tissue type was determined by complete dissection. The vascular space of each tissue (in mL) was then calculated by dividing $C_{perfusate}$ by the product of C_{tissue} and tissue weight. Total blood flow was assumed to equal the blood flow of each tissue type (mL/min/g of tissue) multiplied by the weight of each tissue type in the perfused hindlimb. Each tissue was assumed to be a well-stirred compartment, and the lateral spread of melphalan from one tissue to another was considered negligible. Additionally, each tissue compartment represented a noneliminating organ (except for hydrolysis), so the rate constant in each tissue was the same. Melphalan hydrolysis in the perfusate was assumed to decline monoexponentially, and the corresponding rate constant was calculated from a semilogarithmic plot of concentration versus time. The model adequately described the concentration time profile of melphalan in perfusate outflow and tissues and accurately predicted the experimental data in skin, muscle, and fat during perfusion and washout periods. The model also allowed predictions of perfusate and tissue profiles containing dextran rather than albumin.

11.2.10 Topotecan

Sung *et al.* (1994) developed a PBPK model in nonhuman primates to study the pharmacokinetics of topotecan administered via two different dosing regimens: *i.v.* infusion (10 mg/m^2 over 10 minutes) or intraventricular bolus (0.1 mg). The model consisted of three compartments: cerebrospinal fluid (CSF), plasma, and body. Clearance of topotecan from the CSF occurred by bulk flow and microvascular exchange with the plasma compartment. The plasma compartment communicated with the body compartment by an intercompartmental transport parameter, *K*, which was a function of plasma flow rate and capillary exchange properties. Within each compartment, the lactone form of topotecan interconverted nonenzymatically with the hydroxyl acid form. The model assumed topotecan, in either form, was eliminated from the plasma compartment with the same rate constant. Similarly, the microvascular exchange rate constant between plasma and ventricular CSF was assumed to be the same for both forms of the drug.

Under conditions of *i.v.* infusion, a significant fraction of the AUC in plasma may have occurred during the actual infusion period. Importantly, the PBPK model permitted the estimation of topotecan plasma concentrations during infusion, providing information important in determining total AUC and therefore the overall clearance of the drug. The model indicated that clearance of the total drug was 116 mL/min, which was significantly greater than the estimated monkey glomerular filtration rate of 24 mL/min. Accordingly, the authors suggested that kidney secretion or other metabolic processes may have contributed to elimination of the drug, noting that the model assumption of exclusive plasma elimination may not have been accurate if significant metabolism of the drug occurred in the body compartment. The model also predicted a plasma clearance of topotecan lactone of 26.3 L/hr/m^2, in good agreement with values reported in the literature.

11.2.11 17-(Allylamino)-17-demethoxygeldanamycin

Xu *et al.* (2003) developed a PBPK model for the antiproliferative agent 17-(allylamino)-17-demethoxygeldanamycin (17AAG) and its active metabolite 17-(amino)-17-demethoxygeldanamycin (17AG) using data from mice, and then they applied it to study the disposition of both compounds in tumors of nude mice bearing human breast cancer xenografts. Normal adult female CD2F1 mice and adult female NCR SCID mice bearing MDA-MB-453 human breast cancer xenografts were administered a single dose of 40 mg/kg 17AAG. The model for normal mice consisted of eight compartments (lung, brain, heart, spleen, liver, kidney, muscle, and miscellaneous tissue) and described the distribution of 17AAG using diffusion-limited conditions and that of 17AG using flow-limited conditions. For tumor-bearing mice, the model included an additional tumor compartment; the physiological parameters and drug transport kinetics employed in the tumor model were consistent with those used in the normal mouse model.

The authors also used the PBPK model to develop estimates of intrinsic clearance and clearance-related parameters in the mouse liver, which was assumed to be the primary site of elimination. Intrinsic clearance rates were determined using local models that related systemic clearance to intrinsic clearance of 17AAG and 17AG in the liver. The relationship between intrinsic clearance and systemic clearance of 17AAG was derived from a simplified whole-body model by solving the model equations in Laplace domain assuming linear kinetics and exclusive hepatic elimination. The RBC : plasma partition coefficient was determined from *in vivo* studies in normal mice, while plasma binding constants were estimated using *in vitro* data. The model assumed that exchange of 17AAG and 17AG across the vascular wall occurred between the free fraction in the blood and the total amount in tissue. Concentrations of parent drug and metabolite were assumed to be homogeneously distributed in both interstitial fluid and cells.

For the tumor model, the authors tested several different diffusion-limited models that evaluated (1) vascular and extravascular spaces, (2) extracellular and cellular spaces, or (3) vascular, interstitial fluid, and cellular spaces. Experimental data were best described by the tumor model that included vascular, interstitial fluid, and cellular space compartments within the tumor compartment. In the tumor compartment, unlike the other compartments of the model, diffusion-limited conditions were used to describe the distribution of both 17AAG and 17AG. Parameters of the tumor models were estimated by simultaneously fitting model predictions to measured tumor tissue concentrations of 17AAG and 17AG.

With the exception of 17AG in the brain and lung, predictions generated by the model agreed reasonably well with the experimental data for 17AAG and 17AG in blood and all organs. The authors suggested the observed discrepancy between model output and brain experimental data was due to limited uptake by the brain. The model could not account for the 17AG experimental data in the lung, which exhibited a peak value at 5 minutes, while the peak blood concentration did not occur until 30 minutes. The model also consistently overpredicted 17AAG concentrations at early time points, for which the authors offered several explanations. Since the model did not include uptake of 17AAG by the small intestine, stomach, or pancreas, uptake by these tissues may have significantly affected drug distribution. Alternatively, continuous metabolism of 17AAG in the liver could have accounted for the model overpredictions. Finally, the assumption of a constant fraction of intrinsic clearance of 17AAG related to the formation of 17AG may have been responsible for model overpredictions.

In the tumor model, concentrations of 17AAG and 17AG in tumor cells were over twofold greater than concentrations in the extravascular space of the tumor, indicating preferential uptake of both 17AAG and 17AG in tumor cells. Sustained concentration–time profiles of 17AAG and 17AG in tumor tissue were due to slow diffusion across cell membranes, while in normal organs this process was estimated to be essentially instantaneous. At 2 hours post-dosing, the concentration of 17AG in the blood was generally lower than that of 17AAG, while the concentration of 17AG in the tumor was much higher than that of 17AAG. The model suggested that this phenomenon was due to preferential uptake of 17AG over 17AAG in tumor tissue, perhaps, as the authors suggested, due to more a effective uptake mechanism of 17AG in tumor cells rather than in the interstitial fluid of tumor tissue.

11.3 SUMMARY

PBPK models provide a whole-body approach for considering the combined effects of absorption, blood flow, physicochemical interactions, metabolism, and elimination of antineoplastic agents across species. They facilitate integration of *in vitro* and interspecies data into *in vivo* human models in a reliable and realistic fashion. Because they offer the ability to predict drug levels in tissues not readily accessible for sampling, they enhance the understanding of systemic drug exposure and, more importantly, target tissue doses and associated toxicities. Clinical evaluation of antineoplastic agents can be enhanced by the use of PBPK models which offer a framework for developing and refining clinical therapeutic protocols. Not only can the effects of dose, dose frequency and timing, and route of administration be evaluated in a noninvasive manner to the patient, but also consideration of compromised organ function, a frequent occurrence among cancer patients, is possible. Furthermore, PBPK models can be extended to investigate the effects of multiple drug treatment regimens. As such, these models represent a useful tool that affords valuable insights in both clinical oncology and drug development arenas and ultimately improves the outcome for cancer patients. The examples provided here show the diverse applications of PBPK models over the past 30 years with these compounds. The progress in this area strongly emphasizes the innovative and informative direction imparted to PK studies of antineoplastic compounds by development of these PBPK approaches by the chemical engineering community.

NOTATION

17AAG 17-(allylamino)-17-demethoxygeldanamycin
17AG 17-(amino)-17-demethoxygeldanamycin

Ara-C	1-β-D-arabinofuranosylcytosine
Ara-U	1-β-D-arabinofuranosyluracil
ATDA	2-amino-1,3,4-thiadiazole
AUC	area under the curve
$C_{\text{perfusate}}$	the concentration of chemical in perfusate inflow
CSF	cerebrospinal fluid
C_{tissue}	the concentration of chemical in tissue
DDP	*cis*-dichlorodiammine-platinum (II)
5′DFUR	5′-deoxy-5-fluorourindine
DPD	dihydropyrimidine dehydrogenase
dThdPase	thymidine phosphorylase
5FU	5-fluorouracil
FUH_2	dihydroxyfluorouracil
ILP	isolated limb perfusion
i.p.	intraperitoneal
i.v.	intravenous
k_m	a hybrid rate constant representing rate of parent compound depletion
MTX	methotrexate
Q/V	tissue perfusion rate per unit volume
$V_{\max}$	the maximal velocity of metabolism

REFERENCES

Bischoff, K. B., Dedrick, R. L., and Zaharko, D. S. (1970). Preliminary model for methotrexate pharmacokinetics. *J. Pharm. Sci.* **59**, 149–154.

Bischoff, K. B., Dedrick, R. L., Zaharko, D. S., and Longstreth, J. A. (1971). Methotrexate pharmacokinetics. *J. Pharm. Sci.* **60**, 1128–1133.

Blesch, K. S., Gieschke, R., Tsukamoto, Y., Reigner, B. G., Burger, H. U., and Steimer, J. L. (2003). Clinical pharmacokinetic/pharmacodynamic and physiologically based pharmacokinetic modeling in new drug development: The capecitabine experience. *Invest. New Drugs* **21**, 195–223.

Chan, K. K., Cohen, J. L., Gross, J. F., Himmelstein, K. J., Bateman, J. R., Tsu-Lee, Y., and Marlis, A. S. (1978). Prediction of adrimycin disposition in cancer patients using a physiologic, pharmacokinetic model. *Cancer Treat. Rep.* **62**, 1161–1171.

Collins, J. M., Dedrick, R. L., King, F. G., Speyer, J. L., and Myers, C. E. (1980). Non-linear pharmacokinetic models for 5-fluorouracil in man: Intravenous and intraperitoneal routes. *Clin. Pharmacol. Ther.* **28**, 235–246.

Dedrick, R. L., Forrester, D. D., and Ho, D. H. W. (1972). *In vitro–in vivo* correlation of drug metabolism—Deamination of 1-β-D-arabinofuranosylcytosine. *Biochem. Pharmacol.* **21**, 1–16.

Dedrick, R. L., Forrester, D. D., Cannon, J. N., El Dareer, S. M., and Mellett, L. B. (1973a). Pharmacokinentics of 1-β-D-arabinofuranosylcytosine (Ara-C) deamination in several species. *Biochem. Pharmacol.* **22**, 2405–2417.

Dedrick, R. L., Zaharko, D. S., and Lutz, R. J. (1973b). Transport and binding of methotrexate in vivo. *J. Pharm. Sci.* **62**, 882–890.

Dedrick, R. L., Zaharko, D. S., Lutz, R. J., and Drake, J. C. (1974). Device for controlled drug release—Application to methotrexate infusion in mice. *Biochem. Pharmacol.* **23**, 2457–2461.

Dedrick, R. L., Myers, C. E., Bungay, P. M., and DeVita, V. T. (1978). Pharmacokinetic rationale for peritoneal drug administration in the treatment of ovarian cancer. *Cancer Treat. Rep.* **62**, 1–11.

Evans, W. E., Crom, W. R., Tsiatis, A., Green, A. A., Hayes, F. A., and Pratt, C. B. (1982). Pharmacokinetic modeling of cisplatin disposition in children and adolescents with cancer. *Cancer Chemother. Pharmacol.* **10**, 22–26.

Farris, F. F., King, F. G., Dedrick, R. L., and Litterst, C. L. (1985). Physiological model for the pharmacokinetics of *cis*-dichlorodiammineplatinum (II) (DDP) in the tumored rat *J. Pharmacokinet. Biopharm.* **13**, 13–39.

Gustafson, D. L., Rastatter, J. C., Colombo, T., and Long, M. E. (2002). Doxorubicin pharmacokinetics: Macromolecule binding, metabolism, and excretion in the context of a physiologic model. *J. Pharm. Sci.* **91**, 1488–1501.

Harris, P. A., and Gross, J. F. (1975). Preliminary pharmacokinetic model for adriamycin (NSC-123–127). *Cancer Chemother. Rep.* **59**, 819–825.

King, F. G., and Dedrick, R. L. (1979). Pharmacokinetic model for 2-amino-1 3,4-thiadiazole in mouse, dog, and monkey. *Cancer Treat. Rep.* **63**, 1939–1947.

King, F. G., and Dedrick, R. L. (1981). Physiologic model for the pharmacokinetics of 2′ deoxycoformycin in normal and leukemic mice. *J. Pharmacokinet. Biopharm.* **9**, 519–534.

King, F. G., and Dedrick, R. L. (1992). Physiological pharmacokinetic parameters for cis-dichlorodiammineplatinum(II) (DDP) in the mouse. *J. Pharmacokinet. Biopharm.* **20**, 95–99.

King, F. G., Dedrick, R. L., and Farris, F. F. (1986). Physiological pharmacokinetic modeling of cis-dichlorodiammineplatinum(II) (DDP) in several species. *J. Pharmacokinet. Biopharm.* **14**, 131–154.

LeRoy, A. F., Lutz, R. J., Dedrick, R. L., Litterst, C. L., and Guarino, A. M. (1979). Pharmacokinetic study of cis-dichlorodiammineplatinum(II) (DDP) in the beagle dog: Thermodynamic and kinetic behavior of DDP in a biologic milieu. *Cancer Treat. Rep.* **63**, 59–71.

Li, J., and Gwilt, P. (2002). The effect of malignant effusions on methotrexate disposition. *Cancer Chemother. Pharmacol.* **50**, 373–382.

Lutz, R. J., Dedrick, R. L., Straw, J. A., Hart, M. M., Klubes, P., and Zaharko, D. S. (1975). The kinetics of methotrexate distribution in spontaneous canine lymphosarcoma. *J. Pharmacokinet. Biopharm.* **3**, 77–97.

Lutz, R. J., Galbraith, W. M., Dedrick, R. L., Shrage, R., and Mellett, L. B. (1977). A model for the kinetics of distribution of actinomycin-D in the beagle dog. *J. Pharmacol. Exp. Ther.* **200**, 469–478.

Sonneveld, P., and Mulder, J. A. (1981). Development and identification of a multicompartment model for the distribution of adriamycin in the rat. *J. Pharmacokinet. Biopharm.* **9**, 577–601.

Sung, C., Blaney, S. M., Cole, D. E., Balis, F. M., and Dedrick, R. L. (1994). A pharmacokinetic model of topotecan clearance from plasma and cerebrospinal fluid. *Cancer Res.* **54**, 5118–5122.

Tsukamoto, Y., Kato, Y., Ura, M., Horii, I., Ishitsuka, H., Kusuhara, H., and Sugiyama, Y. (2001a). A physiologically based pharmacokinetic analysis of capecitabine, a triple prodrug of 5-FU, in humans: The mechanism for tumor-selective accumulation of 5-FU. *Pharmacol. Res.* **18**, 1190–1202.

Tsukamoto, Y., Kato, Y., Ura, M., Horii, I., Ishikawa, T., Ishitsuka, H., and Sugiyama, Y. (2001b). Investigation of 5-FU disposition after oral administration of capecitabine, a triple prodrug of 5-FU, using a physiologically based pharmacokinetic model in a human cancer xenograft model: Comparison of the simulated 5-FU exposures in the tumor tissue between human and xenograft model. *Biopharm. Drug Dispos.* **22**, 1–14.

Wu, Z.-Y., Smithers, B. M., and Roberts, M. S. (1997). Tissue and perfusate pharmacokinetics of melphalan in isolated perfused rat hindlimb. *J. Pharmacol. Exp. Ther.* **282**, 1131–1138.

Xu, L., Eiseman, J. L., Egorin, M. J., and D'Argenio, D. Z. (2003). Physiologically-based pharmacokinetics and molecular pharmacodynamics of 17-(allylamino)-17-demethoxygeldanamycin and its active metabolite in tumor-bearing mice. *J. Pharmacokinetic. Biopharm.* **30**, 185–218.

Zaharko, D. S., Dedrick, R. L., and Oliverio, V. T. (1972). Prediction of the distribution of methotrexate in the sting rays Dasyatidae *sabina* and *sayi* by use of a model developed in mice. *Comp. Biochem. Physiol.* A **42**, 183–194.

Zaharko, D. S., Dedrick, R. L., Peale, A. L., Drake, J. C., and Lutz, R. J. (1974). Relative toxicity of methotrexate in several tissues of mice bearing Lewis lung carcinoma. *J. Pharmacol. Exp. Ther.* **189**, 585–592.

PART IV

PBPK MODELING APPROACHES FOR SPECIAL APPLICATIONS

CHAPTER 12

PERINATAL PHARMACOKINETICS

Sun K. Lee

12.1 INTRODUCTION

Frequently, we hear the statement "... Children are not just little people ..." While this statement is true, the implication of the physiological uniqueness of different stages of the perinatal period goes much farther than that. If we consider the first 6 months of gestation, the human conceptus grows from microscopic size (i.e., a single cell) to about 500 g (about 1 pound). This is almost "infinite" growth. After 3 more months of growth, the average fetal weight increases to about 3.5 kg (7.5 pounds), and the organs and physiological systems are sufficiently mature that adaptation to life outside the maternal body is reasonably assured. From birth through adolescence, the rates of physical growth and functional development vary from system to system, organ to organ, and tissue to tissue. Thus, not only do infants and children differ from adults, but also at any point during maturation, the individual differs in structure and function from him/herself at any other age (National Research Council 1993). A reasonable question to ask, at this point, is: How does one study such an ever-changing, complex, dynamic biological process? One efficient and

---

Physiologically Based Pharmacokinetic Modeling, edited by Micaela B. Reddy, Raymond S. H. Yang, Harvey J. Clewell III, and Melvin E. Andersen.

realistic way to approach this issue is through computer simulation from modeling tools built and verified through detailed animal experimentation and the integration of human physiological data. Physiologically based pharmacokinetic (PBPK) modeling is, thus, a promising approach in this very important area of perinatal pharmacokinetics.

Babies and children also differ substantially from adults in their susceptibility to hazardous chemicals. Vulnerability often depends on developmental stage during which an individual is exposed to chemical insults (Rice and Barone 2000). There are critical periods of development during both prenatal and postnatal stages where specific structures or functions will be most sensitive to disruption (Weiss 2000). Since human experimental studies, particularly involving fetuses, neonates, or children, are impractical and have ethical limitations, we must resort to perinatal studies in animals for safety/risk assessment of drugs and chemicals. PBPK modeling, once again, is a very useful tool for extrapolating perinatal pharmacokinetics from laboratory animals to humans.

There are some unique characteristics in PBPK models describing perinatal transfer of chemicals. First, these models should incorporate physiological changes of the mother/dam, fetus, and baby/pup (we consider both human and laboratory animal work) during the perinatal period. The dynamic nature of physiological changes makes it necessary to collect such information. Second, these models should consider the effects of chemicals on the physiology of all perinatal stages including the pregnant mother/dam. Third, these models should incorporate the unique physiological compartments appearing at specific time points such as placenta in gestation and breast milk in lactation. For these characteristics, PBPK models for perinatal transfer can be complex.

Despite the complexity, much progress has been made in developing PBPK models for some chemicals and drugs transferred from mother to fetus/pup during perinatal period. Olanoff and Anderson (1980) presented a PBPK model for tetracycline disposition in the pregnant rat. This is the first PBPK model describing transplacental movement of a drug in a pregnant animal. It incorporated the growth of tissue and body weight during the gestational period. This model came to be the prototype PBPK model for the chemicals transferred through the placenta. Shelly *et al.* (1988) developed a PBPK model for lactational transfer of a number of volatile organic solvents and estimated the chemical concentration in neonates assuming that chemical was only exposed to the mothers by inhalation. Since there was a public health concern for the exposure of volatile organics to the pregnant and lactating mothers in industrial workplace, Shelly *et al.* (1988) used their model for risk assessment. This was the first PBPK model considering lactational transfer of chemicals.

Other physiological changes during perinatal period continued to be incorporated into the PBPK models developed in 1990s. Fisher *et al.* (1989) reported a PBPK model for transplacental movement of trichloroethylene in rats. In this article, they incorporated body weight change and increases of specific tissue sizes during pregnancy. The following year, Fisher *et al.* (1990) published a PBPK model for lactational transfer of trichloroethylene in rats. They extended a previous model to the lactational transport and described the increase of tissue volume as linearly proportional to body weight. In a somewhat different approach, other PBPK models considered nonlinear changes of body weight and tissue volume. For instance, O'Flaherty *et al.* (1992) and Luecke *et al.* (1994) incorporated the nonlinear growth of body weight and tissue volume into their models.

Over 40 PBPK models have been published describing perinatal transfer of chemicals and different life stages (Table 12.1). In addition, there are some excellent review articles summarizing perinatal transfer of chemicals (Clewell and Gearhart 2002; Corley *et al.* 2003). The majority of these PBPK models focused on the disposition of a chemical in a mother or a fetus/pup for a limited time points (e.g., one or two days after exposure rather than the whole perinatal period). In this chapter, we first present an overview of important physiological changes of mother and fetus/pup during the perinatal period and modeling approaches to incorporate these changes into PBPK models. Then, we will discuss each PBPK model developed for perinatal exposure to chemicals.

12.2 PHYSIOLOGICAL AND BIOCHEMICAL CHANGES DURING PREGNANCY

12.2.1 Body Weight Changes and Organ Growth

Fetal weight increases dramatically during gestational period and is affected by maternal age, weight, race, ethnicity, fetal sex, and the number of fetuses (Davis and Dobbing 1981). Placental growth shows a different pattern from fetal growth (Davis and Dobbing 1981). The maternal component of total weight gain comprises the increases in uterine and breast tissue, extracellular fluid, and fat (Goehbloed 1976). The uterus shows a dramatic increase in cavity volume. Extracellular fluid is also increased. The increase of plasma volume accounts for approximately half of that; the remainder is interstitial fluid. Another important component of the maternal weight gain is the accumulation of fat. Maternal fat stores are laid down primarily during the first half of pregnancy and provide an energy store for the third trimester when fetal growth predominates.

12.2.2 Physiological and Biochemical Changes in Pregnant Females

We will briefly review physiological and biochemical changes required to maintain a viable pregnancy with a special emphasis on humans. First, there are significant changes in endocrine system (Fisher 1992). In the placenta, hCG and hPL are produced, which are important in maintaining pregnancy and growth. In addition, maternal hormone levels such as progesterone, prolactin, ACTH, MSH, thyroxine, tri-iodothyronine, cortisol, aldosterone, and estrogen increase throughout the pregnancy. Parathyroid hormone also increases, which will increase calcium absorption by the gut so as to maintain calcium supply to the fetus.

Changes in intermediary metabolism and excretion are also important for pregnancy (Davis and Dobbing 1981). The fasting plasma glucose concentration falls during the first trimester, then rises and falls again towards term. Furthermore, there seems to be a progressive reduction of glucose tolerance following an oral glucose

TABLE 12.1 PBPK Models for Perinatal Transfer

Chemical class	Chemical	Stage	Species	References
Drug	Methadone	Fetus	Rat, human	Gabrielsson *et al.* (1985) Gabrielsson and Groth (1988)
Drug	Morphine	Fetus	Rat, human	Gabrielsson and Paalzow (1983)
Drug	Pethidine	Fetus	Rat, human	Gabrielsson *et al.* (1986)
Drug	PPBA	Fetus	Rat	Kawahara *et al.* (1998)
Drug	Tetracycline	Fetus	Rat	Olanoff and Anderson (1980)
Drug	Theophylline	Fetus	Rat, human	Gabrielsson *et al.* (1984)
Drug	Vitamin A acid	Fetus	Rat, mouse, monkey, human	Clewell *et al.* (1997)
Organic solvents	DMO	Fetus	Rat, mouse	O'Flaherty *et al.* (1992)
Organic solvents	2-Ethoxyethanol	Fetus	Rat, human	Gargas *et al.* (2000b)
Organic solvents	Methanol	Fetus	Rat, mouse	Ward *et al.* (1997) Clarke *et al.* (1993) Terry *et al.* (1995)
Organic solvents	2-Methoxyethanol	Fetus	Rat, mouse, human	Welsch *et al.* (1995) Hays *et al.* (2000) Gargas *et al.* (2000a)
Organic solvents	Solvents[a]	Neonate	Human	Fisher *et al.* (1997) Schreiber (1993)

Organic solvents	Tetrachloroethylene	Neonate	Rat	Byczkowski and Fisher (1994) Byczkowski *et al*. (1994) Byczkowski and Fisher (1995)
Environmental contaminants	2,4-D	Fetus	Rabbit	Kim *et al*. (1996)
Environmental contaminants	DDE	Fetus, neonate	Rat	You *et al*. (1999) Gray (1995)
Environmental contaminants	Methylmercury	Fetus, neonate	Rat, human	Clewell *et al*. (1999) Byczkowski and Lipscomb (2001)
Environmental contaminants	Trichloroethylene	Fetus, neonate	Rat	Fisher *et al*. (1989) Fisher *et al*. (1990)
Environmental contaminants	Perchlorate	Fetus, neonate	Rat, human	R. A. Clewell *et al*. (2001; a,b); Merrill *et al*. (2003)
Environmental contaminants	Isopropanol, vinyl chloride, methylene chloride, perchloroethylene, nicotine, and dioxin	Infants and children	Human	Gentry *et al*. (2003b); Sarangapani *et al*. (2003); Clewell *et al*. (2004)

[a] Solvents include 19 chemicals described in text.

load. The plasma concentration of free fatty acids, glycerol, cholesterol, and phospholipid increases during pregnancy. The significance of this change is not clear, but it should increase placental transfer of important substrates to the fetus. Pregnancy causes a reduction in amino acid catabolism as amino acids are used preferentially for gluconeogenesis, thus resulting in the decrease of plasma amino acid level. Plasma albumin concentration decreases during pregnancy, contributing to the edema seen in late pregnancy. Excretion of glucose, water-soluble vitamins, and amino acids increases partially due to the increase of GFR and diminished ability of reabsorption in kidney.

In addition, changes in cardiovascular and respiratory system occur (Rudolf 1992; Spitzer and Chevalier 1992). Plasma volume increases by approximately 43%, which result in the physiological anemia of pregnancy. Cardiac output increases by approximately 40% during the first trimester because of an increase in both heart rate and stroke volume. Much of the increased cardiac output is directed to the uteroplacental circulation, which has a 10-fold increase in blood flow. Because of fetal oxygen consumption, oxygen demand increases. This can be met by increased red cell mass and minute ventilation.

12.2.3 Physiological Changes in Fetuses

Before birth, the lungs in fetuses are not functional and blood flow is diverted away from them. Oxygen and nutrients are all transferred from the placenta. When the baby takes its first breath, pulmonary blood flow starts (Rudolf 1992). Cardiac output is a meaningless term in fetal life because of the shunting of blood. In the neonate, cardiac output is higher than that in adult. Heart rate rises at birth but falls soon after, reaching adult rates at puberty (Goehbloed 1976). The number of red cells increases dramatically throughout gestation and the size of red cells in fetuses is bigger than that in adults. After birth there is a rapid breakdown of red cells (Rudolf 1992).

The fetus obtains all nutritional requirements from its mother via the placenta. Glucose, amino acids, and fatty acids cross placenta easily, but the passage of some lipids is restricted. Plasma glucose level is half that of adults, but its control is poor. Fetal plasma amino acids are higher, which is believed to involve active transport mechanisms of the placenta. Towards the end of gestation, relatively large amounts of glycogen are stored in fetal muscle, liver, and fat stores to be used for the survival of neonate before the initiation of feeding (Rudolf 1992).

12.2.4 Mechanisms of Chemical Transfer Through Placenta

The placenta, which is vital for normal fetal growth and development, has three main functions. First, the placenta is an important endocrine organ of pregnancy. Second, the placenta protects the fetal allograft from attack by the maternal immune system. Third, the placenta is the interface between maternal and fetal plasma across which maternofetal transfer of nutrients and waste products occurs (Davis and Dobbing 1981).

The placenta shows the greatest morphological and physiological diversity across species of any mammalian organ. For example, pig and sheep placentas have four continuous cell layers separating the maternal and fetal plasma whereas the human placenta has two cell layers (Case and Waterhouse 1994). Therefore, the permeability of sheep and pig placentas to the solutes is lower than that of the human placenta. The two cell layers separating maternal and fetal blood in the human placenta are the syncytiotrophoblast and the fetal capillary epithelium. Solutes may cross these cell layers by the following three mechanisms (Case and Waterhouse 1994).

Passive diffusion. Fick's law of diffusion can be related to placental transfer of hydrophobic molecules. Since fetal side is slightly negative from maternal side, positively charged materials will be transferred to the fetus in addition to hydrophobic solutes.

Carrier-mediated transport. Transcellular transport across the placental cell layer requires two transport proteins, one in the maternal-facing plasma membrane and one in the fetal-facing plasma membrane. An example of this transport is ionized calcium.

Endocytosis/exocytosis. Large molecules (e.g., proteins) may gain access into the placenta by fluid phase endocytosis in which the plasma membrane invaginates to engulf solute and water in the extracellular space. In receptor-mediated endocytosis, the solute binds to specific receptors on the plasma membrane, which eventually invaginates. The syncytiotrophoblast of placenta is well endowed with coated pits and vesicles, which are indispensable for endocytosis. Thus, this may be the primary mechanism of maternofetal transfer of large molecules such as immunoglobulin G.

12.2.5 Mechanisms of Chemical Transfer Through Breast Milk

In late pregnancy, progesterone and estrogen play a role in stimulating the growth of mammary glands so that milk will be secreted (Corley *et al.* 2003). At parturition, a series of dramatic changes leads the cells in mammary glands to secrete milk. Milk is produced and stored in alveolar units in mammary glands (Linzell *et al.* 1975). Removal of milk from alveoli is accomplished by contraction of myoepithelial cells surrounding alveoli. The production, secretion, and composition of milk are determined by endogenous hormones such as progesterone, estrogen, growth hormone, adrenal glucocorticoids, and insulin (Corley *et al.* 2003).

Five distinct pathways are involved in the synthesis and secretion of milk in mammary glands operating in parallel to transform precursors derived from the plasma or interstitial fluid into milk constituents (Oskarsson *et al.* 1998). Even though the composition of milk is different from one species to another, the biochemical mechanisms leading to the secretion of various milk components are similar.

Exocrine pathway. Most aqueous phase of milk components are secreted by this pathway. Major milk proteins are synthesized on ribosomes and transferred to the Golgi system. Secretory vesicles produced in the Golgi system move to the plasma membrane where they fuse and release their contents into the milk space.

Lipid transport pathway. Milk lipids are secreted to the mammary glands via this pathway. Triglycerides are synthesized from precursor fatty acids in the smooth endoplasmic reticulum of the mammary alveolar cell and coalesce into large droplets that are drawn to the apex of the cell. The droplets become enveloped in apical membrane and then extruded from the cell as the complex state called the milk fat globuli.

Transport across the apical membrane. Small molecules including sodium, potassium, chloride, and glucose can pass across the apical membrane although detailed mechanism of this pathway remains to be explored.

Transcytosis. Intact proteins can cross the mammary epithelium from the interstitial fluid by transcytosis. Immunoglobulins are the best studied molecules that pass through transcytosis. The proteins bind to receptors at the basolateral membrane. At the apical membrane the extracellular portion of the receptor is cleaved and secreted with bound protein. Many peptide hormones and proteins are considered to be transferred via this pathway.

The paracellular pathway. During lactation the passage of molecules between alveolar cells is impeded by a gasket-like structure called the tight junction. Sometimes the junctions become leaky, allowing the components of interstitial space to pass into the milk. At the same time, milk components can cross between the cells.

The transport of chemicals into milk may follow the same pathways as do milk components, but this issue has not been extensively studied. In general, passive diffusion across the membrane is believed to be the major transport mechanism of xenobiotics (Corley *et al.* 2003). Unionized and lipophilic chemicals can be transported by this way. In addition, certain type of carrier proteins may be involved in the transfer of toxic metals into the milk (Oskarsson *et al.* 1998).

12.3 PHYSIOLOGICAL FACTORS INCORPORATED INTO PBPK MODELS FOR PERINATAL PHARMACOKINETICS

In this section, the mathematical approaches that investigators have used to model the physiological changes of pregnancy and lactation are presented, including equations developed to describe physiological phenomena. The variable names in the equations have been modified so that a consistent nomenclature could be presented here. Also, in certain instances where coefficients were calculated based on a specific dataset, we present general designations such as a, b, c . . . , and so on, instead of the reported values by the investigators that are specific to their dataset. In other cases where equations have general applicability, the statistically derived coefficients from the original publications are retained.

12.3.1 Body Weight in the Mother

Body weight changes dramatically in both the mother and the fetus/neonate. Therefore, most PBPK models of perinatal stages have incorporated these changes. There are three approaches to incorporating body weight changes in pregnant and lactating mothers into the model. In the first approach, body weight is measured at some time points during pregnancy and mathematical expressions were derived using linear regression analysis. In perinatal transfer model of trichloroethylene in rats developed by Fisher *et al.* (1989, 1990), body weight changes of pregnant and lactating mothers were estimated by subtracting calculated placental/mammary and fetus weight from the measured weight of pregnant/lactating rat. Then, body weight changes were described as linear increase. In doing so, the maternal body weight changes were 10% increase during gestation and 16% increase during lactation.

In the second approach, O'Flaherty *et al.* (1992) described the body weight change as a function of body weight increase of nonpregnant animal and organ weight increase of placenta, mammary gland, and other maternal tissues. Thus, for the construction of a PBPK model describing placental transfer of a weak organic acid, maternal weight changes were described as nonlinear increases of body weight dependent on age with volume increases of the tissues including placenta, mammary gland, fat, and liver. Tissue volume changes were described as nonlinear increases representing published physiological changes.

In the third approach, certain PBPK models describing perinatal transfers do not incorporate body weight changes of pregnant mothers (Gabrielsson and Paalzow 1983; Gabrielsson *et al.* 1985). Since body weight change occurs during whole pregnancy period, it is not a significant factor if the chemical eliminates very rapidly from the body.

12.3.2 Organ Volume and Cardiac Output in the Mother

In general, organ volume is proportional to body weight. However, the organ volume changes in pregnant mother and the fetus/pup are not linearly correlated with body weight changes because every organ has a different growth pattern during the perinatal stages. In the pregnant mother, volumes of fat, placenta, and mammary gland significantly increase. In Fisher's PBPK model (Fisher *et al.* 1989), fat accumulation during pregnancy was expressed as a linear increase from 6% to 12% of body weight during the gestational period. Accumulated fat was depleted during lactational period, which was expressed as a linear decrease from 12% to 6% of body weight (Fisher *et al.* 1990). In O'Flaherty's model (O'Flaherty *et al.* 1992), they also described fat accumulation as linear increase during pregnancy. Comparing with fat accumulation, volume changes in other organs are not so dramatic. Therefore, most PBPK articles did not consider volume changes in specific organs except fat. However, O'Flaherty *et al.* (1992) described volume increase of the liver in mice during pregnancy. They described volume change of liver as linear increase from gestational day (GD) 6.

Luecke *et al.* (1994) tried to express volume changes in maternal organs using regression equation. They used following allometric equation to describe the increase of each tissue size.

$$\ln W_i = \ln a_i + b_i \times \ln \mathrm{BW}_F + c_i \times (\ln \mathrm{BW}_F)^2 \tag{12.1}$$

where W_i is the weight of a given tissue in the mother. BW_F is the body weight of the fetuses, and a_i, b_i, and c_i are constants fitted to data for each of the "*i*" tissues.

Maternal cardiac output and tissue-specific blood flows are affected by volume changes of tissues and water during pregnancy. In nonpregnant rat, cardiac output is defined using the following equation:

$$Q_C = Q_{\mathrm{CC}} \times (\mathrm{BW})^{0.74} \tag{12.2}$$

where Q_C is cardiac output of an animal. Q_{CC} is allometric constant. In most PBPK models, Eq. (12.2) is used to describe maternal cardiac output. Fisher *et al.* (1989) used the same equation for estimating maternal cardiac output during the gestational period, but they observed an increase in maternal cardiac output during the lactational period. They incorporated a linear increase in Q_{CC} during the lactational period. O'Flaherty *et al.* (1992) considered more complex approach to modeling the changes of cardiac output and blood flow during pregnancy. First, they estimated the changes of cardiac output in growing nonpregnant rat as follows:

$$Q_{\mathrm{CG}} = Q_{\mathrm{C}} \times (\mathrm{BW}_G / \mathrm{BW}_M)^{0.7} \tag{12.3}$$

where Q_{CG} is the cardiac output of the growing rat. BW_G is the body weight of the growing rat. BW_M is the weight of young mature animal. To describe cardiac output during pregnancy, they incorporated the changes of cardiac output in the placenta into total cardiac output. Total cardiac output was estimated by the following equation:

$$Q_{\mathrm{CP}} = Q_{\mathrm{CG}} + N \times (Q_{\mathrm{DEC}} + Q_{\mathrm{CAP}}) \tag{12.4}$$

where Q_{CP} is the cardiac output of the pregnant rat, Q_{CG} is the cardiac output of the growing, nonpregnant rat, N is the number of concepti, Q_{DEC} is the change of cardiac output associated with yolk sac placenta, and Q_{CAP} is the change of cardiac output associated with chorioallantoic placenta. Q_{DEC} was described differently based on gestational time. First, the increase of cardiac output in yolksac placenta between GD 6 and GD 10 was expressed as

$$Q_{\mathrm{DEC1}} = 0.55 \times (\mathrm{GD} - 6) \tag{12.5}$$

where GD is gestational day [or EAGE = embryo age in days as in the original publication by O'Flaherty *et al.* (1992)].

Second, the fall in cardiac output associated with disappearance of yolk sac placenta after GD 10 is described as

$$Q_{DEC2} = 2.2 \times \exp[-0.23 \times (GD - 10)] \quad (12.6)$$

The change of cardiac output associated with yolk sac placenta, Q_{DEC}, is the sum of these two equations.

$$Q_{DEC} = Q_{DEC1} + Q_{DEC2} \quad (12.7)$$

The change of cardiac output associated with chorioallantoic placenta, Q_{CAP}, was described as

$$Q_{CAP} = a \times (GD - 12)^{4.36} \quad (12.8)$$

where a is empirical constant.

The total cardiac output of each fetus was modeled with the logistic expression

$$Q_{CF} = Q_{CON} / [1 + 20{,}000 \times \exp(-0.55 \times GD)] \quad (12.9)$$

where Q_{CF} is the cardiac output of each fetus, Q_{CON} is the maternal blood flow to each fetus, and GD is gestational day. O'Flaherty *et al.* (1992) assumed that 25% of Q_{DEC} goes to the fetus. Thus, the maternal blood flow, Q_{CON} was described as

$$Q_{CON} = (0.25/N) \times Q_{DEC} + Q_{CAP} \quad (12.10)$$

where N is the number of fetus in a pregnant mother.

12.3.3 Chemical Transfer Through the Placenta and Mammary Gland

Fisher *et al.* (1989) separated the placenta into maternal and fetal compartments. Chemical transfer through each placenta was considered as flow-limited. Clarke *et al.* (1993) developed a PBPK model in which the transfer of chemicals from maternal placenta to fetal placenta is diffusion-limited. In one study, the placenta was combined to the richly perfused compartment in a PBPK model (Gargas *et al.* 2000a).

In most PBPK papers dealing with lactational transfer, mammary gland is included or connected to milk compartment. The transfer of chemicals from mammary gland to the pups is considered to be dependent upon the rate of milk production (Clewell and Gearhart 2002). One PBPK model showed that mammary gland was combined to fat compartment (Gargas *et al.* 2000a).

12.3.4 Body Weight and Organ Volume in the Fetus/Pup

Body weight increases in fetus/pup are more dramatic than in the mother. The body weight of a fetus increases by a factor of more than 100 during the gestational period (Luecke *et al.* 1994). To simulate weight increases in fetuses and pups, several approaches have been used. First, Fisher *et al.* (1989, 1990) developed two equations to express body weight changes in rat fetuses and pups. For fetal growth, they used the following equation:

$$BW_F = [a(GD - 13)]^3 \tag{12.11}$$

where a is specific growth constant. This equation was based on an investigation of the relationship between maternal age and body weight of the fetus (Huggett and Widdas 1951). For the growth of the pup, Fisher *et al.* (1990) used an exponential equation to describe body weight increase such as

$$BW_P = 0.0045 \times \exp(0.0036 \times PND) \tag{12.12}$$

To parameterize above coefficients, they measured pup's body weight from postnatal day (PND) 3 to PND 23 and described using exponential function.

Second, O'Flaherty *et al.* (1992, 1995) described body weight changes in murine fetuses and pups using a segmented equation. They divided body weight growth into three periods at which they showed different growth patterns according to the following equation:

$$BW_F = EW1 + EW2 + EW3 \tag{12.13}$$

where $EW1 = (0.12 \times GD)^{4.53}$ [from GD 1 to GD 8.6], $EW2 = [1.2 \times (GD - 8.6)]^{2.6}$ [from GD 8.6 to GD 15.8], $EW3 = a \times [\text{birth weight} - EW2]^{1/3.2}$ [from GD 15.8 to parturition]. Using this approach, they were able to describe body weight changes in both rats and mice with slight differences in constants and coefficients.

Third, Luecke *et al.* (1994) used Gompertz equation to describe body weight changes in the human fetus. In this equation, the fetal body weight increase is described as

$$BW_F = BW_0 \times \exp\{A_0/a[1 - \exp(-a \times (t - t_0))]\} \tag{12.14}$$

where BW_0 is initial weight at the time t_0, A_0 is the relative growth at the time t_0, a is the exponential rate of decrease in growth rate, and t is day post-conception. Using this equation, Luecke *et al.* (1994) could simulate successfully body weight increases in fetus of different species using specific coefficients.

Finally, You *et al.* (1999) used a simple approach to describe body weight changes in pups. They measured body weight directly at different time points and developed a relationship between body weight and time using table function included in their simulation software.

Even though there are some studies on fetal and neonatal growth and body weight changes, the changes of individual tissues in fetuses or neonates have not been studied extensively. Luecke *et al.* (1994) described organ weight increase in fetus using the following equation.

$$\ln W_{Fi} = \ln a_i + b_i \times \ln BW_F \tag{12.15}$$

where W_{Fi} is the weight of each tissue in fetus and a_i and b_i are constants fitted to the data. Olanoff and Anderson (1980) described tissue weight changes using the following equation:

$$W_{Fi}^{1/3} = a_i \times (GD - 11) \tag{12.16}$$

where a_i is a constant calculated by fitting the equation to data, and "i" is each ith tissue. The growth of pup's fat during the lactational period was estimated as

$$V_{PF} = 0.0199 \times \mathrm{BW}_P + 1.644 \tag{12.17}$$

by You *et al.* (1999), when V_{PF} is the percent of body weight in the fat compartment.

12.4 PBPK MODELS FOR PERINATAL TRANSFER

In this section, we provide a brief discussion on chemical-specific PBPK modeling related to perinatal pharmacokinetics. The discussion is arranged chronologically starting with the earliest study by Olanoff and Anderson (1980). When there were more than one study, we used the first in the series for chronological arrangement. We chose to provide some highlights of the significant study(ies) for each of the chemical listed rather than detailed discussion of the experimental details. Interested readers are encouraged to consult the original articles for more information.

12.4.1 Tetracycline

Tetracycline is an antibiotic with broad spectrum activity and has been used for the treatment of many infectious diseases. Although tetracycline is known for low toxicity, it can cause acute fatty liver syndrome in the mother and dental defects and abnormalities in the fetus. Olanoff and Anderson (1980) developed a PBPK model to describe the disposition of tetracycline in maternal and fetal tissues of pregnant rats incorporating the growth of mammary glands, uterus, placenta, fetus, and blood in the pregnant rat. Partition coefficients in the pregnant rat and fetus were determined by fitting the tissue and plasma concentration data obtained from an *in vivo* kinetic study. They did not consider the metabolism of tetracycline or the transfer of metabolites to the fetus. This model was the first PBPK model to include the dynamics of growth during pregnancy and served as a default schematic for the subsequent PBPK models describing pregnancy and placental transfer of chemicals.

12.4.2 Morphine

Morphine is used as a potent pain killer. If pregnant mothers use morphine chronically, children born to the mother show higher incidences of withdrawl symptoms. Gabrielsson and Paalzow (1983) developed a PBPK model for the disposition of morphine in rats during pregnancy. They did not consider the growth of maternal and fetal tissues because they were interested in simulating the disposition of morphine during a short time period (hours) in which tissue growth is not a significant factor. All tissue compartments were treated as flow-limited except for the placenta, which was modeled using diffusion-limited uptake. The fetuses were connected to maternal tissues as a single compartment. Parameters including tissue:blood partition coefficients, protein binding constants, and diffusion permeation constants were optimized by fitting the tissue and plasma concentration data obtained from an intravenous infusion study. Metabolic constants were obtained from previous pharmacokinetic studies. To improve the ability of the model to describe experimental data, the model was modified to simulate the sedative effects of morphine by reducing the blood flow rate to the liver. Gabrielsson and Paalzow (1983) used this model for estimating the disposition of morphine in humans after scale-up.

12.4.3 Theophylline

Theophylline is used to treat or prevent the symptoms of bronchial asthma, chronic bronchitis, and emphysema. Since theophylline causes birth defects in animals, it is not reccmmended for use by pregnant women. Gabrielsson *et al.* (1984) developed a PBPK model describing the disposition of theophylline in pregnant rats similar to the morphine model (Gabrielsson and Paalzow 1983). However, there were some differences including physiological parameters (e.g., tissue volumes and blood flow rates) and the transfer mechanism of theophylline to the brain. The transfer of theophylline to maternal brain was considered to be diffusion-limited. Pharmacokinetic and simulation results showed that fetal levels of theophylline were lower than maternal levels. They used the model to estimate the transfer of theophylline in humans where the levels of theophylline in fetus were lower than those in the mother.

12.4.4 Methadone

Methadone is used to prevent morphine withdrawal symptoms. Compared with morphine, methadone is considered to be relatively safe to the pregnant mother and fetus. Gabrielsson *et al.* (1985) and Gabrielsson and Groth (1988) developed two PBPK models for methadone. The first model was developed for pregnant rats (Gabrielsson *et al.* 1985). The model structure and physiological parameters were based on the theophylline model. Mass transfer into all tissue compartments was assumed to be flow-limited with the exception of brain, intestine, and fetus, which were assumed to be diffusion-limited. The renal clearance constant was obtained from previous pharmacokinetic studies, but metabolic constants for the liver were optimized using data obtained from intravenous infusion studies.

Pharmacokinetic behavior of methadone was complex and their first model could not explain completely. Thus, Gabrielsson and Groth (1988) extended their model to include additional maternal tissues (adipose and heart tissues) and separate fetus into several compartments (brain, liver, and remaining carcass). Adipose and heart compartments were added because heart muscle stored more methadone than skeletal muscle and fat had a significant capacity for methadone retention. The fetus was separated to allow the consideration of embryotoxic effects, analgesic activity, and different pharmacokinetics from mother. The extended model was useful to explain the complex pharmacokinetic behaviors of methadone.

12.4.5 Pethidine

Pethidine, which has been used for analgesia during labor, can induce respiratory depression in newborns. Therefore, Gabrielsson *et al.* (1986) developed a PBPK model for the disposition of pethidine in the pregnant rat. The modeling approach was similar to that used for morphine, theophylline, and methadone. All tissue compartments were flow-limited except the placenta which was diffusion-limited. Simulation results suggested that the transfer of pethidine to the fuetus is low.

12.4.6 Trichloroethylene

Trichloroethylene is an environmental contaminant coming from contaminated shower water, household products, and byproducts in chlorinated drinking water. In toxicology studies, trichloroethylene causes liver, kidney, and brain damage and cancer. Fisher *et al.* (1989) developed the first PBPK model for disposition of trichloroethylene in pregnant rats. The description of tissue growth in the maternal and fetal compartments was based on the growth model of Olanoff and Anderson (1980). Mass transfer of chemical to and from all tissue compartments was assumed to be flow-limited except for the transfer of metabolite between the placenta and fetus, which was assumed to be diffusion-limited.

Fisher *et al.* (1990) also developed a PBPK model for lactational transfer of trichloroethylene in rats. The model structure was similar to that of pregnant rat except that chemical transferred between the dam and pups through breast milk. All tissue compartments were assumed to be flow-limited and time-dependent parameters including physiological changes were estimated using a linear regression technique. The unique characteristics of this model were that all partition coefficients, absorption constants, and metabolic constants were measured directly through extensive pharmacokinetic studies rather than by finding the parameter values that resulted in the minimum difference between model output and experimental data. These two models with trichloroethylene, for pregnancy and lactation, are the basis for many subsequent models describing perinatal transfer.

12.4.7 5,5′-Dimethyloxazolidine-2,4-Dione (DMO)

O'Flaherty *et al.* (1992) developed a PBPK model for DMO, a weak acid with teratogenic potential in rats and mice during the gestational period. The primary purpose of this project was to develop a mathematical framework to describe biological changes in the pregnant animal and fetus. DMO is eliminated without protein binding and metabolism. This characteristic considerably simplify model structure required for the description of pharmacokinetic behavior. Mass transfer to all tissue compartments was described as flow-limited except in the placenta where the transfer was described as diffusion-limited. This model incorporated many of the physiological changes that were not incorporated in previous models that occur during the gestational period.

12.4.8 Tetrachloroethylene

Tetrachloroethylene causes liver and kidney cancers in animals and is easily transferred to neonates through breast milk. Thus, some researchers wanted to simulate lactational transfer of tetrachloroethylene and to apply PBPK modeling to risk assessment in neonates. Schreiber (1993) developed a PBPK model for the lactational transfer of tetrachloroethylene in humans. This model was composed of four tissue compartments (i.e., fat, muscle, liver, and remaining carcass). The mammary gland was not included in the model. Rather, the author assumed that the milk concentration of tetrachloroethylene was proportional to that of the fat compartment. He simulated the amount of tetrachloroethylene in babies for various maternal exposure scenarios assuming that babies were exposed to tetrachloroethylene only by breast milk. He concluded that the margin of exposure was relatively small to the exposure levels inducing toxic effects in humans.

Byczkowski *et al.* (1994) also developed a PBPK model for tetrachloroethylene in the lactating rat. The model structure was similar to Schreiber's model (Schreiber 1993) except that a mammary tissue compartment was included. Pup compartments included the lung, GI tract, and remaining carcass. The pup's GI compartment received tetrachloroethylene from the mammary compartment of the mother. Physiological changes in maternal and pup tissues were incorporated based on the model of Fisher *et al.* (1990).

Byczkowski and Fisher (1994) extended this model to simulate human exposures. Human partition coefficients were assessed in nine volunters. Human physiology was adapted from an adult PBPK model published by Ward *et al.* (1988). Kinetic parameters were reoptimized using experimental data from the literature describing human inhalation exposures. They used this model to simulate the concentrations of tetrachloroethylene in blood and exhaled air in occupational exposure scenarios. Byczlowski and Fisher (1995) extended previous models to apply to cancer risk assessment of tetrachloroethylene to breast-fed infants. They simulated a series of exposure scenarios and compared the model output with that of Schreiber's (Schreiber 1993). Their predicted exposures to breast-fed infants were lower than those of Schreiber's because Schreiber's model used peak milk concentrations rather than actual concentrations in milk for each exposure scenarios. Using PBPK model simulations, they calculated cancer risk of breast-fed infants using linearized multistage (LMS) model.

12.4.9 2-Methoxyethanol and Methoxyacetic Acid

2-Methoxyethanol is teratogenic and metabolism of 2-methoxyethanol to methoxyacetic acid is required for teratogenicity. Several PBPK models simulate the dosimetry of 2-methoxyethanol during perinatal period. In the first modeling approach of Clarke *et al.* (1993), a PBPK model for 2-methoxyethanol in pregnant mice was developed based on the model of O'Flaherty *et al.* (1992). Maternal tissue weights were measured directly and used to set parameters of mathematical equations derived from O'Flaherty *et al.* (1992) to simulate body weight changes during gestational period. Weights of individual conceptuses and their tissues were measured at gestational days (i.e., GD 10, 11, 12, 14, 16, and 18). Partition coefficients for 2-methoxyethanol and methoxyacetic acid were determined in gestational day 11 using a vial equilibration and centrifugal ultrafiltration method. Metabolic constants were obtained by *in vitro* data then reoptimized using *in vivo* intravenous study through formal parameter estimation. The pharmacokinetics of the metabolite, methoxyacetic acid, was described using a classical one-compartment model and linked to the placenta compartment for 2-methoxyethanol. The transfer of methoxyacetic acid from the placenta to the embryos was diffusion-limited, while all other uptake processes were described as flow-limited.

Terry *et al.* (1995) refined the model of Clarke *et al.* (1993) to construct some hypothetical model structures to describe the mass transfer of chemicals between dams and embryos for exposures during different days of gestation. For simulation of an exposure on GD 8, a blood-flow limited transfer was adequate to describe the dosimetry of methoxyacetic acid. For exposures on GD 11 and 13, the three different model structures were developed to see which transport process best described the complex dosimetry of methoxyacetic acid in the conceptus: They were (a) pH trapping of ionized methoxyacetic acid, (b) active transport of methoxyacetic acid, and (c) reversible binding of methoxyacetic acid. Among the three, an active transport model best described the dosimetry of methoxyacetic acid at different exposure time points even though the mechanism of transport and accumulation in the conceptus remains to be explored. Welsch *et al.* (1995) extended the previous two models to include rats and humans. Using their model, Welsch *et al.* calculated C_{max} and AUC for methoxyacetic acid in maternal plasma and embryo. These values were compared with those of rats and mice to estimate safe levels in humans. C_{max} was used for as a dose metric for exencephaly; AUC was used as a dose metric for risks of digit malformations.

O'Flaherty *et al.* (1995) developed a PBPK model for methoxyacetic acid in pregnant mice. The model, based on the earlier model for DMO, successfully described the tissue dosimetry of methoxyacetic acid at the lower dose, but underestimated that at the higher dose. Since this model was developed by assuming that the distribution of weak acid was solely dependent on pH, incorporation of further mechanistic information may be needed to describe the distribution of methoxyacetic acid more accurately.

Hays *et al.* (2000) developed a PBPK model for 2-methoxyethanol in the pregnant rat. Both 2-methoxyethanol and methoxyacetic acid have their respective model structures. However, fetal and placental tissues were lumped into richly perfused compartment and the mammary tissue was incorporated into the fat compartment. This simplification was justified by the similarities of partition coefficients between the tissues. All compartments were flow-limited. Using the model, C_{max} and AUC for methoxyacetic acid in maternal plasma and embryos were calculated. Gargas *et al.* (2000a) extended the Hays' model to the human and considered inhalation exposure. C_{max} and daily AUC for methoxyacetic acid in maternal blood were used as a surrogate for extrapolation between species. This model was validated by experimentation on human volunteers. Using traditional PBPK scale-up techniques, the model predicted that pregnant women exposed for 8 hr/day, 5 days/week, for the duration of pregnancy would need to be exposed to 12 or 60 ppm 2-methoxyethanol to produce maternal 2-methoxyacetic acid blood concentrations (C_{max} or average daily AUC) equivalent to those in rats exposed to the NOEL (10 ppm) or LOEL (50 ppm), respectively.

12.4.10 Methylmercury (MeHg)

MeHg is a representative developmental neurotoxicants. That can be easily transferred from mother to fetus/pup through perinatal transfer. Gray (1995) developed a PBPK model for methylmercury in the pregnant rat describing growth using the model of Olanoff and Anderson (1980). The model included the increase of tissue volumes, plasma volume, and blood flow rates that accompany pregnancy. Unlike other PBPK models, there was no incorporation of partition related phenomenon. The model was composed of diffusion-limited transfer (or membrane limited transfer) and protein binding of mercury with protein thiols. Protein binding constants were optimized by *in vivo* studies at pseudo-equilibrium. Simulation results suggested that membrane transport coefficients were important for describing short-term pharmacokinetic behavior of MeHg and the linear binding constants in tissues were important for the long-term disposition of MeHg.

A PBPK model for MeHg in pregnant women was developed by Clewell *et al.* (1999). The model structure, based on the model of Gearhart *et al.* (1995), incorporated Monte Carlo analysis in order to determine the plausible range for a dose conversion factor between hair mercury levels and steady-state intake levels. Most tissue compartments were described as flow-limited with the exception of placenta, brain, and red blood cells, which were described as diffusion-limited. Using simulation results, they calculated reference dose (RfD) and oral minimal risk level (MRL).

This model was later extended by Byczkowski and Lipscomb (2001) to describe lactational transfer of MeHg in women by incorporating parameters involved in lactational physiology. They used same model structure of Clewell *et al.* (1999) for maternal and fetal compartments and incorporated milk and infant compartments. Their model accurately described lactational transfer of MeHg by breast milk in humans and predicted the daily intake of MeHg in infants.

12.4.11 2,4-Dichlorophenoxyacetic Acid (2,4-D)

2,4-D is an organic acid herbicide and accumulates into the brain and cerebrospinal fluid through the active organic ion transport system in the choroid plexus. Kim *et al.* (1996) developed a hybrid PBPK-compartmental model describing 2,4-D disposition in the pregnant rat and rabbit. The maternal brain tissues were subdivided into the hypothalamus, caudate nucleus, hippocampus, forebrain, brainstem, and cerebellum. The brain plasma, cerebrospinal fluid, blood, and placenta were also included to PBPK model. The remaining body was modeled using a classical two-compartment model including central and deep compartment. For fetal tissues the model included brain plasma, cerebrospinal fluid, brain, blood, and the remaining body. The transfer from the maternal blood to the placenta and fetus was treated as flow-limited. Maternal and fetal brains were described as diffusion-limited. Model parameters were obtained by previous intravenous pharmacokinetic studies. This model served as a basis for further investigation of regional brain dosimetry for 2,4-D.

12.4.12 Methanol

Methanol is teratogenic at high concentration exposure. Thus, there is a need to investigate the dose metric relating to developmental defects. Ward *et al.* (1997) developed a PBPK model for methanol in pregnant rats and mice based on the earlier models developed for 2-methoxyethanol and incorporating maternal and fetal tissue growth using the methods of O'Flaherty *et al.* (1992). Most tissue compartments were described as flow-limited; fetus and extraembryonic fluids were described as diffusion-limited. Kinetic parameters and other coefficients were determined in previous studies or measured directly. Interestingly, the diffusion parameters, which they optimized at each gestational day, showed extensive variability. Validation studies for the diffusion parameters demonstrated that methanol limits its own transfer between placenta and fetus by decreasing blood flow to the uterus. Simulation results suggested that methanol levels in fetus are only dependent upon the balance of diffusion between placenta and fetus.

12.4.13 Vitamin A Acid

Vitamin A (i.e., retinoic acid) has been shown to be teratogenic in mice, rats, and monkeys. Clewell *et al.* (1997) developed a PBPK model for vitamin A acid to provide more relevant internal dose estimates for studying human teratogenic risk. The model was composed of 10 maternal compartments including the embryo. Most tissue compartments were described as having flow-limited transfer with the exception of dermal uptake and the transfer between the embryo and placenta, which were described as being diffusion-limited. Model parameters were obtained from diverse pharmacokinetic studies. The model was used to simulate C_{max} and AUC for vitamin A acid and its active metabolites for diverse administered doses in maternal plasma and these were compared with C_{max} and AUC of minimally teratogenic oral dose in mice, rats, and monkeys.

12.4.14 Organic Solvents

Fisher *et al.* (1997) constructed a PBPK model for a variety of organic solvents which are of concern due to occupational exposure. The model was based on their previous models developed for trichloroethylene (Fisher *et al.* 1990) and tetrachloroethylene (Byczkowski and Fisher 1994; Byczkowski *et al.* 1994). Nineteen chemicals (i.e., benzene, carbon tetrachloride, bromochloromethane, chlorobenzene, chloroform, methyl chloroform, diethyl ether, 1,4-dioxane, halothane, *n*-hexane, isoflurane, methylene chloride, methyl ethyl ketone, tetrachloroethylene, styrene, trichloroethylene, tetrachloroethane, toluene, and the *o*-, *m*-, and *p*-xylenes) were selected and simulated for the concentrations in the neonates at occupational exposure scenarios. Based on PBPK model simulations of exposure of a lactating woman to a threshold limit of each of these chemicals, only tetrachloroethylene, bromochloroethane, and 1,4-dioxane exceeded the U.S. Environmental Protection Agency (US EPA) noncancer drinking water ingestion rates for children.

12.4.15 *p*-Phenylbenzoic Acid (PPBA)

PPBA is a nonsteroidal antiinflamatory drug (NSAID). Some research groups investigated pharmacokinetics of PPBA in the pregnant rat because NASIDs can be clinically used to treat inflammation of pregnant women. Kawahara *et al.* (1998) developed a PBPK model for the disposition of PPBA in the pregnant rat. They constructed the model using 10 compartments from mother and eight compartments from fetus. There are two features in this model. One is the exchange of drug between skin and amniotic fluid in the fetus since PPBA in amniotic fluid can pass through fetal skin. The other is fetal blood flow based on the anatomical circulation in the uterus. Drug transfer to each compartment is considered to be flow-limited except several maternal tissues including brain, muscle, fat, and skin. They simulated drug disposition in the pregnant rat and the fetus for three exposure routes (i.e., intravenous, intraumbilical, and intramuscular injection). The simulation results showed that placental transfer clearance from fetus to mother was larger than from mother to fetus, suggesting that some transport mechanism specifically transfers drugs from the fetus into mother's bloodstream.

12.4.16 *p,p′*-Dichloro-2,2-*bis*(*p*-chlorophenyl)ethylene (DDE)

DDE is a metabolite of DDT and can produce estrogenic effects and possibly acts as an androgen receptor agonist in rats. You *et al.* (1999) developed a PBPK model for DDE disposition in both dams and fetuses/neonates during the perinatal period. They developed two different model structures: one for pregnancy, the other for lactation. The pregnancy model was based on the model of O'Flaherty *et al.* (1992). Placenta, fetus, and fat compartment were described as being diffusion-limited. Other tissue compartments were described as being flow-limited. The lactational model was based on the model of Fisher *et al.* (1990). The transfer of DDE from the mammary gland to the milk was diffusion-limited. The diffusion rate was fitted using the data from a cross-fostering study. Other constants including partition coefficients were determined in previous pharmacokinetic studies. Cross-fostering studies were performed to validate these models. For these studies, dams were exposed to DDE at gestational days or postnatal days and tissue concentrations were compared between them. Using these models, they proposed a possible mechanism on the importance of lipoproteins and fatty acids for the transfer of DDE to the fetus and neonate.

12.4.17 2-Ethoxyethanol and Ethoxyacetic Acid

2-Ethoxyethanol can induce similar teratogenic effects with 2-methoxyethanol even though its toxic potency is lower than that of 2-methoxyethanol. Gargas *et al.* (2000b) developed a PBPK model for 2-ethoxyethanol and its metabolite, ethoxyacetic acid in pregnant rat. The model structure for 2-ethoxyethanol was the same as that of 2-methoxyethanol because both 2-ethoxyethanol and 2-methoxyethanol show similar pharmacokinetic behaviors. Partition coefficients for 2-ethoxyethanol and ethoxyacetic acid were assumed to be the same as those for 2-methoxyethanol and methoxyacetic acid. Metabolic constants were obtained from *in vitro* studies using rat and human hepatocytes. As in the 2-methoxyethanol model, placental and fetal tissues were incorporated into the richly perfused compartment, whereas the mammary gland was incorporated into the fat compartment. Growth of the maternal and fetal tissues was based on the models of Fisher *et al.* (1989) and Olanoff and Anderson (1980) for rats and Luecke *et al.* (1994) for human. They used this model to calculate the peak concentration and AUC of ethoxyacetic acid in maternal blood.

12.4.18 Perchlorate

Perchlorate (ClO_4^-) is an inorganic oxidizer often used in missiles and solid rocket boosters. Due to its widespread use, it has also been frequently found as a contaminant in groundwater. Perchlorate contamination is of concern due to its ability to inhibit uptake of iodine into the thyroid, potentially leading to disruption of thyroid hormone homeostasis. Since thyroid hormones play an important role in development, the potential effects of perchlorate exposure during pregnancy and lactation are of particular concern. For this reason, a suite of PBPK models has been developed that demonstrates the capability to accurately simulate the kinetics of perchlorate, and the resulting inhibition of iodine uptake, in the adult rat and human as well as in the pregnant rat/developing fetus and in the lactating rat/neonate (R. A. Clewell *et al.* 2001, 2003a,b; Merrill *et al.* 2003). The models are remarkably comprehensive, providing reasonably accurate simulations of a variety of datasets. One of the most compelling aspects of the suite of models is the high degree of consistency between the model structures and parameters. For the most part, only physiological parameters vary across the models. Of course, it is just these differences in physiology that are responsible for much of the observed differences in kinetics. In cases where different values are used for chemical-specific parameters in the models, the differences are supported by evidence from experimental data for perchlorate or iodine.

These PBPK models were used in US EPA's 2002 (US EPA 2002) risk assessment for perchlorate to support cross-species extrapolation (calculation of human equivalent exposures), as well as to inform concerns regarding sensitivity to the effects of perchlorate during development. Because the PBPK models are grounded in experimental data collected under the conditions of concern for the risk assessment (drinking water exposures in rats and humans) and because of their thorough validation with separate experimental data, there can be high confidence in the dosimetry estimates calculated with the models for the dose metrics used in the risk characterization: blood concentrations of perchlorate and inhibition of iodine uptake. Although it might be possible to obtain similar dosimetry estimates through a combination of classical pharmacokinetic calculations, there would be a much lower level of confidence in such an approach since the assumptions made in such an approach would not be subject to the same level of validation as the PBPK models.

12.5 RISK ASSESSMENT DOSIMETRY MODELS

Another set of publications relating the application of PBPK modeling in reproductive and developmental risk assessment warrants some discussion. A PBPK model for isopropanol and its metabolite acetone in the adult rat and human (H. J. Clewell *et al.* 2001) was subsequently adapted for use in the prediction of dose metrics for exposures during pregnancy. This model was then applied to estimate dose metrics for developmental endpoints in risk assessments for isopropanol (Gentry *et al.* 2002) and acetone (Gentry *et al.* 2003a). The model was further adapted to compare dosimetry during pregnancy and lactation in the human for a number of chemicals: isopropanol, vinyl chloride, methylene chloride, perchloroethylene, nicotine, and dioxin (Gentry *et al.* 2003b). Dosimetry during infancy and childhood for direct exposure to these same chemicals was also evaluated using models of inhalation (Sarangapani *et al.* 2003) and ingestion (Clewell *et al.* 2004) in which the physiological and metabolic parameters were described as age-dependent functions. A review of pharmacokinetic differences between children and adults included case studies demonstrating PBPK modeling of early life exposure to caffeine and theophiline (Ginsberg *et al.* 2004).

12.6 SUMMARY

To date, PBPK models developed for simulating perinatal transfer of chemicals have been used to estimate doses of chemicals to different tissues in the developing fetus and neonate. The success of these models in accurately simulating tissue concentrations of parent compounds and their metabolites is quite variable: Some models perform better than others and, even within the same model applications, there has been variable success across several doses. While some issues with model performance may be related to variability in applying PBPK tools to diverse datasets, a more likely challenge relates to the confidence in our ability to adequately capture all the changes in anatomy and physiology occurring during development. For example, none of the models developed for fetal dosimetry include the intricate changes in blood flow patterns through various shunts that are present during development.

Mathematical expressions for physiological changes, especially body weight and tissue growth during pregnancy and lactational period, are included in these PBPK models describing perinatal transfer. However, other physiological and biochemical changes during perinatal period (e.g., hormonal changes, protein expressions, etc.) are not incorporated into the models. Time-dependent alterations in these parameters may produce variable susceptibility to toxic responses that may also be important for understanding toxic responses of the mother, fetus and neonate. A very significant challenge in perinatal PBPK models involves kinetic control of endogenous compounds that are under feedback control in the adult, but are critical morphogens in specific tissues and organs. An example is the temporal increases in testosterone and its metabolites, estradiol and dihydrotestosterone, that are required for proper formation of epididymis, prostate, and hypothalamic structures, respectively.

In most models, simplifying assumptions about flow or diffusion-limited uptake have been included to account for dynamics of transport across various blood–tissue barriers. In many cases, more specific transport processes exist to ensure transfer of nutrient from mother to offspring in these periods. One area of needed research is in improved characterization of transfer, especially across the placental barriers and from maternal blood into milk during lactogenesis.

Recently, Corley *et al.* (2003) provided a set of research recommendations to improve the development and application of PBPK models for perinatal exposures. They suggested three different focus areas: (1) creating databases for biological information covering pregnancy, lactation, and development; (2) supporting research to reveal strain-specific physiology, ontogeny of metabolic enzymes and transport proteins, mechanisms on chemical transfer, and other important biochemical processes; and (3) developing more comprehensive PBPK models than those exist presently. These areas will have to be advanced if these life-stage specific PBPK models are to continue to make contributions to the risk assessment for neonates and children exposed to various compounds.

NOTATION

$\mathbf{A_0}$	relative growth at the time t_0
ACTH	adrenocorticotrophic hormone
AUC	area under the curve
$\mathbf{BW_0}$	initial weight at the time t_0
$\mathbf{BW}_F$	body weight of the fetuses
$\mathbf{BW}_G$	body weight of growing rat
$\mathbf{BW}_M$	weight of young mature animal
$\mathbf{BW}_P$	pup's body weight
$C_{\max}$	peak blood concentration
2,4-D	2,4-dichlorophenoxyacetic acid
DDE	*p*,*p*′-dichloro-2,2-*bis*(*p*-chlorophenyl)ethylene
DDT	*p*,*p*′-dichloro-2,2-*bis*(*p*-chlorophenyl)trichloroethane
DMO	5,5′-dimethyloxazolidine-2,4-dione
EAGE	embryo age in days
EW1	body weight changes in murine fetuses and pups, Stage 1
EW2	body weight changes in murine fetuses and pups, Stage 2
EW3	body weight changes in murine fetuses and pups, Stage 3
GD	gestational day
GI	gastrointestinal
GFR	glomerular filtration rate
hCG	human chorionic gonadotrophin
hPL	human placental lactogen
LMS	linearized multistage
LOEL	lowest-observable-effect level

MeHg	methylmercury
MRL	oral minimal risk level
MSH	melanocyte stimulating hormone
NSAID	nonsteroidal antiinflamatory drug
NOEL	no-observable-effect level
PBPK	physiologically based pharmacokinetic
PND	postnatal day
PPBA	*p*-phenylbenzoic acid
QC	cardiac output of an animal
QCAP	change of cardiac output associated with chorioallantoic placenta
QCC	allometric constant
QCF	cardiac output of each fetus
QCG	cardiac output of growing rat
QCON	maternal blood flow to each fetus
QCP	cardiac output of pregnant rat
QDEC	change of cardiac output associated with yolk sac placenta
RfD	reference dose
US EPA	U.S. Environmental Protection Agency
$\mathbf{W}_{Fi}$	weight of each tissue in fetus
$\mathbf{W}_{i}$	weight of a given tissue in the mother

REFERENCES

Byczkowski, J. Z., and Fisher, J. W. (1994). Lactational transfer of tetrachloroethylene in rats. *Risk Anal.* **14**, 339–349.

Byczkowski, J. Z., and Fisher, J. W. (1995). A computer program linking physiologically based pharmacokinetic model with cancer risk assessment for breast-fed infants. *Comput. Methods Programs Biomed.* **46**, 155–163.

Byczkowski, J. Z., Kinkead, E. R., Leahy, H. F., Randall, G. M., and Fisher, J. W. (1994). Computer simulation of the lactational transfer of tetrachloroethylene in rats using a physiologically based model. *Toxicol. Appl. Pharmacol.* **125**, 228–236.

Byczkowski, J. Z., and Lipscomb, J. C. (2001). Physiologically based pharmacokinetic modeling of the lactational transfer of methylmercury. *Risk Anal.* **21**, 869–882.

Case, R. M., and Waterhouse, J. M. (1994). *Human Physiology, Age, Stress, and the Environment.* Oxford University Press, New York, 286 pp.

Clarke, D. O., Elswick, B. A., Welsch, F., and Conolly, R. B. (1993). Pharmacokinetics of 2-methoxyethanol and 2-methoxyacetic acid in the pregnant mouse, a physiologically based mathematical model. *Toxicol. Appl. Pharmacol.* **121**, 239–252.

Clewell, H. J, 3rd, Andersen, M. E., Wills, R. J., and Latriano, L. (1997). A physiologically based pharmacokinetic model for retinoic acid and its metabolites. *J. Am. Acad. Dermatol.* **36**(3 Pt 2), S77–S85.

Clewell, H. J., Gearhart, J. M., Gentry, P. R., Covington, T. R., VanLandingham, C. B., Crump, K. S., and Shipp, A. M. (1999). Evaluation of the uncertainty in an oral reference dose for methylmercury due to interindividual variability in pharmacokinetics. *Risk Anal.* **19**, 547–558.

Clewell, H. J., Gentry, P. R., Gearhart, J. M., Covington, T. R., Banton, M. I., and Andersen, M. E. (2001). Development of a physiologically based pharmacokinetic model of isopropanol and its metabolite acetone. *Toxicol. Sci.* **63**, 160–172.

Clewell, H. J., Gentry, P. R., Covington, T. R., Sarangapani, R., and Teeguarden, J. G. (2004). Evaluation of the potential impact of age- and gender-specific pharmacokinetic differences on tissue dosimetry. *Toxicol. Sci.* **79**, 381–393.

Clewell, R. A., and Gearhart, J. M. (2002). Pharmacokinetics of toxic chemicals in breast milk: Use of PBPK models to predict infant exposure. *Environ. Health Perspect.* **110**, A333–A337.

Clewell, R. A., Merrill, E. A., and Robinson, P. J. (2001). The use of physiologically based models to integrate diverse data sets and reduce uncertainty in the prediction of perchlorate and iodide kinetics across life stages and species. *Toxicol. Ind. Health* **17**, 210–222.

Clewell, R. A., Merrill, E. A., Yu, K. O., Mahle, D. A., Sterner, T. R., Fisher, J. W., Gearhart, J. M. (2003a). Predicting neonatal perchlorate dose and inhibition of iodide uptake in the rat during lactation using physiologically-based pharmacokinetic modeling. *Toxicol. Sci.* **74**, 416–436.

Clewell, R. A., Merrill, E. A., Yu, K. O., Mahle, D. A., Sterner, T. R., Mattie, D. R., Robinson, P. J., Fisher, J. W., and Gearhart, J. M. (2003b). Predicting fetal perchlorate dose and inhibition of iodide kinetics during gestation: A physiologically-based pharmacokinetic analysis of perchlorate and iodide kinetics in the rat. *Toxicol. Sci.* **73**, 235–255.

Corley, R. A., Mast, T. J., Carney, E. W., Rogers, J. M., and Daston, G. P. (2003). Evaluation of physiologically based models of pregnancy and lactation for their application in children's health risk assessment. *CRC Crit. Rev. Toxicol.* **33**, 137–211.

Davis, J. A., and Dobbing, J. (1981). *Scientific Foundations of Pediatrics.* Heinemann, London.

Fisher, D. A. (1992). *Endocrinology of Fetal Development.* Saunders, Philadelphia.

Fisher, J., Mahle, D., Bankston, L., Greene, R., and Gearhart, J. (1997). Lactational transfer of volatile chemicals in breast milk. *Am. Ind. Hyg. Assoc. J.* **58**, 425–431.

Fisher, J. W., Whittaker, T. A., Taylor, D. H., Clewell, H. J., III, and Andersen, M. E. (1989). Physiologically based pharmacokinetic modeling of the pregnant rat: A multiroute exposure model for trichloroethylene and its metabolite, trichloroacetic acid. *Toxicol. Appl. Pharmacol.* **99**, 395–414.

Fisher, J. W., Whittaker, T. A., Taylor, D. H., Clewell, H. J., III, and Andersen, M. E. (1990). Physiologically based pharmacokinetic modeling of the lactating rat and nursing pup: A multiroute exposure model for trichloroethylene and its metabolite, trichloroacetic acid. *Toxicol. Appl. Pharmacol.* **102**, 497–513.

Gabrielsson, J. L., and Groth, T. (1988). An extended physiological pharmacokinetic model of methadone disposition in the rat: validation and sensitivity analysis. *J. Pharmacokinet. Biopharm.* **16**, 183–201.

Gabrielsson, J. L., and Paalzow, L. K. (1983). A physiological pharmacokinetic model for morphine disposition in the pregnant rat. *J. Pharmacokinet. Biopharm.* **11**, 147–163.

Gabrielsson, J. L., Paalzow, L. K., and Nordström, L. (1984). A physiologically based pharmacokinetic model for theophylline disposition in the pregnant and nonpregnant rat. *J. Pharmacokinet. Biopharm.* **12**, 149–165.

Gabrielsson, J. L., Johansson, P., Bondesson, U., and Paalzow, L. K. (1985). Analysis of methadone disposition in the pregnant rat by means of a physiological flow model. *J. Pharmacokinet. Biopharm.* **13**, 355–372.

Gabrielsson, J. L., Johansson, P., Bondesson, U., Karlsson, M., and Paalzow, L. K. (1986). Analysis of pethidine disposition in the pregnant rat by means of a physiological flow model. *J. Pharmacokinet. Biopharm.* **14**, 381–395.

Gargas, M. L., Tyler, T. R., Sweeney, L. M., Corley, R. A., Weitz, K. K., Mast, T. J., Paustenbach, D. J., and Hays, S. M. (2000a). A toxicokinetic study of inhaled ethylene glycol monomethyl ether (2-ME) and validation of a physiologically based pharmacokinetic model for the pregnant rat and human. *Toxicol. Appl. Pharmacol.* **165**, 53–62.

Gargas, M. L., Tyler, T. R., Sweeney, L. M., Corley, R. A., Weitz, K. K., Mast, T. J., Paustenbach, D. J., and Hays, S. M. (2000b). A toxicokinetic study of inhaled ethylene glycol ethyl ether acetate and validation of a physiologically based pharmacokinetic model for rat and human. *Toxicol. Appl. Pharmacol.* **165**, 63–73.

Gearhart, J. M., Clewell, H. J., Crump, K. S., Shipp, A. M., and Silvers, A. (1995). Pharmacokinetic dose estimates of mercury in children and dose–response curves of performance tests in a large epidemiologic study. *Water Air and Soil Pollution* **80**, 49–58.

Gentry, P. R., Covington, T. R., Andersen, M. E., and Clewell, H. J. (2002). Application of a physiologically-based pharmacokinetic model for isopropanol in the derivation of an RfD/RfC. *Reg. Toxicol. Pharmacol.* **36**, 51–68.

Gentry, P. R., Covington, T. R. Andersen, M. E., and Clewell, H. J. (2003a). Application of a physiologically based pharmacokinetic model for reference dose and reference concentration estimation for acetone. *J. Toxicol. Environ. Health Pt. A* **66**, 2209–2225.

Gentry, P. R., Covington, T. R., and Clewell, H. J. (2003b). Evaluation of the potential impact of pharmacokinetic differences on tissue dosimetry in offspring during pregnancy and lactation. *Reg. Toxicol. Pharmacol.* **38**, 1–16.

Ginsberg, G., Hattis, D., and Sonawane, B. (2004). Incorporating pharmacokinetic differences between children and adults in assessing children's risks to environmental toxicants. *Toxicol. Appl. Pharmacol.* **198**, 164–183.

Goehbloed, J. F. (1976). The embryonic and postnatal growth of rat and mouse. *Acta Anat.* **95**, 8–33.

Gray, D. G. (1995). A physiologically based pharmacokinetic model for methyl mercury in the pregnant rat and fetus. *Toxicol. Appl. Pharmacol.* **132**, 91–102.

Hays, S. M., Elswick, B. A., Blumenthal, G. M., Welsch, F., Conolly, R. B., and Gargas, M. L. (2000). Development of a physiologically based pharmacokinetic model of 2-methoxyethanol and 2-methoxyacetic acid disposition in pregnant rats. *Toxicol. Appl. Pharmacol.* **163**, 67–74.

Huggett, A. S., and Widdas, W. F. (1951). The relationship between mammalian foetal weight and conception age. *J. Physiol.* **114**, 306–317.

Kawahara, M., Nanbo, T., and Tsuji, A. (1998). Physiologically based pharmacokinetic prediction of *p*-phenylbenzoic acid disposition in the pregnant rat. *Biopharm. Drug Dispos.* **19**, 445–453.

Kim, C. S., Binienda, Z., and Sandberg, J. A. (1996). Construction of a physiologically based pharmacokinetic model for 2,4-dichlorophenoxyacetic acid dosimetry in the developing rabbit brain. *Toxicol. Appl. Pharmacol.* **136**, 250–259.

Linzell, J. L., Peaker, M., and Taylor, J. C. (1975). The effects of prolactin and oxytocin on milk secretion and on the permeability of the mammary epithelium in the rabbit. *J. Physiol.* **253**, 547–563.

Luecke, R. H., Wosilait, W. D., Pearce, B. A., and Young, J. F. (1994). A physiologically based pharmacokinetic computer model for human pregnancy. *Teratology* **49**, 90–103.

Merrill, E. A., Clewell, R. A., Gearhart, J. M., Robinson, P. J., Sterner, T. R., Yu, K. O., Mattie, D. R., Fisher, J. W. (2003). PBPK predictions of perchlorate distribution and its effect on thyroid uptake of radioiodide in the male rat. *Toxicol Sci.* **73**, 256–269.

National Research Council. (1993). *Pesticides in the Diets of Infants and Children.* National Academy Press, Washington, D.C., 386 pp.

O'Flaherty, E. J., Scott, W., Schreiner, C., and Beliles, R. P. (1992). A physiologically based kinetic model of rat and mouse gestation: Disposition of a weak acid. *Toxicol. Appl. Pharmacol.* **112**, 245–256.

O'Flaherty, E. J., Nau, H., McCandless, D., Beliles, R. P., Schreiner, C. M., and Scott, W. J. Jr. (1995). Physiologically based pharmacokinetics of methoxyacetic acid: Dose–effect considerations in C57BL/6 mice. *Teratology* **52**, 78–89.

Olanoff, L. S., and Anderson, J. M. (1980). Controlled release of tetracycline-III: A physiological pharmacokinetic model of the pregnant rat. *J. Pharmacokinet. Biopharm.* **8**, 599–620.

Oskarsson, A., Palminger Hallen, I., Sundberg, J., and Petersson Grawe, K. (1998). Risk assessment in relation to neonatal metal exposure. *Analyst* **123**, 19–23.

Rice, D., and Barone, S., Jr. (2000). Critical periods of vulnerability for the developing nervous system: Evidence from humans and animal models. *Environ. Health Perspect.* **108**(Suppl 3), 511–533.

Rudolf, A. M. (1992). *Pediatrics*, Saunders, Philadelphia.

Sarangapani, R., Gentry, P. R., Covington, T. R., Teeguarden, J. G., and Clewell, H. J. (2003). Evaluation of the potential impact of age- and gender-specific lung morphology and ventilation rate on the dosimetry of vapors. *Inhal. Toxicol.* **15**, 987–1016.

Schreiber, J. S. (1993). Predicted infant exposure to tetrachloroethene in human breastmilk. *Risk Anal.* **13**, 515–524.

Shelley, M. L., Andersen, M. E., and Fisher, J. W. (1988). An inhalation distribution model for the lactating mother and nursing child. *Toxicol. Lett.* **43**, 23–29.

Spitzer, A., and Chevalier, R. L. (1992). *The Developing Kidney and the Process of Kidney.* Raven, New York.

Terry, K. K., Elswick, B. A., Welsh, F., and Conolly, R. B. (1995). Development of a physiologically based pharmacokinetic model describing 2-methoxyacetic acid disposition in the pregnant mouse. *Toxicol. Appl. Pharmacol.* **132**, 103–114.

US EPA. (2002). *Perchlorate Environmental Contamination: Toxicological Review and Risk Characterization*, External Review Draft, NCEA-1-0503, Office of Research and Development, US EPA, Washington D.C.

Ward, K. W., Blumenthal, G. M., Welsch, F., and Pollack, G. M. (1997). Development of a physiologically based pharmacokinetic model to describe the disposition of methanol in pregnant rats and mice. *Toxicol. Appl. Pharmacol.* **145**, 311–322.

Ward, R. C., Travis, C. C., Hetrick, D. M., Andersen, M. E., and Gargas, M. L. (1988). Pharmacokinetics of tetrachloroethylene. *Toxicol. Appl. Pharmacol.* **93**, 108–117.

Weiss, B. (2000). Vulnerability of children and the developing brain to neurotoxic hazards. *Environ. Health Perspect.* **108**(Suppl 3), 375–381.

Welsch, F., Blumenthal, G. M., and Conolly, R. B. (1995). Physiologically based pharmacokinetic models applicable to organogenesis: Extrapolation between species and potential use in prenatal toxicity risk assessments. *Toxicol. Lett.* **82–83**, 539–547.

You, L., Gazi, E., Archibeque-Engle, S., Casanova, M., Conolly, R. B., and Heck, H. D. (1999). Transplacental and lactational transfer of *p,p′*-DDE in Sprague–Dawley rats. *Toxicol. Appl. Pharmacol.* **157**, 134–144.

CHAPTER 13

MIXTURES

Raymond S. H. Yang and Melvin E. Andersen

13.1 INTRODUCTION

Human exposure to chemicals is rarely, if ever, confined to a single compound. Therefore, the study of chemical mixture toxicology has gained a great deal of momentum in the last two decades. Studying chemical mixtures is an extremely complex task because of the astronomical number of possible combinations. Such numbers certainly preclude any systematic experimental assessment of toxicology of all potentially troublesome chemical mixtures. In the past 15 years or so, physiologically based pharmacokinetic/pharmacodynamic (PBPK/PD) modeling has been applied to the toxicological interactions of chemical mixtures. As is generally the case in the evolution of a new area, the progress in the application of PBPK modeling to chemical mixtures has followed different phases from simple binary pharmacokinetic and pharmacodynamic interactions to more and more complex mixtures. First, PBPK modeling of binary chemical mixtures became necessary because of pharmacological or toxicological interactions. Second, as investigators became interested in mechanisms of toxicological interactions, the advances of physiologically based pharmacodynamic (PBPD) modeling formed a natural course of development of this area. Third, when more and more sophistication was incorporated into PBPK modeling, it is inevitable that PBPK modeling of complex chemical mixtures were attempted and novel approaches developed. Since 1996 when the Food Quality and Protection Act was enacted, the United States Environmental Protection Agency (US EPA), under the Congressional mandate, began active

Physiologically Based Pharmacokinetic Modeling, edited by Micaela B. Reddy, Raymond S. H Yang, Harvey J. Clewell III, and Melvin E. Andersen.

consideration of cumulative risk assessment. Out of necessity, toxicological interactions must be taken into consideration. Thus, the application of PBPK modeling in cumulative risk assessment has become an active area of research endeavor (US EPA 2003).

13.2 PBPK MODELING OF CHEMICAL MIXTURES

13.2.1 Earlier Days: PBPK Modeling of Binary Mixtures

Although numerous drug interaction studies have been reported in the scientific literature in past decades (Mozayani and Raymon 2004), the earliest applications of PBPK modeling to chemical mixtures did not occur until the mid-1980s. Because of the necessity for extrapolation from animal experimentation to humans for many toxicants, the advances of PBPK modeling in the area of toxicology has far outpaced its application in pharmacology. An earlier review of PBPK modeling of chemical mixtures (Mumtaz *et al.* 1993) indicated that the "first example" of PBPK modeling of a "chemical mixture" actually involved one chemical, *n*-hexane, and its metabolites, methyl *n*-butyl ketone (MnBK) and 2,5-hexanedione (2,5-HD); thus, it is a kind of "one-chemical mixture." This particular PBPK model for *n*-hexane and its metabolites incorporated three inhibitory interactions: (1) hexane and MnBK are competitive substrates for ω-1 oxidation; (2) MnBK and 2,5-HD are competitive substrates for α oxidation; and (3) 2,5-HD acts as a product feedback inhibitor (Andersen and Clewell 1984). The findings of this modeling study were intriguing and they explained some of the most interesting and complex toxicological and pharmacokinetic behaviors of *n*-hexane in animals (Mumtaz *et al.* 1993). However, this PBPK modeling work was never published in a peer-reviewed journal; the principal reason was that the investigators involved were never happy enough with the PBPK modeling results. Interestingly, after about 20 years, the *n*-hexane PBPK modeling work was revisited (Dennison 2004).

Logically, some of the earliest investigations on PBPK modeling were on binary mixtures. Given below and in Table 13.1 are specific examples on PBPK modeling of binary mixtures following a chronological order. Some aspects of PBPK modeling work of higher order of chemical mixtures are given in later sections.

Specific Example 1: Dibromomethane and Isoflurane (Clewell and Andersen 1985) One of the reasons for the delay in the completion of *n*-hexane work was the difficulties encountered during the experimental and PBPK modeling processes. In those earlier days, as an alternative to mimic the interaction between *n*-hexane and its metabolite, MnBK, Clewell and Andersen (1985) studied, using PBPK modeling, a binary mixture of dibromomethane (DBM) and isoflurane. The rationale, as given later in a review (Mumtaz *et al.* 1993) by Clewell, was that (1) DBM was selected as a "surrogate" of MnBK because of its high tissue solubility and the ease of monitoring its metabolism, and (2) isoflurane was selected as a "surrogate" of *n*-hexane because of its poor tissue solubility and its rapid clearance by exhalation. The Clewell and Andersen (1985) publication was, in general, a review

TABLE 13.1 Compilation of Studies on PBPK Modeling of Chemical Mixtures

Chemicals	Interaction mechanisms	References
Binary Mixtures		
Dibromomethane and Isoflurane	Competitive enzyme inhibition	Clewell and Andersen (1985)
1,1-Dichloroethylene and trichloroethylene	Competitive enzyme inhibition	Andersen *et al.* (1987)
Benzene and toluene	Nonompetitive enzyme inhibition	Purcell *et al.* (1990)
Mirex, phenobartital, or chlordecone and bromotrichloromethane	Enzyme induction or interference with tissue repair	Thakore *et al.* (1991)
Ethanol and trichloroethylene	Competitive enzyme inhibition or enzyme induction	Sato *et al.* (1991)
Toluene and *m*-xylene	Competitive enzyme inhibition	Tardif *et al.* (1993, 1995)
1,3-Butadiene and styrene	Enzyme inhibition	Filser *et al.* (1993)
1,3-Butadiene and styrene; 1,3-butadiene and benzene; 1,3-butadiene and ethanol	Competitive enzyme inhibition; enzyme induction	Bond *et al.* (1994)
Vinyl chloride and trichloroethylene	Competitive enzyme inhibition	Barton *et al.* (1995)
Toluene and dichloromethane	Competitive enzyme inhibition	Krishnan and Pelekis (1995)
1,3-Butadiene and styrene	Competitive enzyme inhibition	Leavens and Bond (1996)
Carbon tetrachloride and Kepone	Inhibition of the repair mechanism	El-Masri *et al.* (1996a)
Trichloroethylene and 1,1-dichloroethylene	Competitive enzyme inhibition	El-Masri *et al.* (1996b,c)
Dichloromethane and toluene	Noncompetitive and uncompetitive enzyme inhibition	Pelekis and Krishnan (1997)
n-Hexane and toluene	Noncompetitive enzyme inhibition	Yu *et al.* (1998)
Toluene and *n*-hexane	Competitive enzyme inhibition	Ali and Tardif (1999)
Methyl chloroform and *m*-xylene	Competitive enzyme inhibition	Tardif and Charest-Tardif (1999)
5-Fluorouracil and sorivudine; Triazolam and erythromycin	"Mechanism-based inhibition" or enzyme inactivation	Ito *et al.* (1998), Kanamitsu *et al.* (2000)[a]
Toluene and trichloroethylene	Competitive enzyme inhibition	Thrall and Poet (2000)
Ethylbenzene and xylenes	Competitive enzyme inhibition	Jang *et al.* (2001)

TABLE 13.1 *continued*

Chemicals	Interaction mechanisms	References
Ternary Mixtures		
Toluene, *m*-xylene, and ethylbenzene	Competitive enzyme inhibition	Tardif *et al.* (1997), Haddad *et al.* (1999a)
Trichloroethylene, tetrachloroethylene, methyl chloroform	Competitive enzyme inhibition	Dobrev *et al.* (2001, 2002)
Toluene, ethylbenzene, and xylenes	Competitive enzyme inhibition	Dennison *et al.* (2004c)
Four-Chemical Mixture		
Benzene, toluene, ethylbenzene, and *m*-xylene	Competitive enzyme inhibition	Haddad *et al.* (1999b)
Five-Chemical Mixture		
Benzene, toluene, ethylbenzene, *m*-xylene, and dichloromethane	Competitive enzyme inhibition	Haddad *et al.* (2000)
Complex Mixtures		
Gasoline and subfractions	Competitive enzyme inhibition	Dennison *et al.* (2003, 2004a,b), Dennison (2004)

[a] The PBPK model in this article is a limited one involving portal vein, systemic blood, and liver; see text for further discussion.

article, although some original experimental and modeling data were apparently included. Many studies described therein utilized the same basic PBPK model of styrene, shown in Fig. 13.1, as a template. A plot of PBPK model simulation versus experimental data showed similar kinetics for the formation of carboxyhemoglobin (HbCO), resulting from the metabolism of DBM to CO, following DBM exposure alone, or in combination with isoflurane in rats. Not much specifics on experimentation and modeling (i.e., animals, exposure regimen, interaction model structure, etc.) were given in this particular publication. Somewhat more details were available in the 1993 review (Mumtaz *et al.* 1993).

Specific Example 2: 1,1-Dichloroethylene and Trichloroethylene (Andersen et al. 1987) The interactive PBPK model of 1,1-dichloroethylene (DCE) and trichloroethylene (TCE), the related discussions on different types of enzyme inhibitions, and the incorporation of competitive inhibition into the liver compartment represent truly the first comprehensive publication (Andersen *et al.* 1987) in the peer-reviewed journal on PBPK modeling of a chemical mixture. Using liver injury [aspartate transaminase (AST)] from DCE as an index, the pharmacokinetics of DCE alone in Fischer 344 rats as well as that in the presence of TCE were compared between PBPK model simulation and experimental results obtained from gas uptake pharmacokinetic studies. For PBPK modeling, Andersen *et al.*

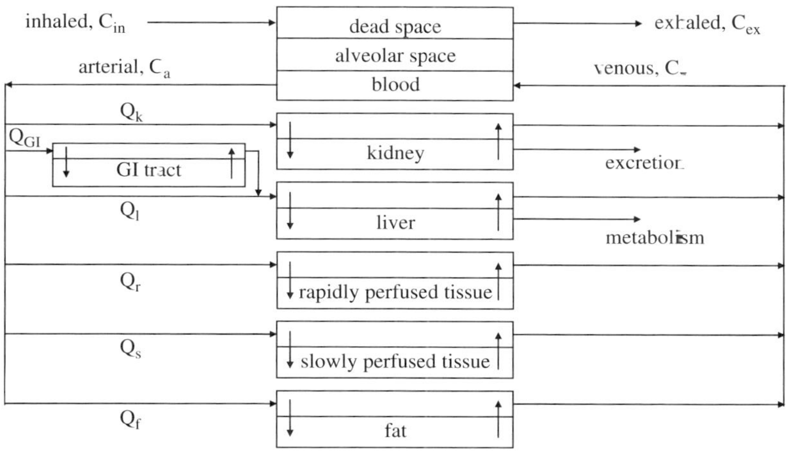

Figure 13.1 A graphical representation of a physiologically based pharmacokinetic (PBPK) model for volatile organic chemicals such as styrene. Modified based on Ramsey, J. C., and Andersen, M. E. (1984). *Toxicol. Appl. Pharmacol.* **73**, 159–175. The variables C_{in} and C_{ex} are the concentrations of chemical in inhaled and exhaled air, respectively, C_a and C_v are the concentrations of chemical in the arterial and venous blood, respectively, and Q_k, Q_{GI}, Q_l, Q_r, Q_s, and Q_f are the flow rate of blood through the kidneys, GI tract, liver, rapidly perfused tissue, slowly perfused tissue, and fat, respectively.

(1987) constructed a PBPK model for each chemical (i.e., DCE or TCE) individually and then linked the two models via the mass balance equation for the liver through enzyme inhibition (see next section for details). The kinetics of each chemical was described by a set of five mass balance differential equations for tissue compartments (fat, muscle/skin, viscera, and liver) and the chamber atmosphere. Thus, the basic template for each PBPK model was again based on that of the styrene model shown in Fig. 13.1 (Ramsey and Andersen 1984). Physiological constants and partition coefficients were either available in the literature (Gargas *et al.* 1986) or, in the case of partition coefficients for TCE, experimentally determined using vial equilibration methods (Sato and Nakajima 1979).

One of the most distinguished features of a PBPK model describing chemical interactions in chemical mixtures in the Andersen *et al.* (1987) article on DCE and TCE, as well as in a large number of subsequent papers on chemical mixtures, is the incorporation of enzyme inhibition as mechanistic basis for interactions. Thus, in many ways, the Andersen *et al.* (1987) article and its related descriptions of different types of enzyme inhibition is the pioneering effort both experimentally and conceptually in the area of PBPK modeling of chemical mixtures. The section below will provide the detailed description of the fundamentals on the mechanistic basis of enzyme inhibition.

Interaction Mechanisms: Enzyme Inhibition Figure 13.2 is a general schematic for multiple mechanisms of enzyme inhibition during coexposure of two substrates (Andersen *et al.* 1987). E is the enzyme in free form; in the case of DCE

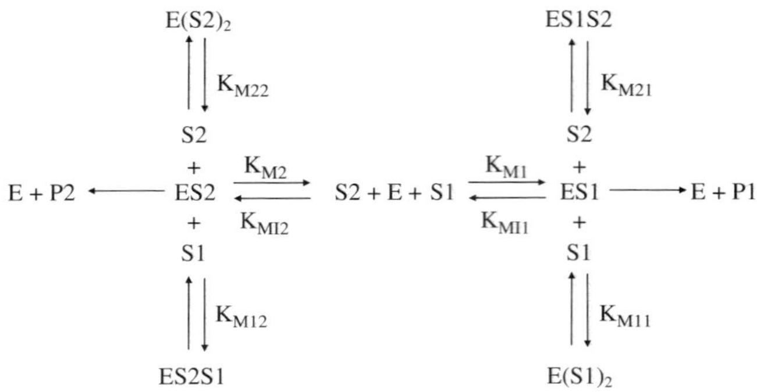

Figure 13.2 A general schematic for multiple mechanisms of enzyme inhibition during exposure to two substrates. E is the free enzyme; S1 and S2 are competing substrates for products P1 and P2; K_{M1} and K_{M2} are the substrate binding constants (they are the same as the inhibitory binding constants K_{MI1} and K_{MI2}). All constants are dissociation equilibrium constants. Redrawn based on Andersen, M. E., Gargas, M. L., Clewell, H. J., III, and Severgn, K. M. (1987). *Toxicol. Appl. Pharmacol.* **89**, 149–157.

and TCE, it is cytochrome P450 2E1 (CYP2E1). S1 and S2 are two substrates; in Example 2 above, they are DCE and TCE. The products formed from S1 and S2, respectively, are P1 and P2. All constants are dissociation equilibrium constants. The enzyme binding constants for substrates 1 and 2, K_{M1} and K_{M2}, are also the inhibitory binding constants when one substrate serves to inhibit the metabolism of a second substrate. In that sense, accordingly, K_{M1} equals K_{MI1} and K_{M2} equals K_{MI2}.

Central to the successful operation of the PBPK model for pharmacokinetic interactions in the binary chemical mixture of DCE and TCE is the incorporation of the set of equations related to different types (competitive, noncompetitive, and uncompetitive) of enzyme inhibition into the liver compartment mass balance differential equation (Andersen *et al.* 1987; note T_1 in the original paper had one term missing) as shown below:

$$V_L \frac{dC_{L1}}{dt} = \frac{d\mathrm{AMT}_{L1}}{dt} = (Q_L C_{a1}) - (Q_L C_{vL1}) - \frac{V_{\max 1} \times C_{vL1}}{K_m(T_1) + C_{vL1}(T_2)} \tag{13.1}$$

$$T_1 = 1 + \frac{C_{vL2}}{K_{MI2}} + \frac{(C_{vL2})^2}{(K_{M12} \times K_{M22})} + \frac{(C_{vL1})(C_{vL2})}{(K_{M12})(K_{M12})} \tag{13.2}$$

$$T_2 = 1 + \frac{C_{vL2}}{K_{M21}} + \frac{(C_{vL1})}{K_{M11}} \tag{13.3}$$

Since this set of equations is very important for the understanding of a large portion of publications involving PBPK modeling of chemical mixtures, it is essential to fully appreciate how these equations came about and what do they mean. Therefore, we will go through the conceptual and algebraic derivation of the above equations.

Based on the general schematic for multiple enzyme inhibition during co-exposure to two substrates in Fig. 13.2, we first write the rate equation for the product of interest (in this case, we concentrate on P1 and S1):

$$v = k_1[\text{ES1}] \tag{13.4}$$

Next, we write the conservation equation for enzyme,

$$\text{E}_{\text{total}} = \text{E}_{\text{free}} + \text{ES1} + \text{ES2} + \text{ES1S2} + \text{ES1S1} + \text{ES2S1} + \text{ES2S2} \tag{13.5}$$

We then solve the conservation equation shown above in terms of ES1, the factor of interest from the rate equation Eq. (13.4), according to the following steps:

We write all the equations for the equilibrium dissociation constants (see Fig. 13.2),

$$K_{\text{M1}} = \frac{(\text{E}_{\text{free}})(\text{S1})}{(\text{ES1})} \tag{13.6}$$

$$K_{\text{M21}} = \frac{(\text{ES1})(\text{S2})}{(\text{ES1S2})} \tag{13.7}$$

$$K_{\text{M11}} = \frac{(\text{ES1})(\text{S1})}{(\text{ES1S1})} \tag{13.8}$$

$$K_{\text{M2}} = \frac{(\text{E}_{\text{free}})(\text{S2})}{(\text{ES2})} \tag{13.9}$$

$$K_{\text{M12}} = \frac{(\text{ES2})(\text{S1})}{(\text{ES2S1})} \tag{13.10}$$

$$K_{\text{M22}} = \frac{(\text{ES2})(\text{S2})}{(\text{ES2S2})} \tag{13.11}$$

Rewrite Eqs. (13.6)–(13.8) in terms of concentrations,

$$(\text{E}_{\text{free}}) = \frac{(\text{ES1})K_{\text{M1}}}{(\text{S1})} \tag{13.12}$$

$$(\text{ES1S2}) = \frac{(\text{ES1})(\text{S2})}{(K_{\text{M21}})} \tag{13.13}$$

$$(\text{ES1S1}) = \frac{(\text{ES1})(\text{S1})}{(K_{\text{M11}})} \tag{13.14}$$

Recasting Eqs. (13.9)–(13.11) is a bit more involved, we start out with rearranging Eq. (13.9), $(\text{ES2}) = (\text{E}_{\text{free}})(\text{S2})/K_{\text{M2}}$, and substitute Eq. (13.12) (i.e., E_{free}) into this equation, we have

$$(\text{ES2}) = \frac{(\text{ES1})(K_{\text{M1}})(\text{S2})}{(\text{S1})(K_{\text{M2}})} \tag{13.15}$$

Similarly, Eqs. (13.10) and (13.11) become

$$(\text{ES2S1}) = \frac{(\text{ES2})(\text{S1})}{(K_{\text{M12}})} \tag{13.16}$$

$$(\text{ES2S2}) = \frac{(\text{ES2})(\text{S2})}{(K_{\text{M22}})} \tag{13.17}$$

Substituting Eq. (13.15) into Eqs. (13.16) and (13.17) completes the process of transforming all forms of enzyme and enzyme/substrate complexes in terms of ES1, we obtain

$$(\text{ES2S1}) = \frac{(\text{S1})(\text{ES1})(K_{\text{M1}})(\text{S2})}{(K_{\text{M12}})(\text{S1})(K_{\text{M2}})} \tag{13.18}$$

$$(\text{ES2S2}) = \frac{(\text{S2})(\text{ES1})(K_{\text{M1}})(\text{S2})}{(K_{\text{M22}})(\text{S1})(K_{\text{M2}})} \tag{13.19}$$

Substituting all these various forms of enzyme and enzyme/substrate complexes, [i.e., Eqs. (13.12)–(13.15), (13.18), and (13.19)] into the conservation equation [i.e., Eq. (13.5)], we get

$$\begin{aligned}
\text{E}_{\text{total}} &= \text{E}_{\text{free}} + \text{ES1} + \text{ES2} + \text{ES1S2} + \text{ES1S1} + \text{ES2S1} + \text{ES2S2} \\
&= \frac{(\text{ES1})(K_{\text{M1}})}{(\text{S1})} + (\text{ES1}) + \frac{(\text{ES1})(K_{\text{M1}})(\text{S2})}{(\text{S1})(K_{\text{M2}})} + \frac{(\text{ES1})(\text{S2})}{(K_{\text{M21}})} + \frac{(\text{ES1})(\text{S1})}{(K_{\text{M11}})} \\
&\quad + \frac{(\text{S1})(\text{ES1})(K_{\text{M1}})(\text{S2})}{(K_{\text{M12}})(\text{S1})(K_{\text{M2}})} + \frac{(\text{S2})(\text{ES1})(K_{\text{M1}})(\text{S2})}{(K_{\text{M22}})(\text{S1})(K_{\text{M2}})}
\end{aligned}$$

Factor out $(E_{\text{S1}})/(\text{S1})$:

$$\text{E}_{\text{total}} = \frac{(\text{ES1})}{(\text{S1})} \left[\begin{aligned} &(K_{\text{M1}}) + (\text{S1}) + \frac{(K_{\text{M1}})(\text{S2})}{(K_{\text{M2}})} + \frac{(\text{S1})(\text{S2})}{(K_{\text{M21}})} + \frac{(\text{S1})^2}{(K_{\text{M11}})} \\ &+ \frac{(\text{S1})(K_{\text{M1}})(\text{S2})}{(K_{\text{M12}})(K_{\text{M2}})} + \frac{(\text{S2})^2 (K_{\text{M1}})}{(K_{\text{M22}})(K_{\text{M2}})} \end{aligned} \right]$$

Now group all the terms in the bracket in terms of S1 and K_{M1}:

$$\text{E}_{\text{total}} = \frac{(\text{ES1})}{(\text{S1})} \left[\begin{aligned} &(\text{S1}) + \frac{(\text{S1})(\text{S2})}{(K_{\text{M21}})} + \frac{(\text{S1})^2}{(K_{\text{M11}})} + (K_{\text{M1}}) \\ &+ \frac{(K_{\text{M1}})(\text{S2})}{(K_{\text{M2}})} + \frac{(\text{S2})^2 (K_{\text{M1}})}{(K_{\text{M22}})(K_{\text{M2}})} + \frac{(\text{S1})(K_{\text{M1}})(\text{S2})}{(K_{\text{M12}})(K_{\text{M2}})} \end{aligned} \right]$$

Now rewrite blocking for S1 and K_{M1}:

$$E_{total} = \frac{(ES1)}{(S1)}\left[(S1)\left(1+\frac{(S2)}{(K_{M21})}+\frac{(S1)}{(K_{M11})}\right) + (K_{M1})\left(1+\frac{(S2)}{(K_{M2})}+\frac{(S2)^2}{(K_{M22})(K_{M2})}+\frac{(S1)(S2)}{(K_{M12})(K_{M2})}\right)\right]$$

Let T_1 and T_2 be the terms, respectively, in the big parentheses above as shown below:

$$T_1 = \left(1+\frac{(S2)}{(K_{M2})}+\frac{(S2)^2}{(K_{M22})(K_{M2})}+\frac{(S1)(S2)}{(K_{M12})(K_{M2})}\right)$$

$$T_2 = \left(1+\frac{(S2)}{(K_{M21})}+\frac{(S1)}{(K_{M11})}\right)$$

Then,

$$E_{total} = \frac{(ES1)}{(S1)}[S1(T_2)+K_{M1}(T_1)]$$

Rearrange to solve for (ES1):

$$ES1 = \frac{E_{total}(S1)}{[S1(T_2)+K_{M1}(T_1)]}$$

Substituting into the original rate equation Eq.(13.4), we obtain

$$v = k_1[ES1] = k_1\frac{E_{total}(S1)}{[S1(T_2)+K_{M1}(T_1)]}$$

Since $k_1 E_{total}$ is V_{max}, highest possible rate, so

$$v = \frac{V_{max}\times(S1)}{[S1(T_2)+K_{M1}(T_1)]} \tag{13.20}$$

Please note that we have successfully derived the last term in Eq. (13.1) and T_1 and T_2 [i.e., Eqs. (13.2) and (13.3)], the equations published in the Andersen *et al.* (1987) article. Note also K_{M2} here equals to K_{M12} in the Andersen *et al.* (1987) paper.

The short statement below is the most important point to extract for the reader—the role of process—in establishing the sets of equations for this or any other interaction model based on enzyme inhibition.

It should be noted that it always follows this process: (1) Write the rate equation; (2) write the conservation equation; (3) derive dissociation constants; (4) solve conservation equation for enzyme species represented in the rate equation.

Even though there are many different mechanisms for enzyme inhibition and an excellent reference (Segal 1975) is available, the final equation above [Eq. (13.20)] carries, simplistically, the following three types of enzyme inhibition:

1. *Competitive Inhibition.* In Eq. (13.20), when $T_2 = 1$ and the inhibitor (in our case, the second substrate) is only affecting K_{M1}, competitive inhibition results. This type of inhibition includes the following scenarios as described in Segal (1975):
   - Substrate (S) and inhibitor (I, or a second substrate) compete for the same binding site.
   - S and I are mutually exclusive because of steric hinderance.
   - S and I share a common binding group on the enzyme.
   - The binding sites for S and I, though distinct, are overlapping.
   - The binding of I to a distinct inhibitor site causes a conformational change in the enzyme that distorts or masks the S binding site, or vice versa.
2. *Noncompetitive Inhibition.* In Eq. (13.20), when the inhibitor is affecting both K_{M1} and S1, noncompetitive inhibition results. In this type of inhibition, S and I (or a second substrate) are not mutually exclusive but ESI (in our case ES1S2) is catalytically inactive. In this case, S and I don't interfere with each other's binding, but the conformational change of the enzyme affect catalytic center (Segal 1975). There are other variations of the above scenario.
3. *Uncompetitive Inhibition.* In Eq. (13.20), when the inhibitor is affecting only S1 (i.e., $T_1 = 1$), uncompetitive inhibition results. In this type of inhibition, I (or a second substrate) only binds to the ES (in our case ES1) complex. When S binds, a conformational change of the enzyme occurs to unmask the I binding site. The resulting ESI (in our case ES1S2) is catalytically inactive.

Other examples involving PBPK modeling of binary chemical mixtures included benzene and toluene (Purcell *et al.* 1990), mirex/phenobarbital/chlordecone and bromotrichloromethane (Thakore *et al.* 1991), ethanol and trichloroethylene (Sato *et al.* 1991), and toluene and *m*-xylene (Tardiff *et al.* 1993). These mixtures and their respective PBPK modeling have been reviewed previously by Krishnan *et al.* (1994a,b); other than the summary in Table 13.1, they will not be further discussed here. Since those reviews by Krishnan *et al.* (1994a,b), there were at least 14 more binary mixtures studied by various investigators using the PBPK modeling approach in one form or another (Table 13.1). With the exception of two binary mixtures involving two drug pairs, 5-fluorouracil/sorivudine and triazolam/erythromycin (Ito *et al.* 1998; Kanamitsu *et al.* 2000), all of the other studies were on volatile organic solvents (VOCs). Since these VOC studies are quite similar to the ones already described above, we chose not to discuss further the individual publications.

Interaction Mechanisms: Enzyme Inactivation or "Mechanism-Based Inhibition" We will specifically discuss the Ito *et al.* (1998) and Kanamitsu *et al.* (2000) articles on two cases of drug–drug interactions involving 5-fluorouracil/sorivudine and triazolam/erythromycin, because: (1) These studies are two of the relatively few PBPK modeling studies on drugs, and it is specifically on drug–drug interactions; (2) the interactions involved a unique mechanism which was implicated in at least 15 human fatalities in Japan; and (3) the PBPK modeling approach in these articles is somewhat limited involving mainly the liver with two related compartments: portal vein and systemic blood.

In 1993 in Japan, 15 patients with cancer and herpes zoster were treated with 5-fluorouracil (5-FU), an anticancer agent, and sorivudine, an antiviral drug, and died from 5-FU toxicity due to a drug–drug interaction. This drug interaction involved a key enzyme, dihydropyrimidine dehydrogenase (DPD), which is a rate-limiting enzyme in the metabolism of 5-FU. It turned out that sorivudine was converted by gut flora to 5-bromovinyluracil, which is then metabolically activated by DPD. The reactive species binds to DPD and renders the enzyme inactive irreversibly. This type of inactivation of the enzyme, unlike competitive or noncompetitive inhibition, is unique mechanistically. Several terms have been used to describe this unique mechanism: "mechanism-based inhibition," "mechanism-based inactivation," "enzyme-activated irreversible inhibition," "suicide inactivation," and "k_{cat} inhibition" (Ito *et al.* 1998). According to Ito *et al.* (1998), this particular "mechanism-based inhibition" has the following characteristics:

- Preincubation time-dependent inhibition of the enzyme (time-dependence).
- No inhibition if cofactors necessary for producing the activated inhibitor are not present in the preincubation medium.
- Potentiation of the inhibition depending on the inhibitor concentration (saturation kinetics).
- Slower inactivation rate of the enzyme in the presence of substrate compared with its absence (substrate protection).
- Enzyme activity not recovered following gel filtration or dialysis (irreversibility).
- 1 : 1 Stoichiometry of the inhibitor and the active site of the enzyme (stoichiometry of inactivation).

Ito *et al.* (1998) also constructed a physiological model for the general "mechanism-based inhibition" of a number of CYP isozymes. This model, shown in Fig. 13.3, consists of three compartments: the liver, systemic blood, and portal vein. The substrate (S) and inhibitor (I) share the same model structure. The mass balance equations are as follows:

For the Substrate

$$V_h \times \left(\frac{dC_h}{dt}\right) = Q \times C_{\text{portal}} - Q \times \frac{C_h}{K_p} - f_b \times CL_{\text{int}} \times \frac{C_h}{K_p}$$

$$CL_{\text{int}} = \frac{V_{\max}}{K_m + f_b \times \dfrac{C_h}{K_p}}$$

$$V_{\max} = V_{\max}(0) \times \frac{E_{\text{act}}(t)}{E_0}$$

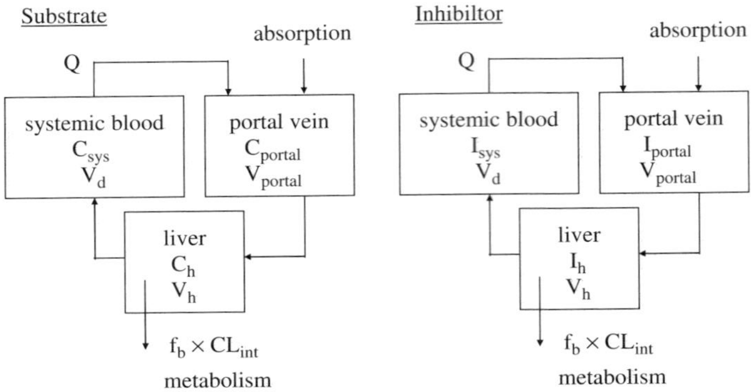

Figure 13.3 A conceptual physiological model for the time profiles of substrate and inhibitor concentrations in the plasma and liver. Q is the blood flow rate; C_{sys} and I_{sys} are substrate and inhibitor concentrations in systemic blood; V_d is the volume of distribution; C_{portal} and I_{portal} are substrate and inhibitor concentrations in portal vein; V_{portal} is the volume of portal vein; C_h and I_h are substrate and inhibitor concentrations in the liver; V_h is the volume of the liver; f_b is the unbound fraction in blood; CL_{int} is the intrinsic metabolic clearance. Redrawn based on Ito, K., Iwatsubo, T., Kanamitsu, S., Ueda, K., Suzuki, H., and Sugiyama, Y. (1998). *Pharmacol. Rev.* **50**, 387–411.

$$V_{portal} \times \left(\frac{dC_{portal}}{dt}\right) = Q \times C_{sys} + V_{abs} - Q \times C_{portal}$$

$$V_{abs} = k_a \times D \times F \times e^{-k_a \times t}$$

$$V_d \times \left(\frac{dC_{sys}}{dt}\right) = Q \times \frac{C_h}{K_p} - Q \times C_{sys}$$

For Inhibitor

$$V_h \times \left(\frac{dI_h}{dt}\right) = Q \times I_{portal} - Q \times \frac{I_h}{K_p} - f_b \times CL_{int} \times \frac{I_h}{K_p}$$

$$CL_{int} = \frac{V_{max}}{K_m + f_b \times \frac{I_h}{K_p}}$$

$$V_{portal} \times \left(\frac{dI_{portal}}{dt}\right) = Q \times I_{sys} + V_{abs} - Q \times I_{portal}$$

$$V_{abs} = k_a \times D \times F \times e^{-k_a \times t}$$

$$V_d \times \left(\frac{dI_{sys}}{dt}\right) = Q \times \frac{I_h}{K_p} - Q \times I_{sys}$$

where Q represents blood flow rate, C_{sys} and I_{sys} represent concentrations of substrate and inhibitors, respectively, in systemic blood, V_d represents the volume of distribution in the systemic blood compartment, C_{portal} and I_{portal} represent concentrations of substrate and inhibitor, respectively, in the portal vein, D represents dose, V_{portal} represents the volume of portal vein, C_h and I_h represent concentrations of substrate and inhibitor, respectively, in the liver, V_h represents the volume of the liver, f_b represents the unbound fraction in blood, CL_{int} represents the intrinsic metabolic clearance, F_a (although the original article did not clarify this, F in the above equations really should have been F_a) represents the fraction absorbed from the gastrointestinal tract, K_m represents the Michaelis constant for the metabolic elimination, V_{max} represents the maximum metabolic rate, and K_p represents the liver-to-blood concentration ratio.

For active and inactive enzymes in the liver, Ito *et al.* (1998) provided the following differential equations:

$$\frac{dE_{act}}{dt} = -\frac{k_{inact} \times E_{act} \times f_b \times \dfrac{I_h}{K_p}}{K_{i,app} + f_b \times \dfrac{I_h}{K_p}} + k_{deg}(E_0 - E_{act})$$

$$\frac{dE_{inact}}{dt} = -\frac{k_{inact} \times E_{act} \times f_b \times \dfrac{I_h}{K_p}}{K_{i,app} + f_b \times \dfrac{I_h}{K_p}} + k_{deg} \times E_{inact}$$

Ito *et al.* (1998) further defined that k_{deg} is the degradation rate constant (i.e., turnover rate constant) of the enzyme. The initial conditions at $t = 0$ are $E_{act} = E_0$ and $E_{inact} = 0$. In the absence of an inhibitor, the enzyme level in the liver is at a steady state and the degradation rate ($k_{deg}E_0$) is equal to the synthesis rate, which is assumed to be unaffected by an inhibitor.

Subsequently, the same research group reported a similar "mechanism-based inhibition" of CYP3A4 by macrolide antibiotics, erythromycin, in a drug–drug interaction with triazolam (Kanamitsu *et al.* 2000). These investigators used the above three-compartment physiological model and obtained quantitative predictions of the erythromycin/triazolam interaction. The predicted increase in triazolam AUC following erythromycin pretreatment was 2.0-fold (from 61.0 to 119 nM·hr) and 2.6-fold (from 61.0 to 156 nM·hr) from model simulation. These model predictions were very close to the actual observed value, *in vivo* in human, of 2.1-fold increase (from 58.6 to 121 nM·hr) reported in the literature.

It should be noted here that a similar "suicide inhibition" phenomenon (i.e., an enzyme biotransforms a substrate to a reactive species which, in turn, "kills" the enzyme—a "suicide" from the perspective of the enzyme) had also been observed in CYP2E1 catalyzed metabolism of some volatile oraganic solvents such as *cis* and *trans* DCE (Andersen *et al.* 1987; Lilly *et al.* 1998).

PBPD Modeling of Binary Mixtures Since the emphasis of this monograph is on PBPK modeling and PBPD modeling of chemical mixtures is still at its infancy, we choose to only briefly discuss PBPD modeling by concentrating on two published examples on binary mixtures. For more details on the basic principles of PBPD modeling and specific information on the two case studies discussed below, the readers are encouraged to consult, respectively, El-Masri *et al.* (1996a,c).

One of our earlier examples was the PBPK/PD modeling of a toxicological interaction between Kepone (also known as chlordecone) and carbon tetrachloride (CCl_4) based on mechanisms of interactive toxicity and the application of computer technology in acute toxicity studies. Briefly, CCl_4 is a well-known hepatotoxin. Following free radical formation through the P450 enzyme system, the toxicity of CCl_4 can be an accumulation of lipids (steatosis, fatty liver) and degenerative processes leading to cell death (necrosis). Kepone is found in the environment as a result of photolytic oxidation of mirex, a pesticide used for the control of fire ants, or as a pollutant from careless and irresponsible discharge. At relatively low levels (e.g., 10 ppm in the diet), even repeated dosing of Kepone in the diet up to 15 days caused no apparent toxicity to the liver. The toxicological interaction between Kepone and CCl_4 was elucidated to be the impairment, by Kepone, of the liver's regeneration process. These mechanistic studies were summarized in a number of publications (Mehendale 1984, 1991, 1994).

El-Masri *et al.* (1996a) constructed a PBPD model based on the mechanism of toxicological interaction between Kepone and CCl_4. This PBPD model was verified by literature information, and it was capable of providing time-course computer simulations of mitotic, injured, and pyknotic (dead) cells after treatment with CCl_4 alone or with Kepone pretreatment. This PBPD model was further linked with Monte Carlo simulation to predict the acute lethality of CCl_4 alone and in combination with Kepone.

The second case study involved PBPK/PD modeling of pharmacodynamic interactions between trichloroethylene (TCE) and 1,1-dichloroethylene (DCE) regarding their binding and depletion of hepatic glutathione (GSH) in relation to the intrinsic hepatic GSH systhesis (El-Masri *et al.* 1996c). A PBPK/PD model was used to identify critical time point at which hepatic GSH is at a minimum in response to both chemicals. PBPK models for interactions leading to depletion of hepatic glutathione had been developed by several investigators (D'Souza *et al.* 1988; Frederick *et al.* 1992). Model-directed gas uptake experiments with DCE revealed that DCE was the only chemical capable of significantly depleting hepatic GSH. TCE exposure higher than 100 ppm to the rats obstructed the ability of DCE to deplete hepatic GSH, indicating metabolic competitive inhibition of DCE biotransformation to reactive metabolites. TCE exposure lower than 100 ppm was ineffective in inhibiting DCE from significantly depleting hepatic GSH. El-Masri *et al.* (1996c) further applied these quantitative analyses in establishing an "interaction threshold" (see discussion below in Section 13.2.2) between TCE and DCE.

13.2.2 More Recent Endeavor: PBPK Modeling of Higher-Order Mixtures

In the above section, we provided a glimpse of the development of chemical mixture toxicology in the "early days." As the field of PBPK modeling grows in parallel with the science of toxicology, the natural progression proceeded in two directions. First, when the toxicology of chemical mixture moves from descriptive work to mechanistic-based research, PBPK modeling transforms into PBPD modeling (discussed above). Second, investigators, driven by innate curiosity and practical need, begin to explore PBPK modeling of more and more complex chemical mixtures. Thus, in the next few sections, we will provide a few examples that reflect the application of PBPK modeling to the toxicology of more complex chemical mixtures.

PBPK Modeling of Ternary and Four-Chemical Mixtures Pioneering efforts in the PBPK modeling of more complex chemical mixtures were from a research group led by Krishnan and various colleagues; two comprehensive reviews of work up to 1994 are available (Krishnan *et al.* 1994a,b). Earlier work from this group concentrated on interactions and PBPK modeling between two chemicals (Tardif *et al.* 1993, 1995; Pelekis and Krishnan 1997). As progress was made, these investigators began to build up the mixtures and devoted their effort to PBPK modeling of more and more complex chemical mixtures (Tardif *et al.* 1997; Haddad *et al.* 1999a,b, 2000).

As shown in Table 13.1, PBPK modeling of a ternary mixture on toluene, *m*-xylene, and ethylbenzene was studied and reported by Tardif *et al.* (1997), and the mechanism involved was competitive inhibition. The details of the conceptual interactive PBPK model and the equations with incorporation of competitive inhibition are provided below under the section for five-chemical mixture. Subsequently, Haddad *et al.* (1999a) applied this interactive PBPK model to the calculation of biological hazard index (BHI). BHI, defined as the biological level tolerable for exposure to mixtures, is traditionally calculated in an analogous way as the hazard index under additivity assumption (Ogata *et al.* 1993; Hadded *et al.* 1999a). However, Haddad *et al.* incorporated toxicological interaction by using PBPK modeling to obtain "simulation concentration" (SC) and modified the BHI calculation according to the following equation:

$$\mathrm{BHI} = \sum_{i=1}^{n} \frac{\mathrm{SC}_i}{\mathrm{BEI}_i}$$

Where BHI and SC were defined before; BEI refers to the concentration or excretion rate of a biomarker in a healthy worker exposed to TLV. In doing so, Haddad *et al.* (1999a) applied interactive PBPK modeling of a chemical mixture into the risk assessment process.

Using the same principle and similar technique, researchers at Colorado State University studied PBPK modeling of two ternary mixtures [trichloroethylene (TCE), tetrachloroethylene (PERC), methyl chloroform (MC) and toluene, ethylbenzene, and xylenes] to respectively enhance the concept of "interaction thresholds" and modify and improve the "Mixture Formula" risk assessment by using an interactive PBPK modeling approach (Dobrev *et al.* 2001, 2002; Dennison *et al.* 2005a).

Haddad *et al.* (1999b) also studied the PBPK modeling of a four-chemical mixture involving benzene, toluene, ethylbenzene, and *m*-xylene. In general, the incorporation of the interaction mechanism, at the level of the liver metabolic enzyme inhibition, is similar to those described above for binary and ternary mixtures, albeit a bit more complicated. We will demonstrate again the principle and techniques involved in a later section on a five-chemical mixture.

The Concept of the "Interaction Threshold" In 1996, El-Masri *et al.* introduced the idea of "interaction thresholds" as the minimal level of change in tissue dosimetry of two or more chemicals associated with a significant health effect (El-Masri *et al.* 1996b). When two or more interactive chemicals are studied together, theoretically there could be infinite interaction thresholds. However, if we specify certain occupational or environmental exposure concentrations for all the other component chemicals in the mixture except one, we may obtain an interaction threshold for that set of exposure conditions. This definition is important because human risk from exposure to multiple chemicals may not always obey the rule of additivity. Dobrev *et al.* (2001) estimated the interaction thresholds of three common volatile organic solvents—TCE, PERC, and MC—under different dosing conditions. First, an interactive PBPK model was built where PERC and MC were competitive inhibitors for TCE, the compound most extensively metabolized among the three. The model was developed and validated by gas uptake pharmacokinetic studies in Fischer 344 rats at relatively high doses of single chemicals, binary mixtures, and the ternary mixture. Using computer simulation to extrapolate from high to low concentrations, Dobrev *et al.* (2001) investigated the toxicological interactions at occupational exposure levels, specifically at around threshold limit value/time-weighted average (TLV/TWA). Since long-term toxicity/carcinogenicity of these three solvents is clearly associated with their metabolism, and TCE is the most extensively metabolized among them, this study focused on changes in internal TCE dose measures related to the mixture co-exposure. Using a 10% elevation in parent compound blood level as a criterion for significant interaction, interaction thresholds were estimated with two of the three chemicals held at constant concentrations. Under the above exposure conditions (i.e., TCE and PERC at their TLVs but varying MC concentrations), the interaction threshold for the ternary mixture was 50, 130, and 25 ppm for TCE, MC, and PERC, respectively. This work was later extended, using computer simulation (i.e., *in silico* toxicology), to human exposure to this three-chemical mixture and the estimation of interaction thresholds for humans (Dobrev *et al.* 2002). Increases in the TCE blood levels led to higher availability of the parent compound for glutathione conjugation, a metabolic pathway associated with kidney toxicity/carcinogenicity. The simulated change in production rates of toxic conjugative metabolites exceeded 17% for a corresponding 10% increase in TCE blood concentration, indicating a nonlinear risk increase due to combined exposures to TCE. This study (Dobrev *et al.* 2002) and the related discussion above reveal that evaluation of metabolic interactions and their thresholds illustrates a unique application of PBPK modeling in risk assessment of occupational exposures to chemical mixtures. It further underscores the importance of incorporating PBPK modeling into the cumulative risk assessment process.

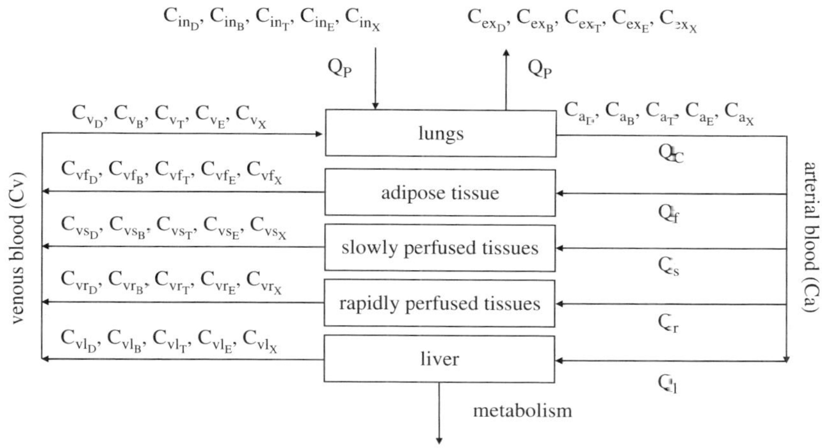

Figure 13.4 Conceptual interactive PBPK model for a five-component chemical mixture. D, dichloromethane; B, benzene; T, toluene; E, ethylbenzene; X, *m*-xylene. Pharmacokinetic interactions occur in the liver as competitive enzyme inhibition. Q_P is alveolar ventilation rate; Q_C is cardiac output; C_v and C_a are venous and arterial blood concentrations of individual component chemicals; C_{vi} is venous blood concentration leaving tissue compartments; Q_i is blood flow to tissues (f, adipose tissue; s, slowly perfused tissues; r, rapidly perfused tissues; l, liver); C_{in} and C_{ex} are inhaled or exhaled concentrations of individual component chemicals. Redrawn based on Haddad, S., Charest-Tardif, G., Tardif, R., and Krishnan, K. (2000). *Toxicol. Appl. Pharmacol.* **167**, 199–209.

Development of a PBPK Model for a Chemical Mixtures Involving Five Components and the Concept of Predicting Pharmacokinetics of Different Chemical Mixtures Based on PBPK Studies of Binary Mixtures An interactive model for 5 chemicals (benzene, toluene, ethylbenzene, *m*-xylene, and dichloromethane) has also been completed (Haddad *et al.* 2000; Krishnan *et al.* 2002), leading to the idea that pharmacokinetic interactions of complex chemical mixtures, regardless of the number of components, may be predicted based on the PBPK modeling of binary mixtures of the component chemicals (Haddad *et al.* 2000; Krishnan *et al.* 2002). Since, to date, their studies represent the most complex interactive PBPK model for a mixture with five reconstituted chemicals, we will provide more detailed discussion below. Also, their concept of predicting the kinetics of complex mixtures based on PBPK studies of binary mixture is interesting and revolutionary, and it warrants further discussion.

As shown in Fig. 13.4, the conceptual PBPK model of the five-chemical mixture (Haddad *et al.* 2000) is similar to the styrene model in Fig. 13.1; the only difference is the incorporation of the individual chemical concentrations in the arterial and venous blood as designated by subscripts D, B, T, E, X, representing the five chemicals. Because the mechanism of interaction was determined to be competitive enzyme inhibition of CYP2E1, the following Michaelis–Menten equations were incorporated into the mass balance differential equations in the liver compartment for the respective chemicals: benzene (B), toluene (T), ethylbenzene (E), *m*-

xylene (X), and dichloromethane (D). RAM here represents rate of metabolism or velocity of the enzyme reaction.

$$\mathrm{RAM}_B = \frac{V_{\max B} \times C_{vlB}}{K_{mB}\left(1+\frac{C_{vlE}}{K_{iEB}}+\frac{C_{vlT}}{K_{iTB}}+\frac{C_{vlX}}{K_{iXB}}+\frac{C_{vlD}}{K_{iDB}}\right)+C_{vlB}}$$

$$\mathrm{RAM}_T = \frac{V_{\max T} \times C_{vlT}}{K_{mT}\left(1+\frac{C_{vlB}}{K_{iBT}}+\frac{C_{vlE}}{K_{iET}}+\frac{C_{vlX}}{K_{iXT}}+\frac{C_{vlD}}{K_{iDT}}\right)+C_{vlT}}$$

$$\mathrm{RAM}_E = \frac{V_{\max E} \times C_{vlE}}{K_{mE}\left(1+\frac{C_{vlB}}{K_{iBE}}+\frac{C_{vlT}}{K_{iTE}}+\frac{C_{vlX}}{K_{iXE}}+\frac{C_{vlD}}{K_{iDE}}\right)+C_{vlE}}$$

$$\mathrm{RAM}_X = \frac{V_{\max X} \times C_{vlX}}{K_{mX}\left(1+\frac{C_{vlB}}{K_{iBX}}+\frac{C_{vlE}}{K_{iEX}}+\frac{C_{vlT}}{K_{iTX}}+\frac{C_{vlD}}{K_{iDX}}\right)+C_{vlX}}$$

$$\mathrm{RAM}_D = \frac{V_{\max D} \times C_{vlD}}{K_{mD}\left(1+\frac{C_{vlB}}{K_{iBD}}+\frac{C_{vlE}}{K_{iED}}+\frac{C_{vlT}}{K_{iTD}}+\frac{C_{vlX}}{K_{iXD}}\right)+C_{vlD}}$$

Experimentally, Haddad *et al.* (2000) did PBPK model simulations of the pharmacokinetics of components under two scenarios: (1) when one of the mixture components was substituted with another such as benzene in the BETX mixture was substituted with dichloromethane and (2) when another chemical was added to an existing four-chemical mixture model such as dichloromethane was added to the existing BTEX model. These investigators also did pharmacokinetic studies and specifically obtained blood kinetic data in rats on all the new binary mixtures with D added; thus, animal experimental data were generated on D-B, D-E, D-T, D-X binary mixtures. When competitive inhibition was incorporated into the interactive PBPK model, the model simulation not only matched the newly obtained binary mixture kinetic data, but also matched earlier data for a variety more complex mixtures beyond binary combinations. Therefore, from their studies emerged the concept that predictability of pharmacokinetic and pharmacodynamic consequences of chemicals in more complex chemical mixtures is possible as long as there is the availability of quantitative data in the literature on binary chemical interactions (Haddad *et al.* 2000; Krishnan *et al.* 2002). So far their approach (Haddad *et al.* 2000; Krishnan *et al.* 2002) has worked for the volatile organic chemicals that they studied. Whether or not this concept has a broader application to mixed classes of chemicals in a mixture remained to be evaluated.

PBPK Modeling of Complex Chemical Mixtures As PBPK modeling of chemical mixtures progresses to involving more and more components, it is a natural course of development that investigators will attempt to tackle the real-world complex chemical mixtures. Verhaar *et al.* (1997) proposed the incorporation of lumping analyses (a chemical engineering technique used in petroleum engineering processes) and QSAR to PBPK modeling. The idea was that each of the three techniques would serve its unique function in the overall goal of predicting some aspects of the chemical mixtures of interest. Thus, QSAR analysis can be used to predict needed physicochemical and toxicological parameters for unknown compounds or for surrogate compounds (from lumping); lumping analysis can drastically reduce the complexity of the description of a mixture; and PBPK/PD modeling can be used to describe the pharmacokinetics, and possibly pharmacodynamics, of an ensemble of compounds or lumped pseudocompounds, including possible interaction effects. A detailed statistical/mathematical treatment on how to minimize errors in lumping was given in an appendix in this article. Verhaar *et al.* (1997) specifically suggested the application of these technologies (i.e., QSAR, lumping analysis, PBPK/PD modeling) to JP-5.

These ideas have now been applied with gasoline as the complex mixture (Dennison *et al.* 2003, 2004a,b, 2005b; Dennison 2004) developing both the PBPK modeling framework and lumping approach. Experimental work involved gas uptake pharmacokinetic studies in male Fischer 344 rats of single, multiple selected target components (benzene, toluene, ethylbenzene, *o*-xylene, and *n*-hexane), and two blends (summer and winter) of gasoline, as well as the volatile fractions of the gasoline. The target components were selected based on the prevalence and toxicological importance; the remainder of the hundreds of component chemicals were lumped into a pseudo chemical (Dennison *et al.* 2003; Dennison 2004). Technological development necessary for this effort included (1) modification of the gas uptake pharmacokinetic chamber system for more efficient incorporation of probes and (2) utilization of more efficient CO_2 absorbent. The PBPK model tracks selected target components and a lumped chemical group representing all nontarget components (Dennison *et al.* 2004a,b, 2005b; Dennison 2004). Competitive inhibition was the principal mechanism of pharmacokinetic interactions among these five selected target single chemicals and a pseudochemical from the lumped components. Computer simulation results from the six-chemical interaction model matched well with gas uptake pharmacokinetic experimental data from single chemicals, five-chemical mixture, and the two blends of gasoline (Dennison *et al.* 2003). Thus, for the first time, we have a PBPK model for a real-world complex chemical mixture.

13.3 FUTURE PERSPECTIVES: SECOND-GENERATION PBPK/PD MODELING

Although the problems and tasks associated with finding a reasonable way to handle, and eventually predict, adverse health effects due to chemical mixtures are complex, human health problems are really the final manifestation of dynamic equilibria of multiple stressors of which chemical mixtures are but one element. Thus, the potential combinations of these multiple stressors may approach infinity. In order to circumvent the study of astronomically large number of combinations, the only logical way is to concentrate on the finite system—in this case, the human body. Dr. Craig Venter of the human genome fame stated: "... If we hope to understand biology, instead of looking at one little protein at a time, which is not how biology works, we will need to understand the integration of thousands of proteins in a dynamically changing environment. A computer will be the biologist's number one tool ..." (Butler 1999). In line with this thinking, we believe that the only efficient way of studying chemical mixtures or multiple stressors is to understand our body in an integrated manner through biologically based computer simulation such as PBPK/PD modeling and very focused experimentation. In essence, this emphasis leads to a systems biology approach with tools from *in silico* toxicology.

In silico toxicology, by our defintion, means integrating computer modeling with focused, mechanistic, animal experimentation such that experiments which are impractical (e.g., too large, too expensive) or impossible (e.g., human experiments with carcinogens) to perform may be conducted by computer simulation. With this type of approach, once we have a "virtual human" in place, multiple stressors and their integrated adverse human health effects are nothing more than the perturbation of certain processes in the normal systems. In that sense, adverse health effects are therefore the manifestation of parameter changes in the computer simulation of the "virtual human." The classical compartmental pharmacokinetic models may be considered an embryonic form of "Virtual Human" (Yang *et al.* 2004b). Indeed, despite the crude nature of these models as compared with the human body, classical pharmacokinetics has contributed very significantly to the field of medicine. The advancement of PBPK modeling, as a result of "delumping" and the incorporation of physiology into the modeling process, results in a better "Virtual Human."

It is our strong belief that utilization of computer modeling is essential in the studies of toxicology of chemicals, chemical mixtures, and multiple stressors. Biology will be well-served by the application of computer technology as an alternative research method to conserve resources and minimize the use of laboratory animals. In the past few years, tools such as "Reaction Network Modeling" and "Gene Network Modeling" have become available to support computer simulation at the molecular interaction level. For more information on the specifics of these newer modeling approaches in biomedical sciences, the readers are referred to a number of publications (Andersen *et al.* 2002; Klein *et al.* 2002; Liao *et al.* 2002; Reisfeld and Yang 2004; Liao 2004). Looking into the future, linkage of PBPK modeling with "Reaction Network Modeling" and/or "Gene Network Modeling" has the potential of providing a computer simulation platform for modeling complex biological systems from the whole organism down to the molecular interaction level.

13.4 SUMMARY

In studying chemical mixtures in the last two decades or so, and the multiple stressors more recently, we reach the inevitable conclusion that it is impossible to seek knowledge on every possible environmental mixture. A reasonable alternative is therefore to begin with careful study of underlying biology. Although the body is itself a complex chemical system, compared to the infinite combinations of external chemical, physical, biological challenges as well as other multiple stressors, the body represents a finite system and a "constant." Thus, if we understand enough of this underlying biology to be able to carry out computer simulations correctly, we should be able to simulate or even predict, by computer modeling, the potential perturbations caused by these infinite number of external stressors.

With the recent advances in computational technology and biomedical research methodologies and past research on the toxicological interactions of chemical mixtures, the integration of computer modeling and focused laboratory experimentation in an iterative manner should improve modeling and risk assessment with various mixtures. For computer modeling and simulation of complex biological processes to be successful and useful, the integration of biology, chemistry, physics, engineering, and computer science is necessary.

NOTATION

AST	aspartate transaminase
BEI	the concentration or excretion rate of a biomarker in a healthy worker exposed to the TLV
BHI	biological hazard index, a parameter estimating the biological level tolerable for exposure to mixtures
C_a	the concentration of chemical in the arterial blood
C_h	concentration of substrate in the liver
C_{ex}	the concentration of chemical in exhaled air
C_{in}	the concentration of chemical in inhaled air
C_{portal}	concentration of substrate in the portal vein
C_{sys}	concentration of substrate in systemic blood
C_v	the concentration of chemical in the venous blood
CCl_4	carbon tetrachloride
CL_{int}	intrinsic metabolic clearance
CYP2E1	cytochrome P450 2E1
CYP3A4	cytochrome P450 3A4
D	dose
DBM	dibromomethane
DCE	1,1-dichloroethylene
DPD	dihydropyrimidine dehydrogenase
F_a	fraction absorbed from the gastrointestinal tract
f_b	unbound fraction in blood
5-FU	5-fluorouracil
GSH	glutathione
HbCO	carboxyhemoglobin
2,5-HD	2,5-hexanedione
I_h	concentration of inhibitor in the liver
I_{portal}	concentration of inhibitor in the portal vein
I_{sys}	concentration inhibitor in systemic blood
K_m	Michaelis constant for metabolic elimination

K_{M1}	substrate 1: enzyme binding constant
K_{M2}	substrate 2: enzyme binding constant
K_{MI1}	substrate 1: inhibitory binding constant
K_{MI2}	substrate 2: inhibitory binding constant
K_p	liver-to-blood concentration ratio
MC	methyl chloroform
MnBK	methyl *n*-butyl ketone
P1	product 1
P2	product 2
PBPD	physiologically based pharmacodynamic
PBPK	physiologically based pharmacokinetic
PD	pharmacodynamic
PERC	tetrachloroethylene
Q	blood flow rate
Q_C	cardiac output
Q_f	the flow rate of blood through the fat
Q_{GI}	the flow rate of blood through the GI tract
Q_k	the flow rate of blood through the kidneys
Q_l	the flow rate of blood through the liver
Q_P	alveolar ventilation rate
Q_r	the flow rate of blood through the rapidly perfused tissue
Q_s	the flow rate of blood through the slowly perfused tissue
S1	substrate 1
S2	substrate 2
SC	simulation concentration
TCE	trichloroethylene
TLV	threshold limit value
TWA	time-weighted average
US EPA	United States Environmental Protection Agency
V_d	volume of distribution in the systemic blood compartment
V_h	volume of the liver
V_{max}	maximum metabolic rate
V_{portal}	volume of the portal vein
VOC	volatile organic solvent

REFERENCES

Ali, N., and Tardif, R. (1999). Toxicokinetic modeling of the combined exposure of toluene and *n*-hexane in rats and humans. *J. Occup. Health* **41**, 95–103.

Andersen, M. E., and Clewell, H. J. (1984). Pharmacokinetic interactions of mixtures. In: *Proceedings of the 14th Conference on Environmental Toxicology*, AFAMRL-TR-83-099, Air Force Aerospace Medical Research Laboratory, Wright-Patterson AFB, OH, pp. 226–238.

Andersen, M. E., Gargas, M. L., Clewell, H. J., III, and Severyn, K. M. (1987). Quantitative evaluation of the metabolic interactions between trichloroethylene and 1,1-dichloroethylene *in vivo* using gas uptake methods. *Toxicol. Appl. Pharmacol.* **89**, 149–157.

Andersen, M. E., Yang, R. S. H., French, C. T., Chubb, L. S., and Dennison, J. E. (2002). Molecular circuits, biological switches and non-linear dose–response relationships. *Environ. Health Perspect.* **110**(Suppl 6), 971–978.

Barton, H. A., Creech, J. R., Godin, C. S., Randall, G. M., and Seckel, C. S. (1995). Chloroethylene mixtures: Pharmacokinetic modeling and *in vitro* metabolism of vinyl chloride, trichloroethylene, and *trans*-1,2-dichloroethylene in rats. *Toxicol. Appl. Pharmacol.* **130**. 237–247.

Bond, J. A., Csanady, G. A., Gargas, M. L., Guengerich, F. P., Leavens, T., Medinsky, M. A., and Recio, L. (1994). 1,3-Butadiene: linking metabolism, dosimetry, and mutation induction. *Environ. Health Perspect.* **102**(Suppl 9), 87–94.

Butler, D. (1999). Computing 2010 from black holes to biology. *Nature* **402**, C67–C70.

Clewell, H. J., III, and Andersen, M. E. (1985). Risk assessment extrapolations and physiological modeling. *Toxicol. Ind. Health* **1**, 111–131.

Dennison, J. E. 2004. Physiologically-Based Pharmacokinetic Modeling of Simple and Complex Mixtures of Gasoline and the Gasoline Components n-Hexane, Benzene, Toluene, Ethylbenzene, and Xylene. Ph.D. dissertation, Colorado State University, 218 pp.

Dennison, J. E., Andersen, M. E., and Yang, R. S. H. (2003). Characterization of the pharmacokinetics of gasoline using PBPK modeling with a complex mixture chemical lumping approach. *Inhalation Toxicol.* **15**, 961–968.

Dennison, J. E., Andersen, M. E., Dobrev, I. D., Mumtaz, M. M., and Yang, R. S. H. (2004a). PBPK modeling of complex hydrocarbon mixtures: Gasoline. *Environ. Toxicol. Pharmacol.* **16**: 107–119.

Dennison, J. E., Andersen, M. E., Clewell, H. J., and Yang, R. S. H. (2004b). Development of a PBPK model for volatile fractions of gasoline using chemical lumping analyses. *Environ. Sci. Tech.* **38**: 5674–5681.

Dennison, J. E., Bigelow, P. L., Mumtaz, M. M., Andersen, M. E., Dobrev, I. D., and Yang, R. S. H. (2005a). Evaluation of potential toxicity from co-exposure to three CNS depressants (toluene, ethylbenzene, and xylenes): A PBPK modeling approach. *Am. Ind. Hyg. Assoc. J.* In press.

Dennison, J. E., Andersen, M. E., and Yang, R. S. H. (2005b). Pitfalls and related improvements of *in vivo* gas uptake pharmacokinetic experimental systems. Submitted for publication.

Dobrev, I., Andersen, M. E., and Yang, R. S. H. (2001). Assessing interaction thresholds for trichloroethylene, tetrachloroethylene, and 1,1,1-trichloroethane using gas uptake studies and PBPK modeling. *Arch. Toxicol.* **75**, 134–144.

Dobrev, I., Andersen, M. E., and Yang, R. S. H. (2002). *In silico* toxicology: Simulating interaction thresholds for human exposure to mixtures of trichloroethylene, tetrachloroethylene, and 1,1,1-trichloroethane. *Environ. Health Perspect.* **110**, 1031–1039.

D'Souza, R. W., Francis, W. R., and Andersen, M. E. (1988). Physiological model for tissue glutathione depletion and increased resynthesis after ethylene dichloride exposure. *J. Pharmacol. Exp. Ther.* **245**, 563–568.

El-Masri, H. A., Thomas, R. S., Sabados, G. R., Phillips, J. K., Constan, A. A., Benjamin, S. A., Andersen, M. E., Mehendale, H. M., and Yang, R. S. H. (1996a). Physiologically based pharmacokinetic/pharmacodynamic modeling of the toxicologic interaction between carbon tetrachloride and Kepone. *Arch. Toxicol.* **70**, 704–713.

El-Masri, H. A., Tessari, J. D., and Yang, R. S. H. (1996b). Exploration of an interaction threshold for the joint toxicity of trichloroethylene and 1,1-dichloroethylene: Utilization of a PBPK model. *Arch. Toxicol.* **70**, 527–539.

El-Masri, H. A., Constan, A. A., Ramsdell, H. S., and Yang, R. S. H. (1996c). Physiologically based pharmacodynamic modeling of an interaction threshold between trichloroethylene and 1,1-dichloroethylene in Fischer 344 rats. *Toxicol. Appl. Pharmacol.* **141**, 124–132.

Filser, J. G., Johanson, G., Kessler, W., Kreuzer, P. E., Stei, P., Baur, C., and Csanady, G. A. (1993). A pharmacokinetic model to describe toxicokinetic interactions between 1,3-butadiene and styrene in rats: predictions for human exposure. *IARC Sci Publ.* **127**, 65–78.

Frederick, C. B., Potter, D. W., Chang-Mateu, M. I., and Andersen, M. E. (1992). A physiologically based pharmacokinetic and pharmacodynamic model to describe the oral dosing of rats with ethyl acrylate and its implications for risk assessment. *Toxicol. Appl. Pharmacol.* **114**, 246–260.

Gargas, M. L., Andersen, M. E., and Clewell, H. J., III. (1986). A physiologically based simulation approach for determining metabolic constants from gas uptake data. *Toxicol. Appl. Pharmacol.* **86**, 341–352.

Haddad, S., Tardif, R., Viau, C., and Krishnan, K. (1999a). A modeling approach to account for toxicokinetic interactions in the calculation of biological hazard index for chemical mixtures. *Toxicol Lett.* **108**, 303–308.

Haddad, S., Tardif, R., Charest-Tardif, G., and Krishnan, K. (1999b). Physiological modeling of the toxicokinetic interactions in a quaternary mixture of aromatic hydrocarbons. *Toxicol. Appl. Pharmacol.* **161**, 249–257.

Haddad, S., Charest-Tardif, G., Tardif, R., and Krishnan, K. (2000). Validation of a physiological modeling framework for simulating the toxicokinetics of chemicals in mixtures. *Toxicol. Appl. Pharmacol.* **167**, 199–209.

Ito, K., Iwatsubo, T., Kanamitsu, S., Ueda, K., Suzuki, H., and Sugiyama, Y. (1998). Prediction of pharmacokinetic alterations caused by drug–drug interactions: Metabolic interaction in the liver. *Pharmacol Rev.* **50**, 387–412.

Jang, J. Y., Droz, P. O., and Kim, S. (2001). Biological monitoring of workers exposed to ethylbenzene and co-exposed to xylene. *Int. Arch. Occup. Environ. Health* **74**, 31–37.

Kanamitsu, S., Ito, K., Green, C. E., Tyson, C. A., Shimada, N., and Sugiyama, Y. (2000). Prediction of *in vivo* interaction between triazolam and erythromycin based on *in vitro* studies using human liver microsomes and recombinant human CYP3A4. *Pharm Res.* **17**, 419–426.

Klein, M. T., Hou, G., Quann, R., Wei, W., Liao, K. H., Yang, R. S. H., Campain, J. A., Mazurek, M., and Broadbelt, L. J. (2002). BioMOL: A computer-assisted biological modeling tool for complex chemical mixtures and biological processes at the molecular level. *Environ. Health Perspect.* **110**(Suppl 6), 1025–1029.

Krishnan, K., Andersen, M. E., Clewell, H. J., III, and Yang, R. S. H. (1994a). Physiologically Based Pharmacokinetic Modeling of Chemical Mixtures. In *Toxicology of Chemical Mixtures: Case Studies, Mechanisms, and Novel Approaches*, R. S. H. Yang, ed., Academic Press, San Diego, CA, pp. 399–437.

Krishnan, K., Clewell, H. J., III, and Andersen, M. E. (1994b). Physiologically-based pharmacokinetic analyses of simple mixtures. *Environ. Health Perspect.* **102**(Suppl 9), 151–155.

Krishnan, K., Haddad, S., Beliveau, M., and Tardif, R. (2002). Physiological modeling and extrapolation of pharmacokinetic interactions from binary to more complex chemical mixtures. *Environ. Health Perspect.* **110**(Suppl 6), 989–994.

Krishnan, K., and Pelekis, M. (1995). Hematotoxic interactions: Occurrence, mechanisms and predictability. *Toxicology* **105**, 355–364.

Leavens, T. L., and Bond, J. A. (1996). Pharmacokinetic model describing the disposition of butadiene and styrene in mice. *Toxicology* **113**, 310–313.

Liao, K. H. (2004). Reaction Network Model for the Prediction of Mammalian Metabolism of Benzo[*a*]pyrene. Ph.D. dissertation, Colorado State University.

Liao, K. H., Dobrev, I., Dennison, Jr., J. E., Andersen, M. E., Reisfeld, B., Reardon, K. F., Campain, J. A., Wei, W., Klein, M. T., Quann, R. J., and Yang, R. S. H. (2002). Application of biologically based computer modeling to simple or complex mixtures. *Environ. Health Perspect.* **110**(Suppl 6), 957–963.

Lilly, P. D., Thornton-Manning, J. R., Gargas, M. L., Clewell, H. J., and Andersen, M. E. (1998). Kinetic characterization of CYP2E1 inhibition *in vivo* and *in vitro* by the chloroethylenes. *Arch Toxicol.* **72**, 609–621.

Mehendale, H. M. (1984). Potentiation of halomethane hepatotoxicity: Chlordecone and carbon tetrachloride. *Fundam. Appl. Toxicol.* **4**, 295–308.

Mehendale, H. M. (1991). Role of hepatocellular regeneration and hepatolubular healing in the final outcome of liver injury. *Biochem. Pharmacol.* **42**, 1155–1162.

Mehendale, H. M. (1994). Mechanism of the interactive amplification of halomethane hepatotoxicity and lethality by other chemicals. In: Yang, R. S. H., ed., *Toxicology of Chemical Mixtures: Case Studies, Mechanisms, and Novel Approaches*, Academic Press, San Diego, CA, pp. 299–334.

Mozayani, A., and Raymon, L. P. (2004). *Handbook of Drug Interactions. A Clinical and Forensic Guide*, Humana Press, Totowa, NJ, 663 pp.

Mumtaz, M. M., Sipes, I. G., Clewell, H. J., and Yang, R. S. H. (1993). Risk assessment of chemical mixtures: Biologic and toxicologic issues. *Fundam. Appl. Toxicol.* **21**, 258–269.

Ogata, M., Fiserova-Bergerova, V., and Droz, P. O. (1993). Occupational exposures to mixtures of industrial chemicals. *Appl. Occup. Environ. Hyg.* **8**, 609–617.

Pelekis, M., and Krishnan, K. (1997). Assessing the relevance of rodent data on chemical interactions for health risk assessment purposes: A case study with dichloromethane–toluene mixture. *Regul. Toxicol. Pharmacol.* **25**, 79–86.

Purcell, K. J., Cason, G. H., Gargas, M. L., Andersen, M. E., and Travis, C. C. (1990). *In vivo* metabolic interactions of benzene and toluene. *Toxicol. Lett.* **52**, 141–152.

Ramsey, J. C., and Andersen, M. E. (1984). A physiological model for the inhalation pharmacokinetics of inhaled styrene monomer in rats and humans. *Toxicol. Appl. Pharmacol.* **73**, 159–175.

Reisfeld, B., and Yang, R. S. H. (2004). A reaction network model for CYP2E1-mediated metabolism of toxicant mixtures. *Environ. Toxicol. Pharmacol.* **18**, 173–179.

Sato, A., and Nakajima, T. (1979). A vial-equilibration method to evaluate the drug-metabolizing enzyme activity for volatile hydrocarbons. *Toxicol. Appl. Pharmacol.* **47**, 41–46.

Sato, A., Endoh, K., Kaneko, T., and Johanson, G. (1991). Effects of consumption of ethanol on the biological monitoring of exposure to organic solvent vapor: A simulation study with trichloroethylene. *Brit. J. Ind. Med.* **48**, 548–556.

Segal, I. H. (1975). Enzyme Kinetics. *Behavior and Analysis of Rapid Equilibrium and Steady-State Enzyme Systems*, John Wiley & Sons, New York, 957 pp.

Tardif, R., and Charest-Tardif, G. (1999). The importance of measured end-points in demonstrating the occurrence of interactions: a case study with methylchloroform and *m*-xylene. *Toxicol. Sci.* **49**, 312–317.

Tardif, R., Lapare, S., Krishnan, K., and Brodeur, J. (1993). A descriptive and mechanistic study of the interaction between toluene and xylene in humans. *Int. Arch. Occup. Environ. Health* **65**(1 Suppl), S135–S137.

Tardif, R., Lapare, S., Charest-Tardif, G., Brodeur, J., and Krishnan, K. (1995). Physiologically-based pharmacokinetic modeling of a mixture of toluene and xylene in humans. *Risk Anal.* **15**, 335–342.

Tardif, R., Lapare, S., Charest-Tardif, G., Brodeur, J., and Krishnan, K. (1997). Physiologically based pharmacokinetic modeling of a ternary mixture of alkyl benzenes in rats and humans. *Toxicol. Appl. Pharmacol.* **144**, 120–134.

Thakore, K. N., Gargas, M. L., Andersen, M. E., and Mehendale, H. M. (1991). PBPK derived metabolic constants, hepatotoxicity, and lethality of $BrCCl_3$ in rats pretreated with chlordecone, Phenobarbital or mirex. *Toxicol. Appl. Toxicol.* **109**, 514–528.

Thrall, K. D., and Poet, T. S. (2000). Determination of biokinetic interactions in chemical mixtures using real-time breath analysis and physiologically based pharmacokinetic modeling. *J. Toxicol. Environ. Health A.* **59**, 653–670.

US EPA. (2003). Physiologically-based Pharmacokinetic/Pharmacodynamic Modeling: Preliminary Evaluation and Case Study for the N-Methyl Carbamate Pesticides: A Consultation, December 11 and 12, 2003 FIFRA Scientific Advisory Panel Meeting, Arlington, Virginia. SAP Minutes No. 2003-06 (http://www.epa.gov/oscpmont/sap/2003/index.htm#121103; accessed March 28, 2004).

Verhaar, H. J. M., Morroni, J. S., Reardon, K. F., Hays, S. M., Gaver, D. P., Carpenter, R. L., and Yang, R. S. H. (1997). A proposed approach to study the toxicology of complex mixtures of petroleum products: The integrated use of QSAR, lumping analysis, and PBPK/PD modeling. *Environ. Health Perspect.* **105**(Suppl 1), 179–195.

Yang, R. S. H., Andersen, M. E., Dennison, J. E., Ou, Y. C., Liao, K. H., and Reisfeld, B. (2004a). Physiologically based pharmacokinetic and pharmacodynamic modeling, in *Mouse Models of Cancer*, E. C. Holland, ed., John Wiley & Sons, New York, pp. 391–405.

Yang, R. S. H., El-Masri, H. A., Thomas, R. S., Dobrev, I., Dennison, Jr., J. E., Bae, D. S., Campain, J. A., Liao, K. H., Reisfeld, B., Andersen, M. E., and Mumtaz, M. M. (2004b). Chemical mixture toxicology: From descriptive to mechanistic, and going on to *in silico* toxicology. *Environ. Toxicol. Pharmacol.* **18**, 65–81.

Yu, X., Johanson, G., Ichihara, G., Shibata, E., Kamijima, M., Ono, Y., and Takeuchi, Y. (1998). Physiologically based pharmacokinetic modeling of metabolic interactions between *n*-hexane and toluene in humans. *J. Occup. Health* **40**, 293–301.

CHAPTER 14

DERMAL EXPOSURE MODELS

Micaela B. Reddy

14.1 INTRODUCTION

The first PBPK modeling article that included the dermal exposure route appeared in 1986. The dermal absorption of dichloromethane (DCM), dibromomethane (DBM), and bromochloromethane (BCM) vapors were studied in the rat using PBPK modeling techniques (McDougal *et al.* 1986) (Table 14.1). Since then, many other PBPK models have included the dermal exposure route. Models have been developed for the dermal absorption of pesticides (e.g., malathion and chlordecone), industrial solvents (e.g., perchloroethylene and toluene), and a variety of other types of compounds (Table 14.2). Primarily, the purposes of the articles have been for conducting risk assessments or for evaluating occupational exposures. Understanding and predicting dermal absorption is important for risk assessment, drug delivery, and many other applications, and the dermal exposure route is relatively underrepresented in the PBPK modeling literature. Of the three exposure routes included in risk assessment (i.e., inhalation, oral and dermal exposures), dermal absorption has been afforded the least attention and has been called the "missing link" of risk assessment (Zartarian and Leckie 1998). This chapter outlines important factors to consider in modeling dermal absorption and the available choices of model structures. Selected experimental methods used to study dermal absorption using PBPK modeling techniques will be presented.

Physiologically Based Pharmacokinetic Modeling, edited by Micaela B. Reddy, Raymond S. H. Yang, Harvey J. Clewell III, and Melvin E. Andersen.

TABLE 14.1 Dermal PBPK Models for Halogenated Methanes

Chemical	Species	Reference
DCM	Rat	McDougal *et al.* (1986)
BCM	Rat	McDougal *et al.* (1986)
	Rat	Jepson and McDougal (1997)
	Rat	Jepson and McDougal (1999)
Chloroform	Human	McKone (1993)
	Human	Chinery and Gleason (1993)
	Human	Georgopoulos *et al.* (1994)
	Human	Roy *et al.* (1994)
	Human	Roy *et al.* (1996b)
	Human	Roy *et al.* (1996a)
	Human	Corley *et al.* (2000)
	Human	Levesque *et al.* (2000)
DBM	Rat	McDougal *et al.* (1986)
	Rat	Bookout *et al.* (1996)
	Rat	Bookout *et al.* (1997)
	Rat	Jepson and McDougal (1997)
	Rat	Jepson and McDougal (1999)
Carbon tetrachloride	Rat	Thrall and Kenny (1996)

14.2 FACTORS TO CONSIDER IN MODELING DERMAL ABSORPTION

Physiologically, skin is a multilayered membrane (Kligman 1964). For many chemicals, the outermost skin layer, the stratum corneum (sc), is the rate-limiting barrier for mass transfer into and through skin. For highly lipophilic chemicals, the second skin layer, the viable epidermis (ve), also contributes a significant resistance to percutaneous penetration. The sc is composed of dead, desiccated, keratinized cells, but the cells of the ve are alive and capable of metabolizing some chemicals. Together, the sc and ve form the epidermis (epi). The dermis, located beneath the epi, is a highly vascularized tissue that usually has sufficient blood flow to clear away all chemical passing through the epi (Scheuplein and Bronaugh 1983). Chemical placed on the skin surface, often presented as a solution or suspension in a vehicle, can partition into and passively diffuse through skin. Both the vehicle and the absorbing chemical can interact with skin, which could alter its physicochemical properties.

A variety of mathematical models describing dermal absorption during various types of exposures are available (Roberts *et al.* 2001), and choosing an appropriate model structure can be difficult. Three types of dermal absorption models (i.e., membrane models and one- and two-compartment models) will be discussed in the following sections. There are often many different ways that dermal absorption can be described. For example, dermal absorption of chloroform has been incorporated into PBPK models using a one-compartment model [e.g., by Corley *et al.*

TABLE 14.2 Dermal PBPK Models for Other Chemicals

Chamical Class	Chemical	Species	Reference
Solvent/disinfectant	Isopropanol	Rat	Clewell *et al.* (2001)[a]
Pesticides	Fluazifop-butyl	Human	Auton *et al.* (1994)
	Chlordecone	Young and adult rat	Heatherington *et al.* (1998)
	Chlorpyrifos	Human	Timchalk *et al.* (2002)[a]
	Isofenphos	Rat	Knaak *et al.* (1994)
	Malathion	Human adult and child	Rabovsky and Brown (1993)
		Human adult and child	Dong *et al.* (1994)
Environmental contaminants	TBDD	Rat	Kedderis *et al.* (1993)
	MTBE	Rat, human	Rao and Ginsberg (1997)
Industrial solvents	2-butoxyethanol	Rat	Shyr *et al.* (1993)
		Rat, human	Corley *et al.* (1994)
	m-Xylene	Human	Loizou *et al.* (1999)
		Rat	McDougal *et al.* (1990)
	o-Xylene	Rat, human	Thrall and Woodstock (2003)
	Hexane	Rat	McDougal *et al.* (1990)
	Benzene	Rat	McDougal *et al.* (1990)
	Trichloroethylene	Rat, human	Poet *et al.* (2000a)
	Perchloroethylene	Rat	McDougal *et al.* (1990)
		Human adult and child	Rao and Brown (1993)
	Toluene	Rat	McDougal *et al.* (1990)
		Rat	Thrall and Woodstock (2002)
		Human	Thrall *et al.* (2002)
	Styrene	Rat	McDougal *et al.* (1990)
Volatile anesthetics	Isoflurane	Rat	McDougal *et al.* (1990)
	Halothane	Rat	McDougal *et al.* (1990)
Water disinfection byproduct	Methyl chloroform	Human	Bogen and Hall (1989)[a]
		Rat, human	Poet *et al.* (2000b)
Industrial compounds	D4	Rat	Sarangapani *et al.* (2003)
	Benzoic acid	Hairless guinea pig	Macpherson *et al.* (1996)

[a] PBPK models for more than one species were developed, but the dermal exposure route was only included in the PBPK model for one species.

(2000)], a two-compartment model [i.e., by Chinery and Gleason (1993)], and a membrane model [e.g., by Georgopoulos *et al.* (1994)].

Modeling dermal absorption of volatile chemicals can present unique modeling considerations. For example, in a PBPK model describing dermal absorption of neat octamethylcyclotetrasiloxane (D4, a semivolatile silicone fluid) in the rat, Sarangapani *et al.* (2003) modeled dermal absorption as a bolus dose to a topical compartment on top of the exposed skin. The model included the volatilization of D4 from the topical compartment because most of the applied chemical evaporated from the skin surface before absorption into cutaneous blood could take place. The choice of a model structure for dermal absorption can depend on both the physical situation that is going to be simulated and the quality of the data available for model development.

After an appropriate model structure for the skin has been selected, this dermal piece must be incorporated in the PBPK model. For situations where a small amount of skin is exposed to a chemical, the chemical absorbed from an exposure is often added to the mixed venous blood, even though the bulk skin compartment is still treated as part of the slowly perfused tissue compartment, as was done for benzoic acid (MacPherson *et al.* 1996) and 2,3,7,8-tetrabromodibenzo-*p*-dioxin, TBDD (Kedderis *et al.* 1993). When the entire animal is exposed to a chemical—for example, in humans exposed to chloroform while swimming (Levesque *et al.* 2000) or during whole-body exposures of humans to *m*-xylene vapor (Loizou *et al.* 1999)—the skin is often split from the slowly perfused compartment and treated as a separate compartment with an absorption term. Other approaches are also possible. In a dermal absorption model for 2-butoxyethanol, Shyr *et al.* (1993) split the skin from the slowly perfused compartment, and treated it as two compartments, with one receiving an exposure and the other not receiving an exposure. In another approach taken by Thrall and Woodstock (2003) in modeling dermal absorption of aqueous *o*-xylene by human volunteers, the majority of the skin was included in the slowly perfused compartment, but a skin compartment representing the exposed skin was included separately. The method chosen for incorporating the dermal absorption model into a PBPK model will depend upon the type of exposure, the surface area of skin exposed, and the dermal absorption model selected.

For a physiologically realistic model of dermal absorption, consideration must also be given to other exposure conditions. For example, if the purpose of the PBPK model development is to understand the risks of exposure to water disinfectant byproducts such as chloroform during showering, both the dermal and inhalation exposure routes must be considered, as was done in a PBPK model simulating chloroform exposures while showering by McKone (1993). Another factor for consideration is the length of the exposure. For certain exposures, the duration of exposure is easy to determine (e.g., a shower usually lasts 10 to 15 minutes), but for other situations it can be harder to estimate (e.g., for exposure to pesticide where the chemical might remain on the skin until it is removed either by showering or rubbing onto clothing). These types of considerations may impact the selection of an appropriate dermal absorption model structure.

14.3 DERMAL ABSORPTION MODELS

14.3.1 Membrane Models

Often, dermal absorption is represented mathematically as passive diffusion through one or more membranes in series (Scheuplein 1978; Cleek and Bunge 1993). For describing dermal absorption of chemicals with low to moderate lipophilicity, a one-layer membrane including the mass transfer resistance of the sc is adequate as long as the absorbing chemical is not metabolized in the ve. For describing dermal absorption of highly lipophilic chemicals, both the sc and ve should be included as separate membranes with distinct properties (Cleek and Bunge 1993). Membrane models employ Fick's Law in describing mass transport of chemical through the skin.

Although these models are considered to be physiologically realistic (Scheuplein and Blank 1971), they can be difficult to implement because they are mathematically more complicated than simpler models with well-mixed compartments. The solutions for compartment models are ordinary differential equations, while the solution to membrane models are partial differential equations, which must be solved using numerical techniques, such as finite difference approximation, that require mathematical skill.

In modeling the dermal absorption and systemic distribution of the lipophilic herbicide fluazifop-butyl (log $K_{o/w} = 4.5$), Auton *et al.* (1994) described the skin as a two-layer membrane including the mass transfer resistances of both the sc and ve. Also, they accounted for metabolism in the ve and chemical removal from the skin surface by desquamation (i.e., the sloughing of the outer skin layer due to epidermal turnover) and rubbing onto clothing. The systemic pharmacokinetic model used by Auton *et al.* (1994) was a simple two-compartment model, possibly due to limitations in the data (i.e., the studies were performed on humans and only concentrations of chemical in the urine and blood plasma were available). In the fluazifop-butyl study, including surface phenomena (e.g., pesticide rubbing onto clothing) was critical to accurately estimate systemic absorption. Although desquamation was not a significant mechanism for elimination of fluazifop butyl from the skin, Reddy *et al.* (2000) showed that desquamation can significantly decrease systemic absorption for chemicals with high lipophilicity (i.e., log $K_{o/w}$ greater than about 4) or high molecular weight (i.e., greater than about 350 Da).

Several investigators (e.g., Georgopoulos *et al.* 1994; Roy *et al.* 1994, 1996a), have demonstrated the use of a membrane model combined with a PBPK model to simulate human dermal exposures to chloroform. In one study, Roy *et al.* (1996b) compared the results of three chloroform PBPK models that are identical in all respects except for including the dermal exposure route modeled in three different ways. The simulation results for two-compartment models for skin and a one-membrane model for skin were compared. Although the authors did find that the membrane model was superior to the compartment models in some ways (e.g., the membrane model predicted a short lag time while the compartment models did not), the authors stated that more data are needed to "clarify the dynamics of the dermal absorption process" to discriminate among models.

14.3.2 Compartment Models

Compartment models (e.g., the one-compartment model shown in Fig. 14.1) have been widely used to describe dermal uptake of compounds. Chemical transfer between the skin and the vehicle or blood compartments depends on four rate constants (k_1, k_{-1}, k_2, and k_{-2}, m^3/sec), the concentrations of chemical in the vehicle, C_v, and systemic blood, C_b, and the average concentration of chemical in the skin, C_{skin}. In physiologically relevant skin models, the rate constants are defined in terms of skin properties (e.g., partition coefficients and permeability coefficients). When combined with a pharmacokinetic model for distribution of chemical throughout the body, compartment models for skin can easily be implemented with the first-order ordinary differential equation solvers that are typically used in PBPK modeling. The

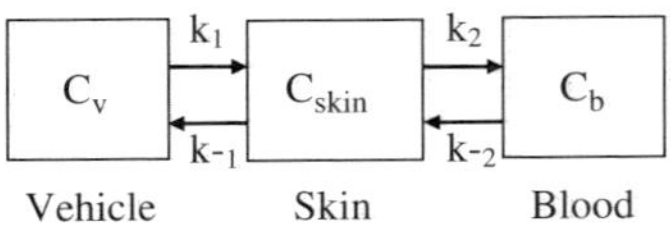

Figure 14.1 A schematic diagram of a one-compartment model for skin. Compound diffuses across the skin barrier and then mixes into skin tissue. This general structure, linked to a PBPK model for distribution of compound in the body, was used by MacDougal *et al.* (1986) and in many subsequent publications by these and other investigators.

many different physiologically relevant skin compartment models have been reviewed by McCarley and Bunge (2001).

One-Compartment Models The earliest PBPK models including the dermal exposure route used one-compartment models for skin that lumped all the skin layers together. In a study of the dermal absorption of DBM, BCM and DCM from the vapor phase in the Fischer 344 rat, McDougal *et al.* (1986) were the first researchers to include the dermal exposure route in PBPK models. In a later study, McDougal *et al.* (1990) used the same description of the dermal exposure route in PBPK models for styrene, toluene, benzene, hexane, isoflurane, perchloroethylene, *m*-xylene, and halothane uptake by skin from the vapor phase. In both studies, the PBPK models were used to calculate permeability coefficients of the absorbing chemicals for the lumped skin compartment by dividing the model-predicted total amount absorbed systemically by the exposed area, the exposure concentration, and the exposure time. Later a similar modeling approach was used by Jepson and McDougal (1997) in their study of the dermal absorption of DBM and BCM from aqueous solutions and by Jepson and McDougal (1999) in their study of the dermal absorption of DBM and BCM from water, mineral oil, and corn oil vehicles.

Many other investigators have used the McDougal *et al.* (1986) dermal absorption model. Corley *et al.* (1994) developed a PBPK model including the intravenous (*i.v.*) infusion, oral, dermal, and inhalation exposure routes for 2-butoxyethanol in rats and humans, using the McDougal *et al.* (1986) skin model to describe the dermal exposure route. They included the metabolite 2-butoxyacetic acid (i.e., the chemical species believed to cause hemolysis, the most sensitive response to this compound) in the model to determine the expected toxicity in humans based on rat study results. Additionally, Corley *et al.* (2000) described a PBPK model for chloroform including the *i.v.* and intraperitoneal (*i.p.*) injection, oral, inhalation, and dermal exposure routes, which also represented dermal uptake using the McDougal *et al.* (1986) model. With the chloroform model, they studied the relative body burden of chloroform resulting from various household exposures to contaminated tap water.

The McDougal *et al.* (1986) skin model was also used in PBPK models developed to estimate permeability coefficients from real-time data (i.e., breath concentrations measured using an ion-trap mass spectrometer, which can detect chemical concentrations in the range of 1–5 ppb) taken after a dermal exposure to volatile chemicals. Thrall *et al.* (2000) used a PBPK model to estimate permeability coefficients of methyl chloroform, trichloroethylene, and benzene in rats, monkeys, and humans by measuring the concentration in exhaled breath following dermal exposures from a saline vehicle using this real-time MS/MS technique. The same method has been used to study the dermal absorption of trichloroethylene and to estimate permeability coefficients for dermal absorption of methyl chloroform from soil and water vehicles for rats and humans (Poet *et al.* 2000a,b).

In a study of the risk associated with the presence of tetrachloroethylene in groundwater during dermal exposures while bathing and showering, Rao and Brown (1993) also used the McDougal *et al.* (1986) skin model for the dermal exposure route. During bathing, people are exposed to tetrachloroethylene in the bath water and in the air; thus two routes of dermal exposure were included: the aqueous and gas phases. The risk associated with bathing and showering in methyl *t*-butyl ether (MTBE)-contaminated groundwater has also been studied (Rao and Ginsberg 1997). The model included the primary metabolite *t*-butyl alcohol, because both MTBE and *t*-butyl alcohol produce central nervous system depression. The dermal exposure route was treated the same as it was for tetrachloroethylene.

The McDougal *et al.* (1986) skin model was a major advance in the PBPK modeling literature because it included the dermal exposure route and described dermal absorption in terms of physiologically relevant skin parameters. This simple skin model, which was intended for use with volatile compounds, described the mass transfer of chemical away from the skin to be flow-limited and described the skin compartment to be well-mixed and in equilibrium with the venous blood. Strictly speaking, these conditions are not typically the case under normal physiological conditions unless an animal is administered a vasoconstrictor or the skin temperature is decreased (Scheuplein and Bronaugh 1983). Due to the potential for a concentration gradient in the skin, after an exposure ends the skin can act as a reservoir for some chemicals, slowly releasing chemical into the bloodstream. However, a model describing the skin as well mixed and in equilibrium with the venous blood cannot describe this reservoir effect (Reddy *et al.* 1998), leading to potential inaccuracies in predictions of the time-dependence of systemic absorption. For the same reason, these simple descriptions cannot predict the time lag for the onset of systemic absorption observed with some chemicals. Models describing the skin as well-mixed can, however, predict the steady-state rate of dermal absorption accurately and have accurately described dermal absorption in several studies. An important consideration in selecting an appriate model structure for the dermal exposure route is determining if a simple model is adequate to describe the data or situation, or if a more complex model structure is necessary.

Because the compartment model equation contains less information than the membrane model equation (i.e., the variation of concentration with position in the skin is unknown for compartment models), compartment model predictions can never exactly match membrane model predictions for all cases (McCarley and Bunge 1998). For example, no compartment model can match the predictions of the membrane model at the limiting conditions of both steady state and equilibrium. As a result, it is possible to develop many different compartment models in which the relationships between the four rate constants and the sc properties are defined to match the membrane model predictions for different limiting cases. McCarley and Bunge (1998) illustrated this procedure by defining 11 different one-compartment models for skin from many more possibilities.

Other compartment models have also been developed to describe dermal absorption and have been implemented in PBPK models. Bookout *et al.* (1996), in analyzing data from rats exposed to dibromomethane (DBM) by skin-only vapor exposures, compared the ability of a homogenous skin model and a two-subcompartment skin model to describe the data. The homogeneous skin model described the skin as a homogeneous, well-stirred compartment, while the two sub-compartment skin model described a sc compartment and a composite dermal compartment containing the ve, dermis and subcutaneous fat. In a similar study, Bookout *et al.* (1997) compared the ability of a homogenous skin model, a parallel subcompartment skin model, and a layered parallel subcompartment skin model to describe the same dibromomethane dataset. The parallel subcompartment skin model included a shunt pathway for mass transport of DBM through appendages in the skin, and the skin was described using an sc compartment and a composite dermal compartment.

Two-Compartment Models Several two-compartment models representing the mass transfer resistances of the sc and ve separately have also been used in PBPK models. Chinery and Gleason (1993) developed a PBPK model for chloroform including a two-layer compartment model for the dermal exposure route and a shower model in order to be able to predict the chloroform concentrations in exhaled breath of a person exposed to chloroform while showering. Although they used a two-compartment pharmacokinetic model including both the sc and ve separately, the ve did not play a rate-limiting role to mass transport for chloroform as would be expected for this moderately lipophilic chemical (log $K_{o/w} = 1.4$). A variety of two-compartment models have been developed to match two-layer membrane model characteristics (e.g., lag time and steady-state flux) (McCarley and Bunge 2000). However, for hydrophilic and moderately lipophilic chemicals, the ve is not expected to provide a significant resistance to dermal absorption and a one-layer compartment model should be adequate.

In a PBPK model describing the distribution of the lipophilic pesticide chlordecone (log $K_{o/w} = 5.4$), Heatherington *et al.* (1998) represented skin using the Guy *et al.* (1985) skin model, which represents the skin as two compartments (i.e., the sc and ve) in series. In this study, *in vivo* and *in vitro* dermal absorption of chlordecone applied in acetone was studied in Fischer 344 rats. Because the acetone evaporates rapidly leaving a residue of chlordecone, the chosen skin model was developed for describing dermal absorption from a solid on the skin surface. In this case, an appropriate parameter for use in the PBPK model would be solubility, instead of using a partition coefficient of the absorbing chemical between the vehicle and the sc. Although the rate constants in the Guy *et al.* (1985) model do have physiological meaning, they were treated as empirical parameters.

Empirical Models Several PBPK models have used an empirical approach to including the dermal exposure route. For example, Shyr *et al.* (1993) developed a PBPK model for 2-butoxyethanol including the dermal exposure route. Unlike Corley *et al.* (1994), they mathematically described skin by fitting a simple first-order cumulative exponential function to the data. Additionally, they included metabolism (i.e., minor glucuronidation) in the skin model at both the exposure site and the skin storage compartment. Kedderis *et al.* (1993), in a study of the tissue distribution and CYP1A enzyme induction of TBDD in rats, used a skin model that described dermal absorption as first-order disappearance from the exposed site.

14.4 EXPERIMENTAL METHODS

The work developing the PBPK model for dermal absorption of vapors (McDougal *et al.* 1986, 1990) relied on the development of an innovative skin-only inhalation chamber for studies with rodents (McDougal *et al.* 1985). Rats were prepared with indwelling jugular cannulas, and fur was removed with electric clippers taking care to avoid nicking or abrading the skin surface. The rats were placed in a specially designed inhalation chamber that prevented inhalation of the vapors by providing clean breathing air at positive pressure through a face mask on each rat while the body surface was exposed to atmospheres at various concentrations of test chemicals. The cannulas were exteriorized from the chamber, and blood samples were taken at multiple times throughout the 4-hour exposures. The overall shape of the accumulation curve in blood demonstrated a significant lag period, providing assurance that there was minimal uptake into the animals by inhalation. The PBPK analyses of these dermal vapor absorption studies were important to assess the contribution that dermal vapor exposure might have in absorption for a worker who wears a respirator but has skin contact with dangerous vapors.

Some investigators have used *in vitro* dermal absorption data in their addition of the dermal exposure route to a PBPK model. For example, in a model for benzoic acid (MacPherson *et al.* 1996), dermal absorption was described using a transepidermal input function derived with *in vitro* dermal absorption data from a flow-through diffusion cell study. Although MacPherson *et al.* (1996) found that varying the parameters for the dermal absorption model improved the model description of the data, the use of *in vitro* data in developing a PBPK model incorporating the dermal exposure route has promise for improving the physiological basis of the PBPK model.

PBPK modeling analyses have been performed to study human exposure to water disinfectant byproducts during showering and bathing exposures. The uptake of chloroform from the monitoring of the concentration in exhaled air of humans exposed while showering (Chinery and Gleason 1993), while bathing (Corley *et al.* 2000), or while swimming (Levesque *et al.* 2000) has been studied using PBPK modeling techniques. For volatile chemicals, monitoring the concentration in exhaled breath provides a noninvasive and relatively simple method for studying dermal absorption. This methodology is useful for assessing rates of dermal uptake for compounds with relatively low blood : air partition coefficients. These compounds have high pulmonary clearance and are rapidly exhaled. Under appropriate experimental conditions the appearance of chemical in air would be an accurate indicator of the kinetic processes controlling uptake across the skin.

Many dermal absorption studies for volatile chemicals have been conducted by applying a dermal dose and then monitoring the concentration of test chemical in exhaled air in real time using a tandem ion-trap mass spectrometer (MS/MS) analysis. The frequency with which the test chemical concentration in exhaled air is measured with this technique provides great resolution on the kinetics of the dermal exposure. For example, *in vivo* dermal absorption of several volatile chemicals [e.g., *o*-xylene (Thrall and Woodstock 2003), methyl chloroform (Poet *et al.* 2000b), and trichloroethylene (Poet *et al.* 2000a)] has been studied by exposing rats or humans to a chemical and quantifying the concentration of chemical in the exhaled air using MS/MS real-time analysis.

A good example of the real-time MS/MS experiment was a study of the absorption of *o*-xylene by rats and humans (Thrall and Woodstock 2003). To study dermal absorption of *o*-xylene in rats, 2 mL of an aqueous solution of *o*-xylene was applied using a glass exposure patch attached to the back of rats, and then rats were placed in a small off-gassing chamber and the concentrations in chamber air were monitored. The concentration of the original dosing solution and the concentration of the remaining solution in the exposure patch at the end of the study were determined so that the total absorbed dose could be measured. To simulate these dermal exposures, the toluene–*m*-xylene PBPK model of Tardif *et al.* (1993) was adapted to include only *m*-xylene and to include the dermal exposure route. The *m*-xylene model was further modified to include partition coefficients calculated from tissue : air partition coefficients measured *in vitro* for *o*-xylene.

For human dermal absorption studies (Thrall and Woodstock 2003), volunteers placed their lower legs in a hydrotherapy tub containing an initial concentration of 500 μg/L *o*-xylene, and the concentration of *o*-xylene in the tub water was determined every five minutes throughout the exposure, which lasted up to 30 minutes. After the dermal exposure, the expired air was monitored for an additional 30 minutes. The rat PBPK model was extended to humans by adjusting the physiological parameters appropriately. Using the PBPK model analysis, it was possible to determine the rate of dermal absorption and the amount absorbed through the skin during these studies based on the amount of test chemical exhaled. This analysis allowed the comparison of the results of the rat and human dermal absorption studies.

14.5 SUMMARY

There are many dermal absorption models available for adding the dermal exposure route to a PBPK model. Membrane models are not commonly used. Compartment modes, which are more compatible with PBPK models and easier to use, are more frequently used. Factors to be considered in developing a dermal PBPK model include the choice of model, the physical situation of the exposure, and whether or not to include a separate skin compartment. Many types of dermal absorption models are available, and careful consideration must be applied in choosing an appropriate model for a given chemical and exposure situation. For volatile chemicals, monitoring the concentration in exhaled breath following a dermal exposure provides a noninvasive and relatively simple method for studying dermal absorption. Perhaps that is why studies of dermal absorption of volatile chemicals (e.g., water disinfection byproducts such as chloroform and industrial solvents such as *o*-xylene) have been common.

NOTATION

BCM	bromochloromethane
C_b	concentration of chemical in the systemic blood
C_{skin}	average concentration of chemical in the skin
C_v	concentration of chemical in the vehicle
D4	octamethylcyclotetrasiloxane
DBM	dibromomethane
DCM	dichloromethane, methylene chloride
epi	epidermis
i.p.	intraperitoneal
i.v.	intravenous
$K_{o/w}$	octanol : water partition coefficient
MTBE	methyl *t*-butyl ether
sc	stratum corneum
TBDD	2,3,7,8-tetrabromodibenzo-*p*-dioxin
ve	viable epidermis

REFERENCES

Auton, T. R., Westhead, D. R., Woollen, B. H., Scott, R. C., and Wilks, M. F (1994). A physiologically-based mathematical-model of dermal absorption in man. *Hum. Exp. Toxicol.* **13**, 51–60.

Bogen, K. T., and Hall, L. C. (1989). Pharmacokinetics for regulatory risk analysis: The case of 1,1,1-trichloroethane (methyl chloroform). *Regul. Toxicol. Pharmacol.* **10**, 26–50.

Bookout, R. L., Jr., McDaniel, C. R., Quinn, D. W., and McDougal, J. N. (1996). Multilayered dermal subcompartments for modeling chemical absorption. *SAR QSAR Environ. Res.* **5**, 133–150.

Bookout, R. L., Jr., Quinn, D. W., and McDougal, J. N. (1997). Parallel dermal subcompartments for modeling chemical absorption. *SAR QSAR Environ. Res.* **7**, 259–279.

Chinery, R. L., and Gleason, A. K. (1993). A compartmental model for the prediction of breath concentration and absorbed dose of chloroform after exposure while showering. *Risk Anal.* **13**, 51–62.

Cleek, R. L., and Bunge, A. L. (1993). A new method for estimating dermal absorption from chemical exposure. 1. General approach. *Pharm. Res.* **10**, 497–506.

Clewell, H. J., Gentry, P. R., Gearhart, J. M., Covington, T. R., Banton, M. I., and Andersen, M. E. (2001). Development of a physiologically based pharmacokinetic model of isopropanol and its metabolite acetone. *Toxicol. Sci.* **63**, 160–172.

Corley, R. A., Bormett, G. A., and Ghanayem, B. I. (1994). Physiologically-based pharmacokinetics of 2-butoxyethanol and its major metabolite, 2-butoxyacetic acid, in rats and humans. *Toxicol. Appl. Pharmacol.* **129**, 61–79.

Corley, R. A., Gordon, S. M., and Wallace, L. A. (2000). Physiologically based pharmacokinetic modeling of the temperature-dependent dermal absorption of chloroform by humans following bath water exposures. *Toxicol. Sci.* **53**, 13–23.

Dong, M. H., Draper, W. M., Papanek, P. J., Ross, J. H., Woloshin, K. A., and Stephens, R. D. (1994). Estimating malathion doses in California medfly eradication campaign using a physiologically-based pharmacokinetic model. *Environ. Epidemiol.* **241**, 189–208.

Georgopoulos, P. G., Roy, A., and Gallo, M. A. (1994). Reconstruction of short-term multiroute exposure to volatile organic compounds using physiologically-based pharmacokinetic models. *J. Expo. Anal. Environ. Epidemiol.* **4**, 309–328.

Guy, R. H., Hadgraft, J., and Maibach, H. I. (1985). Percutaneous absorption in man: A kinetic approach. *Toxicol. Appl. Pharmacol.* **78**, 123–129.

Heatherington, A. C., Fisher, H. L., Sumler, M. R., Waller, C. L., Shah, P. V., and Hall, L. L. (1998). Percutaneous absorption and disposition of [^{14}C]Chlordecone in young and adult female rats. *Environ. Res.* **79**, 138–155.

Jepson, G. W., and McDougal, J. N. (1997). Physiologically based modeling of nonsteady state dermal absorption of halogenated methanes from an aqueous solution. *Toxicol. Appl. Pharmacol.* **144**, 315–324.

Jepson, G. W., and McDougal, J. N. (1999). Predicting vehicle effects on the dermal absorption of halogenated methanes using physiologically based modeling. *Toxicol. Sci.* **48**, 180–188.

Kedderis, L. B., Mills, J. J., Andersen, M. E., and Birnbaum, L. S. (1993). A physiologically-based pharmacokinetic model for 2,3,7,8-tetrabromodibenzo-*p*-dioxin (TBDD) in the rat—tissue distribution and cyp1a induction. *Toxicol. Appl. Pharmacol.* **121**, 87–98.

Kligman, A. M. (1964). The biology of the stratum corneum. In: *The Epidermis*, W. Montagna and W. C. Lobitz, eds., Academic Press, New York, pp. 387–433.

Knaak, J. B., Albayati, M. A., Raabe, O. G., and Blancato, J. N. (1994). Prediction of anticholinesterase activity and urinary metabolites of isofenphos—Use of a percutaneous physiologically-based pharmacokinetic physiologically-based pharmacodynamic model. *Biomarkers of Human Exposure to Pesticides ACS Symposium Series* **542**, 284–300.

Levesque, B., Ayotte, P., Tardif, R., Charest-Tardif, G., Dewailly, E., Prud'Homme, D., Gingras, G., Allaire, S., and Lavoie, R. (2000). Evaluation of the health risk associated with exposure to chloroform in indoor swimming pools. *J. Toxicol. Environ. Health Pt. A* **61**, 225–243.

Loizou, G. D., Jones, K., Akrill, P., Dyne, D., and Cocker, J. (1999). Estimation of the dermal absorption of *m*-xylene vapor in humans using breath sampling and physiologically based pharmacokinetic analysis. *Toxicol. Sci.* **48**, 170–179.

MacPherson, S. E., Barton, C. N., and Bronaugh, R. L. (1996). Use of in vitro skin penetration data and a physiologically based model to predict *in vivo* blood levels of benzoic acid. *Toxicol. Appl. Pharmacol.* **140**, 436–443.

McCarley, K. D., and Bunge, A. L. (1998). Physiologically relevant one-compartment pharmacokinetic models for skin. 1. Development of models. *J. Pharm. Sci.* **87**, 470–481.

McCarley, K. D., and Bunge, A. L. (2000). Physiologically relevant two-compartment pharmacokinetic models for skin. *J. Pharm. Sci.* **89**, 1212–1235.

McCarley, K. D., and Bunge, A. L. (2001). Pharmacokinetic models of dermal absorption. *J. Pharm. Sci.* **90**, 1699–1719.

McDougal, J. N., Jepson, G. W., Clewell, H. J., III, and Andersen, M. E. (1985). Dermal absorption of dihalomethane vapors. *Toxicol. Appl. Pharmacol.* **79**, 150–158.

McDougal, J. N., Jepson, G. W., Clewell, H. J., MacNaughton, M. G., and Andersen, M. E. (1986). A physiological pharmacokinetic model for dermal absorption of vapors in the rat. *Toxicol. Appl. Pharmacol.* **85**, 286–294.

McDougal, J. N., Jepson, G. W., Clewell, H. J. D., Gargas, M. L., and Andersen, M. E. (1990). Dermal absorption of organic chemical vapors in rats and humans. *Fundam. Appl. Toxicol.* **14**, 299–308.

McKone, T. E. (1993). Linking a pbpk model for chloroform with measured breath concentrations in showers—Implications for dermal exposure models. *J. Expo. Anal. Environ. Epidemiol.* **3**, 339–365.

Poet, T. S., Corley, R. A., Thrall, K. D., Edwards, J. A., Tanojo, H., Weitz, K. K., Hui, X., Maibach, H. I., and Wester, R. C. (2000a). Assessment of the percutaneous absorption of trichloroethylene in rats and humans using MS/MS real-time breath analysis and physiologically based pharmacokinetic modeling. *Toxicol. Sci.* **56**, 61–72.

Poet, T. S., Thrall, K. D., Corley, R. A., Hui, X. Y., Edwards, J. A., Weitz, K. K., Maibach, H. I., and Wester, R. C. (2000b). Utility of real time breath analysis and physiologically based pharmacokinetic modeling to determine the percutaneous absorption of methyl chloroform in rats and humans. *Toxicol. Sci.* **54**, 42–51.

Rabovsky, J., and Brown, J. P. (1993). Malathion metabolism and disposition in mammals. *J. Occup. Med. Toxicol.* **2**, 131–168.

Rao, H. V., and Brown, D. R. (1993). A physiologically based pharmacokinetic assessment of tetrachloroethylene in groundwater for a bathing and showering determination. *Risk Anal.* **13**, 37–49.

Rao, H. V., and Ginsberg, G. L. (1997). A physiologically-based pharmacokinetic model assessment of methyl *t*-butyl ether in groundwater for a bathing and showering determination. *Risk Anal.* **17**, 583–598.

Reddy, M. B., McCarley, K. D., and Bunge, A. L. (1998). Physiologically relevant one-compartment pharmacokinetic models for skin. 2. Comparison of models when combined with a systemic pharmacokinetic model. *J. Pharm. Sci.* **87**, 482–490.

Reddy, M. B., Guy, R. H., and Bunge, A. L. (2000). Does epidermal turnover reduce percutaneous penetration? *Pharm. Res.* **17**, 1414–1419.

Roberts, M. S., Anissimov, Y. G., and Gonsalvez, R. A. (2001). Mathematical models in percutaneous absorption [reprinted from *Percutaneous Absorption*, pp. 3–55, 1999]. *J. Toxicol.-Cutan. Ocul. Toxicol.* **20**, 221–270.

Roy, A., Weisel, C. P., Gallo, M. A., and Georgopoulos, P. G. (1994) Studies of multiroute exposure/dose reconstruction using physiologically based pharmacokinetic models. *Hazard. Waste Public Health, Int. Congr. Health Eff. Hazard. Waste*, 284–293.

Roy, A., Weisel, C. P., Gallo, M., and Georgopoulos, P. (1996a). Studies of multiroute exposure/dose reconstruction using physiologically based pharmacokinetic models. *J. Clean Technol. Environ. Toxicol. Occup. Med.* **5**, 285–295.

Roy, A., Weisel, C. P., Lioy, P. J., and Georgopoulos, P. G. (1996b). A distributed parameter physiologically-based pharmacokinetic model for dermal and inhalation exposure to volatile organic compounds. *Risk Anal.* **16**, 147–160.

Sarangapani, R., Teeguarden, J., Andersen, M. E., Reitz, R. H., and Plotzke, K. P. (2003). Route-specific differences in distribution characteristics of octamethylcyclotetrasiloxane in rats: Analysis using PBPK models. *Toxicol. Sci.* **71**, 41–52.

Scheuplein, R. J. (1978). Permeability of the skin: A review of major concepts. *Curr. Prob. Dermatol.* **7**, 172–186.

Scheuplein, R. J., and Blank, I. H. (1971). Permeability of the skin. *Physiol Rev.* **51**, 702–747.

Scheuplein, R. J., and Bronaugh, R. L. (1983). Percutaneous absorption. In: *Biochemistry and Physiology of the Skin.* L. A. Goldsmith, ed. New York, Oxford University Press.

Shyr, L. J., Sabourin, P. J., Medinsky, M. A., Birnbaum, L. S., and Henderson, R. F. (1993). Physiologically based modeling of 2-butoxyethanol disposition in rats following different routes of exposure. *Environ. Res* **63**, 202–218.

Tardif, R., Lapare, S., Krishnan, K., and Brodeur, J. (1993). Physiologically based modeling of the toxicokinetic interaction between toluene and m-xylene in the rat. *Toxicology & Applied Pharmacology* **120**, 266–273.

Thrall, K. D., and Kenny, D. V. (1996). Evaluation of a carbon tetrachloride physiologically based pharmacokinetic model using real-time breath-analysis monitoring of the rat. *Inhal. Toxicol.* **8**, 251–261.

Thrall, K. D., Poet, T. S., Corley, R. A., Tanojo, H., Edwards, J. A., Weitz, K. K., Hui, X., Maibach, H. I., and Wester, R. C. (2000). A real-time in-vivo method for studying the percutaneous absorption of volatile chemicals. *Int. J. Occup. Environ. Health* **6**, 96–103.

Thrall, K. D., Weitz, K. K., and Woodstock, A. D. (2002). Use of real-time breath analysis and physiologically based pharmacokinetic modeling to evaluate dermal absorption of aqueous toluene in human volunteers. *Toxicol. Sci.* **68**, 280–287.

Thrall, K. D., and Woodstock, A. D. (2002). Evaluation of the dermal absorption of aqueous toluene in F344 rats using real-time breath analysis and physiologically based pharmacokinetic modeling. *J. Toxicol. Env. Health Pt. A* **65**, 2087–2100.

Thrall, K. D., and Woodstock, A. D. (2003). Evaluation of the dermal bioavailability of aqueous xylene in F344 rats and human volunteers. *J. Toxicol. Env. Health Pt. A* **66**, 1267–1281.

Timchalk, C., Nolan, R. J., Mendrala, A. L., Dittenber, D. A., Brzak, K. A., and Mattsson, J. L. (2002). A physiologically based pharmacokinetic and pharmacodynamic (PBPK/PD) model for the organophosphate insecticide chlorpyrifos in rats and humans. *Toxicol. Sci.* **66**, 34–53.

Zartarian, V. G., and Leckie, J. O. (1998). Dermal exposure: The missing link. *Environ. Sci. Technol.* **32**, 134–137.

CHAPTER 15

CONCLUSIONS AND FUTURE DIRECTIONS

Melvin E. Andersen, Micaela B. Reddy, Harvey J. Clewell III, and Raymond S. H. Yang

15.1 INTRODUCTION

The PBPK modeling approaches discussed throughout this volume represent an evolution from simpler, compartmental kinetic models toward more realistic, biological descriptions of the determinants that regulate disposition of chemicals and drugs in the body. The development of these PBPK models was frequently met with skepticism. These models introduced a large number of variables for metabolism, transport, binding, and so on. The older classical pharmacokinetic models had a much more sparse set of variables. Concerns were raised about the ability to obtain values for all these parameters. However, many of these constants were identified from experiments separate from collection of time-course studies or were known from physiological or anatomical research. These independently measured parameters could be introduced into structured PBPK models and simulations (predictions) from the model could be compared to data to decide if the model structure were accurate

Physiologically Based Pharmacokinetic Modeling, edited by Micaela B. Reddy, Raymond S. H. Yang, Harvey J. Clewell III, and Melvin E. Andersen.

or in need of re-parameterization or more extensive re-formulation. This iteration of proposed physiological structures for the models and testing them against data is simply a form of adherence to the scientific method: hypothesis generation and testing applied to problems of drug and chemical disposition and persistence.

15.2 A SYSTEMS APPROACH FOR PHARMACOKINETICS

To a large extent, the application of these PBPK models to study the time courses of compounds in the body is simply an integrated systems approach to understanding the biological processes that regulate the delivery of active forms of chemicals and drugs to target sites. PBPK modeling was a pioneering application of a systems approach to ascertain the mechanisms of distribution and interactions of drug or chemical in the body. Modern PBPK modeling was conceived by two chemical engineers, Drs. Kenneth Bischoff and Robert Dedrick, who integrated engineering principles, physiology, chemistry, and biochemistry into a computer modeling platform (Bischoff and Brown 1966; Bischoff *et al.* 1971; Dedrick 1973). The past four to five decades has witnessed a great deal of progress in PBPK modeling; however, the fundamental basis is still the application of mathematics, engineering principles, and computer science to understanding a range of complex biological systems.

Many PBPK models integrate information across multiple levels of organization, especially when describing interactions of compounds with molecular targets, such as reversible binding of ligands to specific receptors, as with models for methotrexate (Bischoff *et al.* 1971) or dioxin (US EPA 2000b), or the adduction of proteins or DNA with reactive metabolites produced by biotransformation in various tissues, as with models for ethylene oxide (Krishnan *et al.* 1992) or acrylonitrile (Gargas *et al.* 1995). In these cases, the PBPK models integrate molecular, cellular, organ level, and organism-level processes to account for the time courses of chemicals, metabolites, and bound complexes within organs and tissues in the body. The system under scrutiny in PBPK models is more the integrated physiological system, appropriate enough for a discipline defined as physiologically based modeling.

As emphasized frequently throughout this monograph, the main goal of these PBPK models is quite simple: to predict the target tissue dose of compound and their metabolites at target tissues and, in some cases, to describe interactions in target tissues. PBPK models, once developed, are extensible. They can be used to extrapolate to various other conditions because of their biological fidelity. While the goal in applying these models is to predict dosimetry, it is important to remember that the overall goal of dose–response assessment activities in toxicology and in safety and risk assessment with drugs is broader than simply estimating tissue dose, regardless of the level of detail provided in the interactions of chemicals with tissue constituents. The goal in the larger context is to understand the relationship between dose delivered to target tissues and the biological sequelae of the exposure of target tissues to compounds.

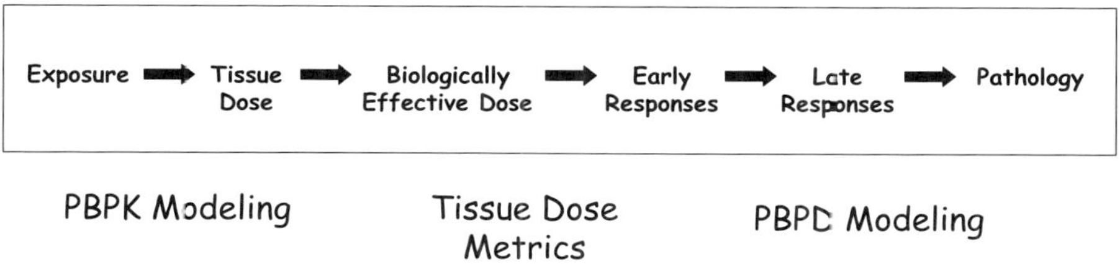

Figure 15.1 Diagram indicating the role played by PBPK and PBPD modeling in studying the relationship between dose and response.

15.3 MODELING BOTH DOSE AND RESPONSE

Over the past 15 years or so, toxicology and risk assessment have emphasized the exposure–dose–response relationships linking the presentation of chemical to an organism with specific toxicological or pathological responses (Fig. 15.1). PBPK models provide more detail on the sequence of steps from exposure through tissue interactions, including reactivity of molecules and their metabolites with biological targets or recognition of chemical structures by binding to cellular targets involved in metabolism or in regulating cell signaling pathways. For many risk assessments, proximate measures of tissue dose, sometimes called dose metrics—that is, that measure of tissue dose believed to be closely associated with the adverse responses—are used as the basis for the dose–response portion of chemical risk assessment. An example here is the US EPA risk assessment for vinyl chloride that relied on a PBPK model for dose-route extrapolations (US EPA 2000a) with calculation of dose of vinyl chloride metabolized to epoxide per tissue volume as the dose metric. The specific steps that lead from these dose metrics to tissue-, organ-, and organism-level responses have usually been considered part of the pharmacodynamic (PD) process.

In general, PD models used in risk assessment have been more empirical, utilizing simple effect compartments correlated with blood or tissue concentrations of active chemical. Sometimes dose metrics are also correlated with more integrated cellular level responses, such as cell death, cell proliferation, or mutation. Two-stage clonal growth models for carcinogenesis and cell-growth-based models for developmental toxicity show some potential for applications in physiologically based pharmacodynamic (PBPD) models for specific adverse responses (Moolgavkar and Luebeck 1990; Leroux *et al.* 1996; Whitaker *et al.* 2003). An example of application of a PBPD model with a two-stage clonal growth model for cancer risk assessment was completed with formaldehyde (Conolly *et al.* 2004). This PBPD model with formaldehyde remains the only true PBPD model applied to cancer risk assessment to date.

Applications of PBPK and PBPD modeling are likely to expand considerably in the years ahead. While many opportunities exist for extending PBPK modeling with environmental chemicals and drugs to study greater numbers of compounds and for refining existing models by creating more extensive datasets for more detailed analysis, major areas of expansion for PBPK modeling are likely to arise from increasing interest within the pharmaceutical community (see Chapter 10 and 11, this volume) and applications of "reaction network" modeling for describing metabolism of xenobiotics with automated PBPK model building for sets of expected metabolites (Liao *et al.* 2002; Chapter 13, this volume). Another inexorable development will be expansion of the systems approaches into the PBPD arena. This latter area will represent a systems biology approach for describing perturbations of biological systems by chemical exposures and the exposure/dose conditions under which these perturbations become sufficiently large to pose significant health risks or sufficiently large to achieve specific therapeutic outcomes. We close this volume with projections of the near-term expansion of PBPK modeling opportunities and of the development of systems biology-oriented approaches within the context of PBPK/PBPD modeling in risk and safety assessments.

15.4 OPPORTUNITIES FOR PBPK MODELING IN THE PHARMACEUTICAL INDUSTRY

It is ironic that PBPK modeling of drugs now appears to lag behind modeling of environmental contaminants (Rowland *et al.* 2004). In fact, the concept of PBPK modeling was first described in the context of drug disposition in the seminal work of Teorell (1937a,b), who elaborated the advantages of using physiological considerations as the basis for a pharmacokinetic description. Unfortunately, at that time the computational resources necessary for solving the systems of simultaneous differential equations were not available. As a result, physiologically accurate descriptions were replaced by simpler compartmental descriptions for which closed form solutions could be derived. However, these simple compartmental pharmacokinetic approaches continued to be used long after the computational resources became available to make PBPK modeling practical. In fact, in the pharmaceutical industry even the compartmental approach has to a large extent been replaced by noncompartmental "model-free" pharmacokinetic analyses, despite the fact that these methods are useful only for summarizing the characteristics of a dataset (e.g., volume of distribution, clearance) and do not provide the advantages inherent in PBPK modeling (hypothesis testing, extrapolation, etc.). Their advantages are that they are rapid and inexpensive to perform.

However, in recent years there has been the increasing pressure on the pharmaceutical industry to accelerate the drug development process, and PBPK modeling has repeatedly been identified as one of the technologies that could prove useful to this end (Peck *et al.* 1992; Charnick *et al.* 1995; Nesterov 2003; Rowland *et al.* 2004). Because the collection of data on the absorption, distribution, metabolism, and elimination (ADME) of a drug is a required element of drug development, much of the data necessary for developing a PBPK model is often available. The PBPK model offers the opportunity to make better use of these data, by serving as a structured repository for quantitative information on the compound, a conceptual framework for hypothesis testing, and a quantitative platform for prediction.

The rapid development of combinatorial chemistry and high-throughput screening has brought increasing attention to the discovery phase of drug development, including growing interest in "discovery toxicology" (van de Waterbeemd and Gifford 2003). PBPK modeling can play a complementary role to two other technologies that are finding increasing use in drug discovery: quantitative structure–activity relationship (QSAR) analysis and genomics. QSAR can be used to estimate chemical-specific parameters for the PBPK model, while genomic data can provide mode of action insights that drive model structure decisions. In this scenario, the PBPK model provides a quantitative biological framework for integrating the physicochemical characteristics of the drug candidate, together with *in vitro* data on its ADME and toxicity, within the constraints of the fundamental physiological and biochemical processes governing chemical behavior *in vivo*.

As the drug development process proceeds, the model can be iteratively informed and refined on the basis of the data being collected, and in turn the model can serve as a platform for *in silico* hypothesis testing and experimental design. As the model becomes more robust, it can also be employed for extrapolation (e.g., selection of first human dose) and for simulation (e.g., Monte Carlo analysis for design of clinical trials). The physiological structure of the PBPK model makes it a perfect platform for conducting evaluations of alternative dosing methods and regimens and for investigating the impact of genetic-, age-, and disease-related differences in physiology and metabolism on drug kinetics. Of particular importance, using a PBPK model, the pharmacodynamic effects of a drug can be investigated more directly, relating the effects to the concentration in the tissue (e.g., the brain) where the compound interacts with the biological system, rather than attempting to elucidate a potentially indirect relationship with a concentration in the central (e.g., plasma) compartment. By obtaining quantitative information on the dose–responses for both the efficacy and toxicity of the compound, the PBPK model can be exercised to evaluate the potential to increase the efficacy/toxicity ratio of the drug through manipulation of dose rate and dose route using novel drug delivery systems. These and other attributes of PBPK models for organizing and interpreting diverse datasets, with the specific goals of understanding efficacy and toxicity, are reviving interest in applying these tools in drug development and evaluation.

15.5 REACTION NETWORK MODELING WITH XENOBIOTICS

Reaction network models arose in petroleum engineering as a tool for predicting the amounts of reactants, intermediates, and products as a function of time for the tens of thousands of coupled chemical reactions in refinery. A Reaction Network Model builder is a tool for the computer generation of a Reaction Network Model. The model builder solves the kinetic equations of interest and generates the reaction mechanisms, rate constants, and reaction equations themselves (Klein *et al.* 2002; Liao *et al.* 2002; Liao 2004; Reisfeld and Yang 2004; Yang *et al.* 2004a,b). The primary applications of this technology were in petroleum process engineering that focused on surface-catalyzed reactions of hydrocarbon in feedstock streams.

In PBPK models for parent chemicals and their metabolites or models for mixtures of chemicals, each compound of interest is described by specifying pertinent compartments and providing equations for tracking compound in the compartments. Reaction network approaches offer promise as tools for rapidly developing more comprehensive PBPK models for mixtures and the suites of metabolites generated by exposures to chemical mixtures. For instance, a parallel can be drawn between an oil refinery and the human body: Individual processing units in the oil refinery are equivalent to specific organs of the human body. Even though a cell or organ may be more complicated than a refinery, the complex biochemical reaction networks for metabolism in each organ could, in principle, be modeled in relation to production of sets of products, much the same way as the individual processing units in a large refinery are linked together.

One element of reaction network modeling is the generation of appropriate reaction rate constants from enzyme kinetic studies and/or quantitative structural–reactivity correlations. This portion of the Reaction Network provides the comprehensive accounting of the suite of compounds that may be produced by metabolism. In this regard, the biological system with multiple enzymatic pathways is considerably more challenging than the refinery reaction pathways. These reaction rates may initially have to be measured with human recombinant enzymes and may eventually be calculated as more sophisticated structure activity models of chemical reaction pathways become available. One of the "omics" of considerable interest today is metabonomics—that is, the identification and measurement of the group of small molecules produced by cellular activity. By analogy, the detailed studies *in vivo* and *in vitro* of the metabolic reaction networks of drugs/chemicals might be considered the metabonome for classes of exogenous compounds.

Reaction network modeling of drugs and chemicals, coupled with detailed enzyme kinetic studies using recombinant human enzymes, may be regarded as a form of metabonomics specifically aiming at a more global, predictive understanding of xenobiotic metabolic reaction networks and their related functional roles *in vivo*. Initial steps in this area have focused on hybrid models that link output of a reaction network generator for specific enzymatic pathways, such as oxidation of exogenous substrates by cytochrome P450 IA1 or cytochrome P450 2E, with generation of PBPK models for each of the predicted metabolites. Theoretical discussions in this area have appeared in the last few years (Klein *et al.* 2002; Liao *et al.* 2002; Reisfeld and Yang 2004; Yang *et al.* 2004a,b). A hybrid PBPK/Reaction Network Model for benzo[*a*]pyrene (Liao 2004) has also recently been described.

15.6 SYSTEMS BIOLOGY AND DOSE–RESPONSE

PBPK models represent a systems approach on the physiological and biochemical level. The ability to apply more integrated systems approaches in biology were simply not possible before the revolutions in molecular biology and various "omics" technologies of the past 20 years. The new, high-throughput, broad coverage technologies—genomics, transcriptomics, proteomics, and metabonomics—provide a comprehensive parts list for cells, tissues, organs, and eventually organisms. High-throughput gain of function screens with transfection of full-length cDNA (Koh *et al.* 2004) and high-throughput loss of function screens using small inhibitory RNA

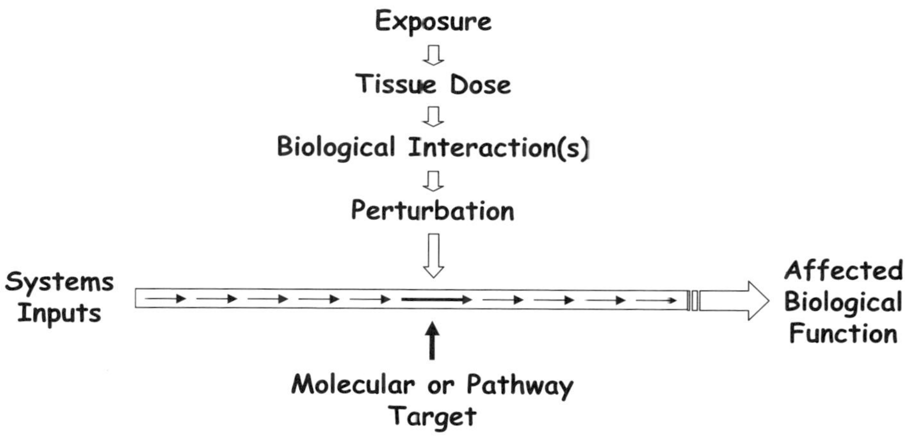

Figure 15.2 Depiction of perturbations of biological function.

molecules (Berns *et al.* 2004) query the effect of each gene on specific biological processes. Systems biology organizes this material to understand how these parts are put together to achieve specific biological functions.

A defining characteristic of systems biology will be the quantitative evaluation, through laboratory experiments and computer modeling, of the manner in which the components of biological systems are organized together to give rise to biological function. Perturbations of biological processes by chemicals lead to either adverse responses (toxicity) or restoration of normal function to a compromised tissue (efficacy). The linear paradigm showing the relationship of exposure, dose, and response for integrating studies of toxicity (Fig. 15.1) needs to be reorganized (Fig. 15.2). This more appropriate schematic representation for toxicology in the twenty-first century focuses on normal biological function and the perturbations associated with chemical exposure. This conceptual organization of the components of toxicological or pharmacological dose–response to chemicals has primary focus on the biology.

Toxicity and efficacy are then defined by an intersection of chemical action with the biological system. Toxicology and pharmacology are disciplines at the interface of chemistry/pharmacokinetics (primarily embedded in the vertical component) and biology/pharmacodynamics (primarily captured by the horizontal chain). Clearly, the main differences in the next generation of systems approaches in PK and PD modeling will be the increasingly detailed descriptions of biology afforded by new technologies and the expansion of modeling tools available for describing the biological signaling processes affected by chemical exposures and drug treatments.

In the engineering community, considerable work has focused on cells as small machines. Simple prokaryotic cells with specific circuit elements (i.e., biological oscillators, switches, amplifiers, etc.) have been produced and examined by laboratory experiments and by computation (Guet *et al.* 2002; Hasty *et al.* 2002; McMillen *et al.* 2002; Tyson *et al.* 2003). Then computational models evaluate the protein networks within cells, the genetic control of these networks, and the logic of cellular responses affected by these networks (Alm and Arkin 2003; Davidson *et al.* 2002; Ferrell 2002). These control networks are likely to be targets for toxic responses of various compounds (Andersen *et al.* 2005). Simulation models of several signaling pathways have also been developed for mammalian cells—including the platelet-derived growth factor (PDGF) pathway in Bhalla *et al.* (2002) and the tumor necrosis factor-alpha (TNF-α) signaling through the nuclear factor kappa B, NF-κB (Cho *et al.* 2003; Hoffmann *et al.* 2002).

These pathway models are similar to the structure of PBPK models. They include multiple reactions within a cell and require evaluation of multiple parameters for multiple biochemical steps embedded within the signaling networks. While there are many signaling modules within any cell, understanding of cellular behaviors is likely to focus on dissecting and modeling individual modules, giving rise eventually to a modular approach to understanding more integrated biological pathways (Hartwell *et al.* 1999). Individual modules likely serve as primary targets of toxic compounds and drugs. The circuitry that controls these target modules are likely to be the primary focus of systems biology approaches with these exogenous compounds.

An interesting aspect of the PDGF signaling (Bhalla *et al.* 2002) is adaptation to persistence of the presence of PDGF through transcriptional control of a phosphatase that blocks the usual activation of mitogen-activated protein kinase (MAPK) responses in the PDGF-signaling network. A more appropriate schematic of the exposure–dose–response paradigm of the past 15 years should include such adaptation (Fig. 15.3) as a central process in regulating dose–response behaviors.

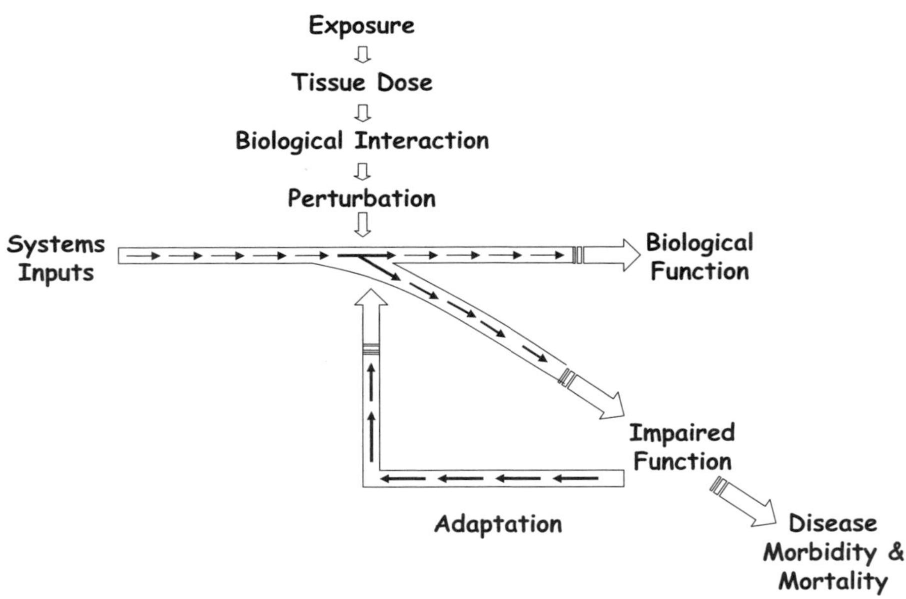

Figure 15.3 A systems approach for safety and risk assessment.

These biological signaling networks become perturbed by exposure of cells within tissues to specific compounds. Dose to tissue can best be described by PBPK approaches and linked through PBPD models of responses of cellular networks. While it is straightforward to consider systems biology modeling of altered signaling with endocrine active compounds that mimic natural hormones, cytotoxic compounds also activate various cell networks, including antioxidant, inflammatory, and DNA-repair pathways that can be associated with the presence of reactive compounds within tissues. The NF-κB pathway, mentioned earlier, is involved in responding to various stress-related responses leading to control of inflammatory responses to tissue damage.

15.7 SUMMARY

In the disciplines of toxicology and chemical risk assessment, PBPK modeling has become a well-established tool for research and analysis. The technologies applied so successfully in developing PBPK models will naturally support their migration to applications in systems biology for describing cell signaling pathways, perturbations of these pathways by exogenous compounds, and adaptation to perturbations with prolonged exposures. The immediate future has many opportunities for the quantitative analysis of various pharmacokinetic and biological studies using computer modeling and systems approaches for kinetics and dynamics. Our hope is that this volume, through the perspective in the Introduction and Conclusions and the broad set of examples provided in the other chapters, will facilitate introduction of these physiologically structured modeling methods to a broader community and spark a new generation of scientists in pharmacology and toxicology to pursue simulation modeling of biological systems in their research and work endeavors.

NOTATION

ADME absorption, distribution, metabolism, and elimination
MAPK mitogen-activated protein kinase
PBPD physiologically based pharmacodynamic
PD pharmacodynamic
PDGF platelet-derived growth factor
QSAR quantitative structure–activity relationship

REFERENCES

Alm, E., and Arkin, A. P. (2003). Biological networks. *Curr. Opinion Struct. Biol.* **13**, 193–202.

Andersen, M. E., Dennison, J. E., and Conolly, R. B. (2005). New directions in incidence–dose modeling. *Trends in Biotech.* **35**.

Berns, K. E., Hijmans, E. M., Mullenders, J., Brummelkamp, T. R., Velds, A., Helmerikx, M., Kerkhoven, R. M., Madiredjo, M., Nijkamp, W., Weigelt, B., Agami, R., Ge, W., Cavet, G., Linsley, P. S.,

Beljersbergen, R. L., and Bernards, R. (2004). A large-scale RNAi screen in human cells identifies new components of the p53 pathway. *Nature* **428**, 431–437.

Bhalla, U. S., Ram, P. T., and Iyengar, R. (2002). MAP kinase phosphatase as a locus of flexibility in a mitogen-activated protein kinase signaling network. *Science* **297**, 1018–1023.

Bischoff, K. B., and Brown, R. H. (1966). Drug distribution in mammals. *Chem. Eng. Prog. Sym. Series* **62**, 33–45.

Bischoff, K. B., Dedrick, R. L., Zaharko, D. S., and Longstreth, J. A. (1971). Methotrexate pharmacokinetics. *J. Pharm. Sci.* **60**, 1128–1133.

Bolouri, H., and Davidson, E. H. (2002). Modeling transcriptional regulatory networks. *BioEssays* **24**, 1118–1129.

Charnick, S. B., Kawai, R., Nedelman, J. R., Lemaire, M., Niederberger, W., and Sat, H. (1992). Perspectives in pharmacokinetics. Physiologically based pharmacokinetic modeling as a tool for drug development. *J. Pharmacokinet. Biopharm.* **23**, 217–229.

Cho, K.-H., Shin, S.-Y., Lee, H.-W., and Wolkenhauer, O. (2003). Investigations into the analysis and modeling of TNFα-mediated NF-κB-signaling pathway. *Genome Res.* **13**, 2413–2422.

Conolly, R. B., Kimbell, J. S., Janszen, D., Schlosser, P. M., Kalisak, D., Preston, J., and Miller, F. J. (2004). Human respiratory tract cancer risks of inhaled formaldehyde: Dose–response predictions derived from biologically-motivated computational modeling of a combined rodent and human data set. *Toxicol. Sci.* **82**, 279–296.

Davidson, E. H., Rast, J. P., Oliveri, P., Ransick, A., Calestani, C., Yuh, C. H., Minokawa, T., Amore, G., Hinman, V., Arenas-Mena, C., Otim, O., Brown, C. T., Livi, C. B., Lee, P. Y., Revilla, R., Rust, A. G., Pan, Z., Schilstra, M. J., Clarke, P. J., Arnone, M. I., Rowen, L., Cameron, R. A., McClay, D. R., Hood, L., and Bolouri, H. (2002). A genomic regulatory network for development. *Science* **295**, 1669–1678.

Dedrick, R. L. (1973). Animal scale-up. *J. Pharmacokinet. Biopharm.* **1**, 435–461.

Ferrell, J. E. (2002). Self-perpetuating states in signal transduction: Positive feedback, double-negative feedback and bistability. *Curr. Opin. Chem. Biol.* **6**, 140–148.

Gargas, M. L., Andersen, M. E., Teo, S. K. O., Batra, R., Fennel, T. R., and Kedderis, G. L. (1995). A physiologically-based dosimetry description of acrylonitrile and cyanoethyleneoxide in the rat. *Toxicol. Appl. Pharmacol.* **134**, 185–194.

Guet, C. C., Elowitz, M. B., Hsing, W., and Liebler, S. (2002). Combinatorial synthesis of genetic networks. *Science* **296**, 1466–1470.

Hartwell, L. H., Hopfield, J. J., Leibler, S., and Murray, A. W. (1999). From molecular to modular cell biology. *Nature* **402**, C47–C52.

Hasty, J., McMillen, D., and Collins, J. J. (2002). Engineered gene circuits. *Nature* **420**, 224–230.

Hoffmann, A., Levchenko, A., Scott, M. L., and Baltimore, D. (2002). The I-κB-NF-κB signaling module: Temporal control and selective gene activation. *Science* **298**, 1241–1245.

Klein, M. T., Hou, G., Quann, R., Wei, W., Liao, K. H., Yang, R. S. H., Campain, J. A., Mazurek, M., and Broadbelt, L, J. (2002). BioMOL: A computer-assisted biological modeling tool for complex chemical mixtures and biological processes at the molecular level. *Environ. Health Perspect.* **110**(Suppl 6), 1025–1029.

Koh, E. Y., Chen, T., and Daly, G. Q. (2004). Genetic complementation of cytokine signaling identifies central role of kinases in hematopoietic cell proliferation. *Oncogene* **23**, 1214–1220.

Krishnan, K., Gargas, M. L., Fennell, T. R., and Andersen, M. E. (1992). A physiologically-based description of ethylene oxide dosimetry in the rat. *J. Toxicol. Ind. Health* **8**, 121–140.

Leroux, B. G., Leisenring, W. M., Moolgavkar, S. H., and Faustman, E. M. (1996). A biologically-based dose-response model for developmental toxicology. *Risk Anal.* **16**, 449–458.

Liao, K. H. (2004). Reaction Network Model for the Prediction of Mammalian Metabolism of Benzo[*a*]pyrene. Ph.D. dissertation, Colorado State University.

Liao, K. H., Dobrev, I., Dennison, Jr., J. E., Andersen, M. E., Reisfeld, B., Reardon, K. F., Campain, J. A., Wei, W., Klein, M. T., Quann, R. J., and Yang, R. S. H. (2002). Application of biologically based computer modeling to simple or complex mixtures. *Environ. Health Perspect.* **110**(Suppl 6), 957–963.

McMillen, D., Kopell, N., Hasty, J., and Collins, J. J. (2002). Synchronizing genetic relaxation oscillators by intercell signaling. *Proc. Natl. Acad. Sci.* **99**, 679–684.

Moolgavkar, S. H., and Luebeck, G. (1990). Two-event model for carcinogenesis biological, mathematical and statistical considerations. *Risk Anal.* **10**, 323–341.

Nestorov, I. (2003). Whole body pharmacokinetic models. *Clin. Pharmacokinet.* **42**, 883–908.

Peck, C. C., Barr, W. H., Benet, L. Z., Collins, J., Desjardins, R. E , Furst, D. E., Hart, J. G., Levy, G., Ludden, T., Rodman, J. H., *et al.* (1992). Opportunities for integration of pharmacokinetics, pharmacodynamics, and toxicokinetics in rational drug development (1992). *Pharm. Res.* **9**, 826–833.

Reisfeld, B., and Yang, R. S. H. (2004). A reaction network model for CYP2E1-mediated metabolism of toxicant mixtures. *Environ. Toxicol. Pharmacol.* **18**, 173–179.

Rowland, M., Balant, L., and Peck, C. (2004). Physiologically based pharmacokinetics in drug development and regulatory science: a workshop report (Georgetown University, Washington, DC, May 29–30, 2002). AAPS Pharm. Sci., 6(1) Article 6, (http://www.aapharmsci.org).

Teorell, T. A. (1937a). Kinetics and distribution of substances administered to the body. I. The extravascular modes of administration. *Arch. Int. Pharmacodyn.* **57**, 205–225.

Teorell, T. A. (1937b). Kinetics and distribution of substances administered to the body. II. The intravascular mode of elimination. *Arch. Int. Pharmacodyn.* **57**, 226–240.

Tyson, J. A., Chen, K. C., and Novak, B. (2003). Sniffers, buzzers, toggles and blinkers: Dynamics of regulatory and signaling pathways in the cell. *Curr. Opin. Cell Biol.* **15**, 221–231.

US Environmental Protection Agency (2000a). *Toxicological review of vinyl chloride.* Appendices A-D. EPA/635R-00/004.

US Environmental Protection Agency (2000b). Dose–response modeling for 2,3,7,8-TCDD. In: *Exposure and Human Health Reassessment of 2,3,7,8-Tetrachlorodibenzo-p-dioxin (TCDD) and Related Compounds. Part II. Health Assessment for 2,3,7,8-Tetrachlorodibenzo-p-dioxin (TCDD) and Related Compounds.* Science Advisory Board Review Draft, NCEA-1-0835, May 2000, Chapter 8.

van de Waterbeemd, H., and Gifford, E. (2003). ADMET *in silico* modeling: Towards prediction paradise. *Nature Rev.* **2**, 192–204.

Whitaker, S. Y., Tran, H. T., and Portier, C. J. (2003). Development of a biologically-based controlled growth and differentiation model for developmental toxicology. *J. Math. Biol.* **46**, 1–16.

Yang, R. S. H., Andersen, M. E., Dennison, J. E., Ou, Y. C., Liao, K. H., and Reisfeld, B. (2004a). Physiologically based pharmacokinetic and pharmacodynamic modeling. In: *Mouse Models of Cancer*, E. C. Holland, ed., John Wiley & Sons, New York, pp. 391–405.

Yang, R. S. H., El-Masri, H. A., Thomas, R. S., Dobrev, I., Dennison, Jr., J. E., Bae, D. S., Campain, J. A., Liao, K. H., Reisfeld, B., Andersen, M. E., and Mumtaz, M. M. (2004b). Chemical mixture toxicology: From descriptive to mechanistic, and going on to *in silico* toxicology. *Environ. Toxicol. Pharmacol.* **18**, 65–81.

INDEX

NOTE: Page numbers followed by f refer to figures, page numbers followed by t refer to tables.

Physiologically Based Pharmacokinetic Modeling, edited by Micaela B. Reddy, Raymond S. H. Yang, Harvey J. Clewell III, and Melvin E. Andersen.